TETRAHEDRON ORGANIC CHEMISTRY SERIES

Series Editors: J. E. Baldwin, FRS & P. D. Magnus, FRS

VOLUME 9

Conjugate Addition Reactions in Organic Synthesis

Related Pergamon Titles of Interest

Full details of all Pergamon publications/free specimen copy of any Pergamon journal available on request from your nearest Pergamon office.

* In preparation

Conjugate Addition Reactions in Organic Synthesis

P. PERLMUTTER

Department of Chemistry, Monash University

PERGAMON PRESS

OXFORD · NEW YORK · SEOUL · TOKYO

U.K.   Pergamon Press Ltd, Headington Hill Hall,
            Oxford OX3 0BW, England

U.S.A.   Pergamon Press, Inc, 660 White Plains Road,
            Tarrytown, New York 10591-5153, USA

KOREA   Pergamon Press Korea, KPO Box 315, Seoul 110-603,
            Korea

JAPAN   Pergamon Press Japan, Tsunashima Building Annex,
            3-20-12 Yushima, Bunkyo-ku, Tokyo 113, Japan

First edition 1992

British Library Cataloguing in Publication Data
A catalogue record for this book is available from the
British Library.

Library of Congress Cataloging in Publication Data
Perlmutter, P.
Conjugate addition reactions in organic synthesis /
P. Perlmutter.
1st ed.
p. cm. -- (Tetrahedron organic chemistry series; v. 9)
Includes bibliographical references and index.
1. Organic compounds--Synthesis. 2. Addition reactions.
I. Title. II. Series.
QD262.P38 1992   547'.2--dc20   92-1266

ISBN 0 08 037066 7 Hardcover
ISBN 0 08 037067 5 Flexicover

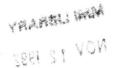

Printed in Great Britain by B.P.C.C. Wheatons Ltd. Exeter

This book is dedicated to the memory of Dr. Geoffrey I. Feutrill

Acknowledgements

This book was begun at Monash University, continued at Denmark's Technical University and Chalmers Technical University (Göteborg, Sweden) and completed back at Monash University. I would like to express my gratitude for having been made feel so welcome by Dr. Søren Jensen and his colleagues (Denmark) and Dr. Christina Ullenius and her colleagues (Sweden). In particular I would like to thank Dr. Frank Eastwood (Monash) who read the entire manuscript and offered numerous valuable criticisms and suggestions. I would also like to thank the following people who read and criticized various portions of the manuscript: Drs. Roger Brown, Mikael Söndahl and Christina Ullenius, Beritte Christenson and Mikael Bergdahl. Also my thanks go to Mrs Kirsten Randolf (Chemistry Librarian, DTU) for her assistance in obtaining several important articles.

TABLE OF CONTENTS

ABBREVIATIONS

18-c-6	18-crown-6
Ac	acetyl
Acac	acetylacetonyl
AcOH	acetic acid
AIBN	azo bis*iso*-butyronitrile
Am	amyl
An	4-anisyl
Aq.	aqueous
BIPY	4,4'-bipyridyl
Bn	benzyl
BTAF	benzyltrimethylammonium fluoride
Bu	butyl
BuLi	butyllithium
Bz	benzoyl
BzC	benzyloxycarbonyl
18-c-6	18-crown-6
Cat.	catalytic
Cp	cyclopentadienyl
Cy	cyclohexyl
DABCO	1,4-diazabicyclo[2.2.2]octane
CSA	camphor sulfonic acid
DBN	1,5-diazabicyclo[4.3.0]non-5-ene
DBU	1,8-diazabicyclo[5.4.0]undec-7-ene
DCA	9,10-dicyanoanthracene
DCE	1,2-dichloroethane
DDQ	2,3-dichloro-5,6-dicyano-1,4-benzoquinone
d.e.	diastereomeric excess
DIA	di(*iso*-propyl)amine
DIEA	di(*iso*-propyl)ethylamine
DIBAL	di-*iso*butylalane
DMAP	4-N,N-dimethylaminopyridine
DME	dimethoxyethane
DMF	N,N-dimethylformamide
DMM	dimethoxymethane
DMPS	dimethylphenylsilyl
DMS	dimethyl sulfide
DMSO	dimethyl sulfoxide
DMTHF	2,5-dimethyltetrahydrofuran
DPMS	diphenylmethylsilyl
e.e.	enantiomeric excess
EEO	1-ethoxy-1-ethyl
EG	ethylene glycol
Eq	molar equivalent
Et	ethyl
FMN	fumaronitrile

HMPA	hexamethylphosphoramide
Hunig's base	N-ethyldiisopropylamine
i-	iso
Im	imidazole
Inv	inverse
LDA	lithium di-*iso*-propylamine
LSA	lithium N-trimethylsilylbenzylamine
Lig	ligand
MCPBA	*meta*-chloroperbenzoic acid
Me	methyl
MEM	methoxyethoxymethyl
MoOPH	Oxodiperoxymolybdenum(pyridine)hexamethylphosphoramide
MOM	methoxymethyl
n-	normal
NBS	N-bromosuccinimide
NCS	N-chlorosuccinimide
NEP	N-ethylpiperidine
NMO	N-methylmorpholine, N-oxide
o/n	overnight
OTf	trifluoromethanesulfonyl
Pent	pentyl
PDMS	phenyldimethylsilyl
Pip.	piperidine
Ra-Ni	Raney nickel
RAMP	(*R*)-(+)-1-amino-2-(methoxymethyl)pyrrolidine
R.t.	room temperature
s	secondary
SAMP	(*S*)-(-)-1-amino-2-(methoxymethyl)pyrrolidine
t-	tertiary
tert-	tertiary
TBAF	tetrabutylammonium fluoride
TBDPS	*tert*-butyldiphenylsilyl
TBS	*tertiary*-butyldimethylsilyl
TBTMG	*ter*-butyltetramethylguanidine
TEAF	tetra-ethylammonium fluoride
TFA	trifluoroacetic acid
TfO	trifluoromethanesulfonyl
Th	2-thienyl
THF	tetrahydrofuran
Thiaz. cat	thiazolium catalyst
TMS	trimethylsilyl
TMSCl	chlorotrimethylsilane
TMSCN	cyanotrimethylsilane
TMSI	iodotrimethylsilane
TMEDA	tetramethylethylenediamine
TMG	tetramethylguanidine
TMS	trimethylsilyl
TMSI	iodotrimethylsilane
Tol	4-tolyl

Tosyl	4-toluenesulfonyl
TPP	triphenylphosphine
Tr	trityl
Trisyl	2,4,6-tris-(isopropyl)benzenesulfonyl
Triton B	N-benzyltrimethylammonium hydroxide
Triton X-100	a non-ionic detergent
p-TSA	4-toluenesulfonic acid
TsOH	4-toluenesulfonic acid

Chapter One   Introduction

1   Organization of the book

This book is divided into seven chapters, each based on a particular class of conjugate acceptor. Chapters 2, 3 and 4 are divided into two major sections, the first covering intermolecular reactions and the second intramolecular reactions. Each chapter is then divided into sections dealing with conjugate additions of individual classes of nucleophiles relevant to the chapter. In some of the larger sections of the book a further division, into various combinations of reaction partners on the basis of chirality, has been necessary. These are:

(a)   The addition of achiral nucleophiles to achiral conjugate acceptors
(b)   The addition of achiral nucleophiles to chiral conjugate acceptors
(c)   The addition of chiral nucleophiles to achiral conjugate acceptors

Where a chiral catalyst or ligand has been used this has been included, somewhat arbitrarily, in part (c).

1.1   History

Conjugate additions have a relatively long history. The first example of a conjugate addition was reported by Komnenos in 1883.[1,2] This publication described the addition of diethyl sodiomalonate (1) to diethyl ethylidenemalonate (2).

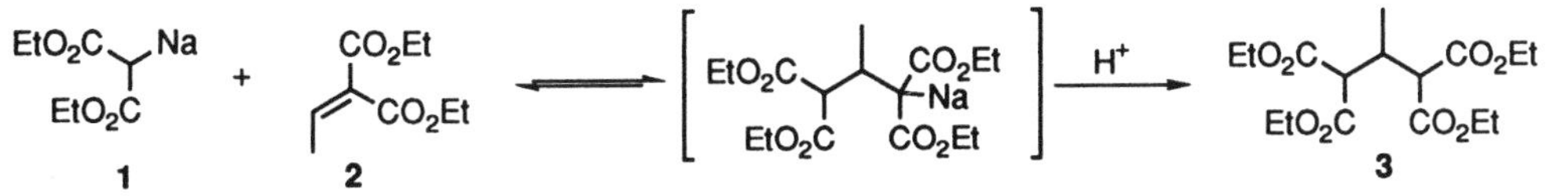

However, the chemistry of conjugate additions really began soon after, with the work of the American,[3] Arthur Michael.[4] His early publications (the first appeared in 1887) on the reaction focussed on the base-promoted additions of the sodium salts of malonates and β-ketoesters to ethyl cinnamate.[5]

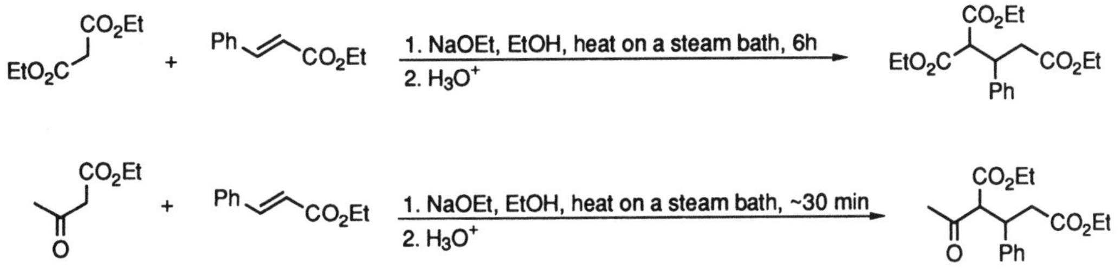

Since that early work was done many stabilized carbanions have been used in the reaction.[6] One of the most important applications of the Michael reaction (using ketone enolates, rather than the more stabilized malonate and acetoacetate carbanions) came with Robinson's introduction of his annulation reaction (see section 2.1.1.1). Since then numerous classes of nucleophiles have been used in conjugate additions. The next major class of nucleophile to be introduced into conjugate addition chemistry, after π-stabilized carbanions, was the (still growing) multitude of organocopper reagents. The success of many total syntheses of important natural products, such as the prostaglandins (section 3.1.2), has hinged on such conjugate additions.

Some of the more significant carbon nucleophiles which have followed include silyl enol ethers, ketene acetals and allyl silanes. There has also been something of a renaissance in the conjugate addition of other organometallic reagents. Organic free radical chemistry is now also an important element in conjugate additions, especially with regard to intramolecular additions. With many of these reactions, high levels of stereocontrol can be achieved by judicious choice of reaction partners. Much is now known about the stereoselectivity of the Michael reaction itself.[7] Because so many of these reactions do show good stereoselectivity, most authors offer their ideas on the likely transition states. Wherever possible, these have been included in this book.

1.2   Nomenclature, classification of reaction types and survey of conjugate acceptors

1.2.1   Nomenclature of reaction types

In searching the literature for reactions relevant to this book it became clear that the use of the terms "Michael addition", "Michael reaction" and "conjugate addition" were often used without any particular system in mind. In this book, the following conventions have been observed:

"Conjugate addition" (or 1,4-addition, see section 1.2.2) refers to the addition of *any* class of nucleophile to an unsaturated system in conjugation with an activating group, usually an electron-withdrawing group. (This is a slightly restricted definition as it does not include the 1,4-addition of, say, bromine across 1,3-butadiene).

"Heteroconjugate addition" is a term coined by Isobe for conjugate additions to alkenes activated by heteroatom-containing functional groups other than carbonyl derivatives.[8] This seems slightly unfortunate as the term could just as easily and, perhaps more appropriately, be applied to the conjugate addition of heteronucleophiles. Consequently, the use of the term has been avoided in this book.

"Michael addition" refers to the addition of *carbanions* to unsaturated systems in conjugation with an activating group.

"The Michael reaction" refers to the addition of *stabilized carbanions* to unsaturated systems in conjugation with a *carbonyl* group.

The term "Michael addition" has not been used in this book. Instead, the term "conjugate addition" has been used throughout and the use of "Michael" has been restricted to the Michael reaction as defined above.

"Cyanoethylation" is often used to describe the conjugate addition of nucleophiles to propenenitrile.[9] In this book the reaction will be found in the appropriate sections of Chapter 4.

1.2.2   1,2 *vs* 1,(2n+2) addition (n = 1,2 etc)

Conjugate addition refers to the addition of a nucleophile to an unsaturated system in conjugation with an activating group, usually an electron-withdrawing group (**A**). Originally, these ideas were applied to additions to α,β-unsaturated carbonyl compounds. As a result, a numbering scheme was developed for these substrates with the numbering beginning at the carbonyl oxygen (C3 and C4 correspond to the α- and β-positions, respectively):

$$RM \quad + \qquad \overset{3}{\diagup}\overset{2}{\diagdown}\underset{O_1}{} \qquad \xrightarrow{\text{1,4 addition}} \qquad R\diagdown\diagup\underset{OM}{\diagup}$$

Conjugate addition may take place at any site 2n atoms distant from the carbonyl carbon, n = 0,1,2.. etc, with n = 1 being by far the most common. As many other **A**'s also activate unsaturated systems towards conjugate addition, this numbering system has been extended and position "2" refers to the atom within **A** directly attached to the unsaturated system. For example, in alkenylsulfoxides the sulfur atom occupies the "2" position:

$$RM \quad + \qquad \overset{3}{\diagup}\overset{2}{\diagdown}\underset{O_1}{S} \qquad \xrightarrow{\text{1,4 addition}} \qquad R\diagdown\diagup S$$

Thus 1,4-addition originally referred to the addition of a carbon nucleophile to positon no. 4 and a metal ion to position no. 1 (and similarly for 1,6- or 1,8-additions). It is also sometimes useful to designate positions using the Greek alphabet:

$$\overset{\gamma \quad \alpha}{\diagup\diagdown\diagup\diagdown}\underset{\beta \quad O}{\diagup}OR$$

Some examples, which display the structural diversity of conjugate acceptors currently so far designed, are given below.

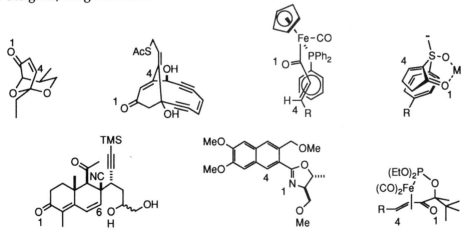

Figure 1.1   Some examples of conjugate acceptors

Since the inception of the Michael reaction many new types of conjugate acceptors have been prepared. Some of these consist of alkenes or alkynes activated by groups other than carbonyls or related functional groups. Often for these molecules the term "1,4" is not relevant as in the following example.

A set of stereochemical descriptors, based on the stereogenicity (or lack thereof) of the reacting centres, for describing all the possible combinations of reaction partners has recently been introduced by Oare and Heathcock.[10]

1.2.3   Survey of conjugate acceptors

There are two components in any conjugate acceptor, (i) the activating substituent (A) and (ii) the unsaturated system. The two archetypal conjugate acceptors are the activated alkene and alkyne structures shown below.

There are many examples of **A**, including:

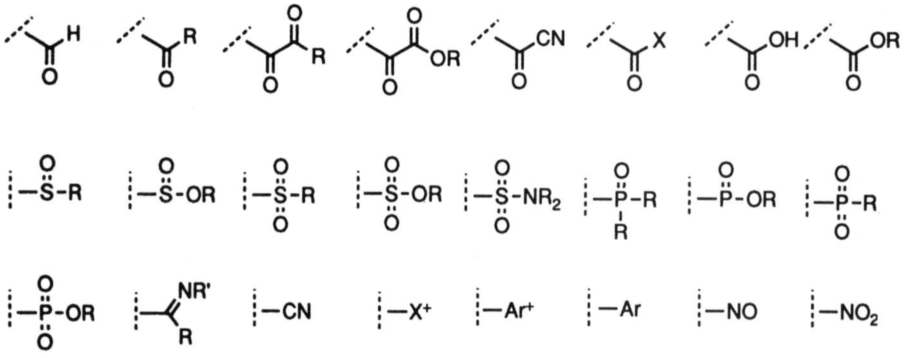

It is also possible for the conjugate acceptor to contain two or more **A** groups acting in concert or opposition. This introduces a more subtle problem of regioselection. Which activating group has the stronger influence?

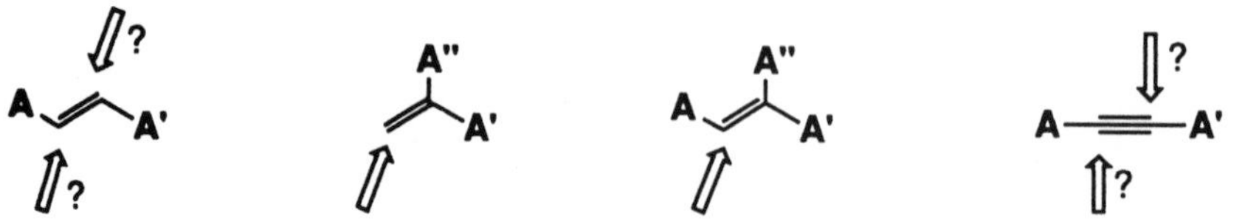

Although of only qualitative value, the relative activating power of these groups can be roughly correlated with their ability to stabilize an adjacent carbanion. In this case, the order is:[11]

$$NO_2 > RCO > CO_2R > SO_2R > CN \sim CONR_2$$

Although there are, of course, many exceptions, this is still quite a useful guide. Also, in general, cyclic derivatives, e.g. cycloalkenones, are more reactive than their acyclic counterparts, e.g. acyclic alkenones.

Some comparisons have been carried out. For example, some generalizations about regiocontrol in enamine additions have been made from such comparisons.[12] A ketone is more powerful than a nitrile and a nitro group is more powerful than an ester group. However, a phenyl group can "beat" a sulfone, depending on the reaction conditions.

A comparison of the ability of the variety of groups mentioned above to activate alkenes and alkynes toward conjugate addition is sometimes complicated by the ambiguity introduced where reversible additions are possible. However, some kinetic studies have been carried out. The following order of reactivity has been put forward by Nagata and Yoshioka for conjugate hydrocyanation reactions:[13]

$$HCO > ArCO > RCO > CN > CO_2R > CONR_2$$

Alkenoyl amides are essentially inert to hydrocyanation unless another activating group is present in the molecule and alkenals suffer from competing 1,2 addition. Alkenylimines possess similar reactivity to their carbonyl analogues. Krief has presented the following reactivity series (in reactions with organometallic reagents) which probably holds true for most nucleophiles.[14]

1.3   Mechanism of conjugate additions

1.3.1   Frontier molecular orbitals and conjugate additions

In a typical conjugate acceptor, say, propenal, the electron deficiency is greater at the carbonyl carbon than at the β-carbon. However, frontier molecular orbital (FMO) analysis reveals that the coefficient of the LUMO is larger at the β-carbon.[15] Thus, soft[16,17] nucleophiles should add to the β-carbon if the additions are under molecular orbital control.[18,232]

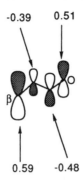

-0.39   0.51

β

0.59   -0.48

Orbital coefficients for the LUMO of propenal

As is evident from the examples to be found in each chapter of this book, many soft nucleophiles, including enolates and other π-stabilized carbanions, copper based reagents, free radicals, enamines and a large variety of heteronucleophiles, efficiently add to the β-carbon of conjugate acceptors.

FMO analysis has been used to account for the transition state structures obtained from *ab initio* calculations of the reaction between a soft nucleophile (an enamine) and an alkenone.[19] Using the 3-21G basis set, the authors found that several structures, including two with significant steric interactions, lay fairly close together in terms of energy. In order to explain this result, it was proposed that some extra stabilization, for the hindered cases, comes from FMO interactions. Examining the two possible molecular orbital combinations, it was concluded that the $HOMO_{enamine}/LUMO_{alkenone}$ combination is largely attractive and also minimizes anti-bonding interactions.

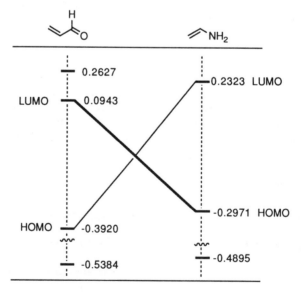

Calculated energies, in au, of the FMOs of
propenal and aminoethene

In addition, they found that the s-*cis* form of the alkenone appears to be less hindered as C2 of the alkenone and C1 of the enamine are not coincident. Thus the chair-like transition structure, with an s-*cis* alkenone, is favoured.

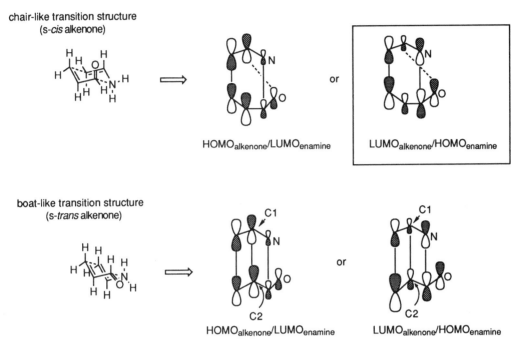

chair-like transition structure
(s-*cis* alkenone)

HOMO$_{alkenone}$/LUMO$_{enamine}$ or   LUMO$_{alkenone}$/HOMO$_{enamine}$

boat-like transition structure
(s-*trans* alkenone)

C1   C1

or

C2   C2

HOMO$_{alkenone}$/LUMO$_{enamine}$ LUMO$_{alkenone}$/HOMO$_{enamine}$

This model was then extended to account for the asymmetric induction found in reactions involving chiral enamines (see section 2.1.5.1).

1.3.2   Thermodynamics and kinetics

As exemplified by the numerous successful reported examples, thermodynamics very often favour a conjugate adduct over the starting materials. Overall, for each reaction, one bond in each of the starting materials is broken and two new bonds are formed in the product. For the addition of heteronucleophiles, such as amines, an analysis, using average bond energies,[20] is rather straightforward and there is a clear gain in energy.

$$RNH_2 \quad + \quad \diagup\!\!\!\diagup CO_2Me \quad \longrightarrow \quad RHN\diagup\!\!\!\diagdown_{CO_2Me}^{H}$$

E_{NH} = ~390 kJmol^{-1} + $E_{CC(\pi)}$ = ~250 kJmol^{-1} Total: ~640 kJmol^{-1}

E_{CN} = ~280 kJmol^{-1} + E_{CH} = ~415 kJmol^{-1} Total: ~695 kJmol^{-1}

Total energy gained = ~55 kJmol^{-1}

Overall, in base-promoted *Michael* reactions,[7] the difference in energy corresponds to the difference in energy between a carbon/carbon single bond (formed in the product) and a carbon/carbon double bond (broken in the starting material). This represents a net gain in energy of ~60 to 90 kJmol[-1]. Where a preformed enolate is added to, say, an alkenoate, it may be sufficient to compare the relative energies of the two enolates. This is only possible in the absence of any electrophiles sufficiently strong to react with the product enolate. This is, of course, also useful when considering *retro*-Michael reactions. However, with many reactions the situation is complicated by the other components in the reaction. Often these additions are carried out in protic media or under conditions where second (and even third) reactions can occur. Often, a Lewis acid is present. Consideration of the thermodynamics of these processes will necessarily be more complicated.

In general, for π-stabilized carbanions, conjugate addition is favoured over carbonyl addition by the use of:

(a)   polar solvents, e.g. HMPA[21]
(b)   larger counterions
(c)   increasing steric hindrance in either reactant (except at the β-carbon of the conjugate acceptor)
(d)   increasing delocalization of the carbanion
(e)   higher temperatures
(f)   longer reaction times

Most of these factors are related to providing means for the reaction to reach equilibrium. This equilibrium usually strongly favours the conjugate adduct. Thus longer reaction times, higher temperatures, polar solvent and larger counterions all enable any 1,2-adduct that is formed to equilibrate with the starting materials and thus find a way through to the more stable conjugate adduct (see, especially, sections 2.1.1 and 3.1.1).

A conceptual model, which accounts for the influence of reaction conditions on the kinetic 1,2- *versus* 1,4-addition of sulfur-stabilized allylic organolithiums, to alkenones, focusses on the role of ion-pairing.[22] The model requires that a rapid equilibrium exists between a contact ion pair (CIP) and a solvent-separated ion pair (SSIP). Solvent-separated ion pairs are assumed to undergo only conjugate addition.

Therefore, according to this model, any factors which promote the formation of solvent-separated ion pairs should also promote conjugate addition over carbonyl addition. The model does seem to explain the role of solvent polarity as an increase in polarity should favour SSIPs.[23] (It is important to distinguish, here, between the role of a polar solvent, such as hexamethylphosphoramide, in promoting a reversal of 1,2-additions and its role in encouraging the formation of solvent-separated ion pairs).

Perhaps the most striking prediction from this theory is that low temperatures favour kinetic 1,4- over 1,2-addition. This has been borne out by a study of the addition of some sulfur-stabilized carbanions to cycloalkenones.[24] What is unclear, at present, is the exact "degree"of stabilization necessary to fit the model. For organolithiums with fewer sulfurs attached to the carbanion, such as 2-lithio-1,3-dithianes, 1,2-addition predominates even at low temperatures (see sections 2.1.1.3 and 3.1.1.3). Ketone and ester enolates, which are quite soft nucleophiles, also give significant 1,2-addition at low temperature (see section 2.1.1.1).

As mentioned above, in an early transition state, the FMOs should control regioselectivity. In a late transition state, the relative stabilities of an enolate ion (from conjugate addition) *versus* an unconjugated oxyanion (from carbonyl addition) will govern the outcome. This should also favour the enolate. However, the traditional view of these (Michael) additions that is assumed in most mechanistic discussions, i.e. that the transition state lies roughly at the mid-point between reactants and products along the reaction pathway, has been challenged.[25] The alternative view, based on an interpretation of the kinetics of addition of cyanide to 1,1-diaryl-2-nitroethenes, is that the transition state is radical-anionic in nature. In other words, there is significant charge transfer in the transition state.

The mechanism of the base-catalyzed addition of *heteronucleophiles* consists of a slow, rate-determining addition of the conjugate base, followed by rapid protonation of the intermediate carbanion.[26] Many studies of the kinetics of these additions (e.g. for alcohols,[27,28,29] amines,[30] thiols[31,32,33]) support this view. The relative reactivity of the conjugate base of some heteronucleophiles was given as[29]: PhS⁻ >> EtO⁻ > HO⁻ > PhO⁻. It was also concluded that polarizability and/or solvation are more important in determining the reactivity of nucleophiles (in additions to aryl vinyl sulfones).

"Anti"-conjugate additions

Very few examples of 1,3- ("anti-") addition have been reported. Perhaps the most remarkable is in the addition of phenoxide and thiophenoxide to benzoyl(trifluoromethyl)ethyne.[34] The additions, which are under kinetic control, are completely regio- and stereoselective. (Reaction at 150°C produced approximately equal amounts of both the conjugate and anti-conjugate products).

$$F_3C-\!\!\!\equiv\!\!\!\overset{O}{\underset{Ph}{\diagup}} \quad \overset{\text{1. PhXH, 1 eq., KO}t\text{-Bu, cat}}{\underset{\text{2. H}_2\text{O, 90\%}}{\overset{\longrightarrow}{\underline{\text{EtOH, 25°C, 6h}}}}} \quad F_3C\overset{O}{\underset{XPh}{\diagdown\!\!\!\diagup Ph}} \quad X = \text{S or O}$$

MNDO molecular orbital calculations show that, for the model system, trifluoromethylpropynal, the electron density at C2 is lower than that at C3 and the LUMO has a larger coefficient at C2 than at C3. These are, of course, ground state properties of (a model of) the conjugate acceptor and to be important in the reaction would require that the addition is under kinetic control. As mentioned above, this is indeed the case.[35]

1.3.3   Conjugate addition of organocopper reagents

Much has been written on the actual mechanism of organocopper additions. In an influential review, House proposed that organocopper reagents add to alkenones via an SET (single electron transfer) mechanism.[36,37] He showed that there is a good correlation between the success of many addition reactions and the redox potentials of the reactants, implying that these nucleophiles could act as one-electron reducing agents.[145] Both an SET mechanism and a direct nucleophilic addition of copper would lead to a copper(III) intermediate as shown below.

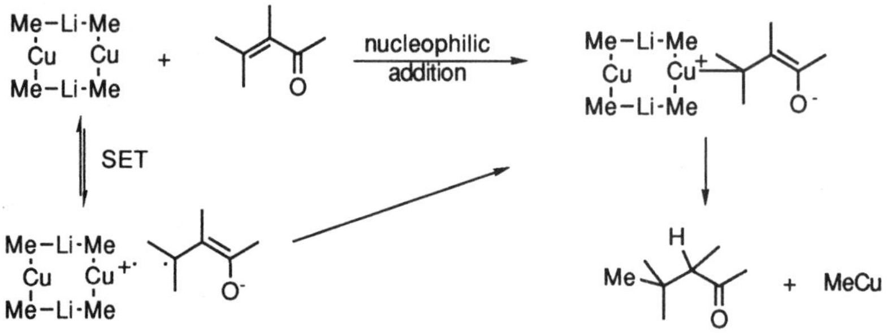

Some evidence exists which argues against a free radical mechanism,[38] and attempts to provide concrete evidence in favour of these processes have so far been unsuccessful.   Most of the early work centred on additions to alkenones, attempting to establish whether or not these additions involved one- or two-electron processes. Mixing dimethylcopperlithium with cholest-4-en-3-one in ether at -100°C generates an orange solution (π-complex?).[39] Warming the mixture to -80°C led to rapid conjugate addition, however no e.s.r. signal was observed.

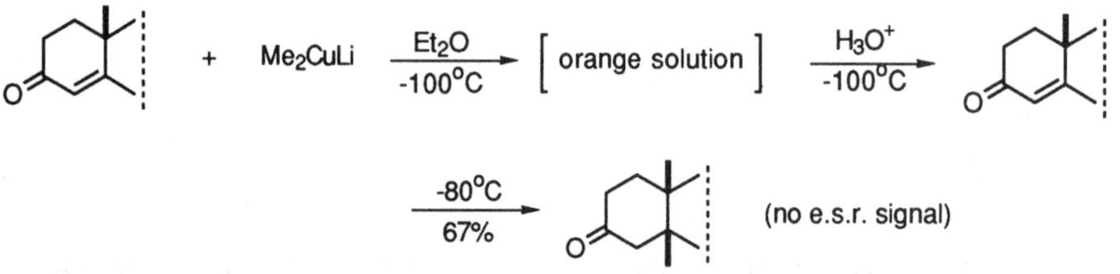

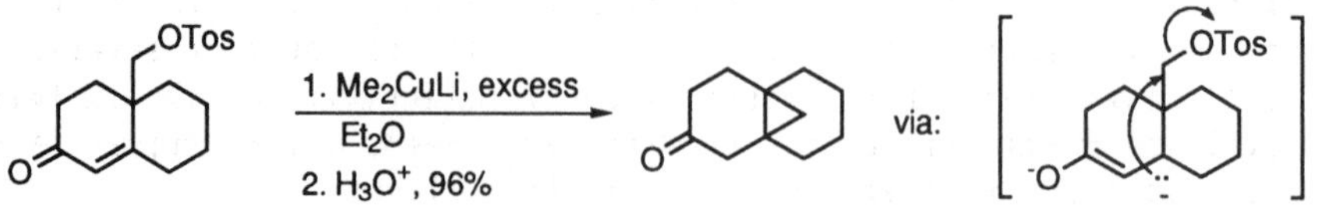

By employing an intramolecular "trap" Hannah and Smith showed that an intermediate with typical enolate reactivity was generated.[36] These authors later proposed a mechanism involving the formation of a Cu(III) intermediate via electron transfer within a charge transfer complex.[40] But this could also be interpreted as proceeding via a radical anion.[41] However, it seems probable that radical anions are not involved. In an important study, Casey showed that the addition of dimethylcopperlithium to a labelled β-cyclopropylcycloalkenone proceeded stereospecifically.[38] Had the reaction mechanism

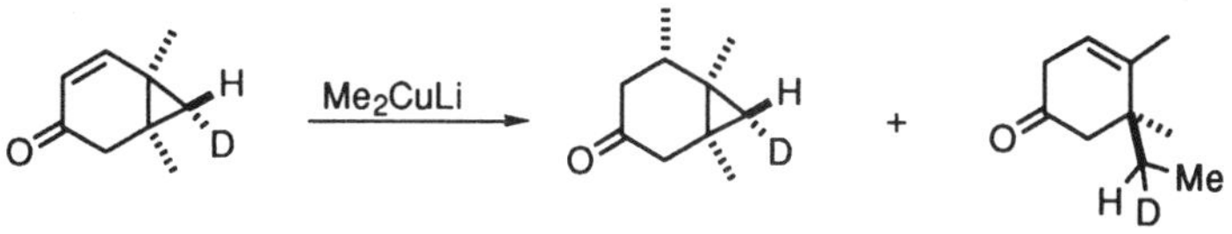

involved a radical anion then the stereochemistry at the labelled carbon would have been scrambled. The result implied that some sort of direct displacement had occurred:

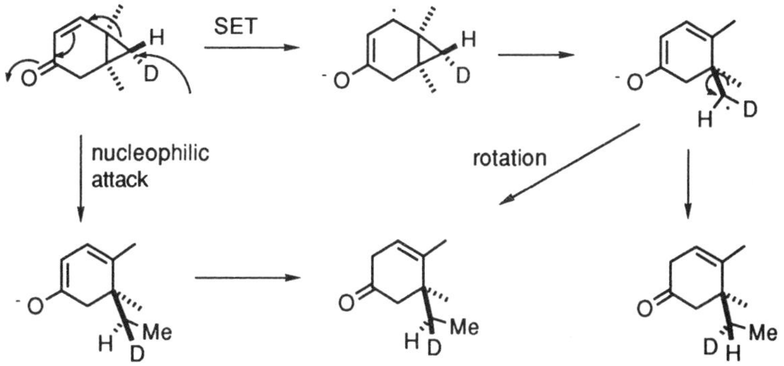

Corey has proposed that a d,π*-complex is formed, with the copper acting as a d^{10}-base and the alkenone acting as a π-acid. He used this idea to account for the stereoselectivity observed in additions to 5-alkyl-2-cyclohexenones and suggested that one of the two complexes (5) would suffer severe pseudo 1,3-diaxial interactions and be disfavoured.[42]

More recently, two groups have provided evidence for the intermediacy of cuprate-alkene π-complexes in additions of LO (lower order, see section 1.4.2.1) cuprates to alkenoates and cycloalkenones which will now be discussed. Ullenius's group have studied the addition of dimethylcopperlithium to a variety of cinnamates.[43] They found that by mixing the reactants at temperatures below about -70°C, the reactants were converted to a new species. The ^{13}C n.m.r. spectrum of this new species showed that the alkene carbons C2 and C3 were shifted strongly upfield. The magnitude of the shifts (~55 and ~73 p.p.m. for C2 and C3 respectively), are similar to those observed for alkenes bound to several other transition metals.[44] Above -50°C, irreversible product formation occurred. On the basis of this, and other evidence, the authors formulated the new species to be a π-complex between copper(I) (from the dimeric cuprate) and the alkenoate double bond.

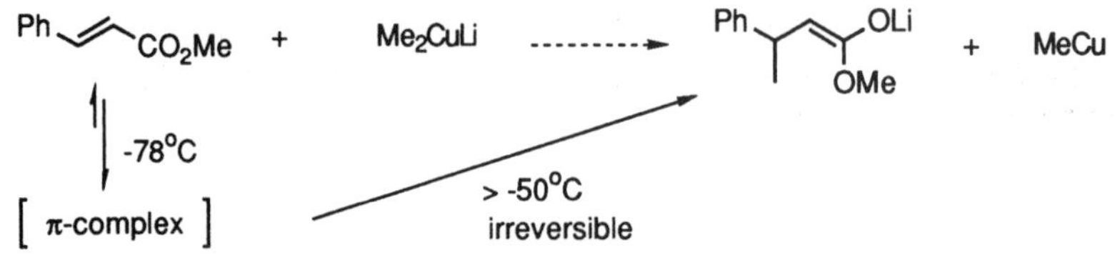

The authors point out that this π-complex corresponds to the previously unidentified intermediate proposed in an earlier study on the kinetics of cuprate additions to alkenones.[45] Bertz's group have studied additions to cycloalkenones.[46] Thus mixing salt-free dimethylcopperlithium with 10-methyl-$\Delta^{1,9}$-2-octalone in diethyl ether-d_{10} at -78°C, they were able to observe several new species in solution. Structural assignments were made on the basis of ^{13}C n.m.r. chemical shifts. Significant amounts of both the lithium-carbonyl (8) and the cuprate-alkene π-complex (7) were present in the solution. By varying the ratio of reagents, several other species were also observed. Two of these were proposed to be lithium-coordinated cuprate-alkene π-complexes (9).[47]

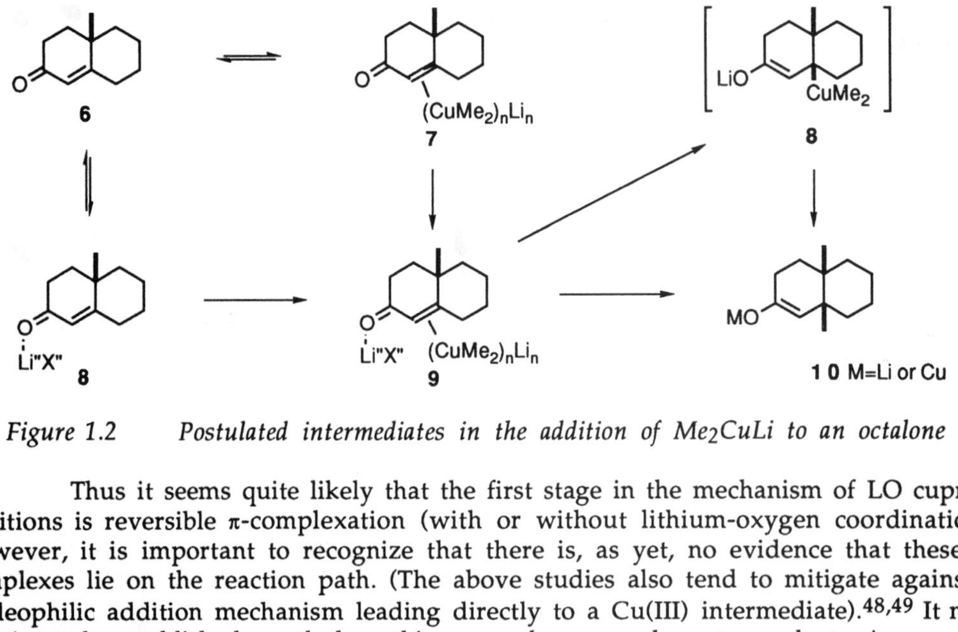

Figure 1.2   Postulated intermediates in the addition of Me$_2$CuLi to an octalone

Thus it seems quite likely that the first stage in the mechanism of LO cuprate additions is reversible π-complexation (with or without lithium-oxygen coordination). However, it is important to recognize that there is, as yet, no evidence that these π-complexes lie on the reaction path. (The above studies also tend to mitigate against a nucleophilic addition mechanism leading directly to a Cu(III) intermediate).[48,49] It now remains to be established exactly how this π-complex proceeds on to products. A summary of the current position is as follows:

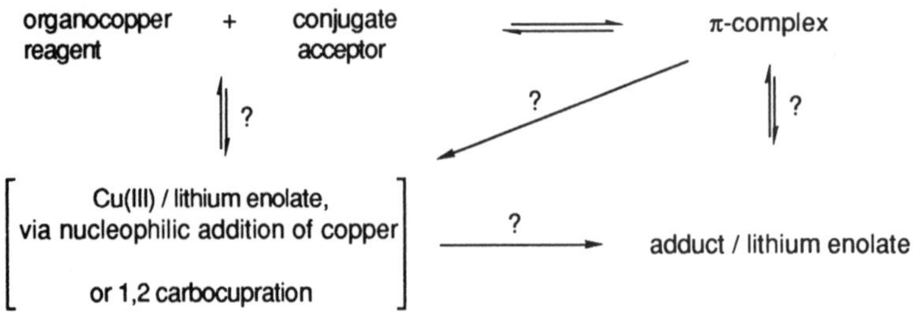

Ab initio calculations

Hartree-Fock calculations, using the 3-21G basis set, have been made on the addition of methylcopper to propenal[50] and both E- and Z-2-butenal.[51] On the basis of both theoretical[52] and experimental[53,54] studies, it was assumed that, in each case, the alkenal conformation was planar, s-*trans* (for propenal this is ~7.1 kJ mol[-1] lower in energy than the s-*cis* form). For the butenals, one C-H bond of the terminal methyl group eclipses the C=C bond.[55]

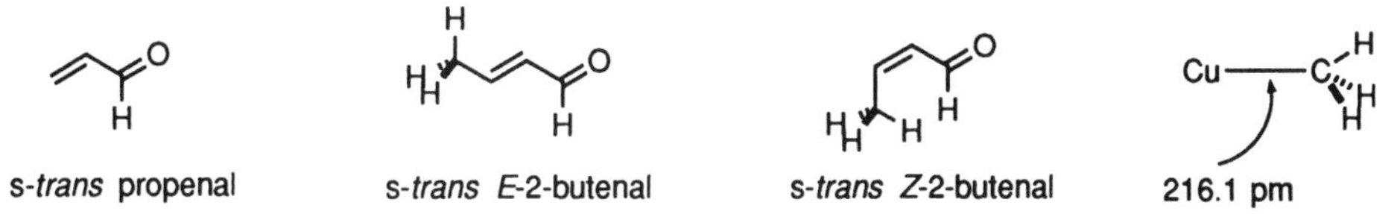

| s-*trans* propenal | s-*trans* E-2-butenal | s-*trans* Z-2-butenal | 216.1 pm |

Several important points emerged from this study. Firstly, the additions are probably concerted but asynchronous. At the transition state, the Cu-O bond length (for the E- and Z-isomers, 201 and 203pm, respectively) is close to that of the enolate (188 pm). On the other hand, the newly-formed C-C bond is still very long (for the E- and Z-isomers, 230 and 232 pm, respectively). Second, the angle of attack (115° and 116°, respectively), is larger than that for nucleophilic additions to carbonyl groups.[56] This is probably due to appreciable bonding being maintained between the methyl group and the copper in the transition state. However, the most significant difference between the transition structures for the E- and Z-isomers is found in the conformation of their terminal methyl groups. For the E-isomer, the terminal methyl group can adopt an almost completely staggered conformation with respect to the incoming methyl group. However, for the Z-isomer, such a conformation would lead to serious steric interactions with the aldehyde moiety, as well as the copper ion. As a result, one hydrogen remains eclipsed with respect to the C=C bond. This becomes significant when one of the protons attached to the terminal methyl group is replaced by another group, such as an alkyl or alkoxy group (see section 1.3.7.1).

1.3.4   Addition/elimination

Many conjugate additions to acceptors bearing a leaving group at the β-position are followed by expulsion of the leaving group. The mechanism of these reactions has been intensively studied.[57,58] Many of these addition/elimination reactions give almost complete retention for either E- or Z- starting alkene stereochemistry. A few examples are given below, all of which gave complete retention.[59]

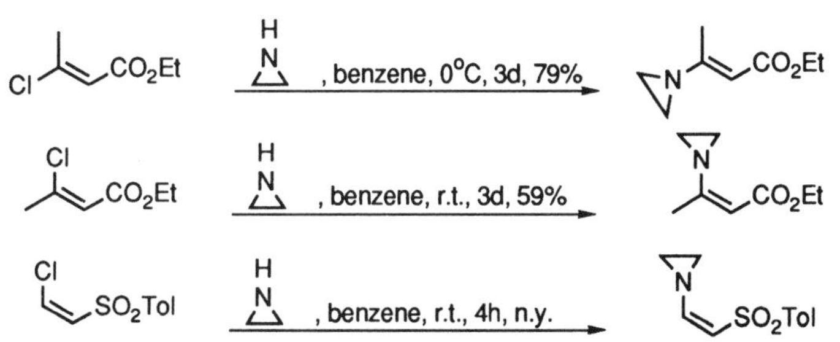

Figure 1.3   Addition/eliminations reactions with overall retention of stereochemistry

Similar results have been obtained for reactions of thioethoxide with 3-chloropropenenitrile,[60] thio(4-nitro)phenoxide with 2-bromoethenyl aryl sulfoxides,[61] methoxide and azide with 2-chloroethenyl aryl sulfones[62] and β-chloroethenyl phenyl ketones.[63]

Quite recently, the first direct observation of an intermediate in a nucleophilic vinylic substitution was reported.[64] Mixing a 50% aqueous solution of DMSO containing (E)-β-methoxy-α-nitrostilbene with a variety of thiolates led to the formation of an intermediate species in each case which slowly converted into the vinylic substitution product (E)-β-alkylthio-α-nitrostilbene.[65]

observable by UV-VIS

Three conditions are necessary for such an intermediate to be observable.

(i)   The equilibrium of the first step must be favourable, i.e., $K_1[RS^-] > (>>) 1$
(ii)   The decay of the intermediate must be slower than its formation, i.e., $k_1[RS^-] > (>>) k_2$
(iii)   The absolute value of k_2 must be low enough to allow detection by suitable techniques (in this case UV-VIS in a conventional or stopped-flow spectrophotometer)

Conjugate addition to a system containing a leaving group at C3 occasionally does not lead to elimination of the leaving group. For example, Cahiez has shown that copper-catalyzed addition of organomanganese compounds to 3-alkoxy-2-cyclohexenone gave clean addition and no elimination of alkoxide (see section 3.1.2). It is interesting to speculate that the intermediate in such an addition may also be observable by spectroscopy. Such a study might shed more light on the nature of the intermediates in organocopper conjugate additions.

1.3.5   Regioselection of additions

A major study of the regio- and stereoselectivity of conjugate additions of preformed ketone and ester enolates has recently been published.[66] With regard to 1,2 *vs* 1,4 addition, it was found that for ketone enolates only 1,4 adducts were observed at low temperatures, where 1,2 addition is normally favoured, *vide infra*. However, 1,2 adducts (aldolates) may be transient intermediates as their corresponding aldols have been prepared and shown to be unstable when converted to the aldolate at low temperatures.[67] On the other hand, ester enolates were found to give significant amounts of 1,2 adducts at low temperatures. Warming these reaction mixtures to room temperature converted them to the isomeric 1,4 adducts. Thus all these conjugate additions are reversible.

The reversibility of additions of α–hetero-substituted ester (and nitrile) enolates to alkenals and alkenones has been investigated by several groups. Schultz and Lee studied the addition of preformed α–hetero-substituted ester enolates to 2-cyclohexenone.[68] Their results are summarized below. Clearly, 1,2 addition is favoured at low temperature. Allowing the reaction mixtures to warm to room temperature leads to high yields of 1,4 adducts. These results also clearly showed that the additions are reversible. Similar results were obtained using α-hetero-substituted acetonitriles.[69] Slight changes in product ratios were observed if the additions were carried out in diethyl ether instead of tetrahydrofuran.[69]

X	1,4	1,2	1,4	1,2
OPh	8	88	84	-
OMe	12	75	62	5
SPh	75	-	86	-
SMe	7	63	85	-
Me	5	88	83	7

Morgans and Feigelson found that the kinetic, 1,2 adduct (**11**) from the reaction of 2-lithio-2-(phenylthio)hexanenitrile and 2-methylpropenal could be smoothly transformed into the 1,4 adduct (**12**) by treatment with lithium diisopropylamide at low temperature followed by warming to zero degrees.[70]

Heathcock's group have examined the regio- and stereoselectivity in the conjugate addition of preformed amide, thioamide, lactam and thiolactam enolates to alkenones.[71] With regard to regiochemistry, several trends emerged:

1. Increase in steric bulk in the enolate leads to more conjugate addition
2. Lactam enolates (which are, effectively, enforced E enolates) show a greater propensity for 1,2 addition than do acyclic Z enolates
3. Increasing the size of the carbonyl ligand decreases the preference for 1,2 addition
4. Enlarging the β-substituent of the alkenone increases the the propensity for 1,2 addition
5. Softer enolates (thioamides and thiolactams) have a greater propensity for conjugate addition then do harder enolates (oxoamides).

These additions also proved to be reversible. Thus warming mixtures of 1,2- and 1,4-adducts from -78°C to room temperature led, in almost all cases, to excellent conversion to the corresponding conjugate adducts. For example:

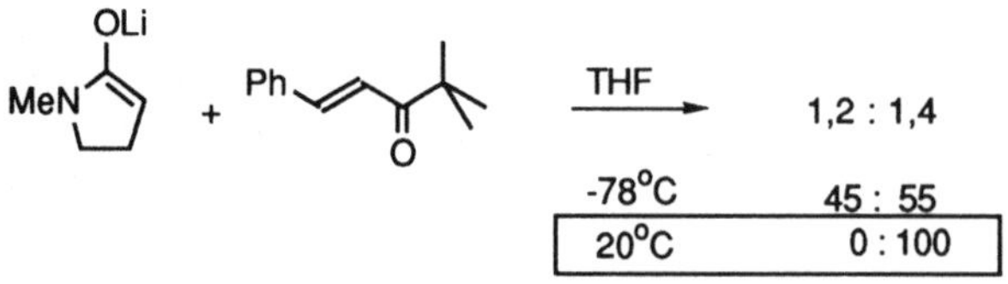

-78°C	45 : 55
20°C	0 : 100

Many other studies, similar to those just cited, using a host of different α-heterosubstituted ester-derived enolates have been published. Some of these are included in the relevant sections throughout the book.

Both reagent and substrate control can be used to direct nucleophiles to add in a conjugate fashion. For example, with nitrogen anions it is possible to adjust the reactivity, and hence regioselectivity, by prior silylation of the nitrogen. In this way, the (overall) conjugate addition of amines to alkenoates has been developed into a powerful new method for the preparation of carbapenam-related azetidinones (see section 4.1.7.1). This is an example of reagent control. Substrate control may be achieved by attenuating the reactivity of the carbonyl group, as shown below.

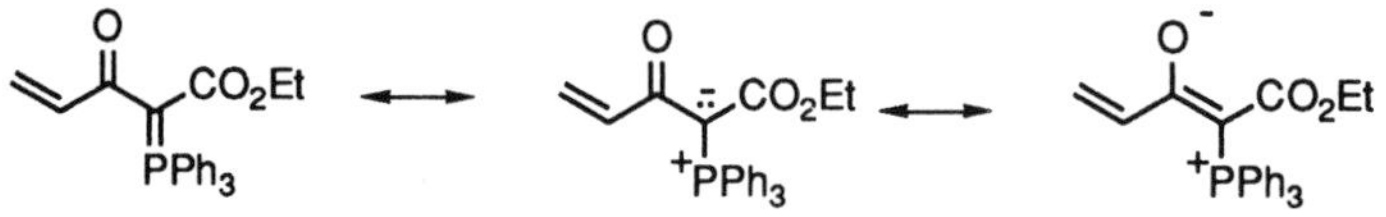

For example, addition of hard nucleophiles, such as non-stabilized organolithium reagents, to α,β-unsaturated acyl ylids leads to clean 1,4 addition.[72] The intermediate enolate can be alkylated and removal of the ylid is quantitative.

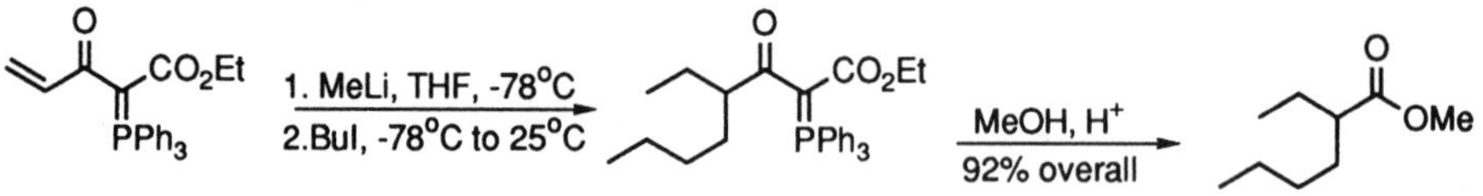

Reactions with more extended conjugate acceptors, e.g. 1,6-acceptors, have been far less studied. However, most nucleophiles add in a 1,6 fashion.[73] Organocopper reagents are sometimes an exception to this. Enamines add in a 1,6 fashion, although it is difficult to distinguish between a conjugate addition and cycloaddition mechanism.

1.3.6   Stereoselection at the nucleophile

1.3.6.1   *Chiral cyclic enolates*

In addition to the control of regioselectivity in enolate formation (discussed in section 1.3.4.1) there is also the issue of stereoselectivity *at the nucleophile* during reaction with an acceptor. This is analogous to the stereoselective alkylation of chiral enolates, an area which has received considerable attention and which has been reviewed recently by Evans.[74] In this review several transition state control elements were identified which were relevant to π-facial selectivity during the alkylation of enolates containing one or more asymmetric centres. The predominant control elements are apparently steric in nature with stereoelectronic effects often having a subordinate role.

In general, exocyclic enolates of 5- and 6-membered rings react at the face *opposite* to that bearing the substituent.

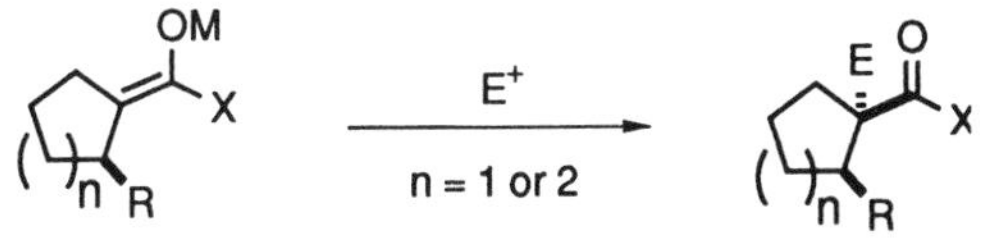

Endocyclic enolates of 4,[75] 5- and 6-membered rings are also usually predictable, although the reasons for the diastereoselectivity are not always so clear.

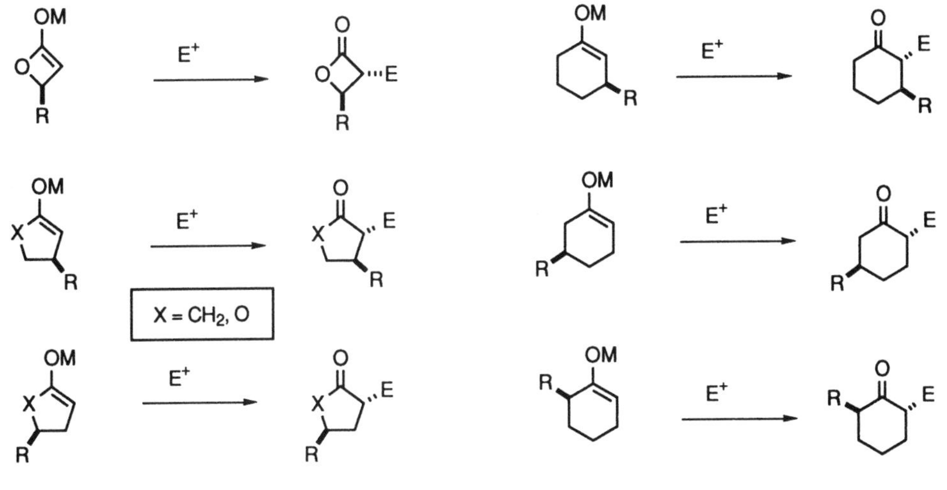

Figure 1.4   Alkylation of chiral, cyclic enolates

In the case of 4-substituted 6-membered endocyclic ketone enolates, the nature of the substituent and/or the electrophile can have a dramatic influence on the stereochemical outcome of the alkylation:

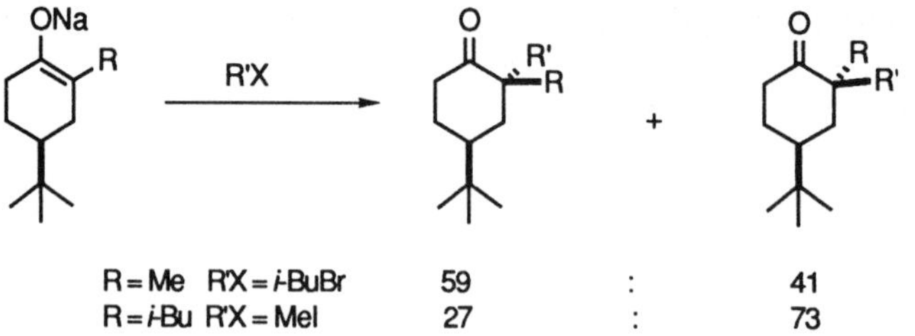

R = Me R'X = *i*-BuBr   59   :   41
R = *i*-Bu R'X = MeI   27   :   73

The situation is less clear with δ-lactone enolates although 3,4-*syn*-disubstituted δ-lactones, e.g., **13**, display quite high levels of diastereoselectivity during alkylation.[76]

(1:1)

13

For all the above systems, the situation becomes, understandably, more complicated when these various enolates are fused to one or more rings. Related to cyclic enolates are the carbanions derived from bislactim ethers. These are, in fact, lithiated cyclic enamines. Deprotonation occurs selectively at the less hindered site as the propyl group effectively blocks access to the adjacent acidic proton.

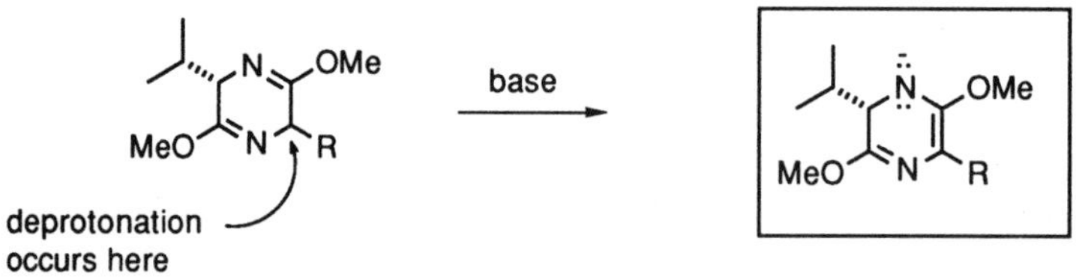

base

deprotonation
occurs here

The propyl group also appears to influence the stereoselection of subsequent alkylations as attack occurs almost exclusively at the face opposite to that bearing the propyl group. This constitutes a very powerful method for preparing enantiomerically pure amino acid derivatives.[77]

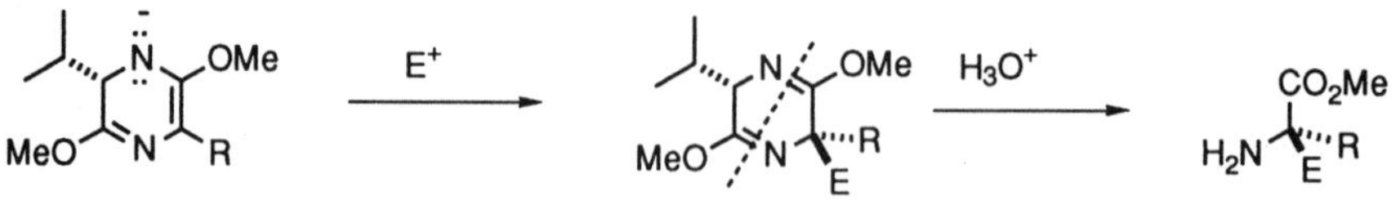

1.3.6.2   *Chiral acyclic enolates*

The last decade has seen the design and synthesis of a variety of chiral ester and amide auxiliaries. Many of these undergo highly stereoselective base-catalyzed alkylations, including additions to conjugate acceptors. The usefulness of these auxiliaries is that they can be simply attached to carboxylic acids producing a chiral ester or amide. These may then be deprotonated and stereoselectively alkylated (the relevant electrophile for our purposes being a conjugate acceptor) and then removed. This last step means that the auxiliary can usually be recycled making the overall process quite efficient. This is summarized below:

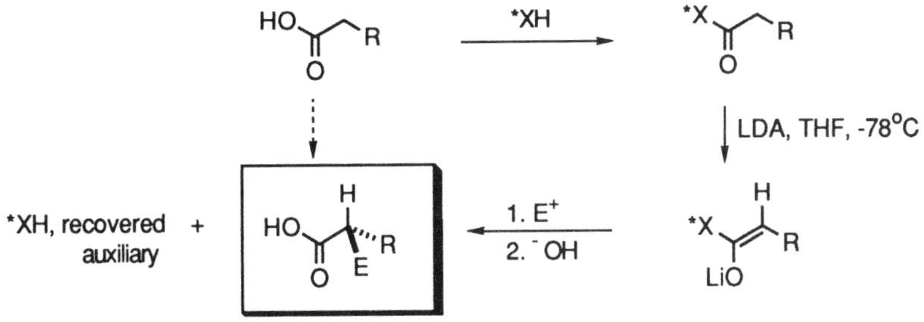

Some of these auxiliaries are shown in the following figure. In each case only one enantiomer is shown although both are usually available.

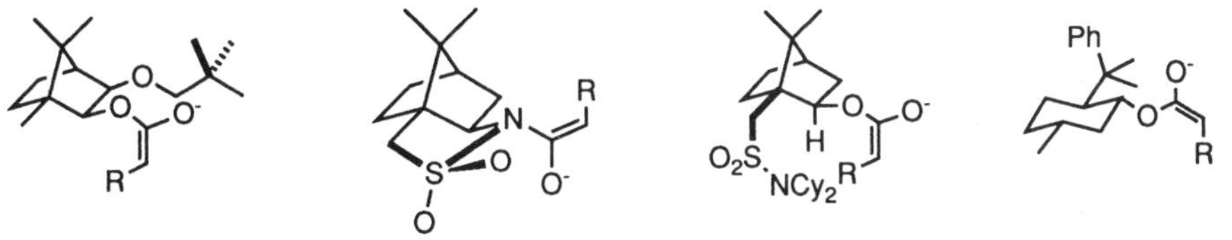

Figure 1.5   Some examples of chiral enolates

1.3.7   Stereoselection at the conjugate acceptor

1.3.7.1   Acyclic systems

Stereoselectivity in conjugate additions may occur during the addition step and/or during reaction (quenching) of the intermediate. Some typical situations are shown below and numerous examples of these can be found throughout this book.

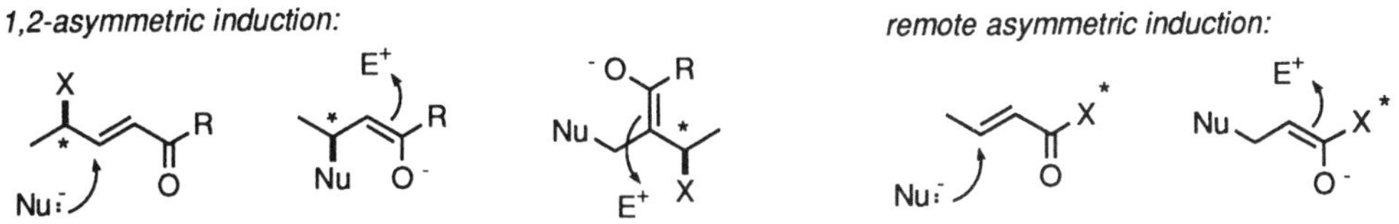

Mohrig's group has shown that under conditions where alkene isomerization is not significant, the base-catalyzed nucleophilic conjugate addition of *tert*-butSD to ethyl *E*-butenoate is highly stereoselective.[230] This is an example of 1,2-asymmetric induction although the relative importance of steric and stereoelectronic factors is as yet unclear. Thus quenching of an enolate generated after the initial addition is often stereoselective, as would be expected from reactions of other chiral enolates (see section 1.3.6.2). Similar selectivity is also found in reactions involving related chiral free radicals (see section 4.1.4).

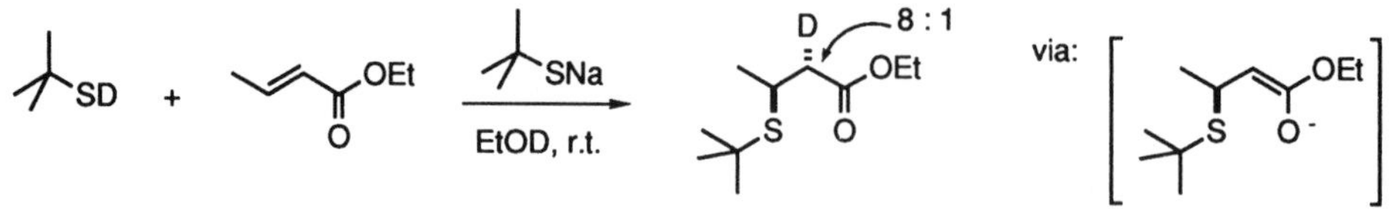

Chiral auxiliaries

      Homochiral alcohols and amines have proved popular as chiral auxiliaries in conjugate additions to alkenoic acid derivatives. Some of these are given below.

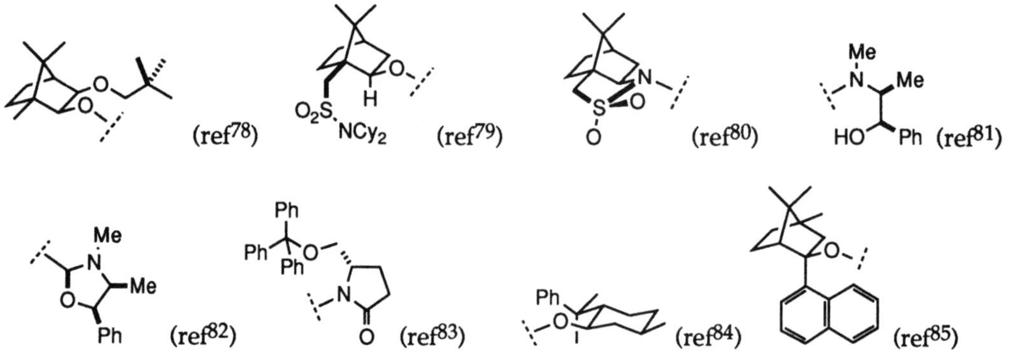

Figure 1.6   Some examples of chiral auxiliaries for alkenoic acids

      Many alkenoate derivatives of these alcohols undergo highly stereoselective conjugate additions with a variety of nucleophiles and examples of these may be found in Chapter 4. Each of the groups that has developed a particular auxiliary has proposed the reactive conformation of the alkenoate derivatives. Fleming has noted that sometimes a conjugate acceptor and a nucleophile will have "mis-matched" conformational preferences.[231] For example, it is becoming clear from Fleming's work and that of others, that silylcuprates or organocopper reagents in the presence of a halosilane, often add to the s-*cis* conformation of alkenoates (and alkenones). However, most crotonate alkenoates tend to react via their s-*trans* conformations. This combination can lead to significant losses in diastereoselectivity (see section 4.1.7.3).

γ-Substituted alkenoyl compounds

      It is now well established that excellent stereocontrol can often be achieved in the addition of nucleophiles to γ-substituted alkenoyl compounds, including alkenals,[86] alkenones[87] and alkenoates.[88] Attempts to understand the stereoselection observed for such additions have all been based on an extension of the largely successful Felkin-Anh transition state models for nucleophilic additions to α-substituted carbonyl compounds. Where a consideration of steric effects is the overriding factor, transition state **A** is the lowest in energy and successfully predicts the outcome of many such additions. On the other hand, when one of the α-substituents is highly electronegative (e.g. oxygen or a halogen) electronic factors may predominate. This leads to a preference for transition states **A** (L= electronegative atom) or **B** (M=electronegative atom).

One problem which arises when extending these ideas to alkenoyl compounds is deciding which of the "inside" or "outside positions" (with respect to the double bond) is the more sterically crowded. Thus a *cis*-substituent, as opposed to the carbonyl oxygen's lone pair, may well render both **A** and **B** much higher in energy than, say, **C** or **D**.

So far, theoretical treatments of conjugate additions to γ-substituted alkenoates have been restricted to analyses of those cases where R=H, i.e. systems containing a β-proton. Both Bernardi's[89] and Morokuma's[86] groups have studied conjugate additions to γ-substituted systems. For E-alkenoates, with a non-polar γ-substituent, minimization of steric repulsion with the incoming nucleophile appears to be the overriding factor and the lowest energy conformer has the γ-proton outside and the largest substituent *anti*. This successfully predicts the outcome of organocuprate additions to such alkenoates.[90]

For similarly substituted Z-alkenoates the steric requirements of the CO_2R group dominate and the inside position appears to be substantially more hindered than for the corresponding α-substituted carbonyl compounds mentioned above. Thus the lowest energy conformers, leading to either the *syn* or *anti* products are those with the small group in the inside position. However, the difference in energies is small. Increasing the size of the nucleophile should lead to an increased selection for the *syn* product. This is indeed observed for sterically demanding organomagnesium reagents[91] as well as dialkylcopperlithiums.[90]

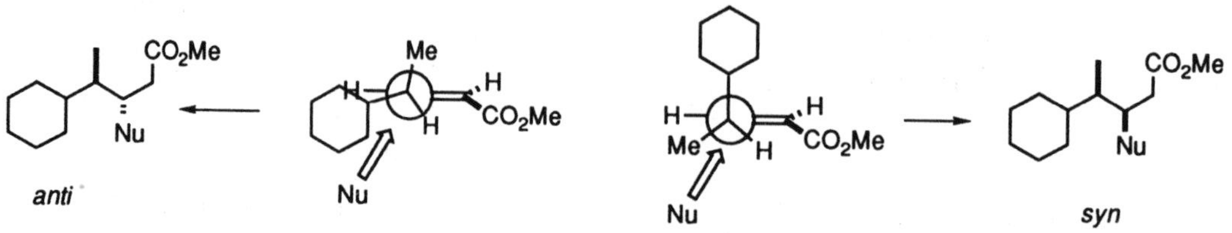

When the γ-substituent is oxygen, e.g. methoxy, then a Felkin-Anh model has been used, i.e. adapted from either **A** or **B**. As shown below the model predicts *syn* selectivity. This holds for organolithium[92] and alkoxide[93] additions. However it does not explain the *anti*-selectivity observed in alkylcopper[90] and cuprate[87] additions.

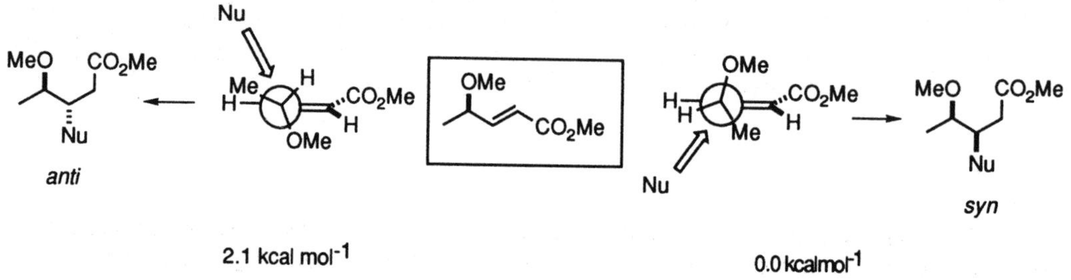

Naturally, these models will improve once more is understood about other factors, such as solvation and chelation. Some examples of the stereochemical trends in these additions, with a variety of nucleophiles is given in the scheme below. For more specific details, see the appropriate sections in the book.

γ-Substituted *E* alkenoates

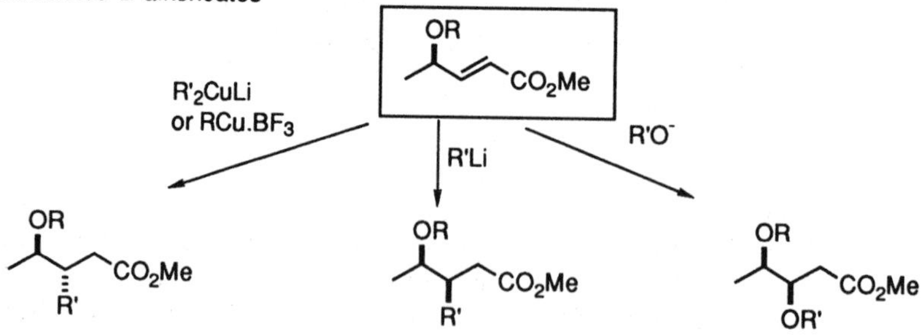

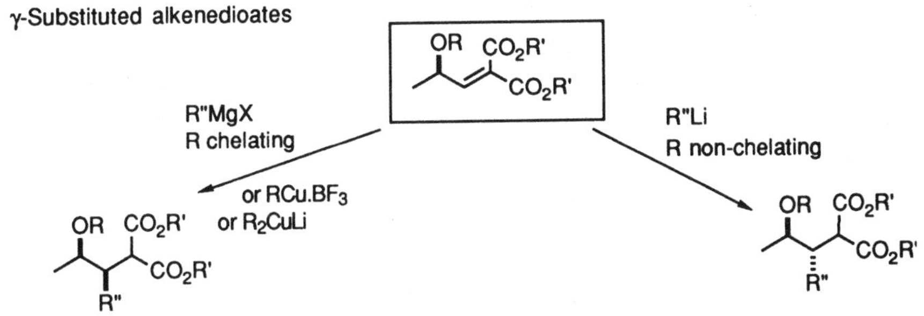

γ-Substituted alkenedioates

R"MgX
R chelating

or RCu.BF$_3$
or R$_2$CuLi

R"Li
R non-chelating

When the γ-substituent = NR$_2$, **E** is apparently the reactive conformation.[94]

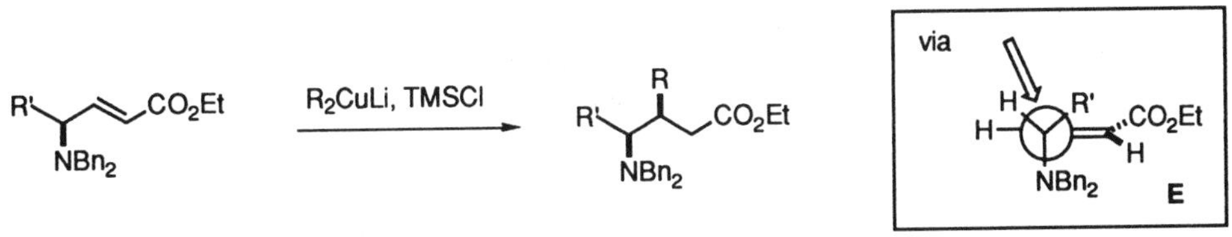

R$_2$CuLi, TMSCI

via

E

Alkenylsulfoxides

Hehre and Kahn have developed a model for chemical reactivity which suggests that product stereochemistries of many reactions, on the basis of electrostatic interactions, are determined early along the reaction coordinate. This model has been used to explain nucleophilic additions to alkenylsulfoxides, and a reactive conformer, with an *s-cis* conformation was proposed.[95,96]

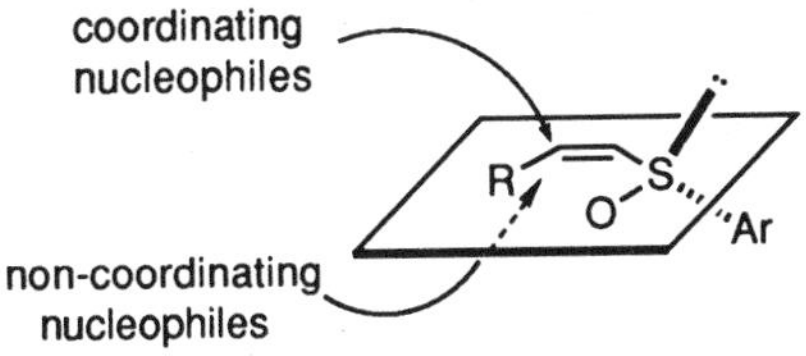

coordinating
nucleophiles

non-coordinating
nucleophiles

The main thrust of this approach is that nucleophiles may be categorized according to whether or not they incorporate an electropositive metal. Those that do may be further divided into those which have either an accessible or inaccessible coordination site. Using the electrostatic model of reactivity it follows that nucleophiles which do not contain an accessible coordination site (e.g. (*i*-PrO)$_3$TiMe or "metal-free" nucleophiles, such as amines) will approach the alkenylsulfoxide from the face remote from the sulfur lone pair (the "electron-rich olefin face"). By contrast, those nucleophiles whose metal counterion is able to coordinate to the sulfur lone pair will approach from the "electron-rich olefin face", i.e. from the same face from which the sulfur lone pair protrudes. Some experimental results were provided to support their conclusions and are reproduced in the Table below. However, some of the results may also be rationalized using steric screening of approach to one face as a consequence of chelation (see section 3.1.3.2). Where examples of these additions have been included in the following chapters, the reacting conformation suggested by the authors of that particular work is given.

Table 1.1 Stereochemical results from conjugate additions to selected alkenylsulfoxides

	Reaction	Side of nucleophilic attack	Selectivity
1[97]	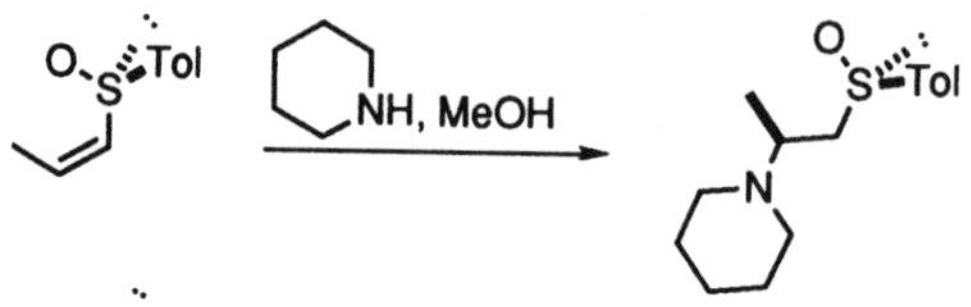 NH, MeOH	Tolyl	3:1
2[98]	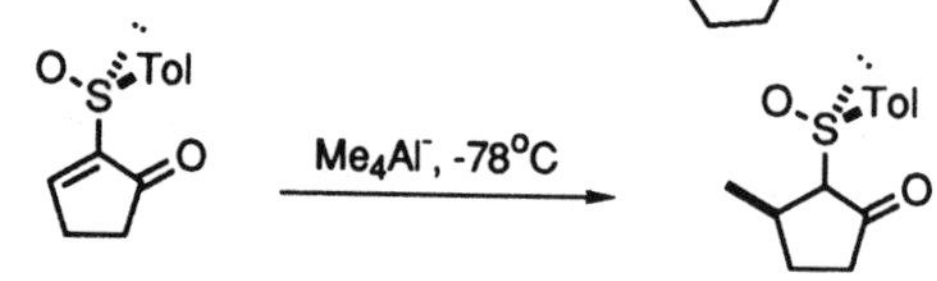 Me$_4$Al$^-$, -78°C	Tolyl	19:1
3[99]	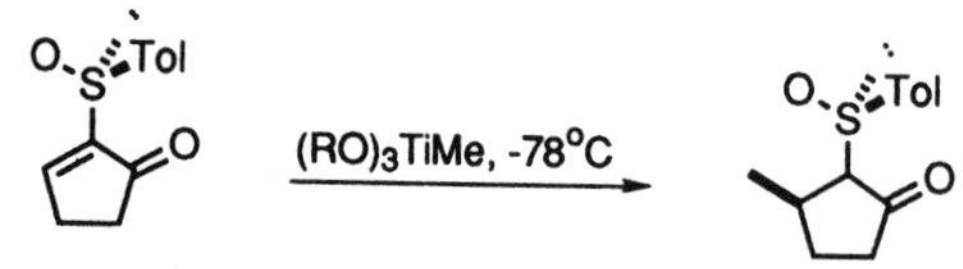 (RO)$_3$TiMe, -78°C	Tolyl	14 to 49:1
4[100]	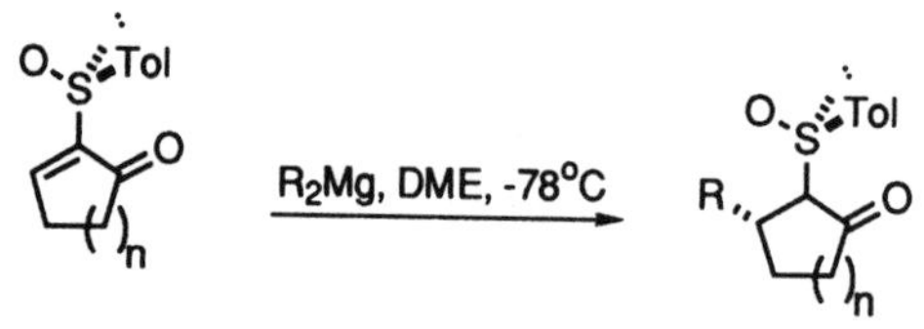 R$_2$Mg, DME, -78°C	Lone pair	3 to 49:1
5[101]	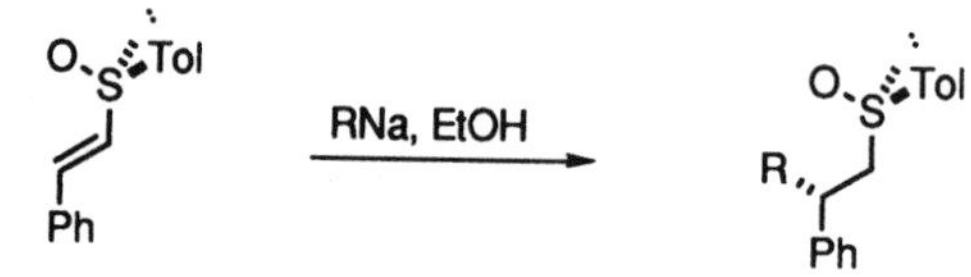 RNa, EtOH	Lone pair	4 to 49:1

The same authors also pointed out the resemblance between their transition structures for the addition of "coordinating" nucleophiles to alkenylsulfoxides and enolate additions to carbonyl compounds.

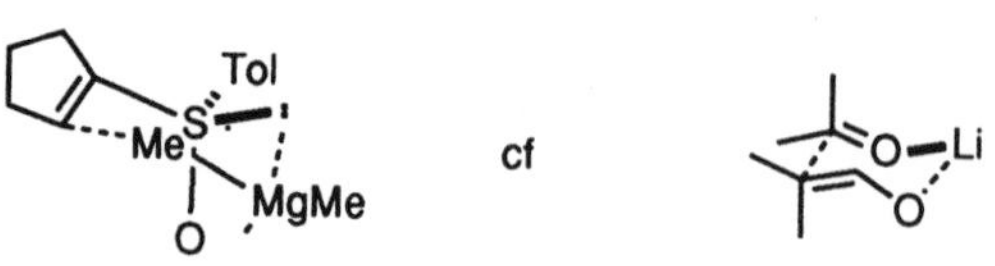

1.3.7.2   *Monocyclic systems*

5-membered rings

For 5-membered rings, the substituent at C4 (or C5 for γ-lactones) generally screens the face from which it protrudes. As a result, conjugate additions usually, and often exclusively, occur from the opposite face. In additions to 4-substituted 2-cyclopentenones, this sort of control has found wide use in prostaglandin synthesis (see section 3.1.2).

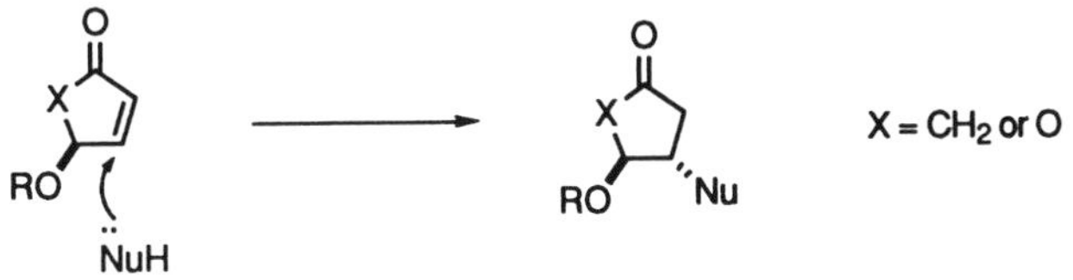

X = CH₂ or O

Additions to 5-substituted 2-cyclopentenones can also be extremely stereoselective. However, this selectivity is largely dependent upon reaction conditions and the nature of the nucleophile (see section 3.1.2).[102]

6-Membered rings

2-Cyclohexenone exists, principally, as a rapidly exchanging mixture of two envelope conformations.[103] Conjugate addition of a nucleophile can, in principle, occur to either face of the alkenone. In each case either "parallel" or "anti-parallel" (with respect to the axial hydrogen at C4, the bold bond in the diagrams below) attack is possible.[104]

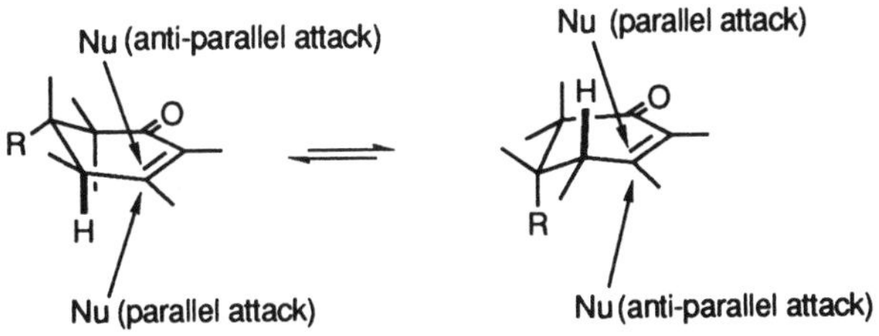

Anti-parallel attack leads to a chair-like intermediate, whereas parallel attack leads to a boat-like intermediate in each case. (Unfortunately, it is quite difficult to see this in a two dimensional projection. Therefore, it is important to build models to convince yourself! A simple guide is the following: in parallel attack, the newly introduced nucleophile forms a *syn diaxial* arrangement with the adjacent C4 hydrogen. Such syn

other factors intervene, then anti-parallel attack is favoured as this leads to a lower energy intermediate (and, by inference, passes through a lower energy transition state).

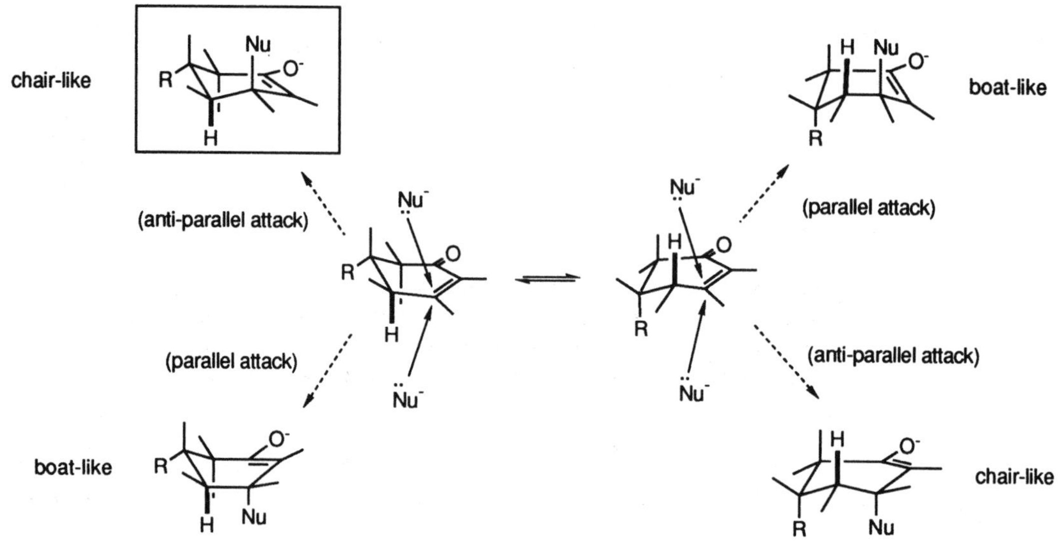

Figure 1.6   Parallel vs anti-parallel nucleophilic addition to cyclohexenones

These arguments were first put succinctly by Allinger and Riew, in their study of the copper-catalyzed addition of iodomethylmagnesium to 5-methyl-2-cyclohexenone.[105] They argued that two factors combined to produce the 3,5-*trans* isomer. Firstly, they assumed that either of the two anti-parallel additions would be preferred. Second, of these two, addition to the conformation with the methyl subsituent (R in the above diagram) *axial* will involve a severe steric interaction compared to addition to the alternative conformation which involves an interaction with the axial hydrogen at C5. Therefore addition should give mainly the 3,5-*trans* product and this is indeed the case.

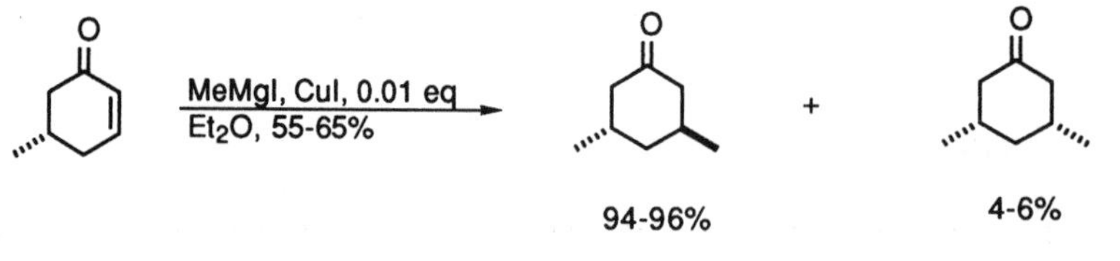

Where R is large, the system is locked in one conformation (R equatorial), and often only one diastereomer is produced. However, steric factors frequently override any stereoelectronic effects. In fact, conjugate addition to monosubstituted cyclohexenones occurs generally from the face opposite to that bearing the substituent, even though this means the reaction may proceed via parallel attack (see Chapter 3).

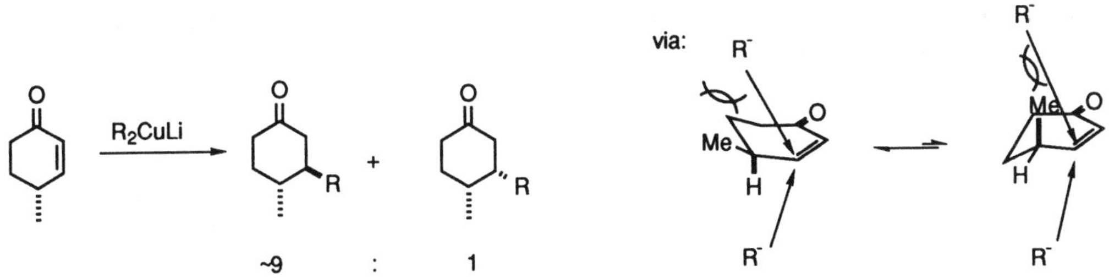

1.3.7.3   Bicyclic systems

In conjugate additions to bicyclic systems, steric effects often play a crucial role in determining stereoselectivity. For example, the addition of nucleophiles to various isomeric octalones is usually very stereoselective but sometimes depends on the particular class of nucleophile being used. This may reflect a change in mechanism.

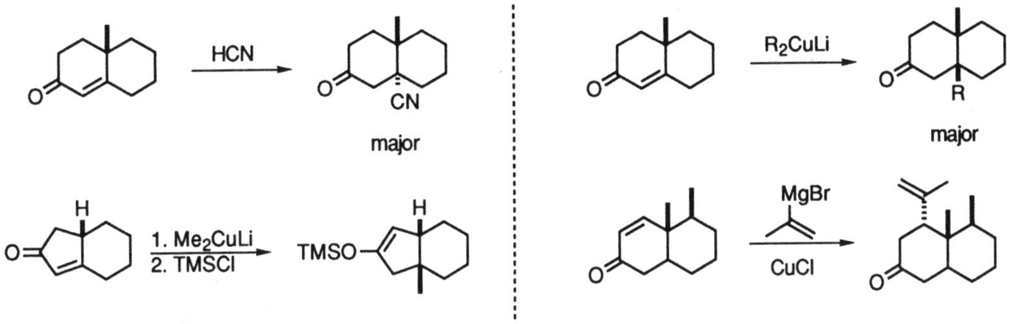

Kinetic 1,6-additions to bicyclic systems are often remarkably stereoselective, as shown below.[106,107]

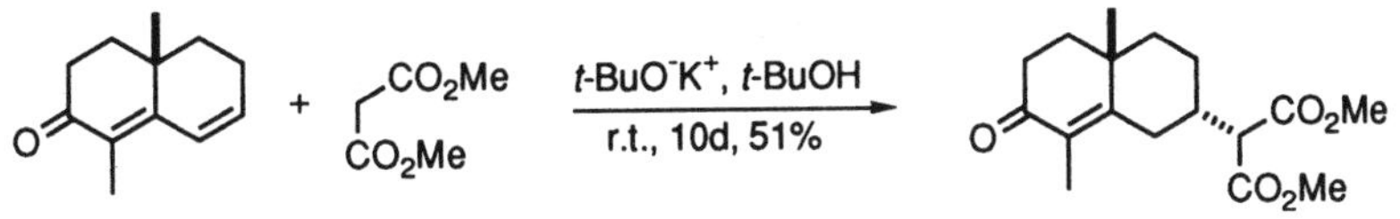

1.3.8   Chiral catalysis

Quite a number of catalysts have been used in attempts to obtain useful levels of asymmetric induction in conjugate additions of enolates to alkenones and alkenoates.[108] The early examples relied upon the the use of chiral natural products such as the alkaloids quinine and quinidine. Extensive studies with these alkaloids and their derivatives have resulted in the development of several highly efficient asymmetric Michael reactions.

The original Michael reaction involved conjugate addition in a protic solvent (e.g. ethanol) using a catalytic amount of a base (an alkoxide or trialkylamine). Many attempts have been made to induce enantioselectivity under these reaction conditions using chiral bases. For example, Wynberg's group have examined enantioselection in Michael reactions using chiral bases derived from the *Cinchona* alkaloids (see, in particular, section 2.1.1.2 for a discussion of the results). Alternatively, a chiral ligand may be added to a reaction. This can also lead to excellent results.

Some examples of the chiral catalysts and ligands which have been used in enantioselective conjugate additions are shown below.

Alkaloid based catalysts

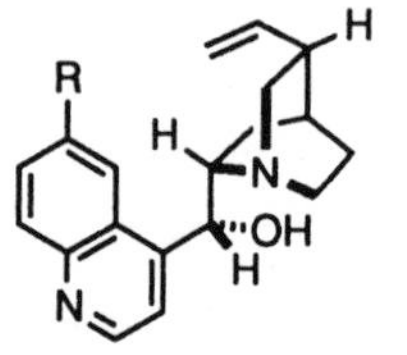

Quinine   R = MeO
Cinchonidine R = H

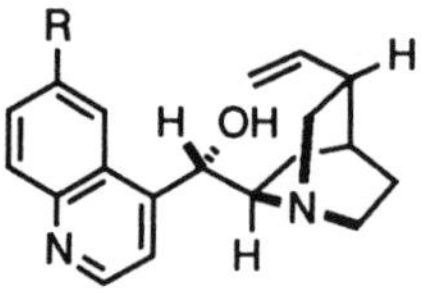

Quinidine   R = MeO
Cinchonine R = H

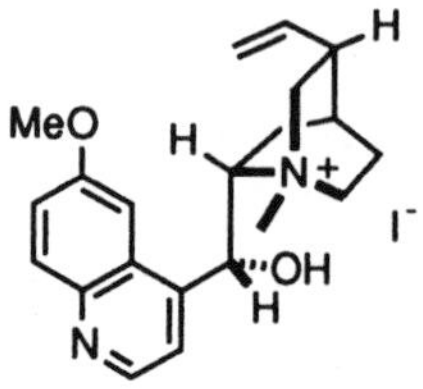

Quinine methiodide

Chiral crown ethers

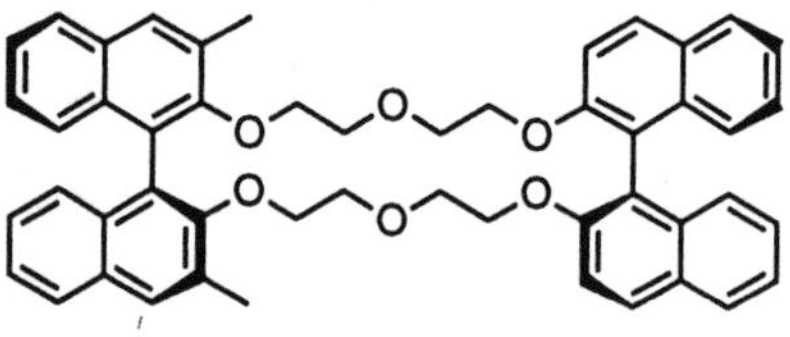

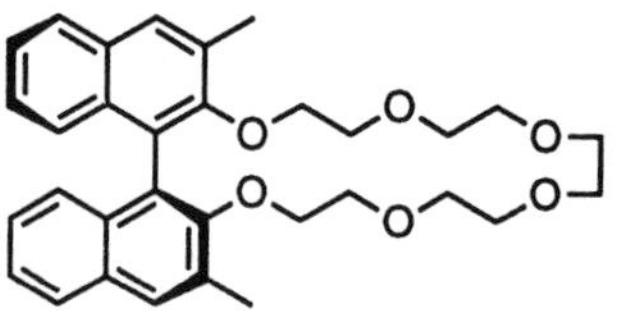

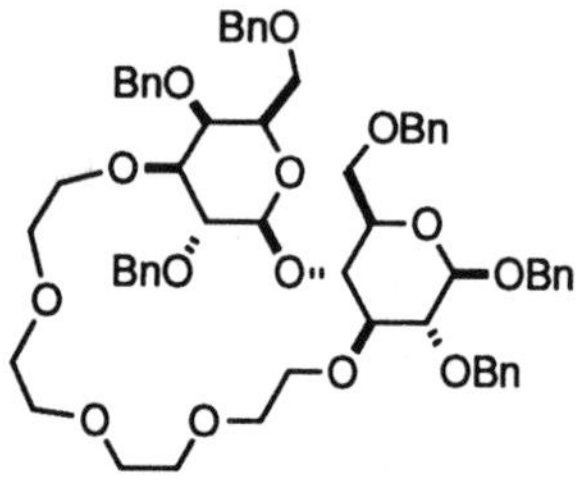

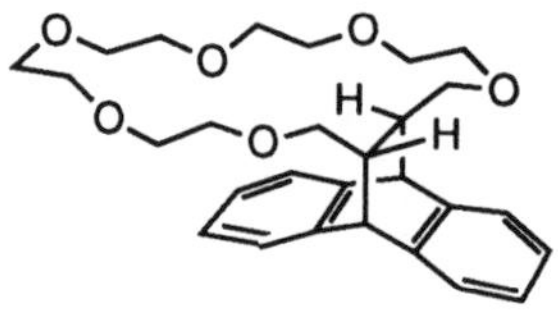

Chiral ligands

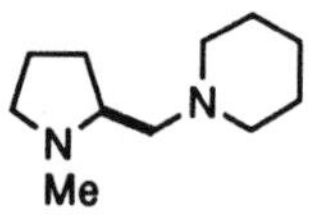

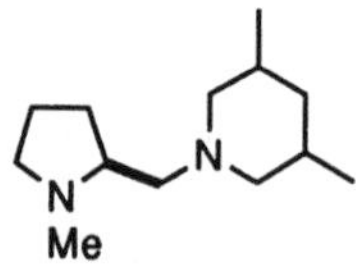

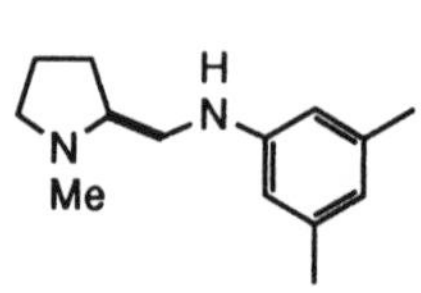

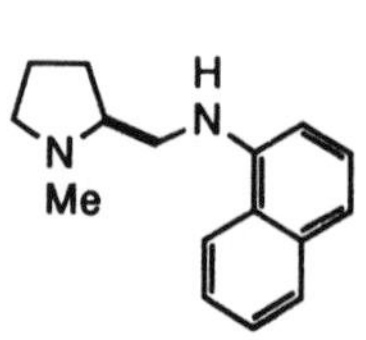

Me   Me   Me   Me

1.4   The nucleophiles

1.4.1   Stabilized carbanions

Stabilized carbanions can be divided into three broad groups:[109]

1.   Carbanions stabilized by π-conjugation with one heteroatom
2.   Carbanions stabilized by π-conjugation with more than one heteroatom
3.   Carbanions stabilized by one or more α-heteroatoms

Because of the thousands of conjugate additions that have been carried out over the years (there are approximately five thousand citations in Chemical Abstracts®) it is more than likely that most existing bases have been used at some time. The original Michael reactions used an alkoxide in the corresponding alcohol as catalyst. However, since then, the wide range of basic amines, especially tertiary amines, have found much use. The powerful lithium amide bases are usually used in stoichiometric amounts generating preformed enolates. Other hindered bases, including DBU and DBN, have also been used. The metal hydrides are more often used stoichiometrically than catalytically.

Phase transfer catalysis (PTC) has become the method of choice for catalyzing the addition of many nucleophiles. A number of catalysts which contain a tetraalkylammonium halide, especially fluoride, have been used. The basicity of fluoride is still a contentious issue. Kuwajima estimated the basicity of fluoride ion in aprotic solvents by comparing the rate of addition of benzyl mercaptan to methyl propenoate, catalyzed by three different bases.[110] In this study it was assumed that the rate of addition of the thiolate ion, once formed, would be very similar irrespective of the counterion. Thus the rate of reaction is directly proportional to the rate of formation of the thiolate and hence to the basicity of the catalyst. From the results shown below, in tetrahydrofuran, fluoride ion is slightly more basic than trialkylamines but much less basic than alkoxides.

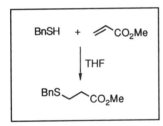

Catalyst	No. of equiv.	Conditions	Yield (%)
NaOEt	0.05	0°C, 5 min	96
TBAF	0.03	~10°C, 20h	87
Oct₃N	0.05	~10°C, 10d	98

Typical reaction conditions which are used for generating each type of carbanion for use in conjugate additions will now be briefly reviewed.

1.4.1.1   *Carbanions stabilized by π-conjugation with one heteroatom*

As the enolate geometry is often crucial to stereoselection in conjugate additions, it is important to have methods which reliably produce the desired geometry.

(a)   Ketone enolates

Oare and Heathcock have demonstrated that pure Z-enolates of unsymmetrical ketones may be generated when one of the carbonyl substituents is non-enolizable and much larger than the other.[7,66,111,112] However, a more general method is to generate the

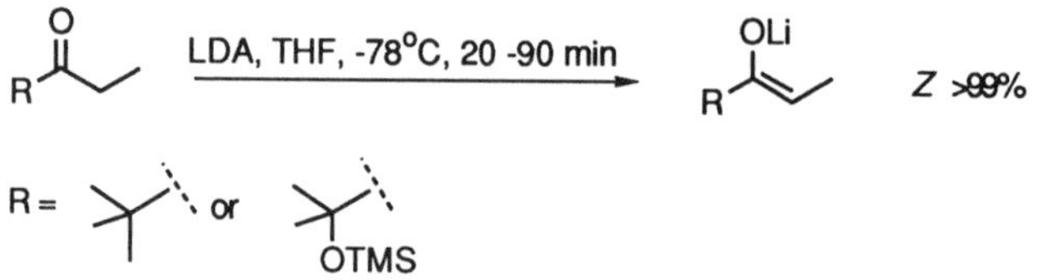

enolate from the previously prepared enol silane, using the procedures of Stork[113] and House.[114,115]

Ireland's group have demonstrated the effect of hexamethylphosphoramide on the enolization of 3-pentanone.[116] Deprotonation with lithium diisopropylamide at -78°C yields predominantly the E-enolate. Inclusion of hexamethylphosphoramide prior to deprotonation leads to the Z-enolate.

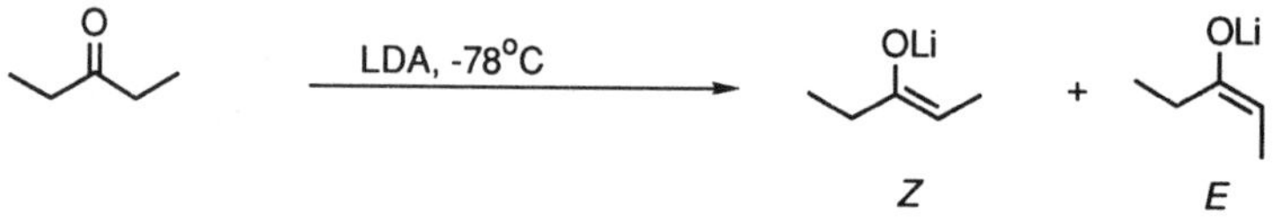

THF	23	:	77
THF/HMPA	95	:	5

The authors rationalized their results (and the corresponding results from the enolization of esters, see below) by assuming that the transition state shown below leading to the E enolate is lower in energy due to the absence of an unfavourable pseudo 1,3-diaxial interaction.

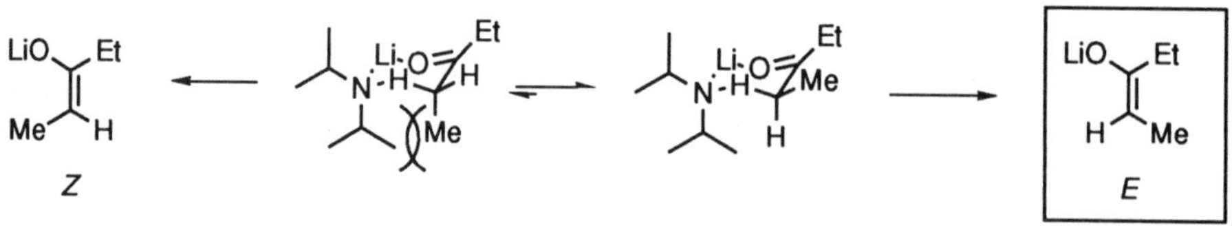

The actual manner in which hexamethylphosphoramide reverses the selectivity in this reaction is still debatable. Corey and Gross have shown that the selectivity in the presence of hexamethylphosphoramide is the result of equilibration to the thermodynamically more stable Z-enolate.[117] In the same paper they also demonstrated the value of using a more hindered base than lithium diisopropylamide, namely lithium *tert*-octyl *tert*-butylamide (LOBA). Furthermore they showed that, rather than quenching the enolate with chlorotrimethylsilane after the enolate had been generated, it was preferable to include the chlorotrimethylsilane in the reaction from the start (this requires that lithium dialkylamides don't react appreciably with chlorosilanes at -78°C, which is indeed the case). However, the use of LOBA has not often been cited since this work was described.

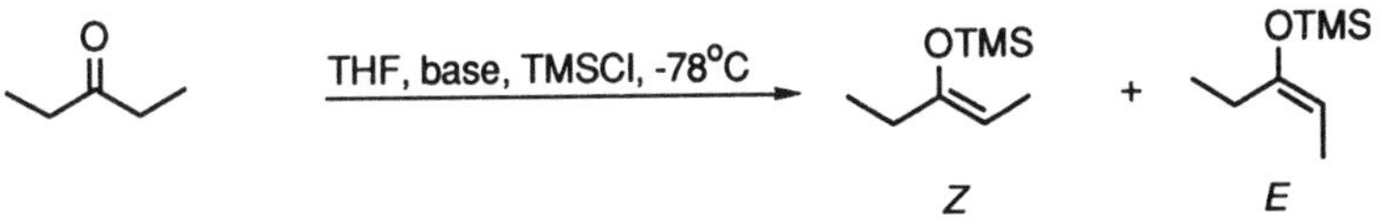

		Z		E
LOBA		2	:	98
LDA		23	:	77

(b)   *Ester enolates*

Ireland's group have also developed reaction conditions which reliably generate either the *E*-or *Z*-enolate selectively from esters.[116] Thus in tetrahydrofuran, deprotonation of propionate esters with lithium diisopropylamide (and other lithium dialkylamides) generates the *E*-enolate with between 87 and 95% selectivity. On the other hand, deprotonation in tetrahydrofuran containing 23% hexamethylphosphoramide produces the *Z*-enolate, usually somewhere between 81 and 97% selectivity.

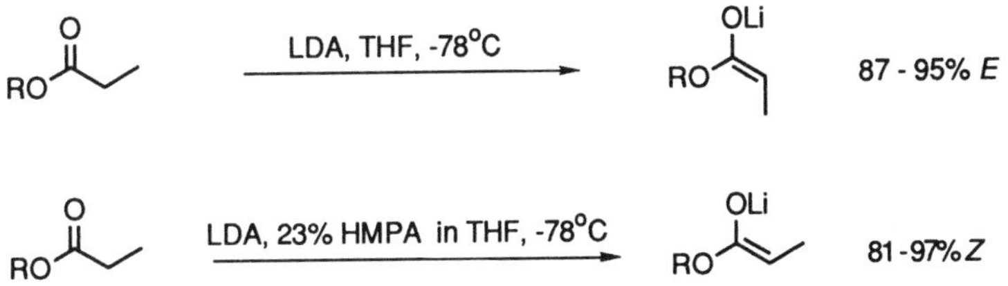

The selectivity for the *E*-enolate has been rationalized in a similar manner to that described above for ketone deprotonation.

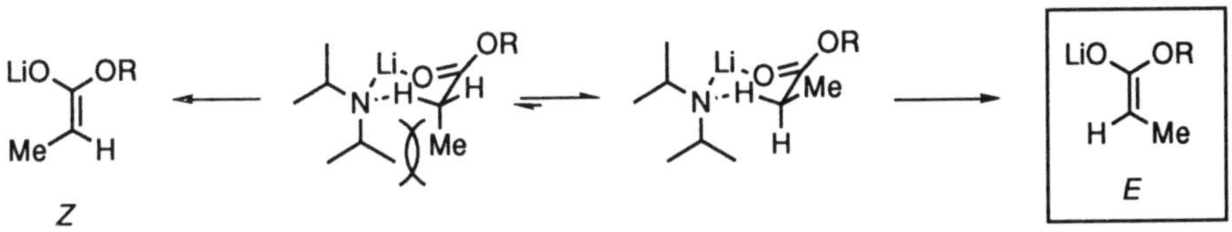

Similar deprotonation of the corresponding propionyl dithioesters leads to disappointing selectivity (*Z/E* ratios of ~1:1 to 3:1).[118] However, in the same paper, a method for the selective formation of the *Z*-enolate, from a mixed anhydride, was briefly described.

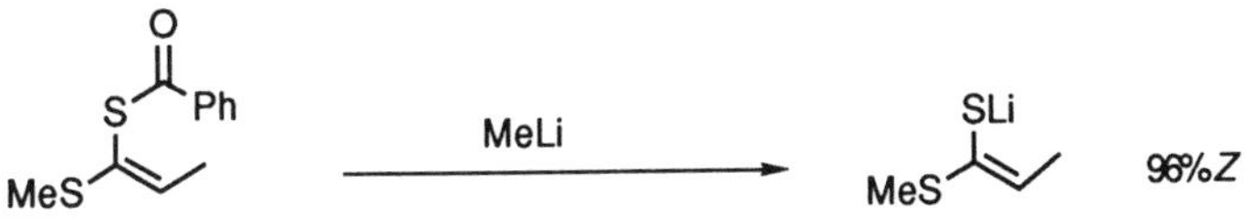

Kinetic deprotonation of longer chain dithioesters is quite selective for the Z-enethiolate.

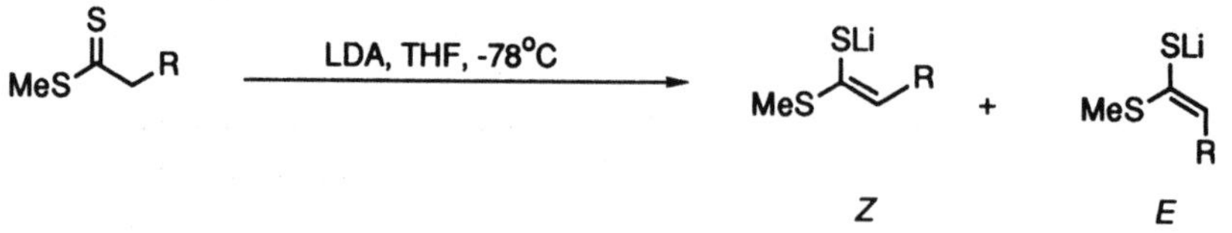

	Z	:	E
R = Me	76	:	24
R=(CH$_2$)$_4$Me	85	:	15
R=CH$_2$CH=CH$_2$	95	:	5

Thioester enolates may be obtained with good selectivity under kinetic or thermodynamic conditions similar to those described above for esters.[119,120]

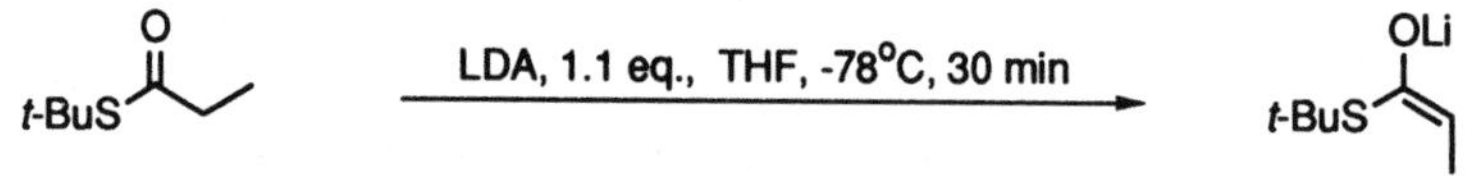

90 - 95% Z

93 - 95% E

(c)   *Amide enolates*

Amide enolates are routinely generated using strong bases, such as lithium diisopropylamide or *n*-butyl lithium. In general, kinetic deprotonation of N,N-dialkylamides yields the Z-enolate.[121,122]

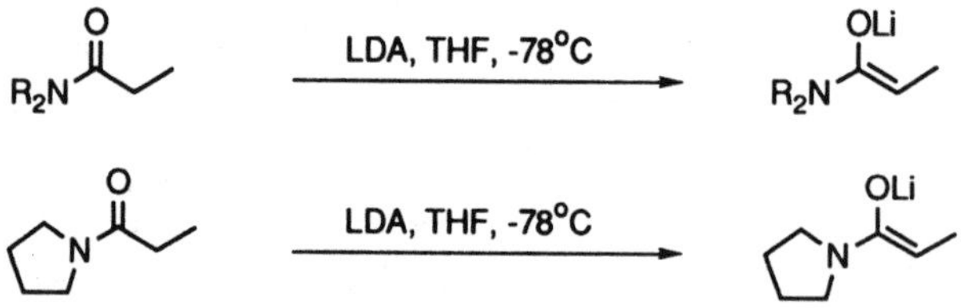

In fact, there are currently no methods available for the generation of E-enolates of amides.[122,123] Where the E-enolate is required, usually for stereochemical studies, N-alkyl lactams can be used. Their enolates, of course, must have E geometry.

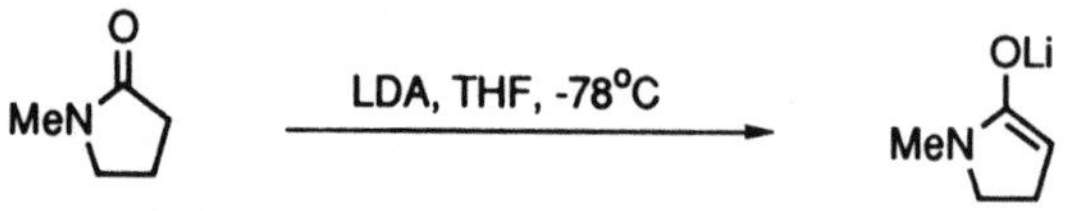

The corresponding thioamides so far studied also give Z-enethiolates with high selectivity. These are generated using *n*-butyllithium or sodium or potassium hexamethyldisilazide.[124] As before, studies which require an *E* enethiolate have used thiolactams as precursors.[122]

Amide dienolates are also produced stereoselectively by deprotonation of enamino- esters (vinylogous carbamates) with a strong base at low temperature.[125,126]

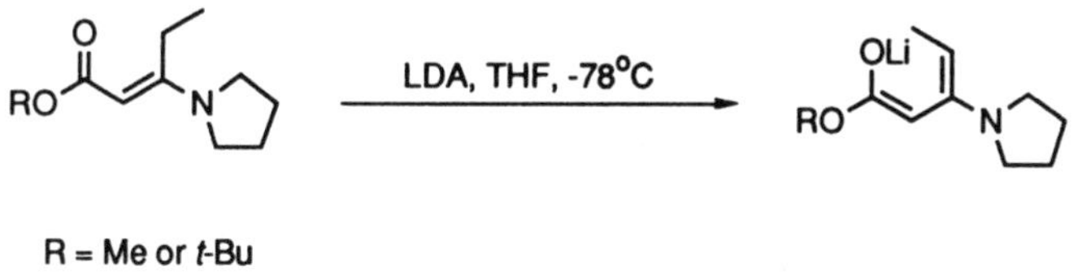

R = Me or *t*-Bu

(d)   *Hydrazones and imines*

Both the *E*-and *Z*-azaenolates of aldehyde N,N-dialkylhydrazones are available using the same approach outlined above for esters.[127,128] Thus low temperature deprotonation leads to virtually exclusive formation of the *E* isomer, whereas inclusion of hexamethylphosphoramide provides the *Z*-isomer with lower, but still useful, selectivity.

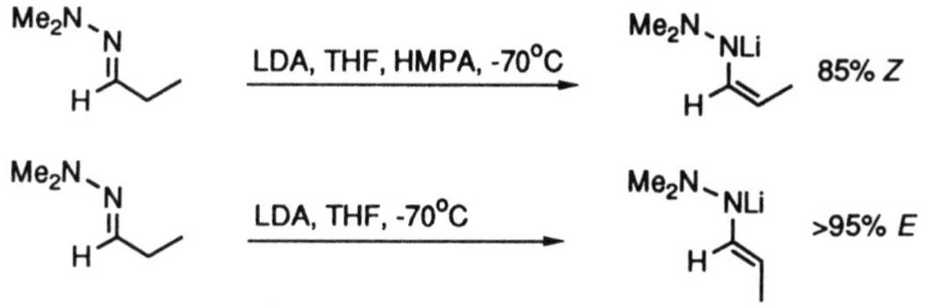

Methods for preparing the corresponding enolates derived from N-arylimines have also been developed.[129] Kinetic deprotonation of the anil of 3-pentanone gives predominantly the *E*-isomer. The *Z*-isomer may be obtained by equilibrating the enolate mixture with an excess of the strongest acid catalyst tolerable in this solution, namely the anil of 3-pentanone.

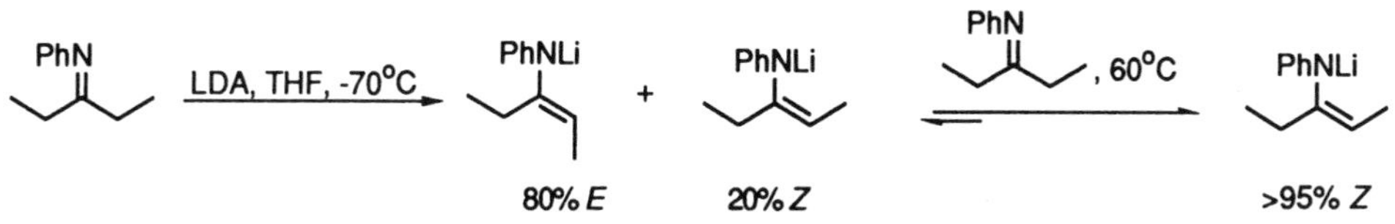

(e)   *Nitro-stabilized carbanions*

Because of their excellent reactivity, these are very popular nucleophiles in studies of additions to new conjugate acceptors. Many bases, including hindered amines such as DBN and TBTMG as well as the powerful LDA and NaH, have been used to deprotonate nitroalkanes.

Clark et al have examined a variety of fluoride catalysts, including supported catalysts, in the reaction of nitroethane with 3-buten-2-one.[130] Some of their results are summarized in Table 1.2. The best loading was found to be ~2.5 molecules (of the fluoride salt) nm^{-2} of well-dried surface material. It was also found that oxygen needed to be excluded in order to avoid oxidation of the (presumed) aci-complex. Hence the major products resulting from stirring CsF in nitroethane were the dimer, 2,3-dinitrobutane and caesium hydrogen fluoride.

$$\text{NO}_2 + \text{CsF} \rightleftharpoons \overset{\text{OH·F}^-\text{Cs}^+}{\underset{\text{O}^-}{\text{N}^+}} \xrightarrow{\text{O}_2} \text{NO}_2 + \text{CsHF}_2 \longrightarrow \underset{\text{NO}_2}{\overset{\text{NO}_2}{\bigvee}}$$

Table 1.2   Conjugate addition of nitroethane to 3-but-en-2-one with F⁻ catalysis

Fluoride catalyst	t_{50}(min)[a]
CsF-alumina	1
KF-alumina	2
CsF-EtOH	3
CsF(wet)	7
CsF(dry)	12
Alumina(dry)	30
KF-18-crown-6	50
KF(dry)	1000

[a] Time for 50% reaction as measured by gas chromatography

(f)   *Nitrile-stabilized carbanions*

One of the important classes of nitrile-stabilized carbanions is that based on cyanohydrins. Their importance lies in their use as masked acyl anions.[131] Again, these require the use of fairly powerful bases, typically LDA, for their generation.

Cyanohydrin based acyl anion synthons

$$\underset{R}{\overset{\text{OR'}}{NC\overset{|}{\underset{|}{C}}H}} \qquad \begin{array}{l}\text{R = alkyl, aryl, alkenyl}\\ \text{R' = TMS, EE, THP}\end{array} \qquad \underset{R}{\overset{\text{NR'}_2}{NC\overset{|}{\underset{|}{C}}H}} \qquad \begin{array}{l}\text{R = alkyl, aryl, alkenyl}\\ \text{R' = H, alkyl, aryl}\end{array}$$

1.4.1.2   *Carbanions stabilized by π-conjugation with more than one heteroatom*

This group of carbon acids constitutes the most acidic of those to be discussed here. As opposed to some other areas of carbanion chemistry, where high acidity of the carbanion precursor can often be correlated with low reactivity (e.g. in S$_N$2 type reactions), these very "soft" nucleophiles react very successfully with a range of conjugate acceptors. Because of their relatively high acidity, very mild conditions are usually employed. A good example of this is the addition of 2-methylcyclopentane-1,3-dione to 3-buten-2-one. No added catalyst is required as the dione is sufficiently acidic to catalyse its own addition.[132] In fact, base-catalysis of this reaction leads to another product being formed. This type of side reaction, caused by base- or acid-catalysed cleavage of the initial adduct, occurs in other systems, including cyclohexane-1,3-dione[133] and 2-methylcyclohexane-1,3-dione.[134,135,136] (See also Section 2.1.1.1).

1.4.1.3   Carbanions stabilized by one or more α-heteroatoms

These anions are almost invariably generated using very strong bases, usually *n*-butyllithium.

Acetal based acyl anion synthons

1.4.2   Carbanions without heteroatom stabilization

1.4.2.1   Copper-based reagents

The first organocopper compound to be reported in the literature was phenylcopper (PhCu), prepared by Reich in 1923.[137] The first study of the potential for organocoppers in synthesis came from Henry Gilman's group in 1936.[138] They reported the successful preparation of an *alkyl*copper reagent, ethylcopper (EtCu), produced from ethylmagnesium iodide and cuprous iodide. Five years later, in the second paper in its series on "factors determining the course and mechanisms of Grignard reactions", Kharasch's group provided unequivocal evidence for the involvement of copper halides in promoting conjugate addition.[139] In particular, they examined the addition of methylmagnesium bromide to isophorone and discovered that, in the absence of any added metal salts, methylmagnesium iodide only gave products arising from 1,2-addition. However, addition of 1.0 mole per cent of cuprous chloride led to almost exclusive conjugate addition in high yield.

43%   48%

CH₃MgBr
CuCl, 0.01 eq
Et₂O

83%   7%

In 1966, House and his group demonstrated that organocuprates were the reactive species responsible for conjugate addition to E-3-penten-2-one and also were, by analogy, the reactive species involved in Kharasch's earlier studies.[140] Over the last twenty years or so there has been something of an explosion in the number and type of organocopper-based reagents which have been prepared. Because of their importance to the subject of conjugate addition, in this section we will group organocopper-based reagents into structural types and briefly examine the methods by which they are prepared.

(a)   *Mono-organocoppers (RCu) and heterocuprates*

Preparation of reagents

Mono-organocoppers are generally prepared, and reacted, *in situ*, according to the following metathesis reaction:

$$RM + CuX \xrightarrow{\text{solvent, low temp}} RCu + MX$$

The most common method of preparation is treatment of either a suspension (in the case of Cu(I) halides) or solution (in the case of CuOAc or soluble Cu(I) salt-ligand complexes) of the copper salt, usually with an organolithium or organomagnesium reagent,[141] in diethyl ether or tetrahydrofuran, at low temperature. Very recently it has been shown that mono-organocopper reagents prepared in neat dimethyl sulfide (in the presence of an equivalent of a lithium or magnesium salt) are both more stable and more reactive than when prepared in diethyl ether or tetrahydrofuran. [142]

The wider use of organocopper chemistry was retarded somewhat by the lack of reproducible results. One of the contributing factors was the use of impure copper salts. This situation no longer exists as several copper salts may be prepared in pure form as their triphenylphosphine,[143] trimethylphosphite[144] or dimethyl sulfide[145,146] complexes. Recrystallization affords Cu(I) salt-ligand complexes pure, free from any impurities. It also renders these salts soluble in solvents such as diethyl ether and tetrahydrofuran. Reaction of any of these salts with either an organolithium or organomagnesium then generates the organocopper reagent.

Reactivity

Although several mono-organocoppers, free of salts, ligands or metal contaminants, can be prepared, there is little synthetic value in isolating the reagents prior to use. They are thermally unstable, often explosive solids, which are also air and moisture sensitive. It has also been demonstrated that, in conjugate addition reactions, salt-free RCu may be less reactive[147] than other organocopper complexes or totally inactive.[148]

Thus mono-organocoppers are generally reacted in the presence of a metal salt (and often a solubilizing ligand-see below). Both House[149], and Riviere[147] demonstrated that addition of either LiX or $MgBr_2$ to inactive, pure organocoppers restores the latter's ability to participate in conjugate additions. For example:

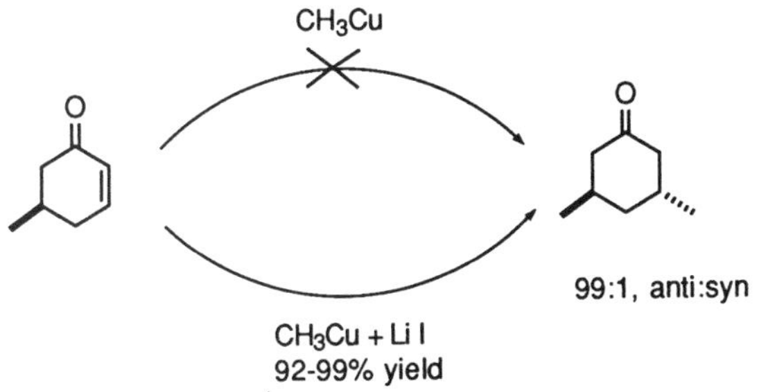

CH$_3$Cu

CH$_3$Cu + Li I
92-99% yield

99:1, anti:syn

These observations, of course, cast strong doubts on the role of RCu in conjugate additions.[150] It is more likely that some type of complex between the organocopper and, at least, the halide is the reactive species. The extreme case would be the following equilibrium, although, to date, there is little evidence for such species being formed.

$$RCu + MX \rightleftharpoons [RCuX]^- M^+$$

At present it is more usual to represent the reactive species as RCu.MX. The dot "denotes a weak interaction between RCu and MX".[151]

A very useful study of the reactivity of mono-organocoppers in several solvents has been reported.[151] It was found that dimethyl sulfide was superior in all cases. Some of the findings are given in the Table below for the following reaction:

Yields of conjugate adduct (%)

Copper (I) salt	DMS	THF		Et$_2$O	
	-78°C	-78°C	0°C	-78°C	0°C
CuI	99	24	40	71	95
CuBr.SMe$_2$	63^a	14	38	36	65
CuOTf	21^b	3	2	1	< 1
CuCN	79	4	60	76	73

a plus 37% of dimeric by-product
b plus 38% of dimeric by-product

The same group also found that mono-organocopper reagents are more stable in dimethyl sulfide than other solvents at higher temperatures, especially >-75°. This was interpreted as being due to the dimethyl sulfide occupying the vacant orbital required for hydride transfer during decomposition via β-hydride elimination.[152]

As with other classes of organocopper reagents, the reactivity of mono-organocoppers is dramatically improved by including chlorotrimethylsilane with a polar additive, in this case TMEDA, in the reaction.[153] Even polymeric methylcopper gives reasonable results under these conditions. As only one equivalent of organic ligand is necessary, these reagents are more economical than organocuprates under conventional conditions. However, little stereoselectivity was obtained with acyclic alkenones. An alternative, which also works well, is to use hexamethylphosphoramide (or 4-(N,N-dimethylamino)pyridine) instead of TMEDA.[154] Lipshutz's group have shown that a

E:Z 1:3.2

reagent, previously formulated as RCu.BF$_3$, contains LiI as an integral component.[155]

Heterocuprates (see next section) may be formed by using strongly coordinating ligands, such as PhS⁻, Ph$_2$P⁻ and R$_2$N⁻. The earlier heterocuprates tended to be thermally unstable and could only be used at low temperatures (though this often proved quite satisfactory). However, those prepared from Ph_2P^- and R_2N^- are quite stable and very reactive (much more so than mixed alkynyl homocuprates too).[156]

(b) Lower order organocuprates (LO cuprates) including copper-catalyzed reactions

LO cuprates have the general structure shown below.

R'(R)CuM   M = Li, MgX

There are several catagories of LO cuprates:

(i)   homocuprates   R = R' = alkyl, alkenyl, alkynyl or aryl

e.g.   Me$_2$CuLi

(ii)   mixed homocuprates   R ≠ R'   R, R' = alkyl, alkenyl, alkynyl or aryl

e.g.[157] $\left(MeO \diagdown \diagup \!\!\!\equiv\!\!\! \diagup \right) CuMeLi$

(iii) heterocuprates   R = alkyl, alkenyl, alkynyl or aryl
                               R' = CN or inorganic

e.g.   Me(Ph$_2$P)CuLi

These are also, most likely, the reactive species involved in copper-*catalysed* organolithium additions where the copper salt is a halide. Lipshutz et al[158] have provided spectroscopic evidence that, in diethyl ether, even with an excess of organolithium only *lower* order cuprates are formed, i.e. :

3 RLi + CuI ⟶ R$_2$CuLi + RLi + LiI   [<u>not</u> R$_3$CuLi$_2$ + LiI]

(However, in tetrahydrofuran, the situation is quite different and higher order cuprates are formed[159] - see page 46). It has also been suggested that it is more accurate to represent these LO cuprates as R$_2$CuLi.LiI or R$_2$Cu(I)Li$_2$.[159] Hence, copper-catalyzed reactions, which by their nature probably always generate homocuprates, will are included in this section.

Preparation of reagents

Homocuprates are most commonly formed simply by the addition of *two* equivalents of an organolithium or organomagnesium reagent to a copper salt in diethyl ether or tetrahydrofuran at low temperature. Mixed homocuprates are formed by the sequential addition of molar equivalents of the two different organolithiums or organomagnesiums to a copper salt:

$$CuX \xrightarrow{\text{RLi}} RCuLiX \xrightarrow{\text{R'Li}} \boxed{R'(R)CuLi} \;+\; LiX$$

In the cases where one of the organic ligands is an alkynyl group, metal-hydrogen exchange is also used:

$$R\!=\!\!=\!\!-H \;+\; R'_2CuLi \longrightarrow R\!=\!\!=\!\!-Cu(R')Li \;+\; R'H$$

Alkenyl(cyano)cuprates may be prepared from trialkyl(alkenyl)stannanes.[160]

Residual (non-transferable) ligands

Because homocuprates only deliver one of their two organic ligands, an inactive ligand was sought which would be cheap and remain attached to the copper during the reaction. (If the latter does not apply, then symmetrization will occur which may well lead to a change in the course of the reaction).[161,162,163] This led to the testing of many different groups, focussing on ligands which were known to bond strongly to copper. The resultant, mixed homocuprates contain two different ligands, often designated R_R and R_T (Residual ligand - also called the "dummy" ligand and Transferable ligand, respectively) yielding reagents with the general formula $R_R R_T CuM$.

Some examples of R_R in $R_R R_T CuM$

The purity of the organolithium used to generate an LO cuprate is also extremely important. For example, it has been shown that the use of solutions of alkyllithiums, from bottles which had already been opened and resealed, led to loss of stereoselectivities of up to 90%! This was apparently due to alkoxide impurities which formed highly reactive, non-selective cuprate species.[169]

Aggregation

Aggregation has been recognized as an important influence in enolate reactions.[170] Consideration of its role in organocopper chemistry is an extremely complex issue as not much is known about the structure of most of these reagents. However, this situation is slowly improving (see discussion below). Dieter and Tokles found that the change from an LO cuprate (containing a chiral ligand) to an HO cuprate, by the inclusion of an extra ligand, changed the sense of asymmetric induction in conjugate additions to alkenones. They concluded that this was due to geometrical and structural changes in the cuprate cluster rather than any subtle change in the orientation of the reacting components.[171]

The influence of crown ethers on reactivity is also well-known.[172] The inclusion of the crown ether, 18-crown-6, in a reaction between an LO cuprate and some cycloalkenones prevents any conjugate addition occurring. This is most likely due to the crown ether sequestering lithium ion from the dimer, producing a completely unreactive ion (see page 44). Addition of lithium iodide restores the reactivity, presumably as this effectively competes for the crown ether. In turn, this allows the reformation of the reactive dimeric cuprate.

Lewis acids

In recent years several groups have demonstrated that the inclusion of various Lewis acids, in particular $BF_3.Et_2O$ and trialkylhalosilanes, in cuprate additions to alkenones, alkenals and alkenoates can lead to enormous increases in reaction rates as well as significant improvement in 1,4/1,2 addition ratios. Lipshutz et al have studied the effects of $BF_3.Et_2O$ on reactions involving LO cuprates.[173] Mainly on the basis of low temperature n.m.r. spectroscopic studies, it was concluded that the BF_3 modifies the originally formed cuprate. This appears to be a consequence of the BF_3 sequestering one RLi from R_2CuLi (which is probably dimeric,[174] i.e. $R_4Cu_2Li_2$) forming R_3Cu_2Li, which is the actual reactive nucleophilic species. A second role of the BF_3, naturally, is substrate activation through complexation to the carbonyl oxygen.[175] The same group has also studied the effect of eleven different Lewis acids on the conjugate addition of an HO cuprate to isophorone.[176] Only $BF_3.Et_2O$ provided excellent yields (98%), whereas most of the others proved to be incompatible with the cuprate resulting in no addition products at all.

Trialkylhalosilanes with or without polar additives

The role of trialkylhalosilanes appears to vary with the nature of the cuprate. LO cuprates react very slowly with trialkylhalosilanes,[177,178] at low temperature, permitting the latter to trap the initial enolate as its silyl enol ether.[179,180] The rate of addition is dramatically increased, as exemplified by the following results for the addition of an LO cuprate to 3-methyl-2-cyclohexenone.[154] Importantly, only a stoichiometric amount of *ligand* is necessary.

without TMSCl	1h	28% (+ 70% recovered starting material)
with TMSCl, 2 eq.,	5 min	99%
n-Bu₂CuLi, 0.6 eq., THF, -78°C / TMSCl, 2 eq., HMPA, 2 eq., 3h		87% (plus 13% recovered starting material)

These additives have also been studied in copper-catalyzed additions.[181] Again little reaction occurs, at low temperature, without the additives. Only with both the chlorosilane and hexamethylphosphoramide do the yields become satisfactory. The use of either cuprous iodide as the catalyst and/or diethyl ether as solvent leads to much poorer results.

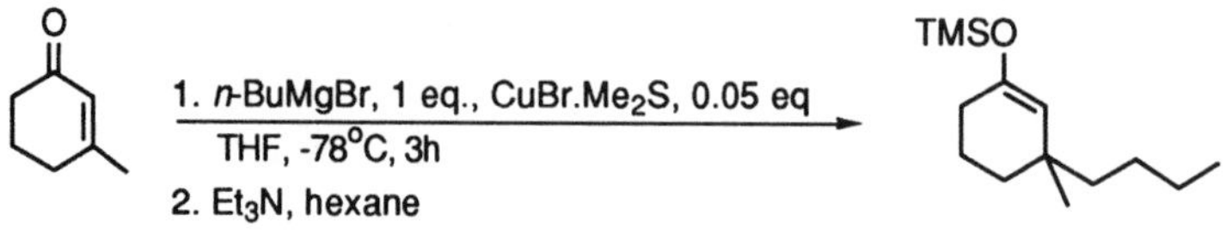

without additives	little reaction
with TMSCl, 2 eq	30-40% plus ~10% 1,2 addition
with TMSCl, 2 eq. and HMPA, 2 eq	99% 1,4 : 1,2 >200:1

The use of these additives in additions to alkenals gives, perhaps, the most impressive improvements, providing high yields and excellent stereoselectivity with both LO cuprate[182] and copper-catalyzed[181] reactions.

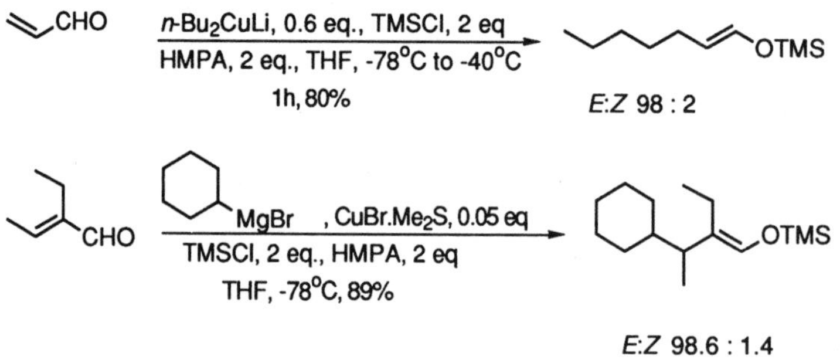

Also, conjugate addition sometimes occurs *only* when a trialkylhalosilane is included. Often the yield is improved significantly. Conjugate additions to alkenoic acid esters and amides also benefit from the inclusion of a trialkylhalosilane. Alkenyl nitriles react well in the presence of a trialkylhalosilane, but the products are those of double alkylation. Alkenones show better reactivity under these conditions, however 1,2 addition is still a significant problem.

Ph $\diagdown$ CO_2Et | 1. Me₂CuLi, 1.2 eq.,TMSCl, 1.2 eq → Et₂O, -78°C, 1h then warm to +20°C over 2h | Ph $\diagdown$ CO_2Et

1. Me₂CuLi, 1.2 eq.,TMSCl, 1.2 eq

Et_2O, -78°C, 1h then warm to +20°C

over 2h

2. Aq. NH₄Cl

Ph $\diagup\diagdown$ CO_2Et

97% with TMSCl
38% without TMSCl

CO_2Me

1. Ph₂CuLi, 1.2 eq., TMSCl, 1.2 eq

Et_2O, -78°C, then 48h 20°C

2. Aq. NH₄Cl

Ph $\diagup\diagdown$ CO_2Me

75% with TMSCl
18% without TMSCl

Ph $\diagdown$ $CONMe_2$

1. Me₂CuLi, 1.2 eq., TMSCl, 1.2 eq

Et_2O, -78°C, 1h then warm to +20°C

over 2h

2. Aq. NH₄Cl

Ph $\diagup\diagdown$ $CONMe_2$

95% with TMSCl
0% without TMSCl

The inclusion of a trialkylhalosilane can sometimes completely change the stereochemical course of a reaction (see section 3.1.2).

Metal salts

Metal salts are nearly always present in the reaction mixtures containing LO cuprates. As pointed out above, n.m.r. studies have suggested that the metal salt should be included in the formulation of, at least, some LO cuprates. LiI has been shown to promote the isomerization of allenolates[183] and equilibria between (Me₂CuLi)₂ and Me₃Cu₂Li + MeLi in tetrahydrofuran/diethyl ether.[184]

(c)   *Higher order organocuprates (HO cuprates)*[185]

Preparation of reagents

Higher order organocuprates are formed when two or more equivalents of an organolithium or organomagnesium reagent are added to copper cyanide or copper thiocyanate. (There has been some debate over the actual existence of HO (cyano)cuprates[186]).

2 RM + CuX $\xrightarrow{\text{Et}_2\text{O or THF}}_{\text{low temp}}$ $\left[R_2Cu(X)M_2 \right]$ R = alkyl, aryl or alkenyl
X = CN, SCN
M = Li or MgY

As mentioned in the previous section, the combination of RLi and a copper halide appears to give only LO cuprates. It remains to be seen whether or not this is also true for organomagnesiums. In the case of HO cuprates a halosilane appears to modify the initial cuprate. For example, even at low temperatures, $Me_2Cu(CN)Li_2$ is converted into the LO cuprate Me_2CuLi:

$$Me_2Cu(CN)Li_2 + TMSCI \xrightarrow{\text{THF, } \leq -78°C} Me_2CuLi + TMSCN + LiCl$$

Thus the TMSCl sequesters lithium cyanide from the HO cuprate generating, as well as the LO cuprate, TMSCN and LiCl. It seems likely that the TMSCN plays an important role in the enhanced selectivity observed in 1,2 additions involving this combination of reagents.[187]

Bertz's group has demonstrated that the formation of HO cuprates can be solvent-dependent.[159] Thus in tetrahydrofuran or diethyl ether, three equivalents of phenyllithium and cuprous iodide yields only Ph_2CuLi and PhLi. Switching to dimethyl sulfide as solvent leads to the generation of a new HO cuprate, Ph_3CuLi_2. This HO cuprate reacts with 2-cyclohexenone much faster than phenyllithium. This is the first HO cuprate, *not* derived from CuCN or CuSCN, so far reported.

(d)   Structure

The first crystal structure studies of Cu(I)-activated alkene complexes have been reported.[188,189] These show copper almost equidistant from the 2 alkene carbons, with little change in bond length of the double bond. This suggests mainly donation from the alkene to the copper with little back-bonding from the copper d-orbitals to π^* of the alkene. The first structure of an organocuprate to be determined, showed that the Cu_2Li_2 unit is almost planar.[190] Each of the R-Cu-R units are almost linear. A similar conclusion, for the structure of dimethylcopperlithium, was reached using MO calculations.[191] Corey and Boaz have isolated a solid which they believe to be a π-complex.[192] So far no crystal structures have been obtained on simple lithium dialkylcuprates. However, several crystallographic studies of lithium di*aryl*cuprates have been reported. These include $[\{Li(OEt)_2\}(CuPh_2)]_2$[193] and $Li_3Cu_2Ph_5(SMe_2)_4$.[194] In the former, the ring formed by the copper and lithium is not planar and is folded at the copper atoms.

Ullenius and Christenson propose that the change in reactivity often associated with a change in solvent reflects a change in the structure of the reacting species.[195] This can be correlated with the Lewis basicity of the solvent (ligand). Thus a progression in solvents from non-coordinating, to weakly coordinating and finally strongly coordinating, would be accompanied by the following structural changes:

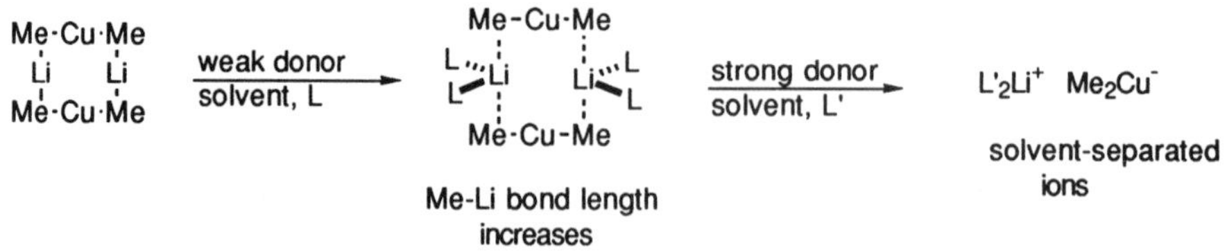

Figure 1.7   Proposed structural dependence of an LO cuprate on the nature of the solvent

Support for the formation of solvent-separated ions comes form the crystal structure of dimethylcopperlithium and 12-crown-4 which shows well-separated cations, $(Li(12-c-4)_2)^+$ and linear anions of Me_2Cu^-.[196]

1.4.2.2   *Other organometallic reagents*

Many organometallic reagents, other than organocopper reagents, add in a 1,4 fashion to conjugate acceptors. In fact, the first example of a 1,4-addition of an organomagnesium reagent was reported as early as 1905. Given the intense interest in the development and application of organocopper chemistry, it is fascinating to note how useful other organometallic reagents can be in conjugate additions. Often these reagents have given poor results in the past because of impurities present in the metal used to prepare the reagent. For example, Munch-Petersen's group demonstrated that, by triply subliming magnesium before using it, the yield of the addition of *n*-butylmagnesium bromide to *sec*-butyl cinnamate improved from 30-35% to 83%.[197] More recently methods for obtaining magnesium in highly divided (and therefore highly reactive) form have been developed, although there has since been little mention of the use of such high grade magnesium in conjugate additions.

Often these hard nucleophiles add well, in a conjugate fashion, if the acceptor is appropriately designed or where 1,2-addition is not possible. Many of these reagents, namely organolithiums and organomagnesiums, are commercially available as solutions in several different solvents. Preparation of organometallic reagents which are not available commercially is most commonly achieved by metal-halogen or metal-hydrogen exchange. However, many other methods exist.[109]

Some work has been done on conjugate additions with triorganozincates. So far, they have only been added successfully to alkenones. There are some parallels with organocopper chemistry in that some residual (non-transferable) ligands have been identified. They also catalyze conjugate additions, giving comparable yields and stereoselectivity to reactions using stoichiometric amounts of the reagents. They may be prepared in one of three ways, as outlined below (see sections 2.1.3 and 3.1.3 for some specific examples and references). The use of the TMEDA-zinc chloride complex is particularly promising as the crystalline complex is air-stable for many months.[198]

$$R_2Zn + R'M \xrightarrow[\text{or -78°C, Et}_2\text{O}]{\text{0°C, THF}} R_2R'ZnM$$

$$ZnCl_2 + 3RM \xrightarrow{\text{-78 to 0°C, Et}_2\text{O}} R_3ZnM + 2MCl$$

$$ZnCl_2\text{.TMEDA} + 3RM \xrightarrow{\text{-78 to 0°C, THF}} R_3ZnM + 2MCl$$

Unlike organocoppers, however, organozincates are not effective in transferring hindered ligands such as *tert*-butyl and *neo*-pentyl.[199] The most popular residual ligand appears to be methyl,[199,200] although *tert*-butoxide is also effective.[201] Indeed, the only way in which methyl can be transferred from an organozincate is by using a transition metal catalyst. A large number of these were screened and Co(acac)$_3$ was found to be the best.[199]

The case of alkynylmetal reagents is particularly important. It is well established that alkynylcopper reagents almost never transfer the alkynyl ligand. In fact, this property makes alkynyls ideal as residual (i.e. non-transferable, see section 1.4.2.1 above) ligands in LO and HO cuprates. However, alkynylboron, aluminium and nickel (and with specially-designed acceptors, lithium) reagents do successfully add in a conjugate fashion.

1.4.3   Free radicals

By far the most common method for generating organic free radicals for conjugate addition reactions has been from an organic halide and AIBN. However several other methods, including the use of Vitamin B$_{12a}$ and its synthetic equivalents or ultrasonication, exist.[202]

1.4.4   Enamines, enol ethers and silyl ketene acetals

1.4.4.1   *Enamines*

The first enamine was prepared by Meyer and Hopf in 1921.[203] They prepared an enamine of acetaldehyde by pyrolyzing choline. However the first general method was developed by Mannich and Davidsen and involved heating a mixture of an aldehyde or

ketone with a secondary amine in the presence of a dehydrating agent such as potassium carbonate.[204] In their early work on enamines, Stork's group found that this method worked well for monosubstituted acetaldehydes but that removal of water by azeotropic distillation with benzene, toluene or xylene was the preferred procedure for cyclic ketones and disubstituted acetone derivatives.[205]

Enamines are, in essence, enol or enolate equivalents. However, one of the virtues of enamines is that they are often sufficiently reactive to add to many conjugate acceptors by simply mixing and (sometimes) heating the two together in solution. No added catalyst is necessary. As a consequence many acid or base sensitive groups may be present in either reactant. This is to be contrasted with the reaction conditions usually required of conventional enol/enolate chemistry (see section 1.4.1). In general, this also means that mono alkylation is more easily achieved. A second virtue of enamines is that they usually form the less substituted isomer from unsymmetrical ketones. For example, 2-methylcyclohexanone forms 3-methyl-2-pyrrolidinocyclohexene rather than the more substituted isomer, 2-methyl-1-pyrrolidinocyclohexene. This is probably due to the methyl substituent in the latter case preventing the amine lone pair from overlapping effectively with the alkene π orbitals.

Metalloenamines, have also found applications in conjugate additions. Methods for their preparation are discussed in section 1.4.1.1.

1.4.4.2   Enol ethers and silyl ketene acetals

Enol ethers most commonly used in conjugate additions are silylenol ethers. These may be prepared by quenching specifically generated enolates, whose preparation is outlined in section 1.4.1.1 or by conjugate addition in the presence of a halosilane (see section 1.4.2.1). This also applies to ketene acetals which are, formally, enol ethers of esters. Their preparation is also outlined in section 1.4.1.1. Trialkylsilyl ketene acetals of thioesters may be prepared with high selectivity for either isomer.[119]

1. LDA, 1.1 eq., THF, -78°C, 30 min	R=*t*-Bu, R'=Me, X=OTf	> 95% *Z*
2. R$_2$R'SiX, 1.1 eq., warm to r.t.	R=R'=Me, X=Cl	90% *Z*
3. Phosphate buffer, pH 7		

1. LDA, 1.1 eq., 25% HMPA in THF, -78°C	R=*t*-Bu, R'=Me, X=Cl	95% *E*
2. R$_2$R'SiX, 1.1 eq., warm to r.t.	R=R'=Me, X=Cl	93% *E*
3. Phosphate buffer, pH 7		

1.4.5   Cyanide

Hydrocyanation of conjugate acceptors is quite a general process. It is often an excellent method for the regio- and stereoselective addition of a C1 unit to a conjugated system.[206] Its only real drawback is that often the cyanide group, once introduced, is difficult (if not impossible) to manipulate further. Where 1,2 addition is also possible, e.g. additions to alkenones, the reverse process is much faster than the reverse conjugate addition. Therefore, the conjugate addition product usually rapidly accumulates as the major, often sole, product. The earliest examples of conjugate hydrocyanations, reported by Bredt and Kallen, described the reaction of diethyl benzalmalonate and diethyl ethylidenemalonate with potassium cyanide and hydrochloric acid.[207]

1. KCN, 1.1 eq., EtOH, H$_2$O	
2. Add conc. HCl with cooling	
3. R.t., 24h	

Over the next few decades the mechanism of addition was studied in detail. It was recognized, early on, that cyanide ion was the reactive species[208] and that the rate of reaction is directly proportional to the cyanide concentration. Since those early studies several sets of conjugate hydrocyanation procedures have been developed. They include:

(i)   KCN, 2 eq., AcOH, H_2O, EtOH[209]

(ii)   KCN or NaCN, 2 eq., NH_4Cl, 1.5 eq., 90% aq. DMF, 100°C[210]

(iii) $HCN_{liq.}$, KCN (or other bases), 40-80°C[211]

(iv)   Acetone cyanohydrin, base, aq. alcohol, reflux[212]

(iii)   HCN, AlR_3, THF[213,214]

(iv) R_2AlCN, THF or other polar aprotic solvent[215,213]

HCN is a weak acid, pK_a 9.21, and, therefore, in general does not add to conjugate acceptors. Thus all the reaction conditions given above involve either acid or base catalysis. A very useful chart, listing hydrocyanations of a number of classes of conjugate acceptors with both the reported and "promising" methods is given in Nagata and Yoshioka's review.[216] This should be consulted when choosing a particular hydrocyanation procedure.

1.4.6   Heteronucleophiles

1.4.6.1   Nitrogen and phosphorus

Most intermolecular additions of nitrogen nucleophiles (e.g. amines and imines) involve use of the free base. Reactions are sometimes carried out neat, although it is more common to use a solvent. Both base and acid catalysis can be used. Preformed nitrogen anions have only occasionally been used as these usually give, exclusively, 1,2-addition. Addition to iron acyls is an exception as are the additions of lithium N-trimethylsilylbenzylamine to alkenoates. Intramolecular additions have mostly involved carbamates. These have been carried out using quite strong bases, such as potassium *tert*-butoxide. Iodine or mercury salts have been used to promote amine additions as well. Phosphorus nucleophiles may react either in their neutral form, or, quite often in their anionic form. For a discussion of these, see sections 2.1.7.1 and 4.1.7.1).

1.4.6.2   Oxygen, sulfur and selenium

Oxygen nucleophiles are mostly added as their alkoxide salts. On the other hand, phase transfer catalyzed (PTC) addition of sulfur nucleophiles has proved to be a very popular method for adding these intrinsically soft nucleophiles. Dialkylaluminium thiolates also deliver sulfur nucleophiles efficiently.[217] Mild acids, such as acetic acid promote the addition of selenium nucleophiles.

1.4.6.3   Silicon and tin

Silyl and stannyl anions may be prepared in several ways. Very often they are prepared by deprotonation of the silane or stannane with *n*-butyllithium at low temperature. Metal-halogen exchange is also used with halosilanes. Trimethylstannyl cuprate can be prepared by generating trimethylstannane *in situ*, thereby avoiding the direct use of this highly volatile and toxic compound.[218]

1.4.6.4   Iodine

Only additions of iodide have been included in this book. Iodotrimethylsilane is an excellent reagent for the conjugate addition of iodide to alkenones, often in the presence of a Lewis acid.

1.5   Intramolecular reactions

<u>1.5.1   Baldwin's rules for ring closure</u>

Not all ring closures proceed with equal facility. In fact, some are known to be difficult or impossible to effect. Partly as a consequence of his studies on the following transformation, Baldwin recognized that there may be geometrical constraints on the closure of various sized ring systems.

Out of these studies came Baldwin's rules for ring closure.[219] These rules describe the relative ease with which various sizes of rings may be formed from different types of ring closures. The term used to describe the ring-forming process is divided into three parts. The first is a number which denotes the size of the ring being formed. The second is a descriptor denoting whether the bond being broken is exocyclic to the smallest so formed ring (*exo*) or whether the bond being broken is endocyclic to the smallest so formed ring (*endo*). The third term describes the geometry at the centre onto which the ring is closed. Hence for a tetrahedral centre the term is *tet*, a trigonal centre, *trig* and a digonal centre, *dig*. Finally, X must be a first row element.

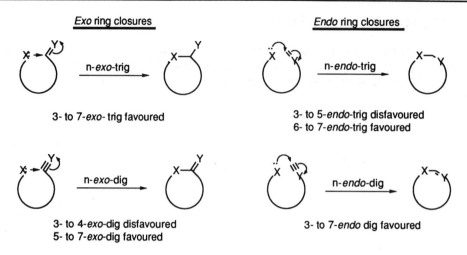

Exo ring closures   Endo ring closures

n-*exo*-trig

3- to 7-*exo*- trig favoured

3- to 5-*endo*-trig disfavoured
6- to 7-*endo*-trig favoured

n-*exo*-dig

n-*endo*-dig

3- to 4-*exo*-dig disfavoured
5- to 7-*exo*-dig favoured

3- to 7-*endo* dig favoured

Figure 1.8   Baldwin's rules for ring closure

These processes are relevant to intramolecular conjugate additions. For example, the hydroxyalkenoate shown below could, in principle, close in either a 5-*exo*- or 5-*endo*-

5-*endo*-trig   5-*exo*-trig

base   base

fashion, leading to the γ-lactone and cyclic ether respectively. In practice, the intramolecular conjugate addition, the 5-*endo*-trig process, is not observed and only the 5-*exo*-trig process (lactonization) occurs.[220] Thus it is important to keep these rules in mind when incorporating intramolecular conjugate additions into synthetic schemes.[221]

Baldwin has extended these rules to consider cyclizations involving enolates as the nucleophile. A new descriptor, for the type of enolate, is added to the term describing the process. Thus if the enolate is exocyclic to the new ring being formed it is described as enol-*exo*, whereas if it is endocyclic to the new ring being formed it is described as enol-*endo*.[222] These processes are particularly relevant to base-catalyzed intramolecular Michael reactions.

Enolate ring closures

n-[enol-*exo*]-*exo*-trig   n-[enol-*endo*]-*exo*-trig

3- to 7-[enol-*exo*]-*exo*- trig favoured   3- to 5-[enol-*endo*]-*exo*-trig disfavoured
6- to 7-[enol-*endo*]-*exo*-trig favoured

1.5.2   Stereoselective ring closures

An interesting extension of the idea of "parallel" *versus* "anti-parallel" attack (see section 1.3.7.2) can be made to include ring closures.[223] Hence it becomes rather easy to see why *cis*-fusion is such a common outcome of these ring closures (see, for example, section 3.2.2).

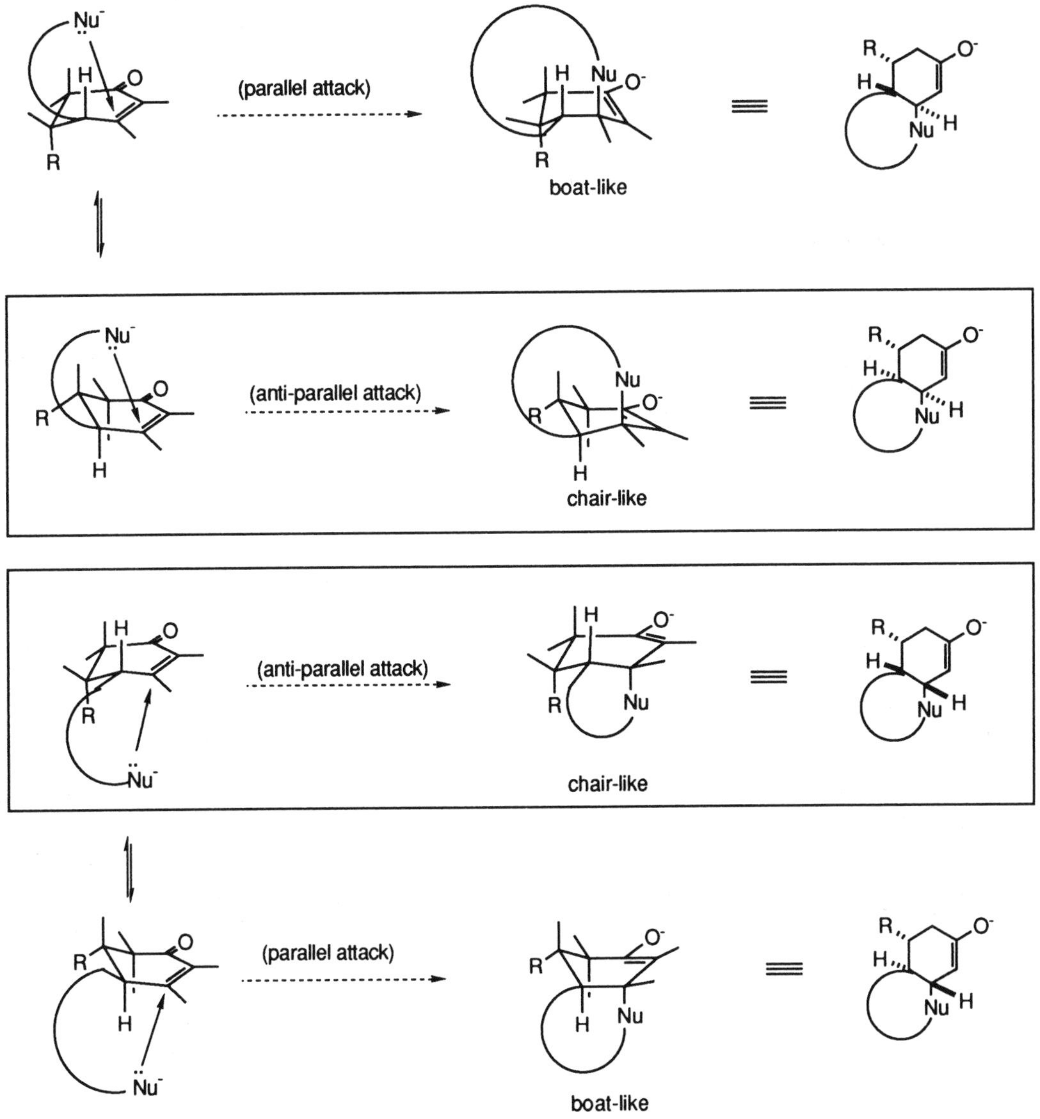

Figure 1.9   *Parallel and anti-parallel attack in ring closures*

Cis-fused products also predominate in ring closures of cyclic enolates onto conjugate acceptors and similar arguments to those just outlined above may be used to account for this as well.[224]

1.6 Tandem reactions

There are many combinations of events which can be linked to a conjugate addition. Some of these are described below.

Conjugate addition followed by:

Quenching with a proton - this can occur either during the reaction or during workup, e.g.[225]

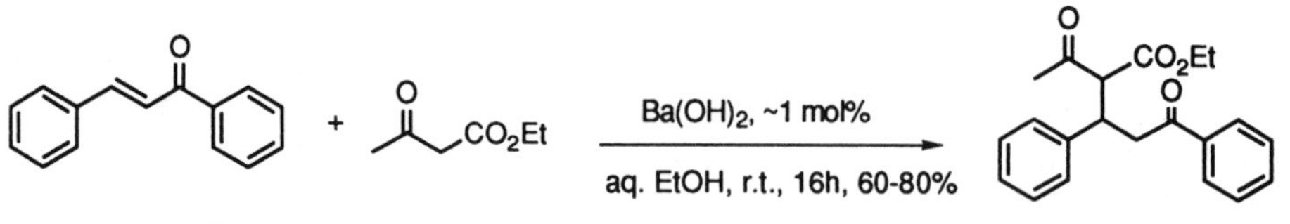

O
O
CO_2Et
+
$Ba(OH)_2$, ~1 mol%
aq. EtOH, r.t., 16h, 60-80%
O
CO_2Et
O

Quenching with a haloalkane

- intramolecularly: cyclization[226]

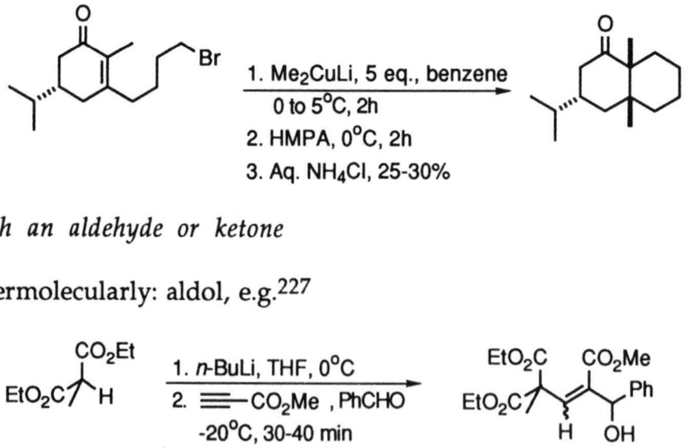

O
Br
1. Me_2CuLi, 5 eq., benzene
0 to 5°C, 2h
2. HMPA, 0°C, 2h
3. Aq. NH_4Cl, 25-30%
O

Quenching with an aldehyde or ketone

- intermolecularly: aldol, e.g.[227]

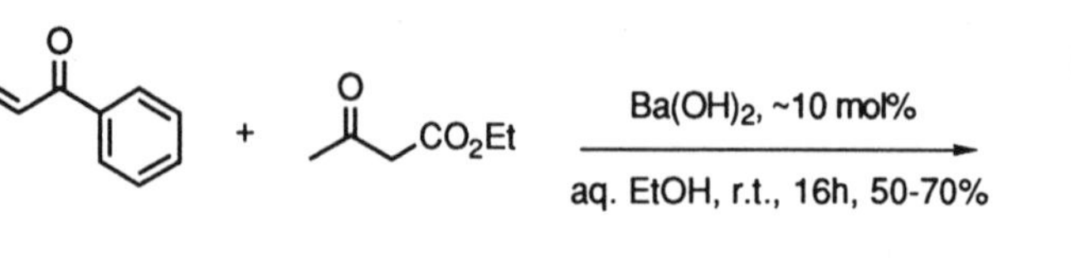

CO_2Et
EtO_2C
H
1. n-BuLi, THF, 0°C
2. ≡—CO_2Me , PhCHO
-20°C, 30-40 min
3. Aq. NH_4Cl, -10°C
47%
EtO_2C
CO_2Me
Ph
EtO_2C
H
OH

- intramolecularly: cyclization (MIRC: Michael-ring closure) e.g.[225]

O
O
+
CO_2Et
$Ba(OH)_2$, ~10 mol%
aq. EtOH, r.t., 16h, 50-70%
O
EtO_2C
HO

often followed by dehydration, e.g.

Quenching with a second conjugate acceptor - often leading to a multicomponent annulation, e.g. an MIMIRC[228]

Elimination of a leaving group - often with retention, e.g.[229]

References

1.   T. Komnenos, *Liebigs Ann. Chem.*, 1883, **218**, 145
2.   For another early example, see L. Claisen, *J. Prakt. Chem.*, 1887, **35**, 413
3.   Michael's nationality is emphasized here only because it is a widely held misconception that Michael was German. Many of his early publications were in German journals largely because he studied, in 1873, with Hofmann in Berlin, then with Bunsen in Heidelberg and once again with Hofmann from 1875 to 1878. He spent 1879 working firstly with Wurtz in Paris and Mendeleev in Russia before returning to America in 1880
4.   For brief biographies of Michael's life and career, see E. W. Forbes, L. F. Fieser and A. B. Lamb, *Harvard University Gazette*, 1943, **38**, 246; A. B. Costa, *J. Chem. Ed.*, 1971, **48**, 243 and D. S. Tarbell and A. T. Tarbell, "Essays on the History of Organic Chemistry in the United States, 1875-1955", Folio, Nashville, 1986, 45
5.   A. Michael, *J. Prakt. Chem.*, 1887, **3**, 349 and *Am. Chem. J.*, 1887, **9**, 112
6.   One major review of the Michael reaction has been published: E. D. Bergmann, D. Ginsburg and R. Pappo, *Org React.*, 1959, **10**, 179
7.   For an excellent review, see D. A. Oare and C. H. Heathcock, in E. L. Eliel and S. H. Wilen (eds), "Topics in Stereochemistry", Wiley, New York, 1989, 227
8.   M. Isobe. M. Kitamura and T. Goto, *Tetrahedron Lett.*, 1980, **21**, 4727
9.   For reviews on cyanoethylation, see H. A. Bruson, *Org. React.*, 1949, **5**, 79 and R. J. Harper, Jr., in H. F. Mark, D. F. Othmer, C. G. Overberger and G. T. Seaborg

      (eds), "Kirk-Othmer Encyclopaedia of Chemical Technology", Wiley, New York, 1979, 370

10.   D. A. Oare and C. H. Heathcock, in "Topics in Stereochemistry", E. L. Eliel and S. H. Wilen (eds), Wiley, New York, 1989, 227

11.   J. March, "Advanced Organic Chemistry", Wiley-Interscience, New York, 1985, 155

12.   G. H. Alt and A. G. Cook in A. G. Cook, "Enamines", Marcel Dekker, New York, 1988, Chapter 4

13.   W. Nagata and M. Yoshioka, *Org. React.*, 1977, **25**, 255

14.   A. Krief, *Tetrahedron*, 1980, **36**, 2531

15.   Ian Fleming, "Frontier Orbitals and Organic Chemical Reactions", Wiley, London, 1976, 70

16.   R. G. Pearson, *J. Amer. Chem. Soc.*, 1963, **85**, 3533

17.   T. L. Ho, "Hard and Soft Acids and Bases Principle in Organic Chemistry", Academic Press, New York, 1977

18.   A. Loupy and J. Seydenne-Penne, *Tetrahedron Lett.*, 1978, 2571

19.   A. Sevin, J. Tortajada and M. Pfau, *J. Org. Chem.*, 1986, **51**, 2671

20.   J. March, "Advanced Organic Chemistry", Wiley-Interscience, New York, 1985, 23

21.   For the first report of this effect, see T. M. Dolak and T. A. Bryson, *Tetrahedron Lett.*, 1977, 1961

22.   See T. Cohen, W. D. Abraham and M. Meyers, *J. Amer. Chem. Soc.*, 1987, **109**, 7923 and references cited therein

23.   This is the case with hexamethylphosphoramide (HMPA). However, tetramethylethylenediamine (TMEDA), another strong complexing agent for lithium cations, is not effective in promoting the formation of solvent-separated ion pairs, apparently due to the difficulty of accommodating eight methyl groups (from two TMEDA molecules) around a lithium ion - see W. N. Setzer and P. R. von Schleyer, *Adv. Organomet. Chem.*, 1985, **24**, 353

24.   M. R. Myers and T. Cohen, *J. Org. Chem.*, 1989, **54**, 1290

25.   Z. Gross and S. Hoz, *J. Amer. Chem. Soc.*, 1988, **110**, 7489

26.   C. K. Ingold, "Structure and Mechanism in Organic Chemistry", Cornell University Press, Ithaca, 1969, 1015

27.   W. G. Davies, E. W. Hardisty, T. P. Nevell and R. H. Peters, *J. Chem. Soc. B*, 1970, 998

28.   B. A. Feit and A. Zhilka, *J. Org. Chem.*, 1963, **28**, 406

29.   P. De Maria and A. Fini, *J. Chem. Soc., Perkin Trans. II*, 1973, 1773

30.   For a review, see S. I. Suminov and A. N. Kost, *Russ. Chem. Rev.*, 1969, **38**, 884

31.   P. De Maria and A. Fini, *J. Chem. Soc. B*, 1971, 2335

32.   P. De Maria and M. Falzone, *Chim. e Ind.*, 1972, **54**, 791

33.   M. Friedman, J. F. Cavins and J. S. Wall, *J. Amer. Chem. Soc.*, 1965, **87**, 3672

34.   C. L. Bumgardner, J. E. Bunch and M-H. Whangbo, *J. Org. Chem.*, 1986, **51**, 4083

35.   For other examples of anti-Michael reactions, see T. Holm, I. Crossland and J. Ø. Madsen, *Acta Chem. Scand., B*, 1978, **32**, 754 and G. W. Klumpp, A. J. C. Mierop, J. J. Vrielink, A. Brugman and M. Schakel, *J. Amer. Chem. Soc.*, 1985, **107**, 6740

36.   H. O. House, *Acc. Chem. Res.*, 1976, **9**, 59

37.   See also Y. Yamamoto, S. Nishii and T. Ibuka, *J. Amer. Chem. Soc.*, 1988, **110**, 617

38.   C. P. Casey and M. Cesa, *J. Amer. Chem. Soc.*, 1979, **101**, 4236

39.   D. J. Hannah and R. A. J. Smith, *Tetrahedron Lett.*, 1975, 187

40.   R. A. J. Smith and D. J. Hannah, *Tetrahedron*, 1979, **35**, 1183

41.   J. A. Marshall and R. A. Ruden, *J. Org. Chem.*, 1972, **37**, 659

42.   E. J. Corey, F. J. Hannon and N. W. Boaz, *Tetrahedron*, 1989, **45** , 545

43.   C. Ullenius and B. Christenson, *Pure Appl. Chem.*, 1988, **60**, 57

44.   P. W. Jolly and R. Mynott, *Adv. Organometal. Chem.*, 1981, **19**, 257

45.   S. R. Krauss and S. G. Smith, *J. Amer. Chem. Soc.*, 1981, **103** , 141

46.   S. H. Bertz and R. A. J. Smith, *J. Amer. Chem. Soc.*, 1989, **111**, 8276

47.   See also J. Berlan, J.-P. Battioni and K. Koosha, *Bull Soc. Chim. Fr.*, 1979, II-183

48.   C. R. Johnson and G. A. Dutra, *J. Amer. Chem. Soc.*, 1973, **95**, 7777 and 7783

49.   C. P. Casey and M. C. Cesa, *J. Amer. Chem. Soc.*, 1979, **101**, 4236

50.   A. E. Dorigo and K. Morokuma, *J. Amer. Chem. Soc.*, 1989, **111**, 4635

51.   A. E. Dorigo and K. Morokuma, *J. Amer. Chem. Soc.*, 1989, **111**, 6524

52.   R. J. Loncharich, T. R. Schwartz and K. N. Houk, *J. Amer. Chem. Soc.*, 1987, **109**, 14

53.   M. S. DeGroot and J. Lamb, *Proc. R. Soc. London, Ser. A*, 1957, **242**, 36

54.   C. E. Blom, R. P. Müller and H. H. Gunthard, *Chem. Phys. Lett.*, 1980, **73**, 483

55.   W. J. Hehre, J. A. Pople and A. J. P. Devaquet, *J. Amer. Chem. Soc.*, 1976, **98**, 664 and *ibid.* 1980, **102**, 2253

56.   H. B. Burgi and J. D. Dunitz, *Acc. Chem. Res.*, 1983, **16**, 153

57.   For reviews, see B. A. Shainyan, *Usp. Khim.*, 1986, **55**, 942; Z. Rappoport, *Recl. Trav. Chim. Pays-Bas*, 1985, **104**, 309 and Z. Rappoport, *Acc. Chem. Res.*, 1981, **14**, 7

58.   See also, G. Lodder, J. W. J. van Dorp, B. Avramovitch and Z. Rappoport, *J. Org. Chem.*, 1989, **54**, 2574 and references cited therein

59.   W. E. Truce and M. L. Gorbaty, *J. Org. Chem.*, 1970, **35**, 2113

60.   J. Biougne, F. Théron and R. Vessiere, *Bull Soc. Chim. Fr.*, 1975, 2703

61.   F. Montanari and A. Negrini, *Gazz. Chim. Ital.*, 1959, **89**, 1543

62.   G. Modena and P. E. Todesco, *Gazz. Chim. Ital.*, 1959, **89**, 866 and L. Maioli and G. Modena, *ibid.*, 1959, **89**, 854

63.   E. Angeletti and F. Montanari, *Bull. Sci. Fac. Chim. Ind. Balogna*, 1958, **16**, 140

64.   C. F. Bernasconi, R. B. Killion, Jr., J. Fassberg and Z. Rappoport, *J. Amer. Chem. Soc.*, 1989, **111**, 6862

65.   The assignment of *E* stereochemistry for both the starting material and product in each case is so far unsubstantiated

66.   D. A. Oare and C. H. Heathcock, *J. Org. Chem.*, 1990, **55** , 157

67.   J. Bertrand, L. Gorrichon and P. Maroni, *Tetrahedron*, 1984, **40**, 4127 and references cited therein

68.   A. G. Schultz and Y. K. Lee, *J. Org. Chem.*, 1976, **41**, 4045

69.   N-y. Wang, S-s. Su and L-y. Tsai, *Tetrahedron Lett.*, 1979, 1121

70.   D. Morgans Jr. and G. B. Feigelson, *J. Org. Chem.*, 1982, **48**, 1131

71.   D. A. Oare, M. A. Henderson, M. A. Sanner and C. H. Heathcock, *J. Org. Chem.*, 1990, **55**, 157

72.   M. P. Cooke, Jr. and R. Goswami, *J. Amer. Chem. Soc.*, 1977, **99**, 642

73.   For an early study, see E. P. Kohler and F. R. Butler, *J. Amer. Chem. Soc.*, 1926, **48**, 1041

74.   D. A. Evans in J. D. Morrison (ed.), "Asymmetric Synthesis", Volume 3, Academic Press, New York, 1984, pp. 50 to 73

75.   J. Mulzer, A. Chucholowski, O. Lammer, I. Jibril and G. Huttner, *J. Chem. Soc., Chem. Commun.*, 1983, 869. The structures of the major products shown in this paper have the incorrect relative configuration

76.   R. Lawrence and P. Perlmutter, *J. Org. Chem.*, 1990, **55**, submitted for publication

77.   R. M. Williams, "Synthesis of Optically Active α-Amino Acids, Pergamon Press, Oxford, 1989

78.   W. Oppolzer, C. Chapuis, M. D. Guo, D. Reichlin and T. Godel, *Tetrahedron Lett.*, 1982, **23**, 4781

79.   W. Oppolzer, C. Chapuis and G. Bernardi, *Tetrahedron Lett.*, 1984, **25**, 5885

80.   M. Vandewalle, J. Van der Eycken, W. Oppolzer and C. Vullioud, *Tetrahedron*, 1986, **42**, 4035

81.   N. Iwasawa and T. Mukaiyama, *Chem. Lett.*, 1981, 913

82.   P. Mangeney, A. Alexakis and J. F. Normant, *Tetrahedron Lett.*, 1983, **24**, 373

83.   K. Tomioka, T. Suenaga and K. Koga, *Tetrahedron Lett.*, 1986, **27**, 369

84.   H. Herzog and H. Scharf, *Synthesis*, 1986, 420

85.   P. Somfai, D. Tanner and T. Olsson, *Tetrahedron*, 1985, **41**, 5973

86.   For a recent theoretical treatment, see A. E. Dorigo and K. Morokuma, *J. Amer. Chem. Soc.*, 1989, **111**, 6524

87.   X= OR, reaction with divinylcopperlithium, W. R. Roush and M. B. Lesur, *Tetrahedron Lett.*, 1983, **24**, 2231; reaction with alkoxide, J. Mulzer, M. Kappert, G. Huttner and I. Jibril, *Angew. Chem. Int. Ed. Engl.*, 1984, **23**, 704

88.   X = OR, see ref 87 and M. Larchevegue, G. Tamagnan and Y. Petit, M. Hirama, *J. Chem. Soc., Chem. Commun.*, 1989, 31; T. Iwakuma and S. Ito, *J. Chem. Soc., Chem. Commun.*, 1987, 1523; M. Isobe, Y. Ichawa, Y. Funabashi, S. Mio and T. Goto, *Tetrahedron*, 1986, **42**, 2863; J. Mulzer and M. Kappert, *Angew. Chem. Int. Ed. Engl.*, 1983, **22**, 63; X = NR$_2$, M. T. Reetz and D. Röhrig, *Angew. Chem. Int. Ed. Engl.*, 1989, **28**, 1706

89.   A. Bernardi, A. M. Capelli, C. Gennari and C. Scolastico, *Tetrahedron Asymmetry*, 1990, **1**, 21

90.   Y. Yamamoto, S. Nishii and T. Ibuka, *J. Chem. Soc., Chem. Commun.*, 1987, 1572

91.   D. Kruger, A. E. Sopchik and C. A. Kingsbury, *J. Org. Chem.*, 1984, **49**, 778

92.   K. Tatsuta, Y. Amemiya, Y. Kanemura and M. Kinoshita, *Tetrahedron Lett.*, 1981, **22**, 3997

93.   J. Mulzer, M. Kappert, G. Huttner and I. Jibril, *Angew. Chem. Int. Ed. Engl.*, 1984, **23**, 704

94.   M. T. Reetz and D. Röhrig, *Angew. Chem. Int. Ed. Engl.*, 1989, **28**, 1706

95.   S. D. Kahn, K. D. Dobbs and W. J. Hehre, *J. Amer. Chem. Soc.*, 1988, **110** , 4602 and references cited therein

96.   The Hehre view has been disputed, see T. Koizumi, Y. Arai and H. Takayama, *Tetrahedron Lett.*, 1987, **28**, 3689

97.   D. J. Abbott, S. Colonna and C. J. M. Stirling, *J. Chem. Soc., Chem. Commun.*, 1971, 471

98.   G. H. Posner, J. P. Mallamo, M. Hulce and L. L. Frye, *J. Amer. Chem. Soc.*, 1982, **104**, 4180

99.   G. H. Posner, L. L. Frye and M. Hulce, *Tetrahedron*, 1984, **40**, 1401

100.   G. H. Posner and L. L. Frye, *Tetrahedron Lett.*, 1984, **25**, 379

101.   G. Tsuchihashi, S. M. Mitamura, S. Inouye and S. Oguro, *Tetrahedron Lett.*, 1973, 323

102.	A. B. Smith III, N. K. Dunlap and G. A. Sulikowski, *Tetrahedron Lett.*, 1988, **29**, 439 and, A. B. Smith III and P. K. Trumper, *ibid*, 1988, **29**, 443
103.	P. Chamberlain and G. H. Whitham, *J. Chem. Soc., Perkin Trans. II*, 1972, 130
104.	This nomenclature was introduced in the following paper: J. Valls and E. Toromanoff, *Bull Soc. Chem. France.*, 1961, 758
105.	N. L. Allinger and C. K. Riew, *Tetrahedron Lett.*, 1966, 1269
106.	M. Yanagita, S. Inayama, M. Hirakura and F. Seki, *J. Org. Chem.*, 1958, **23**, 690
107.	Organocuprates also add with very high stereoselectivity in a 1,6 fashion, see J. A. Campbell and J. C. Babcock, *J. Amer. Chem. Soc.*, 1959, **81**, 4069 and J. A. Marshall and H. Roebke, *J. Org. Chem.*, 1966, **31**, 3109
108.	For a brief review, see D. A. Oare and C. H. Heathcock, in "Topics in Stereochemistry", E. L. Eliel and S. H. Wilen (eds), Wiley, New York, 1989, 227
109.	J. C. Stowell, "Carbanions in Organic Synthesis", Wiley-Interscience, New York, 1979
110.	I. Kuwajima, T. Murobushi and E. Nakamura, *Synthesis*, 1978, 602
111.	D. A. Oare and C. H. Heathcock, *Tetrahedron Lett.*, 1986, **27**, 6169
112.	C. H. Heathcock, C. T. Buse, W. A. Kleschick, M. C. Pirrung, J. E. Sohn and J. Lampe, *J. Org. Chem.*, 1980, **45**, 1066
113.	G. Stork and P. F. Hudrlik, *J. Amer. Chem. Soc.*, 1968, **90**, 4462 and 4464
114.	H. O. House, L. J. Czuba, M. Gall and H. D. Olmstead, *J. Org. Chem.*, 1969, **34**, 2324
115.	See also, G. W. Spears, C. E. Caufield and W. C. Still, *J. Org. Chem.*, 1987, **52**, 1226
116.	R. E. Ireland, R. H. Mueller and A. K. Willard, *J. Amer. Chem. Soc.*, 1976, **98**, 2868. The nomenclature used here to assign *E*- or *Z*-enolate stereochemistry uses the convention suggested by Evans in his review on the alkylation of chiral enolates (see ref. 74). Thus, the metal enolate oxygen always takes priority over the other geminal substituent.
117.	E. J. Corey and A. W. Gross, *Tetrahedron Lett.*, 1984, **25**, 495
118.	K. Kpegba, P. Metzner and R. Rakotonirina, *Tetrahedron Lett.*, 1986, **27**, 1505
119.	C. Gennari, M. G. Beretta, A. Bernardi, G. Moro, C. Scolastico and R. Todeschini, *Tetrahedron*, 1986, **42**, 893
120.	D. A. Evans and L. R. McGee, *Tetrahedron Lett.*, 1980, **21**, 3975
121.	D. A. Evans, J. M. Takacs, L. R. McGee, M. D. Ennis, D. Mathre and J. Bartroli, *Pure Appl. Chem.*, 1981, **53**, 1109
122.	D. A. Oare, M. A. Henderson, M. A. Sanner and C. H. Heathcock, *J. Org. Chem.*, 1990, **55**, 132
123.	D. A. Oare and C. H. Heathcock, in "Topics in Stereochemistry", E. L. Eliel and S. H. Wilen (eds), Wiley, New York, 1989, 321
124.	See Y. Tamaru, Y. Furukawa, M. Mizutani, O. Kitao and Z. Yoshida, *J. Org. Chem.*, 1983, **48**, 3631 and references cited therein
125.	R. H. Schlessinger, M. A. Poss and S. Richardson, *J. Amer. Chem. Soc.*, 1986, **108**, 3112
126.	R. H. Schlessinger, P. Lin, M. A. Poss and J. Springer, *Heterocycles*, 1987, **25**, 315
127.	For reviews, see P. W. Hickmott, *Tetrahedron*, 1982, **38**, 3363
128.	K. G. Davenport, H. Eichenauer, D. Enders, M. Newcomb and D. E. Bergbreiter, *J. Amer. Chem. Soc.*, 1979, **101**, 5654
129.	R. Knorr and P. Löw, *J. Amer. Chem. Soc.*, 1980, **102**, 3241
130.	J. H. Clark, D. G. Cork and H. W. Gibbs, *J. Chem. Soc., Perkin Trans. I*, 1983, 2253
131.	For a review, see J. D. Albright, *Tetrahedron*, 1983, **39**, 3207

132.   U. Eder, G. Sauer and R. Wiechert, *Angew. Chem. Int. Edn. Eng.*, 1971, **10**, 496;
        Z. G. Hajos and D. R. Parrish, *J. Org. Chem.*, 1974, **39**, 1612. See also, L. F. Tietze
        and T. Eicher, "Reactions and Syntheses in the Organic Chemistry Laboratory",
        University Science Books, Mill Valley, 1989, 169
133.   K. Balasubramanian, J. P. John and S. Swaminathan, *Synthesis*, 1974, 51
134.   I. N. Nazarov and S. I. Zav'yalov, *Izvest. Akad. Nauk. SSSR Otd. Khim. Nauk* ,
        1952, 300
135.   N. L. Wender and H. L. Slatey, U. S. pat. 2,542,223; [*Chem. Abstr.*, 1951, **45**, 7599]
136.   S. Ramachandran and M. S. Newman, *Org. Synth.*, 1973, **Coll. Vol. V**, 486
137.   J. Reich, *Compt. Rend.*, 1923, **177**, 322
138.   H. Gilman and J. M. Straley, *Rec. Trav. Chem*, 1936, **55**, 821
139.   M. S. Kharasch and P. O. Tawney, *J. Amer. Chem. Soc.*, 1941, **63**, 2308
140.   H. O. House, W. L. Respess and G. M. Whitesides, *J. Org. Chem.*, 1966, **31**, 3128
141.   There is also a renewed interest in organozinc-derived copper reagents. See, for
        example, M. C. P. Yeh, P. Knochel, W. M. Butler and S. C. Berk, *Tetrahedron Lett.*,
        1988, **29**, 6693. For earlier work, see K. H. Thiele and J. Kohler,
        J. Organometal. Chem., 1968, **12**, 225
142.   S. H. Bertz and G. Dabbagh, *Tetrahedron*, 1989, **45**, 425
143.   J. B. Siddall, M. Biskup and J. H. Fried, *J. Amer. Chem. Soc.*, 1969, **91**, 1853
144.   H. O. House and W. F. Fischer, Jr., *J. Org. Chem.*, 1968, **33**, 949
145.   G. H. Posner, "An Introduction to Synthesis Using Organocopper Reagents",
        Wiley, New York, 1980
146.   H. O. House, C-Y. Chu, J. M. Wilkins and M. J. Umen, *J. Org. Chem.*, 1955, **20**, 1460
147.   N. T. Luong-Thui and H. Riviere, *Compt. Rend.*, 1968, **267**, 776
148.   See N. T. Luong-Thui and H. Rivière, *Tetrahedron Lett.*, 1970, 1583 and
        references cited therein
149.   H. O. House and W. F. Fischer Jr., *J. Org. Chem.*, 1968, **33**, 949
150.   For the preparation of salt-free methylcopper, see B. H. Lipshutz, J. A. Kozlowski
        and C. M. Breneman, *J. Amer. Chem. Soc.*, 1985, **107**, 3197
151.   S. H. Bertz and G. Dabbagh, *Tetrahedron*, 1984, **45**, 1234
152.   G. M. Whitesides, E. J. Panek and E. R. Stedronsky, *J. Amer. Chem. Soc.*, 1972, **94**,
        232
153.   C. R. Johnson and T. J. Marren, *Tetrahedron Lett.*, 1987, **28**, 27
154.   E. Nakamura, S. Matsuzawa, Y. Horiguchi and I. Kuwajima, *Tetrahedron Lett.*,
        1986, **27**, 4029
155.   B. H. Lipshutz, E. L. Ellsworth and S. H. Dimock, *J. Amer. Chem. Soc.*, 1990, **112**,
        5869
156.   S. H. Bertz and G. Dabbagh, *J. Amer. Chem. Soc.*, 1982, **104**, 5824
157.   E. J. Corey, D. Floyd and B. H. Lipshutz, *J. Org. Chem.*, 1978, **43**, 3418
158.   B. H. Lipshutz, J. A. Kozlowski and R. S. Wilhelm, *J. Amer. Chem. Soc.*, 1985, **107**,
        3197
159.   S. H. Bertz and G. Dabbagh, *J. Amer. Chem. Soc.*, 1988, **110**, 3668
160.   E. Piers and V. Karunaratne, *J. Chem. Soc., Chem. Commun.*, 1983, 935
161.   E. J. Corey and D. J. Beames, *J. Amer. Chem. Soc.*, 1972, **94**, 7210
162.   G. M. Whitesides, W. F. Fischer, Jr., J. San Filippo, Jr., R. W. Bashe and H. O.
        House, *J. Amer. Chem. Soc.*, 1969, **91**, 4871
163.   H. O. House, D. G. Koespell and W. J. Campbell, *J. Org. Chem.*, 1972, **37**, 1003
164.   C. R. Johnson and D. S. Dhanua, *J. Chem. Soc., Chem. Commun.*, 1982, 358

165.   H. Malmberg, M. Nilsson and C. Ullenius, *Tetrahedron Lett.*, 1982, **23**, 3823

166.   B. H. Lipshutz, J. A. Kozlowski, D. A. Parker and S. L. Nguyen, *J. Organomet. Chem.*, 1985, **285**, 437

167.   H. Malmberg, M. Nilsson and C. Ullenius, *Acta Chem. Scand. B.*, 1981, **35**, 625

168.   S. H. Bertz, G. Dabbagh and G. Sundararajan, *J. Org. Chem.*, 1986, **51**, 4953

169.   E. J. Corey, R. Naef and F. J. Hannon, *J. Amer. Chem. Soc.*, 1986, **108**, 7115

170.   D. Seebach, *Angew. Chem. Int. Ed. Engl.*, 1988, 1624

171.   R. K. Dieter and M. Tokles, *J. Amer. Chem. Soc.*, 1987, **109**, 2040

172.   C. Ouannes, G. Dressaire and Y. Langlois, *Tetrahedron Lett.*, 1977, 815

173.   B. H. Lipshutz, E. L. Ellsworth and T. J. Siahaan, *J. Amer. Chem. Soc.*, 1989, **111**, 1351

174.   See G. Van Koten, J. T. B. H. Jastrezebski, F. Muller and C. H. Stam, *J. Amer. Chem. Soc.*, 1985, **107**, 697 and references cited therein

175.   M. T. Reetz, M. Hullmann, W. Massa, S. Berger, P. Rademacher and P. Heymanns, *J. Amer. Chem. Soc.*, 1986, **108**, 2405

176.   B. H. Lipshutz, *Synthesis*, 1987, 325

177.   C. Chuit, J. P. Foulon and J. F. Normant, *Tetrahedron*, 1980, **36**, 2305 and 1981, **37**, 1385

178.   M. Bourgain-Commerçon, J. P. Foulon and J. F. Normant, *J. Organomet. Chem.*, 1982, **228**, 321

179.   E. J. Corey and N. W. Boaz, *Tetrahedron Lett.*, 1985, **26**, 6015 and 6019

180.   A. Alexakis, J. Berlan and Y. Besace, *Tetrahedron Lett.*, 1986, **27**, 1047

181.   Y. Horiguchi, S. Matsuzawa, E. Nakamura and I. Kuwajima, *Tetrahedron Lett.*, 1986, **27**, 4025

182.   Y. Horiguchi, S. Matsuzawa, E. Nakamura and I. Kuwajima, *Tetrahedron Lett.*, 1986, **27**, 4029

183.   N. Krause, *Tetrahedron Lett.*, 1989, **30**, 5219

184.   See reference 150

185.   For reviews of HO cuprates, see B. H. Lipshutz, *Synthesis*, 1987, 325 and B. H. Lipshutz, R. S. Wilhelm and J. A. Kozlowski, *Tetrahedron*, 1984, **40**, 5005

186.   See S. H. Bertz, *J. Amer. Chem. Soc.*, 1990, **112**, 4031 (for the prosecution) and B. H. Lipshutz, S. Sharma and E. L. Ellsworth, *ibid.*, 1990, **112**, 4032 (for the defence)

187.   B. H. Lipshutz, E. L. Ellsworth T. J. Siahaan and A. Shirazi, *Tetrahedron Lett.*, 1988, **29**, 6677

188.   S. Andersson, M. Håkansson, S. Jagner, M. Nilsson, C. Ullenius and F. Urso, *Acta Chem. Scand.*, 1986, A40, 58

189.   S. Andersson, M. Håkansson, S. Jagner, M. Nilsson and F. Urso, *Acta Chem. Scand.*, 1986, A40, 194

190.   G. van Koten, J. T. B. H. Jastrzebski, F. Muller and C. H. Stam, *J. Amer. Chem. Soc.*, 1985, **107**, 697

191.   K. R. Stewart, J. R. Lever and M-H. Whangbo, *J. Org. Chem.*, 1982, **47**, 1472

192.   E. J. Corey and N. W. Baoz, *Tetrahedron Lett.*, 1985, **26**, 6015

193.   N. P. Lorenzen and E. Weiss, *Angew. Chem. Int. Ed. Engl.*, 1990, **29**, 300

194.   M. M. Olmstead and P. P. Power, *J. Amer. Chem. Soc.*, 1989, **111**, 4135

195.   C. Ullenius and B. Christenson, *Pure Appl. Chem.*, 1988, **60**, 57

196.   H. Hope, M. M. Olmstead, P. P. Power, J. Sandell and X. Xu, *J. Amer. Chem. Soc.*, 1985, **107**, 4337

197.   S. R. Jensen, A-M. Kristiansen and J. Munch-Petersen, *Acta Chem. Scand.*, 1970, **24**, 2641

198.   R. A. Watson and R. A. Kjonaas, *Tetrahedron Lett.*, 1986, **27**, 1437

199.   W. Tuckmantel, K. Oshima and H. Nozaki, *Chem. Ber.*, 1986, **119**, 1581

200.   R. A. Watson and R. A. Kjonaas, *Tetrahedron Lett.*, 1986, **27**, 1437

201.   J. F. G. A. Jansen and B. L. Feringa, *Tetrahedron Lett.*, 1988, **29**, 3593

202.   B. Geise, "Radicals in Organic Synthesis: Formation of Carbon-Carbon Bonds", Pergamon Press, Oxford, 1986

203.   K. H. Meyer and H. Hopf, *Chem. Ber.*, 1921, **54**, 2277

204.   C. Mannich and H. Davidsen, *Chem. Ber.*, 1936, **69**, 2106

205.   G. Stork, A. Brizzolara, H. Landesman, J. Szmuszkovicz and R. Terrell, *J. Amer. Chem. Soc.*, 1963, **85**, 207

206.   For a review, see W. Nagata and M. Yoshioka, *Org. React.*, 1977, **25**, 255

207.   J. Bredt and J. Kallen, *Ann. Chem.*, 1896, **293**, 338. It is possible that some earlier hydrocyanations involved additions to alkenenitrile intermediates. See A. Claus, *Ann. Chem.*, 1873, **170**, 126 and *ibid*, 1878, **191**, 33

208.   A. Lapworth, *J. Chem. Soc.*, 1903, **83**, 995

209.   A. C. O. Hann and A. Lapworth, *J. Chem. Soc.*, 1904, **85**, 1335

210.   W. Nagata, S. Hirai, H. Itazaki and K. Takeda, *J. Org. Chem.*, 1961, **26**, 2413

211.   P. Kurtz, *Liebig's Ann. Chem.*, 1951, **572**, 23

212.   I. N. Nazarov and S. I. Zav'yalov, *J. Gen. Chem. USSR (Engl. Transl.)*, 1954, **24**, 475

213.   W. Nagata, M. Yoshioka and S. Hirai, *Tetrahedron Lett.*, 1962, 461

214.   For a mechanisitic study of this reagent see W. Nagata, M. Yoshioka and M. Murakami, *J. Amer. Chem. Soc.*, 1972, **94**, 4644

215.   W. Nagata and M. Yoshioka, *Tetrahedron Lett.*, 1966, 1913

216.   W. Nagata and M. Yoshioka, *Org. React.*, 1977, **25**, Table B, pp 320-323

217.   C. Paulmier, "Selenium Reagents and Intermediates in Organic Synthesis", Pergamon Press, Oxford, 1986

218.   B. H. Lipshutz and D. C. Reuter, *Tetrahedron Lett.*, 1989, **30**, 4617

219.   J. E. Baldwin, *J. Chem. Soc., Chem. Commun.*, 1976, 734 and 738; J. E. Baldwin, R. C. Thomas, L. I. Kruse and L. Silberman, *J. Org. Chem.*, 1977, **42**, 3846 and J. E. Baldwin, in "Further Perspectives in Organic Chemistry", CIBA Foundation Symposium 53, Elsevier-North Holland, Amsterdam, 1978

220.   J. E. Baldwin, *J. Chem. Soc., Chem. Commun.*, 1976, 736

221.   An *ab initio* study of the *6-endo* trig process revealed that it is ~25 kJ mol^{-1} higher in energy than the intermolecular counterpart. The transition state is distorted from the optimum geometry for addition and the extra energy is the cost paid for the compromise in the stereoelectronic requirements for the reaction. See C. I. Bayly and F. Greim, *Can. J. Chem.*, 1989, **67**, 2173

222.   J. E. Baldwin and L. I. Kruse, *J. Chem. Soc., Chem. Commun.*, 1977, 233 and J. E. Baldwin and K. J. Lusch, *Tetrahedron*, 1982, **38**, 2939

223.   D. A. Oare and C. H. Heathcock, in E. L. Eliel and S. H. Wilen (eds), "Topics in Stereochemistry", Wiley, New York, 1989

224.   D. A. Oare and C. H. Heathcock, in E. L. Eliel and S. H. Wilen (eds), "Topics in Stereochemistry", Wiley, New York, 1989, 372-374

225.   A. Garcia-Raso, J. Garcia-Rosa, J. V. Sinisterra and R. Mestres, *J. Chem. Ed.*, 1986, **63**, 443

226.   G. H. Posner, J. J. Sterling, C. E. Whitten, C. M. Lentz and D. J. Brunelle, *J. Amer. Chem. Soc.*, 1975, **97**, 107

227.   A. Bury, S. D. Joag and C. J. M. Stirling, *J. Chem. Soc., Chem. Commun.*, 1986, 124

228.   G. H. Posner, S.-B. Lu, E. Asirvatham, E. F. Silversmith and E. M. Shulman, *J. Amer. Chem. Soc.*, 1986, **108**, 511

229.   B. Bernet, P. M. Bishop, M. Caron, T. Kawamata, B. L. Roy, L. Ruest, G. Sauve P. Soucy and P. Deslongchamps, *Can. J. Chem.*, 1985, **63**, 2810

230.   J. R. Mohrig, S. S. Fu, R. W. King, R. Warnet and G. Gustafson, *J. Amer. Chem. Soc.*, 1990, **112**, 3665

231.   I. Fleming and N. D. Kindon, *J. Chem. Soc., Chem. Commun.*, 1987, 1177

232.   See also S. S. Wong, M. N. Paddon-Row, Y. Li and K. N. Houk, *J. Amer. Chem. Soc.*, 1990, **112**, 8679

233.   For an extensive discussion on additions to enones see "The Chemistry of Enones", S. Patai and Z. Rappoport (eds), Wiley, Chichester, 1989

2   Introduction

Acyclic alkenones are one of the most important classes of conjugate acceptors. Much of the early work on these additions focussed on developing the annulation reaction invented by Robinson (see the first section of this chapter). The need to develop an enantioselective version of this reaction led to the investigation of numerous combinations of chiral ligands and catalysts. Some of these are outlined in the sections that follow. The failure of organometallic reagents to add to alkenones (albeit, originally, cycloalkenones) reproducibly, in a conjugate fashion, also provided the impetus for determining the reasons for this. This led to the discovery of the crucial role of copper in these reactions.

2.1   Intermolecular reactions

2.1.1   Stabilized carbanions

2.1.1.1   *Carbanions stabilized by π-conjugation with one heteroatom*

(a)   *Addition of achiral carbanions to achiral alkenones*

(i)   *Ketone enolates*

Ketone enolates give only products of conjugate addition, even at low temperatures. However, this is probably due to the extended reaction times, as it has been observed previously that aldolates do form, reversibly, at these temperatures.[1] Remarkably, the stereoselectivity of the conjugate additions both at low temperature and after allowing the reaction mixtures to warm up is often excellent.[2] Furthermore, the stereostructure of the adducts appears to correlate strongly with the enolate geometry. In general, Z enolates gave *anti* adducts whereas E enolates gave *syn* adducts.

E enolates

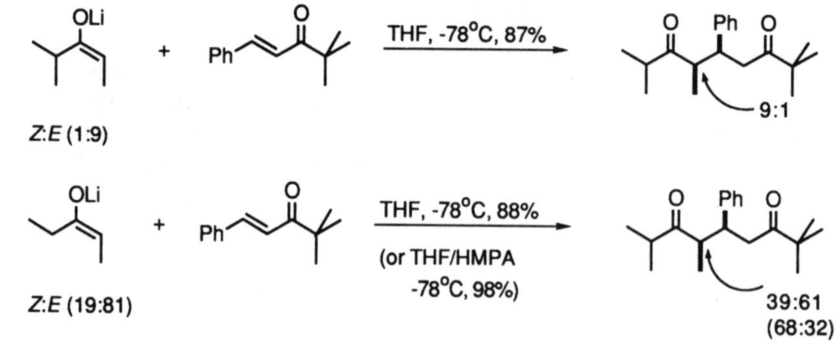
OLi
+
Ph
O
THF, -78°C, 87%
O
Ph
O
9:1
Z:E (1:9)
OLi
+
Ph
O
THF, -78°C, 88%
(or THF/HMPA
-78°C, 98%)
O
Ph
O
Z:E (19:81)
39:61
(68:32)

By observing the stereochemical outcome caused by varying the steric demand of substituents in both the enolate and alkenone, Heathcock's group were able to develop a transition state model for these additions. As shown below the model is based upon an eight-membered, chelated transition state. Everything else being equal, transition states **A** and **C** are preferred as these minimize any steric interactions between R^1 and R. However, the *E* enolate-*syn* correlation breaks down when R is small.

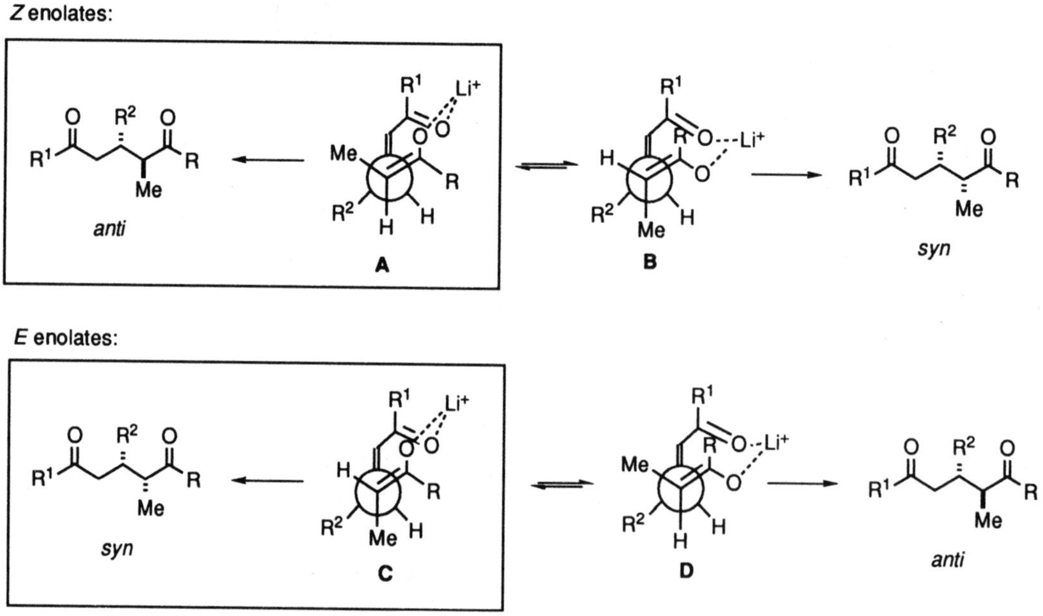
Z enolates:
anti
A
B
syn
E enolates:
syn
C
D
anti

Figure 2.1 Proposed transition state structures for ketone enolate additions to acyclic alkenones

The Robinson annulation[3]

Of all the applications of the Michael reaction to organic synthesis, the Robinson annulation is undoubtedly the best known.[4] The first reaction, reported by Robinson and Rapson,[5] involved adding 4-phenyl-3-buten-2-one (benzalacetone) to an ice-cooled solution of the sodium enolate of cyclohexanone (1) producing the octalone (3) in 43% yield.

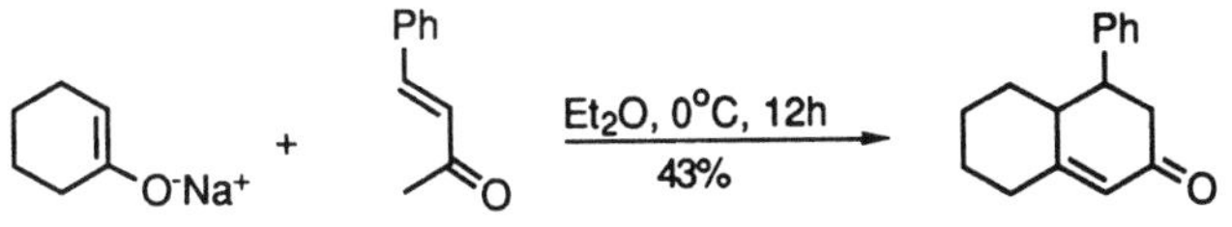

In the following paper on this topic, Robinson and his co-workers reported that all attempts to apply this method to reactions involving either simple alkenones such as 3-buten-2-one (methyl vinyl ketone) or β-chloroketones failed.[6] Successful annulation was finally achieved by generating the alkenones *in situ*, using "Mannich salts" such as the methiodide of 1-(diethylamino)butan-3-one:[300]

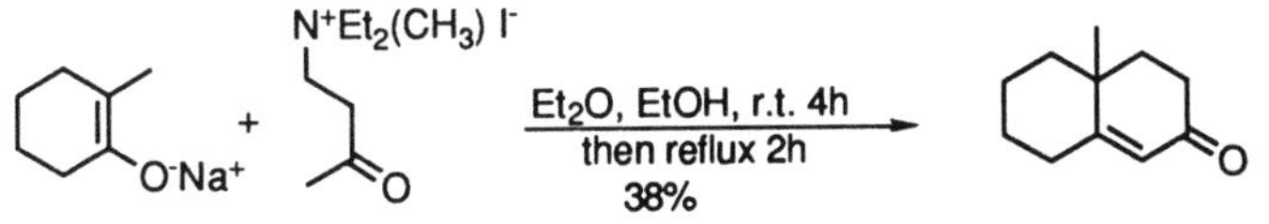

This was the first synthetic method for preparing octalone systems bearing an "angular" methyl group, a characteristic feature of the corresponding portion of the steroid nucleus. A reasonable mechanism is provided in the following scheme.

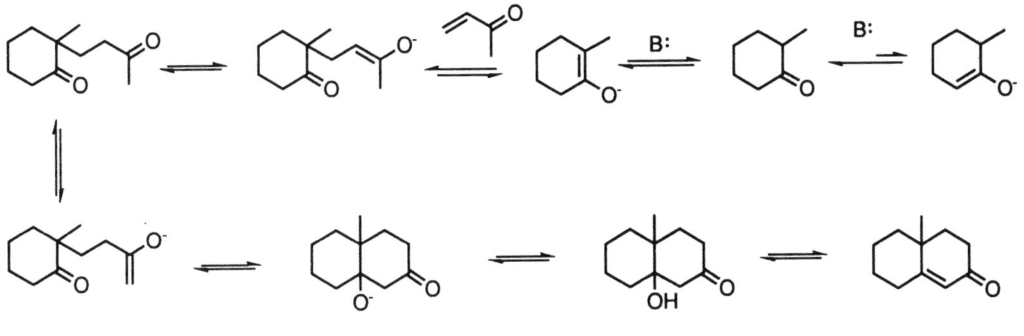

The Robinson annulation suffered from several limitations, each of which has been addressed over the years, and for which solutions have been found, with varying degrees of success. These include:

1.   Selectivity of enolate formation
2.   Michael addition/competing polymerization
3.   Equilibration of intermediate enolates
4.   Ring closure (aldol)
5.   Dehydration

Only the first two of these issues are directly relevant to this book.

Enolate formation

A general discussion of enolate formation is included in Chapter 1 (carbanions stabilized by π-conjugation with one heteroatom). However, there are some aspects which are of particular relevance to the Robinson annulation. In general, it has been found that, for the case of unsymmetrical monoketones, reaction usually occurs at the more substituted α-carbon, i.e. via the thermodynamic enolate, as in the example of 2-methylcyclohexanone shown above. However, steric constraints can reverse this selectivity,[7] e.g.:

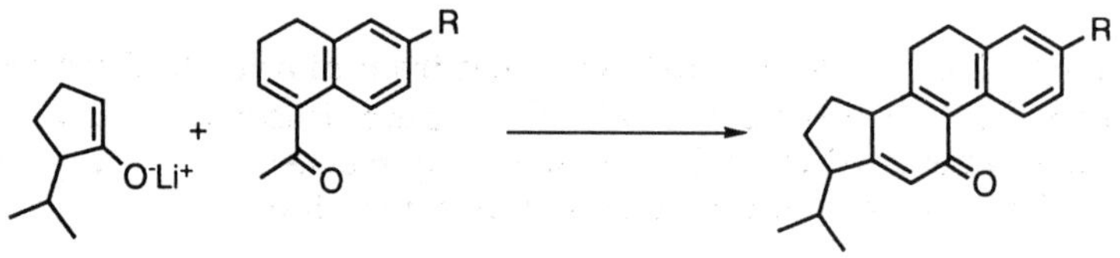

A remarkably simple and stereoselective Robinson annulation has been reported by Scanio and Starrett.[8] By simply varying the solvent, either *cis* or *trans* 4,10-dimethyl-1(9)-octal-2-one may be obtained in ≥ 95% isomeric purity. The authors suggested that the

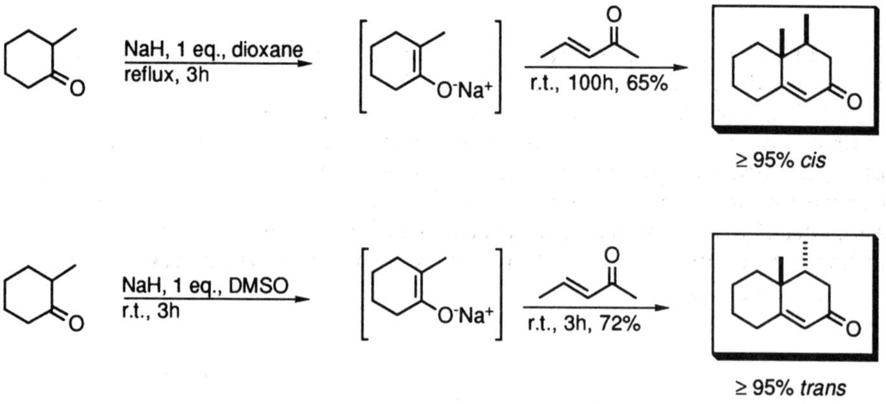

stereoselection obtained in the first case occurs during the initial conjugate addition. A more intriguing mechanism was offered for the second case. As proton transfer is rapid in polar solvents,[9] it is possible that 2-methylcyclohexanone is regenerated and the resulting enolate (**4**) (the butenone serves as the acid) adds to the cyclohexanone. Dehydration, followed by electrocylic rearrangement gives the product with the correct relative stereochemistry. Unfortunately, no supporting mechanistic study has appeared since this work was published.

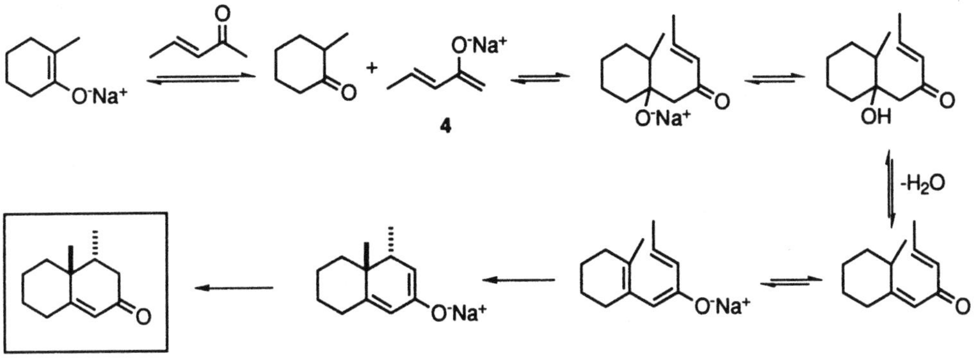

Takagi's group has also found that switching from dioxane to dimethylsulfoxide changes the ratio of product isomers significantly.[10] The pure *cis* isomer is actually obtained in best yield by reacting 3-penten-2-one with a temporarily blocked 2-methylcyclohexanone, (5) in the presence of a catalytic amount of potassium *tert*-butoxide.

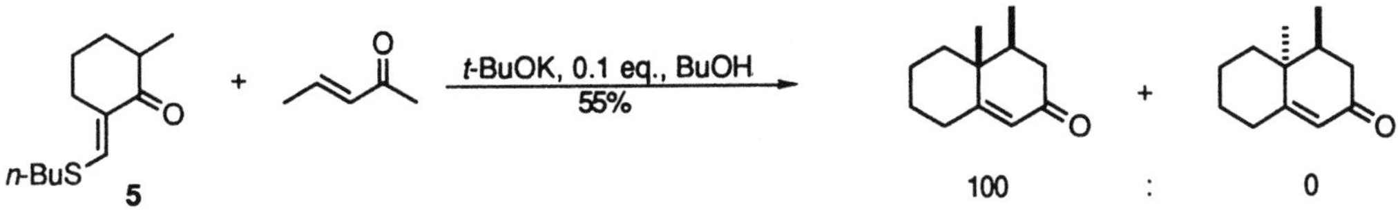

In order to avoid equilibration, the generation and reaction of a specific enolate must be carried out under kinetic control. (An alternative to this approach is to use the corresponding enamine. Enamines of unsymmetrical ketones generally form at the less substituted α-carbon, i.e. the equivalent to a kinetic enolate, see section 2.1.5). This generally requires the use of aprotic conditions. Even under these circumstances, the rate of conjugate addition still needs to be greater than that of proton transfer from any carbon acid intermediates if reasonable yields of products are to be achieved. The yields can be good, as in the low temperature deprotonation of carvomenthone, which was used in a synthesis of 7-hydroxycalamenene.[11]

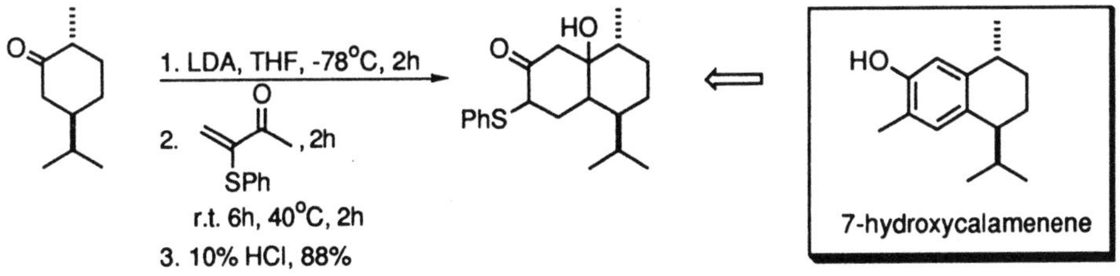

An alternative is to generate the enolate from a trimethylsilylenol ether:[12]

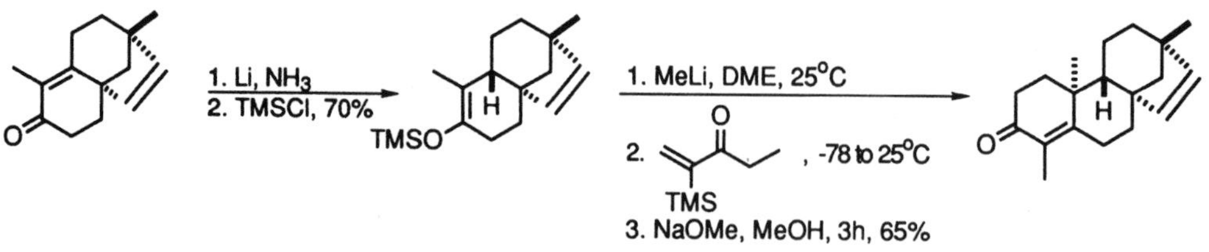

Various stabilizing groups attached to the α-carbon of (usually cyclic) alkenones ensure that anion generation is regiospecific (also see section 2.1.1.2). An example of the use of a stabilizing group (in this case there is no alternative acidic site!) comes from Woodward's total synthesis of cholesterol.[13]

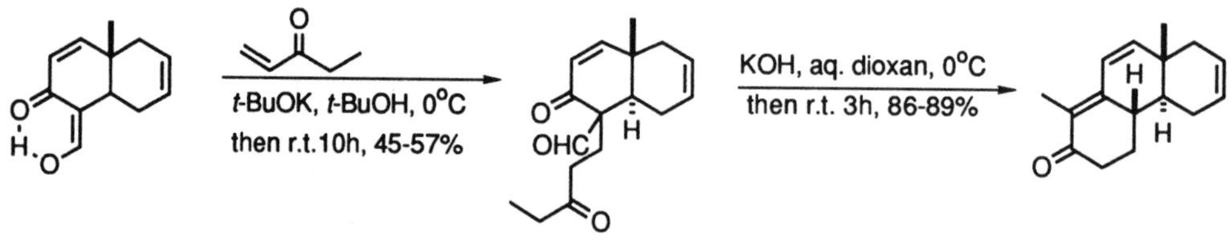

Temporary α-blocking groups

Alternatively, temporary blocking groups can be used to direct enolate formation to the required α-carbon. For example, in his synthesis of (±)-nootkatone, Takagi used a thioalkylidene group to direct enolate formation to the more substituted site of 2-methylcyclohexanone.[14] Better selectivity for (±)-nootkatone could be achieved under more forcing conditions using benzyl in place of *n*-butyl, however the yield was low (30% total).

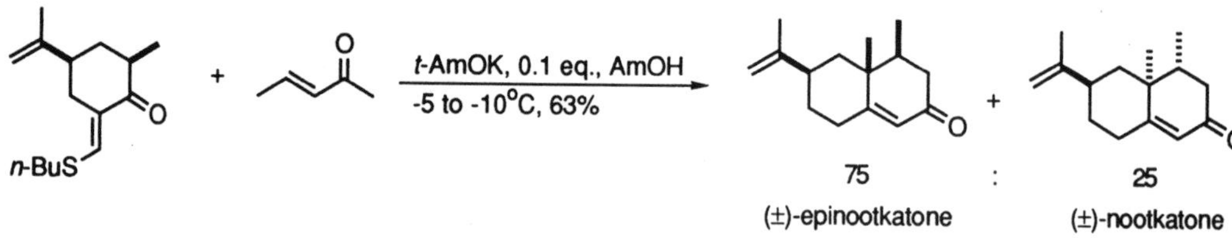

Other types of blocking groups include dithioacetals, benzylidene and other heteroalkylidene groups. α,β-Unsaturation is often a necessary feature of the target molecule and it is therefore economical to incorporate it early in the synthesis, as in Roy's synthesis of (+)-β-cyperone.[15]

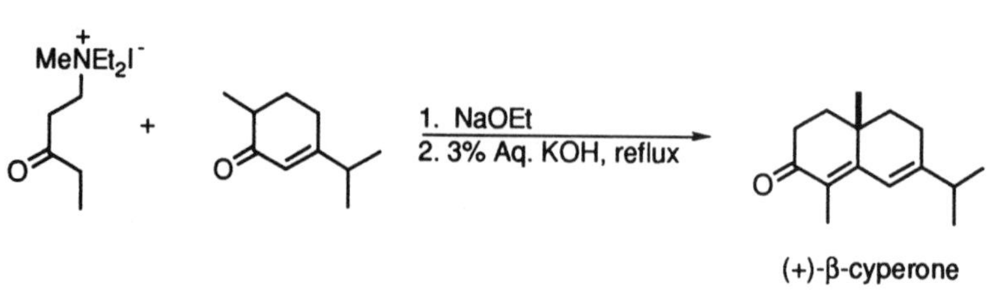

However, it must be kept in mind that γ-deprotonation can compete in some cases. This was used in a total synthesis of the quassinoid ring system.[16]

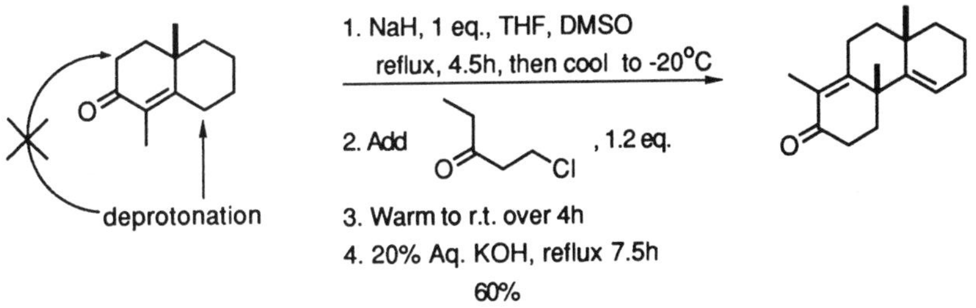

1. NaH, 1 eq., THF, DMSO
   reflux, 4.5h, then cool to -20°C

2. Add [structure] , 1.2 eq.

deprotonation

3. Warm to r.t. over 4h
4. 20% Aq. KOH, reflux 7.5h
   60%

The Nazarov modification of the Robinson annulation, which uses ethyl 3-oxo-4-pentenoate (6) in place of 3-buten-2-one to introduce an α-ethoxycarbonyl group, originally suffered from rather poor yields.[17] Because the reagent is quite acidic, only enolates formed from highly acidic compounds, typically 1,3-diketones, could react successfully under the (basic) reaction conditions. An added complication is the propensity of the reagent to polymerize under these conditions. However, it occasionally does react in very good yields, as in the following example.[18]

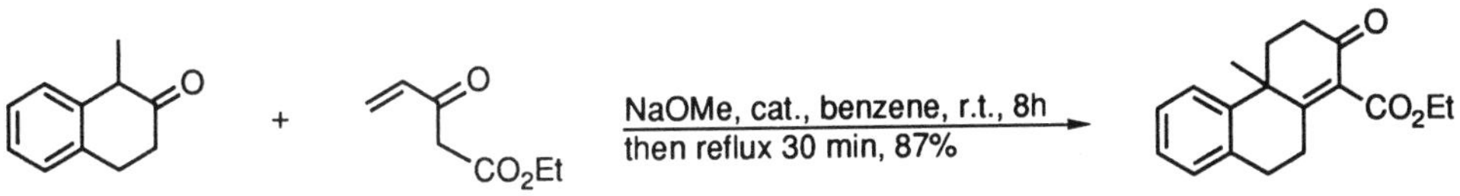

+

NaOMe, cat., benzene, r.t., 8h
then reflux 30 min, 87%

CO$_2$Et

6

Fluoride also catalyzes the addition.[19]

+   CO$_2$Et

KF, MeOH, r.t., 15h, 30-50%

HO   CO$_2$Et

Wenkert has introduced a reagent, 1,4-dimethoxy-2-butanone (7) for use in preparing α-methoxy derivatives.[20] In the following example, the long reaction time was necessary as shorter times led to the isolation of the ring-closed, hydrated intermediate. (An alternative to this is to carry the reaction out under acidic conditions, see later this section).

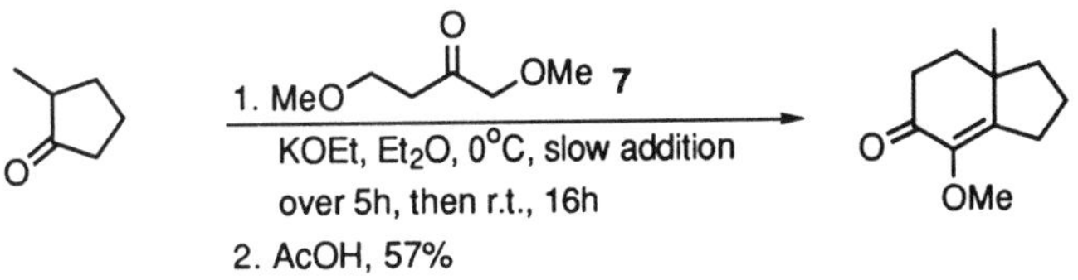

1. MeO   OMe 7

KOEt, Et$_2$O, 0°C, slow addition
over 5h, then r.t., 16h

2. AcOH, 57%

OMe

Enolates of aldehydes also participate in Robinson annulations. For example, base-catalyzed addition of 2-methylpropanal to 3-buten-2-one produces the useful synthetic intermediate 4,4-dimethylcyclohexenone.[21]

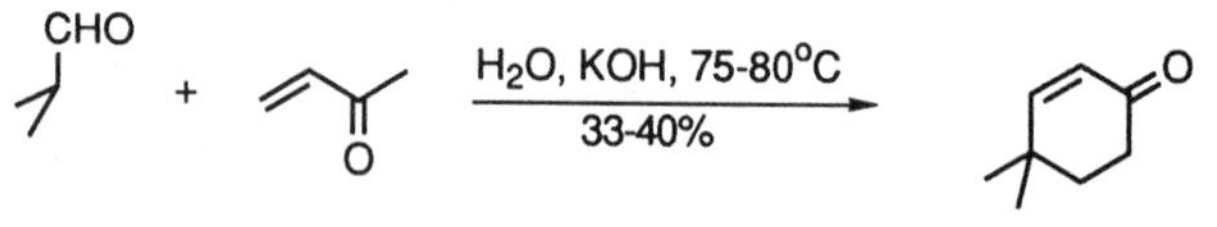

The use of 1,3-dicarbonyl compounds removes any problems of regioselection during enolate formation and is also attractive because of the very mild conditions which can be employed. For example, 2-methyl-1,3-cyclohexanedione reacts with 3-buten-2-one to give the annulated product in 64% yield (even higher yields are possible, see below).[22,23]

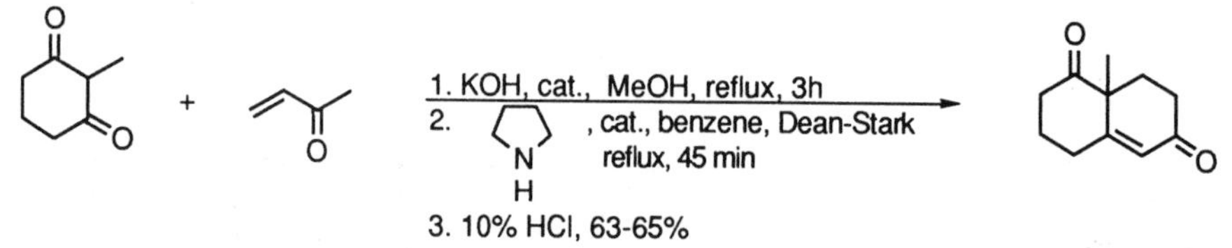

However the use of 1,3-dicarbonyl compounds sometimes leads to the formation of unusual products due to either base- or acid-catalyzed cleavage of the tricarbonyl intermediate.[24,25,26,27] (For more examples, see section 2.1.1.2).

The addition of 2-methyl-1,3-cyclohexanedione to 3-buten-2-one is of particular significance as it offers direct access to optically pure Wieland-Miescher ketone, an important intermediate in the synthesis of a variety of natural products. This method has evolved over a number of years and combines a conjugate addition followed by a catalytic asymmetric cyclization. To date, the simplest procedure reported is that of Harada,[28] which is based on much earlier work.[29,30,31]

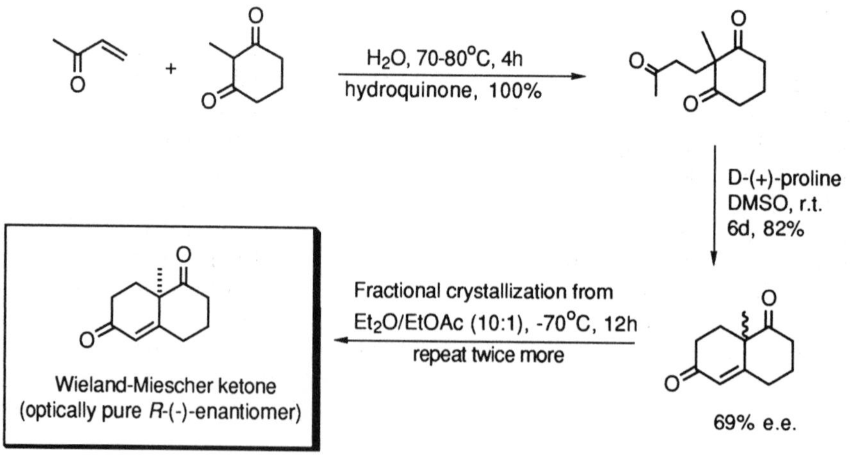

Wieland-Miescher ketone
(optically pure R-(-)-enantiomer)

An interesting procedure, involving a magnesium chelated dianion (8), has recently been reported by Rathke et al.[32] However, it suffers from a lack of regioselectivity in the carboxylation of unsymmetrical ketones.

Robinson annulations have been carried out on arene chromium tricarbonyl complexes. However, although the addition went well, subsequent cyclization of the major diastereomeric adduct (not shown) gave a mixture of products. The unexpected products arose from competitive deprotonation at the benzylic site. The minor diastereomer (9) cyclized normally.[33]

Michael Reaction

Probably the most difficult problem to overcome in the Robinson annulation is that of competing polymerization of the alkenone, especially under aprotic conditions. In 1973 Stork and Ganem introduced the use of α-silylalkenones in place of simple alkenones.[34] These reagents react successfully with enolates under either protic or aprotic conditions. In addition the silyl substituent is subsequently easily removed, usually during cyclization of the initial adduct. The role of the silyl substituent is two-fold. Firstly it provides steric hindrance to further reactions at the newly formed enolate - the most important of these being addition to more alkenone (and hence polymerization). Second, it stabilizes the charge of the newly-formed enolate. By taking care rigorously to remove any protic impurities, using procedures which are now routine in the handling of air and moisture sensitive reactions, excellent regiocontrol is maintained with respect to the starting enolate.[35,36] One example involves the first use of a carbohydrate-derived enolate.[37]

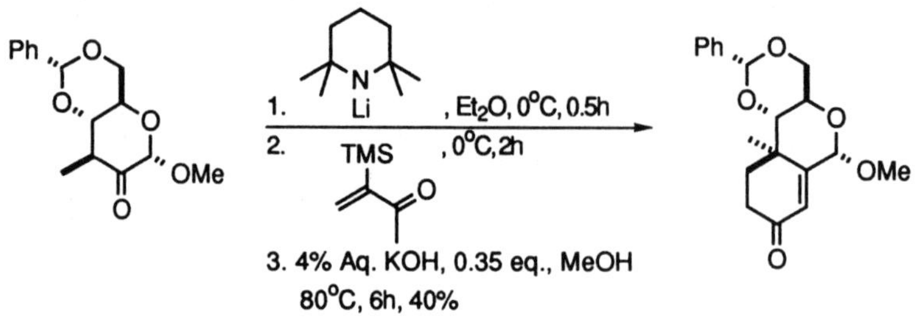

3. 4% Aq. KOH, 0.35 eq., MeOH
   80°C, 6h, 40%

Other stabilizing groups which have been used include alkoxycarbonyl[38,39] and transition metal complexes.[40]

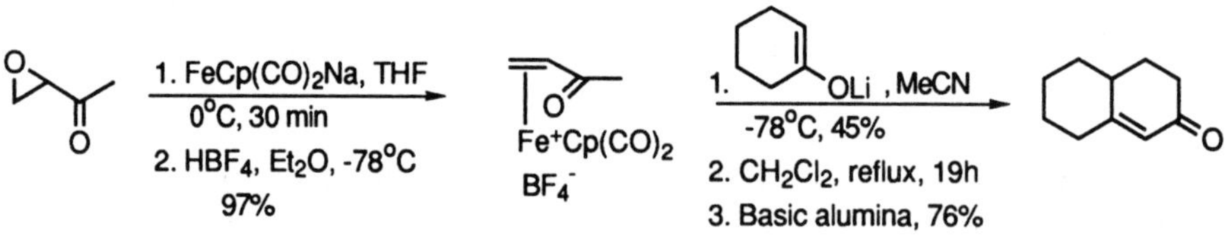

Another useful application of the Robinson annulation is in the synthesis of spirocyclic compounds. For example Pesaro's group required a spirocycle in their syntheses of (-)-acorenone and (-)-acorenone B.[41] This was prepared via a rather low-yielding Robinson spiroannulation:

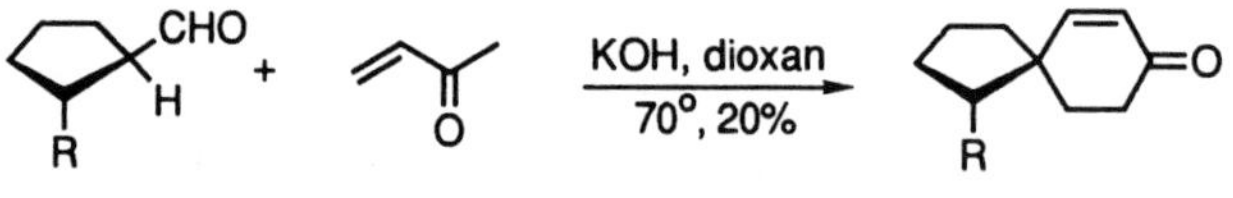

Corey's group have also employed a Robinson spiroannulation, in their total synthesis of (±)-aphidicolin.[42] DeBoer's method of slowly adding gaseous 3-buten-2-one to the reaction mixture was used in this synthesis. However no yield was reported.[43]

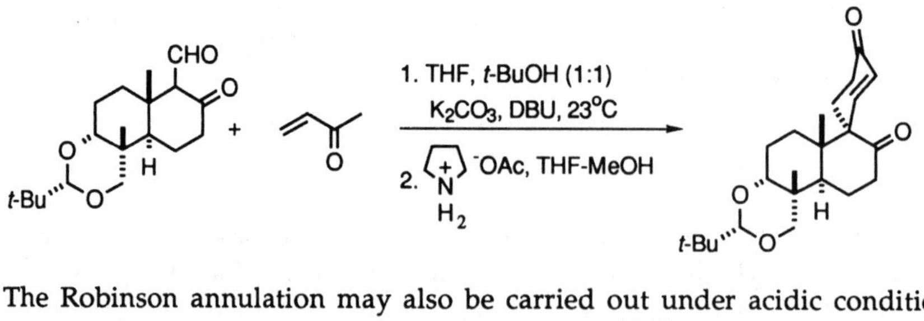

The Robinson annulation may also be carried out under acidic conditions. This method, introduced by Heathcock and McMurry in 1971,[44,45] simply consists of heating a benzene solution of the two reactants in the presence of a catalytic amount of concentrated sulfuric acid. Zoretic has applied this to a highly stereoselective synthesis of cis-5,10-dimethyl-1(9)-2-octalone.[46] Although the yield is not particularly high, it is very direct, involving only one step from the dimethylcyclohexanone. However, the stereoselectivity in this reaction as well as its method of estimation (measurement by [13]C n.m.r spectroscopy) have been questioned.[47] The revised ratio was put at 2.8:1 (as opposed to 9:1 as reported by Zoretic's group).

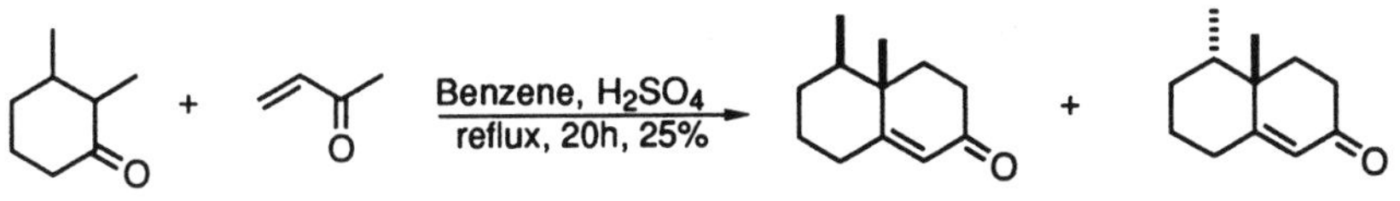

2.8   :   1

By running similar reactions at 0°C, the adduct can be isolated in ~50% yield.[48]

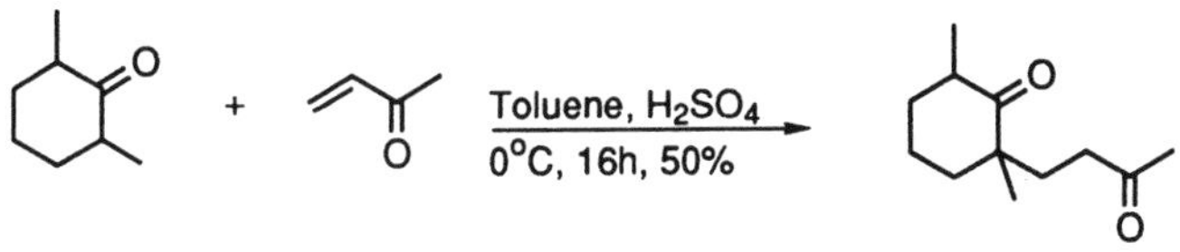

The Nazarov-Robinson reaction may also be carried out under acidic conditions. Significantly, this procedure also succeeds (although in modest yield) with monoketones.[49,50]

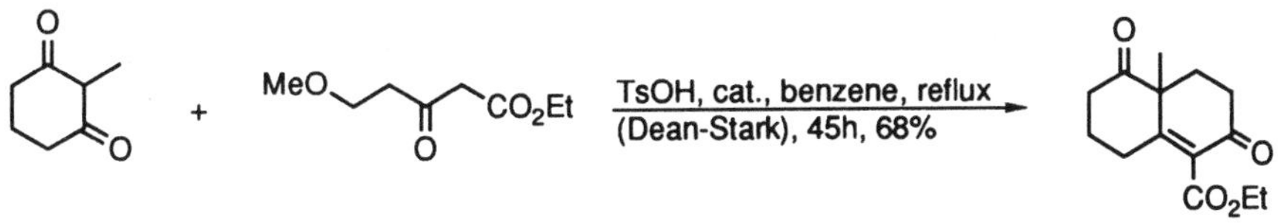

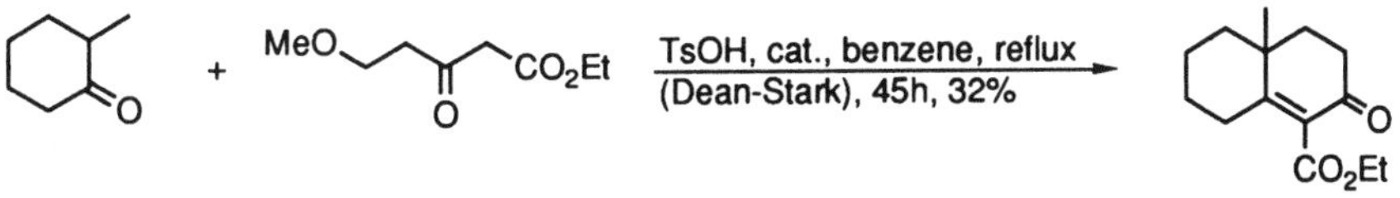

The Robinson annulation has been used to synthesize many natural products. Some of these include the eremophilanes,[51] (+)-progesterone,[52] (+)-testosterone,[53] acorenones,[54] (+)-aphidocolin,[55] and cholesterol[13].

(ii)   Ester enolates

Ester enolates give both 1,2- and 1,4-adducts at low temperature. Significant conversion of 1,2- to 1,4-adducts occurs as the temperature is increased beyond -40°C. The issue of stereoselectivity in these reactions is similar to that for the ketone enolate additions discussed in part (i) above. The postulated transition states are also similar. Some examples are given on the next page.[56]

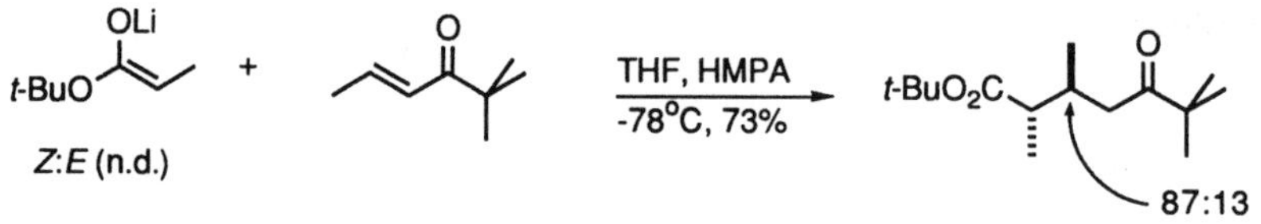

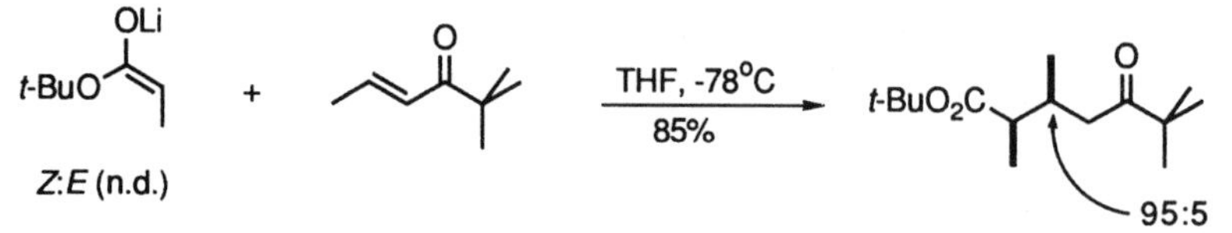

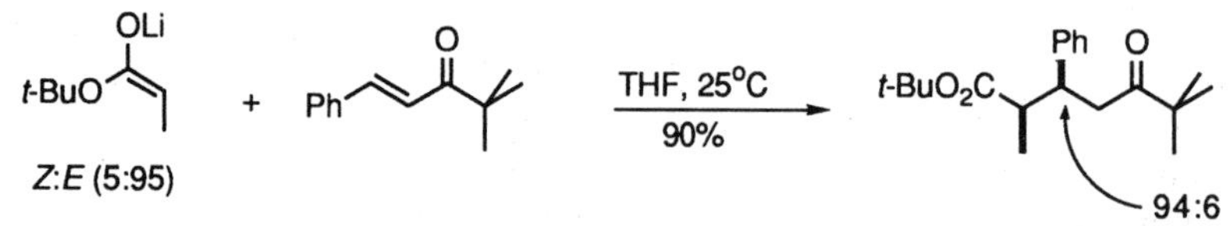

87:13

95:5

94:6

Where the enolate stereochemistry has been determined, a strong correlation between the stereochemistry of the enolate and that of the adduct(s) was again evident. Z enolates give *anti* adducts and E enolates give *syn* adducts. In their review Oare and Heathcock concluded that "the degree of diastereoselectivity obtained is limited only by the ability to generate enolates in an isomerically homogeneous manner".[1]

The situation with amide and lactam enolate additions is a little more complex.[57] Again, initally 1,2 addition is favoured at low temperatures. However, in most cases, warming to room temperature or above equilibrates the inital adducts to, often exclusively, the conjugate adducts. Good stereoselectivity is also often obtained.

Z enolates

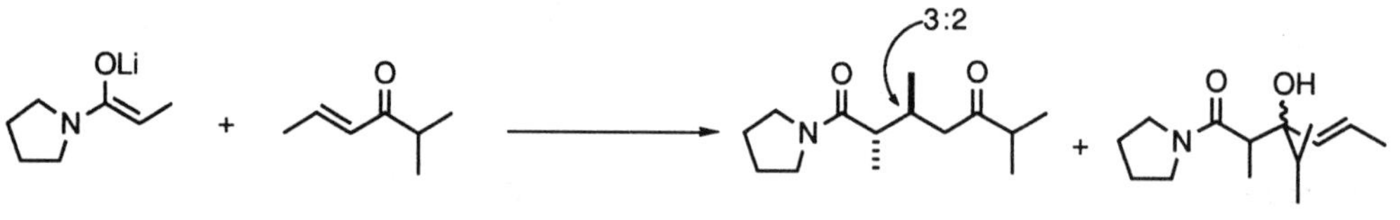

| | THF, -78°C, 48 min, 84% | 71 | : | 29 |
| | THF, 25°C, 12h,   85% | >97 | : | <3 |

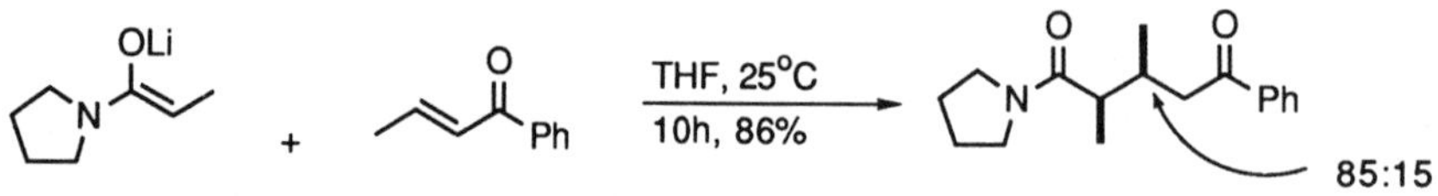

85:15

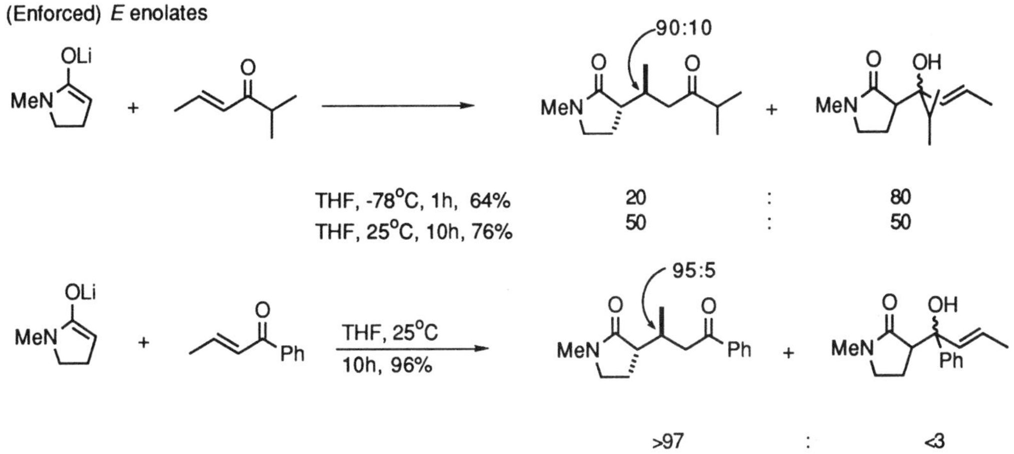

(Enforced) *E* enolates

THF, -78°C, 1h, 64%	20	:	80	
THF, 25°C, 10h, 76%	50	:	50	

>97   :   <3

Several trends emerged from the study. The stereoselectivity of lactam additions did not alter significantly by varying the β-substituent in the alkenone. Propionamide enolates, however, were sensitive to the size of the alkenone β-substituent, with better *anti* selectivity for increasing substituent size. The addition of hexamethylphosphoramide to the reaction medium can also strongly influence stereoselectivity in some cases, especially where the additions are relatively slow. Finally, for lactam enolates (which favour *syn* adducts), the sodium and potassium enolates enhance *syn* selectivity. By contrast, such metal substitutions favour *anti* selectivity for propionamide enolates.

Amide enolates:

Lactam enolates:

[X = O,S]

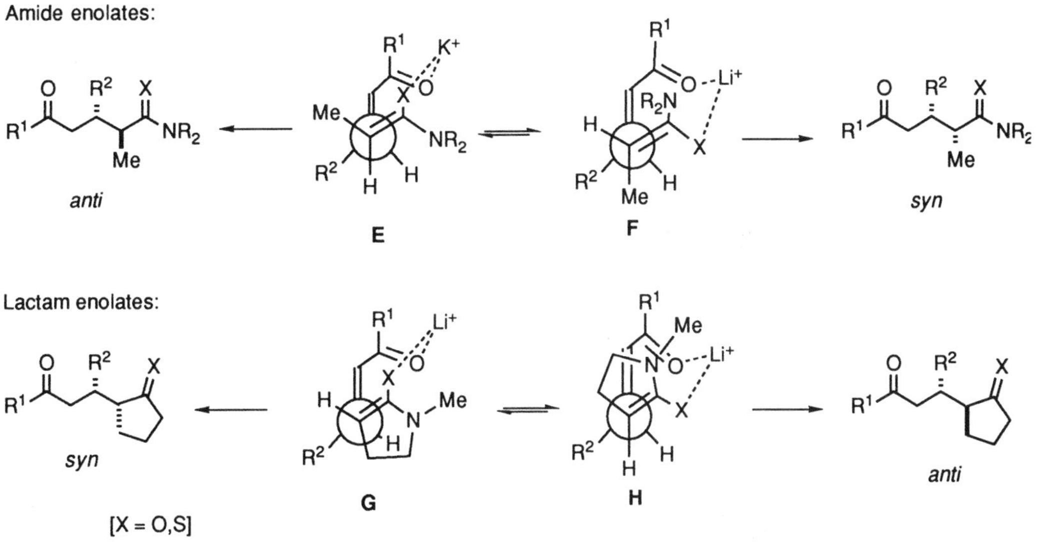

Metzner's group has also shown that it is possible to obtain *syn* products from the addition of enethiolates to 2,3-disubstituted alkenones.[58] As the stereochemistry of the major diastereomer is the opposite to that expected for a protonation under allylic control,[59] it was proposed that protonation occurs intramolecularly. (Quenching the reaction mixture with methyl iodide showed that protonation occurs during the reaction, as the methylation product was a ketene dithioacetal).

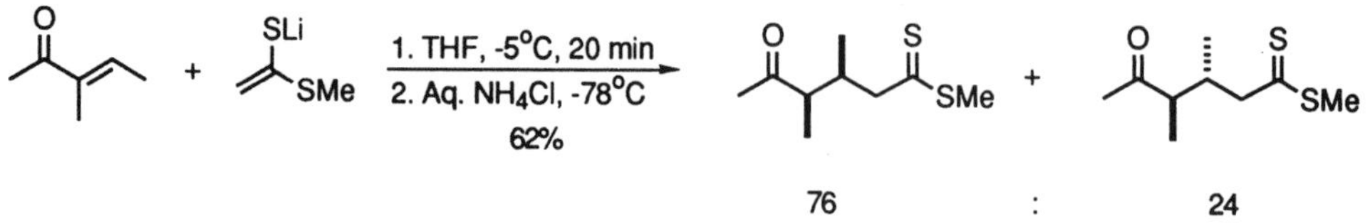

| 76 | : | 24 |

Stetter has developed two important, related processes for the addition of aldehydes to conjugate acceptors.[60,61] Both depend upon the *in situ* generation of a π-stabilized carbanion (10) produced by the addition of a nucleophilic catalyst to the aldehyde carbonyl group.

The first system is analogous to the benzoin reaction in that it uses cyanide as the catalyst. However, whereas the benzoin reaction is carried out in protic media and all the steps are reversible, Stetter's system uses cyanide in aprotic media and the last step is irreversible as shown below. Although this works well for aromatic aldehydes, it fails for aliphatic aldehydes because the latter undergo cyanide-catalyzed aldol condensation instead.

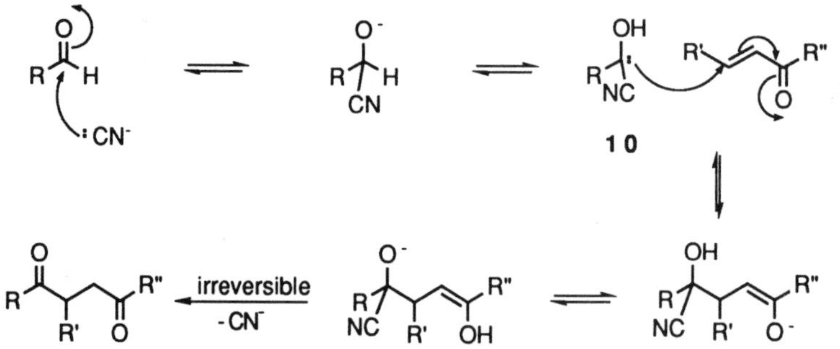

The second system, which works well for both aromatic and aliphatic aldehydes, is also based on analogy, this time with the action of vitamin B$_1$, which is known to convert aliphatic aldehydes into acyloins in buffered aqueous solution.[62,63] Stetter examined a series of thiazolium salts and settled on the cheap, commercially available quaternary salts of 3-alkyl-5-(2-hydroxyethyl)-4-methyl-1,3-thiazole. Reactions can be carried out in both protic or aprotic solvents and usually require heating to between 60 and 80°C for several hours. Interestingly, for both the catalyst systems, benzoins may be used in place of the aromatic aldehyde. However, acyloins could not be used in place of aliphatic aldehydes. Some examples of conjugate additions to alkenals and alkenones are given in Table 2.1.

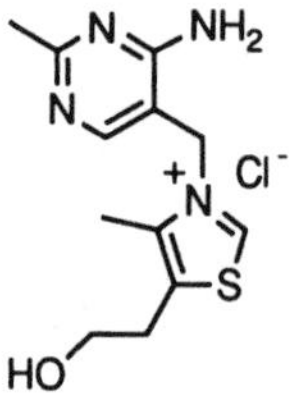

thiamine

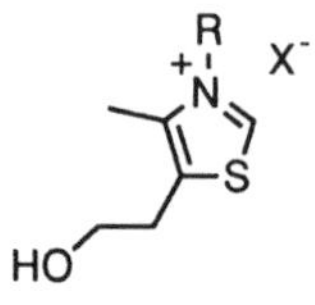

thiazolium salts

R = Me, Et or Bn
X = I, Br or Cl

Table 2.1 Cyanide- and thiazolium-catalyzed conjugate additions of aldehydes to alkenones

164	(propanal)	+ (methyl vinyl ketone)	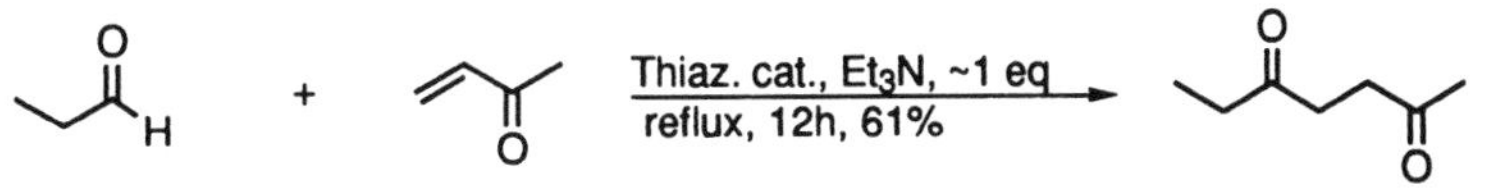 Thiaz. cat., Et₃N, ~1 eq reflux, 12h, 61%	(1,4-diketone product)

164		+	Thiaz. cat., Et₃N, ~1 eq reflux, 12h, 61%	
265		+	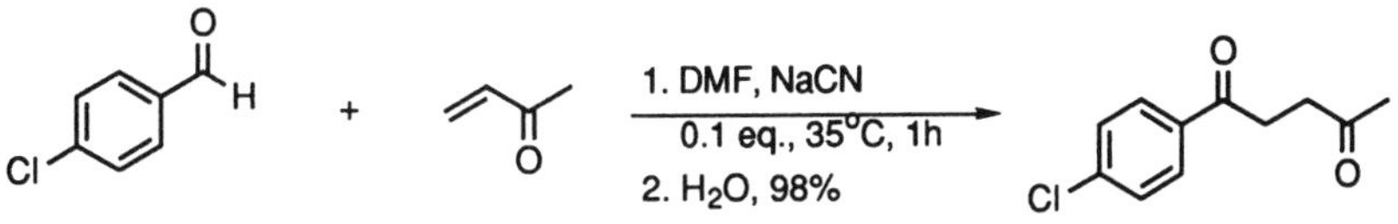 1. DMF, NaCN 0.1 eq., 35°C, 1h 2. H₂O, 98%	
366		+ Ph	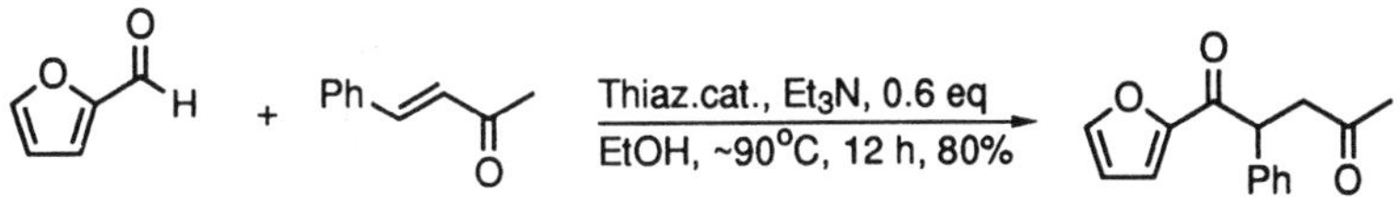 Thiaz.cat., Et₃N, 0.6 eq EtOH, ~90°C, 12 h, 80%	
465		+ Ph	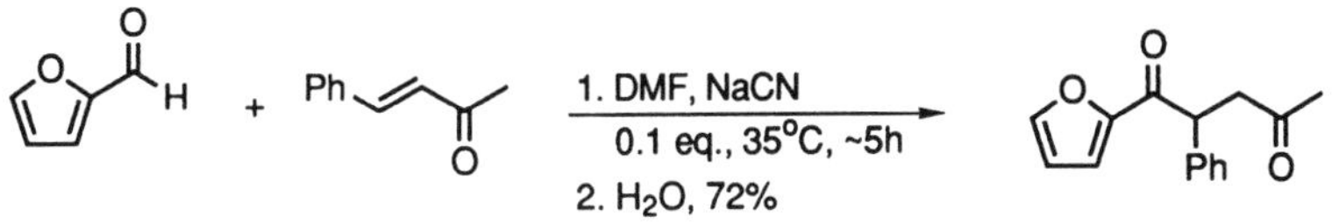 1. DMF, NaCN 0.1 eq., 35°C, ~5h 2. H₂O, 72%	
567		+	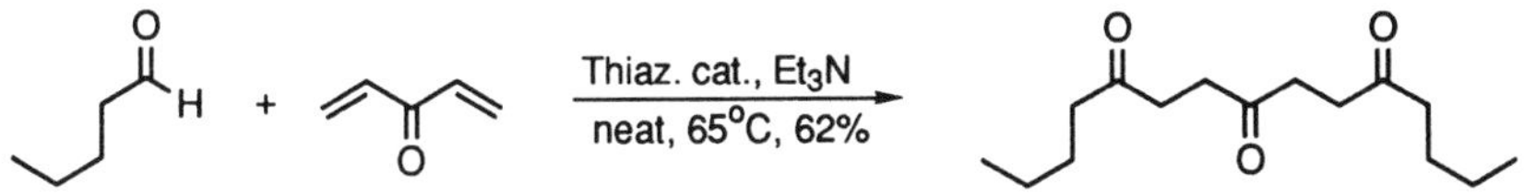 Thiaz. cat., Et₃N neat, 65°C, 62%	

Ahlbrecht's group has added carbanions of α-cyanoamines to alkenones. Hydrolysis of the adduct produces a 1,4-diketone product.[68]

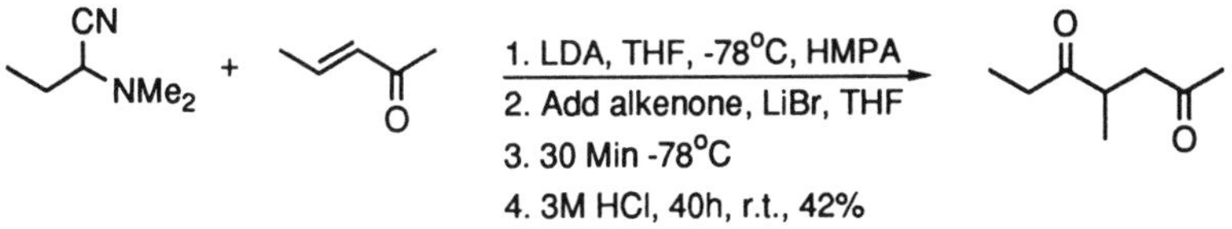

CN
NMe₂ + (alkenone)

1. LDA, THF, -78°C, HMPA
2. Add alkenone, LiBr, THF
3. 30 Min -78°C
4. 3M HCl, 40h, r.t., 42%

It is also possible to trap the intermediate enolate with electrophiles, as in the following example.[69]

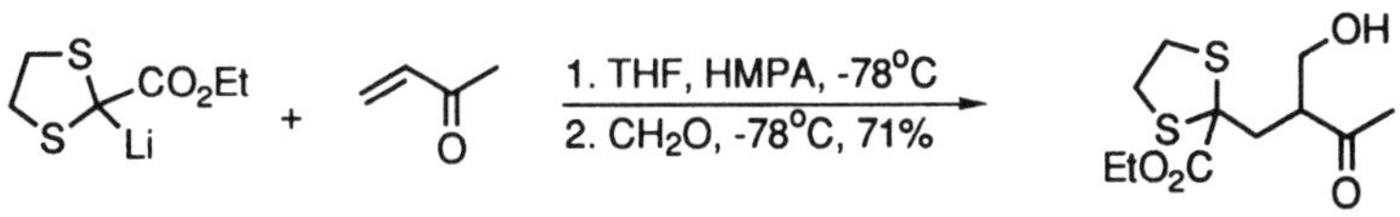

(dithiolane)-CO₂Et
Li +

1. THF, HMPA, -78°C
2. CH₂O, -78°C, 71%

(iii)   *Nitro-stabilized carbanions*

Because of their relatively high acidity, nitroalkanes are very useful sources of carbon nucleophiles which add to conjugate acceptors. Quite a variety of conventional bases work well, including DBU (as shown in the example below),[70] TMG,[71] HCO_2Na,[72] $KOAc$,[73] and PTC.[74,75]

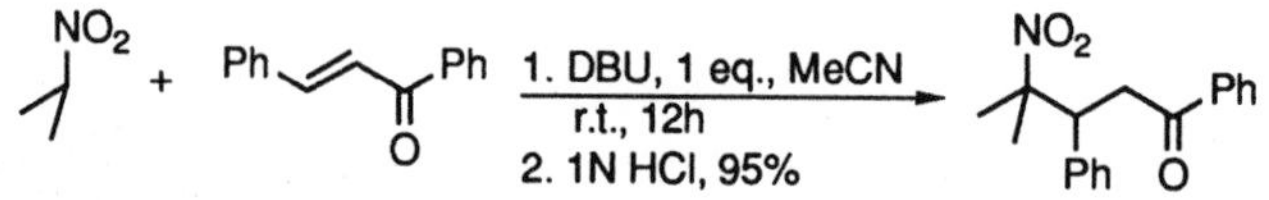

The combination of an addition of a nitroalkane to an alkenone or alkenal with either a Nef reaction[76,77] or ozonolysis leads to another synthesis of 1,4-dicarbonyl compounds. For example, conjugate addition of nitropropane to 3-buten-2-one yields 5-nitroheptan-2-one. Treatment with methoxide generates the corresponding nitronate which is then ozonolyzed *in situ* to give 2,5-heptanedione.[78] Alternatively, the adduct may

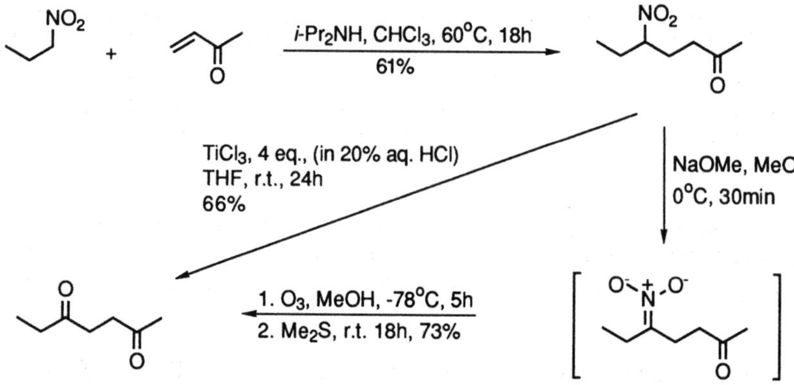

be subjected to a Nef reaction under reductive conditions, using titanium(III)chloride in aqueous acid.[79]

Another 1,4-diketone synthesis, similar to those above has been developed but, rather than a Nef reaction or ozonolysis, the final step is a silica-supported permanganate oxidation.[80]

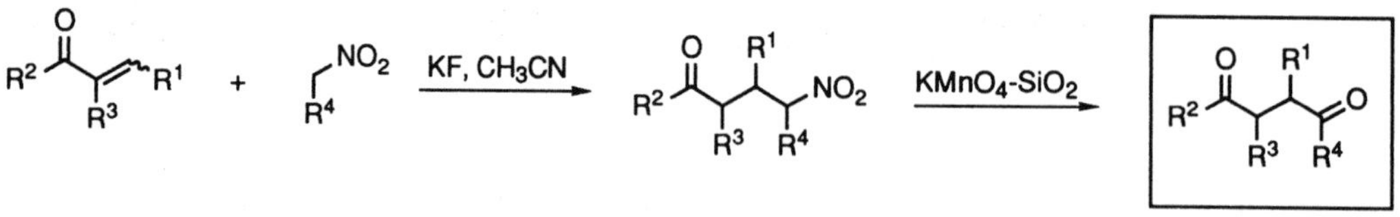

Ballini's group has demonstrated the usefulness of alumina as catalysts for these additions.[81]

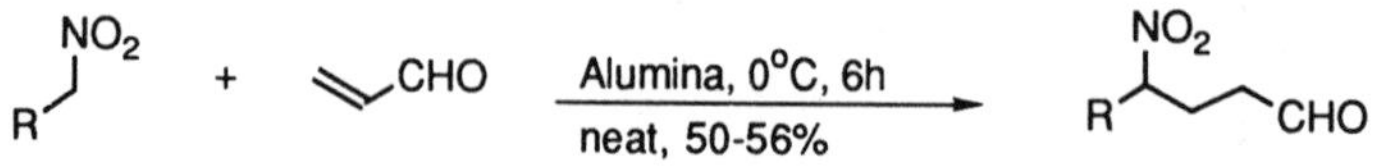

As nitro groups are easily removed reductively, even in the presence of such sensitive functional groups as aldehydes, they show potential as temporary activating groups for the conjugate addition of alkyl groups, e.g.:[82]

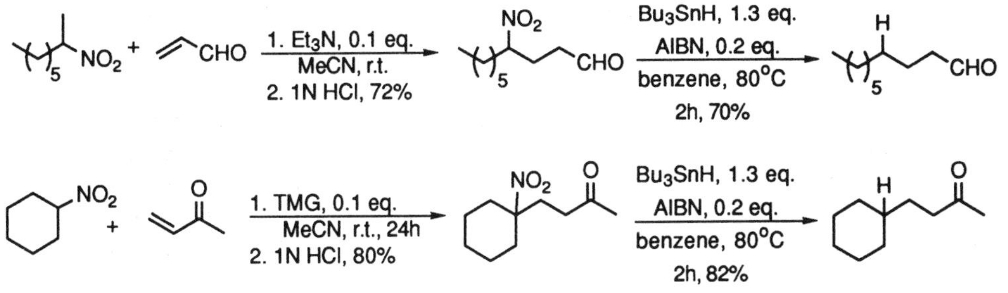

It is sometimes useful to trap, *in situ*, the initial adduct as its acetal.[83]

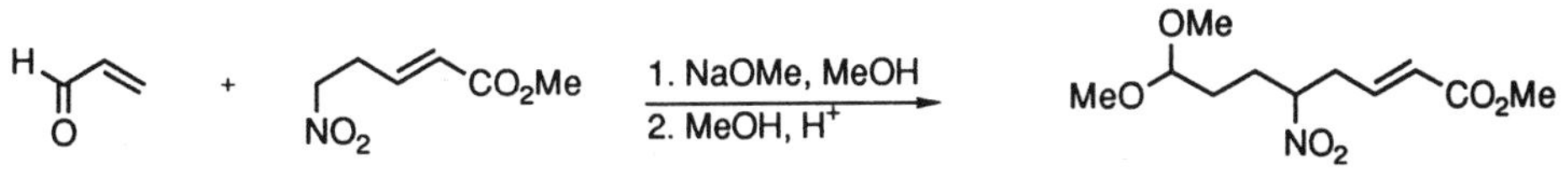

Reductive cleavage works well for tertiary nitro compounds, however the corresponding secondary nitroalkanes sometimes give problems. This can be overcome in those cases where one of the substituents attached to the secondary carbon is either an aryl or carbonyl group (see section 2.1.1.2).

Stevens' group incorporated the conjugate addition of a nitro-stabilized carbanion into their synthesis of (±)-monomorine 1, one of the trail pheromones of the Pharaoh ant (*Monomorium pharaonis* L.).[84]

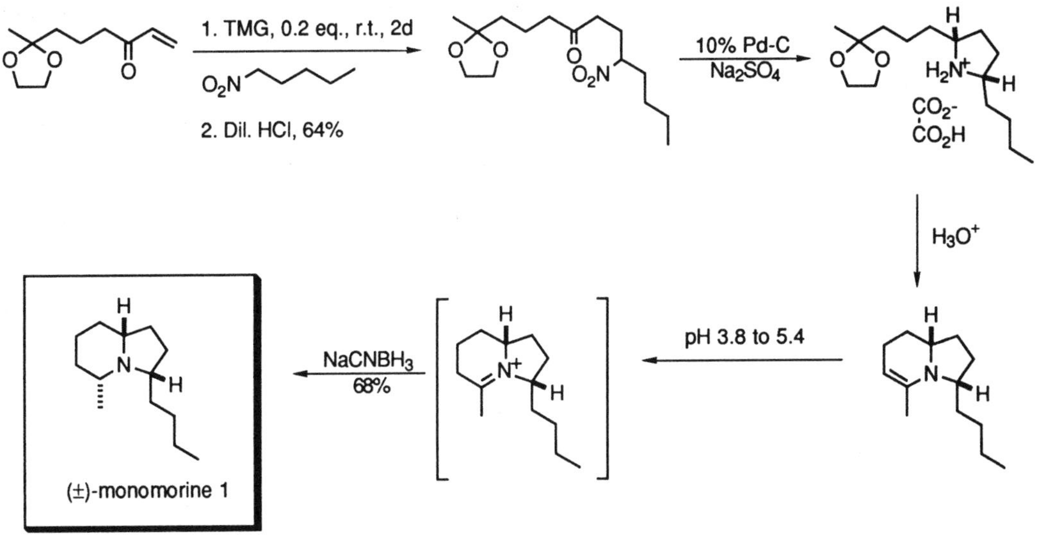

(±)-monomorine 1

An excellent example of the ability of nitro-stabilized carbanions to add to highly hindered alkenones comes from Battersby's total synthesis of (±)-bonellin dimethyl ester, a green pigment from an echurian worm, *Bonella viridis*.[85]

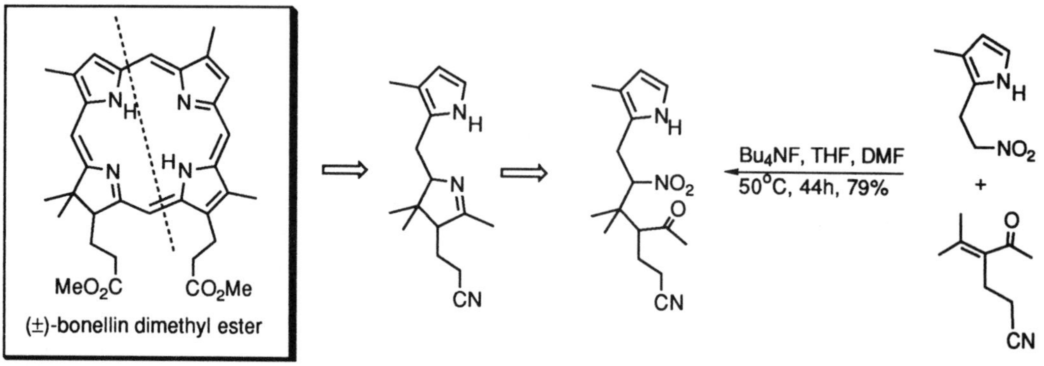

(±)-bonellin dimethyl ester

Bu₄NF, THF, DMF
50°C, 44h, 79%

(iv)   Sulfonyl-stabilized carbanions

α-Thiosulfoxides, e.g. (11) may be used as acyl anion equivalents.[86] This procedure was used to advantage in a synthesis of methylenemycin B.[87] It is interesting to note that conjugate addition occurred without accompanying elimination of thiomethoxide. This can be a problem where alkoxide is the leaving group.[88]

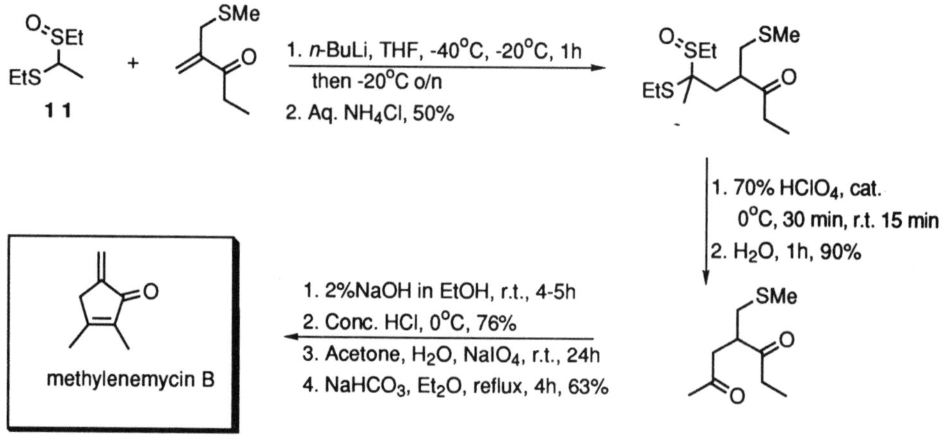

1. *n*-BuLi, THF, -40°C, -20°C, 1h
   then -20°C o/n
2. Aq. NH₄Cl, 50%

1. 70% HClO₄, cat.
   0°C, 30 min, r.t. 15 min
2. H₂O, 1h, 90%

1. 2%NaOH in EtOH, r.t., 4-5h
2. Conc. HCl, 0°C, 76%
3. Acetone, H₂O, NaIO₄, r.t., 24h
4. NaHCO₃, Et₂O, reflux, 4h, 63%

methylenemycin B

(b)   Addition of achiral carbanions to chiral alkenones

A rather remarkable new chiral alkenone system shows great potential for asymmetric synthesis.[89] In this system a phosphite formed from a hydroxyketone is first complexed to an iron dicarbonyl group. Conjugate addition, followed by hydrolysis, led to an enantiomeric excess in the product of greater than 99%.

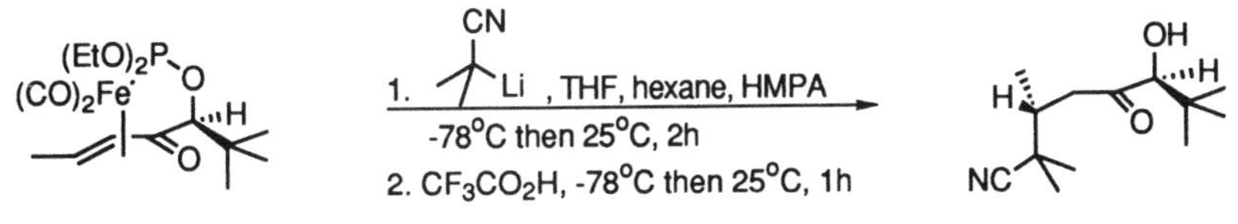

3-(S) **1 2**   >99% e.e.   3-(S), 6-(R)

The authors proposed that addition occurred to the conformation (12) shown above, which is the sterically least congested.

(c)   *Addition of chiral carbanions to achiral alkenones*

Mukaiyama's group has screened a large number of chiral diamines in an effort to find ones which induce high levels of asymmetric induction in both tin and silicon triflate promoted enolate additions. The best ligand so far found is 2-(1-naphthylaminomethyl)-N-methylpyrrolidine (13).[90] An "activator" was found to be necessary. Chlorotrimethylsilane was tried first, but no product was formed. Switching to trimethylsilyl triflate immediately gave good yields of the required adducts.

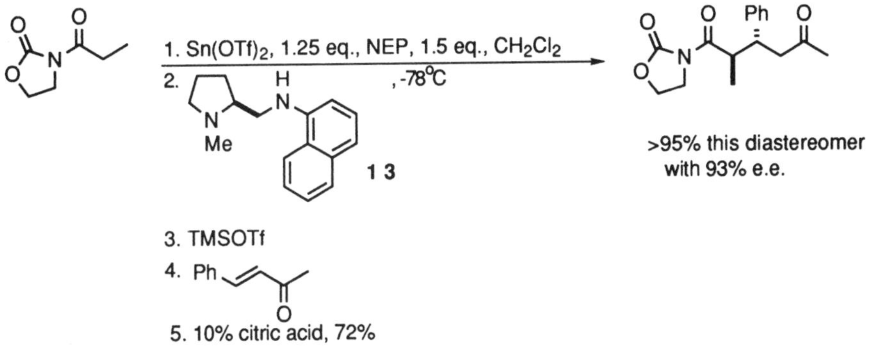

The reaction also works with dithioesters, although the enantioselectivities are not as high. However, a more significant development in these studies was the incorporation of these reactions into a catalytic cycle.[91] By slowly adding a trimethylsilyl enethiolate to a solution containing the alkenone and both the tin triflate and chiral ligand, enantiomeric excesses as high as 70% were obtained. Slow addition was necessary in order to ensure that the chiral tin triflate was the reacting species as the corresponding silyl triflate gives much poorer enantioselection.

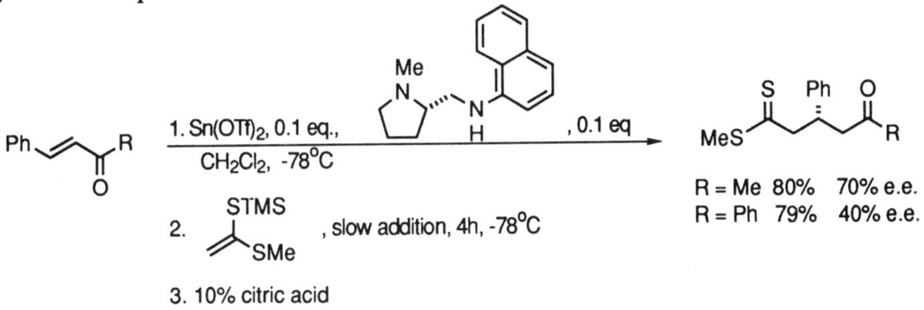

The authors proposed the following catalytic cycle.

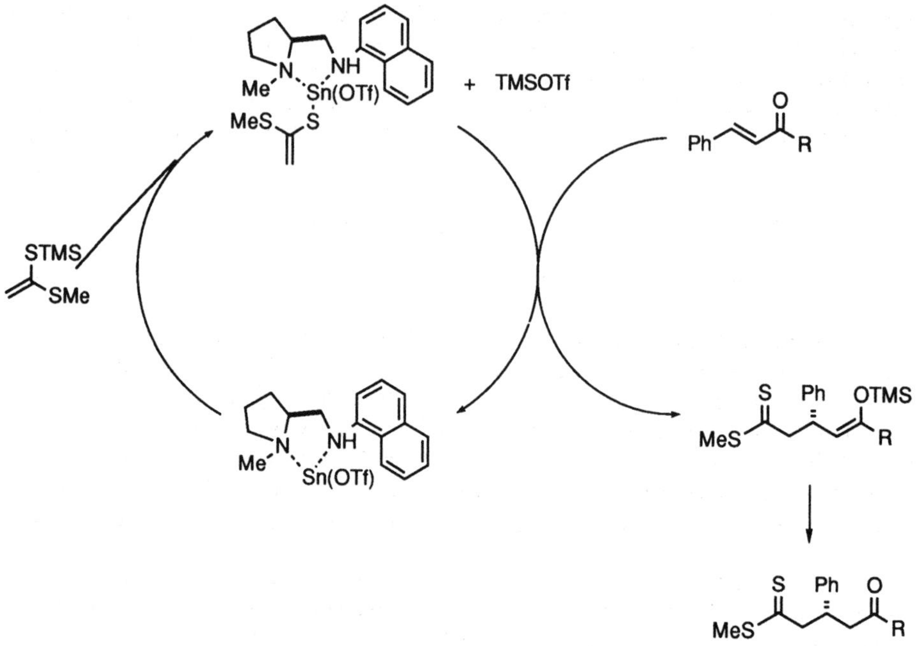

Figure 2.2 Proposed catalytic cycle for tin triflate catalyzed additions of trimethylsilyl enethiolates to alkenones

Highly stereoselective addition of ketone enolates may be achieved by first converting the carbonyl into a chiral imine or hydrazone. For example, the imines obtained from 2-substituted cyclohexanones undergo efficient conjugate additions to 3-buten-2-one.[92,93] In these cases addition is thought to occur via the enamine conformation shown below.

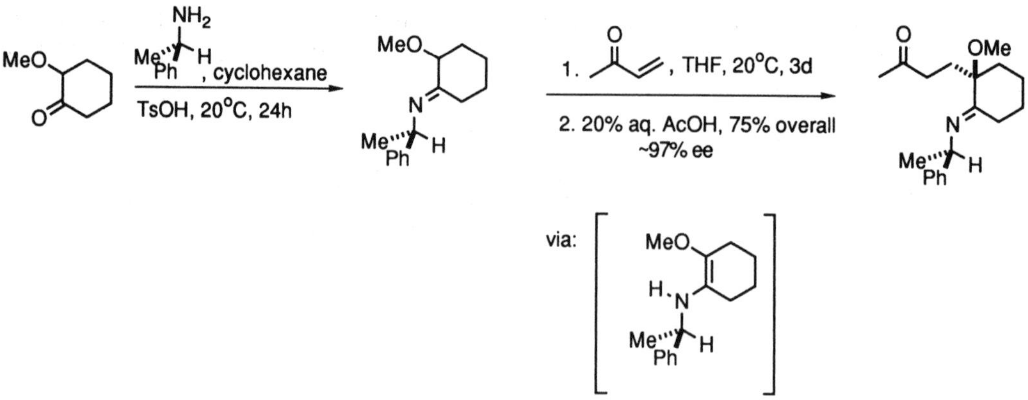

2.1.1.2   Carbanions stabilized by π-conjugation with more than one heteroatom

These highly-stabilized carbanions virtually always add in a conjugate fashion. Hence, regioselectivity is rarely a problem. In addition, several of the stabilizing groups are removable. Therefore, these groups often serve two purposes. Firstly, they ensure conjugate addition occurs and, second, they may be removed or modified after addition. In a way, they may be viewed as "conjugate auxiliaries". Some examples are given in the following section (see the corresponding sections in other chapters for more examples).

(a)   *Addition of achiral carbanions to achiral alkenones*

(i)   *1,3-Dicarbonyls and related compounds*

1,3-Diketones tend to be so acidic that they can sometimes catalyze their own addition to alkenals and alkenones. For example, entries 1 and 2 in Table 2.2, show that simply mixing 2-methyl-1,3-cyclopentanedione with a conjugate acceptor in water is sufficient to obtain excellent yields of adduct. In fact, 2-methyl-1,3-cyclopentanedione has been subjected to a variety of catalysts in its conjugate additions including pyridine in toluene[94], potassium hydroxide in methanol[95] or triethylamine in ethyl acetate.[96] Nickel(II) catalysis has also been successfully used with 1,3-dicarbonyl compounds.[97]

Where the nucleophile is crowded, as in entry 7, an improvement in yields is often achievable using high pressure reaction conditions. Thus the same reactants, but with triethylamine as base, give a much better yield (67%) at 10 kbar.[98,99]

A simple method for adding 1-nitro-1-toluenesulfonylalkyl groups to alkenones was developed by Zeilstra and Egberts. Although the yield is low, the pure product crystallizes from solution.[100]

Heating a mixture of ethyl 3-oxobutanoate, 4-methyl-3-penten-2-one and boron trifluoride dietherate provides isophorone in 27% yield (37% yield, based on unrecovered starting material).[101]

Table 2.2 Conjugate addition of highly-stabilized carbanions to alkenals and alkenones

1[102]			H_2O, r.t., 100%	
2[103]			H_2O, r.t., 5d, 82%	
3[104]			H_2O, 70-80°C, 4h hydroquinone, 100%	
4[105]			NaH (cat), Et_2O 100%	
5[105]			CH_3OH, NaOH, 20°C 74%	
6[106,107]			$RuH_2(PPh_3)_4$, 0.03 eq THF, r.t., o/n, 72%	
7[108]			1. TBAF, 0.6 eq THF, r.t., 3d 2. H_2O, 28%	

Tsuji's group has developed a synthetic strategy which involves sequential addition to the conjugate acceptor moiety of 3-oxo-1,8-nonadiene, followed by a palladium-catalyzed hydration of the terminal alkene (the Wacker reaction).[109]

The resulting 1,6-diketone may then be used in a variety of ways, including an intramolecular aldol reaction. Such a process, which is, effectively, a type of Robinson annulation, has been applied to the total synthesis of such steroids as (+)-19-nortestosterone.[110]

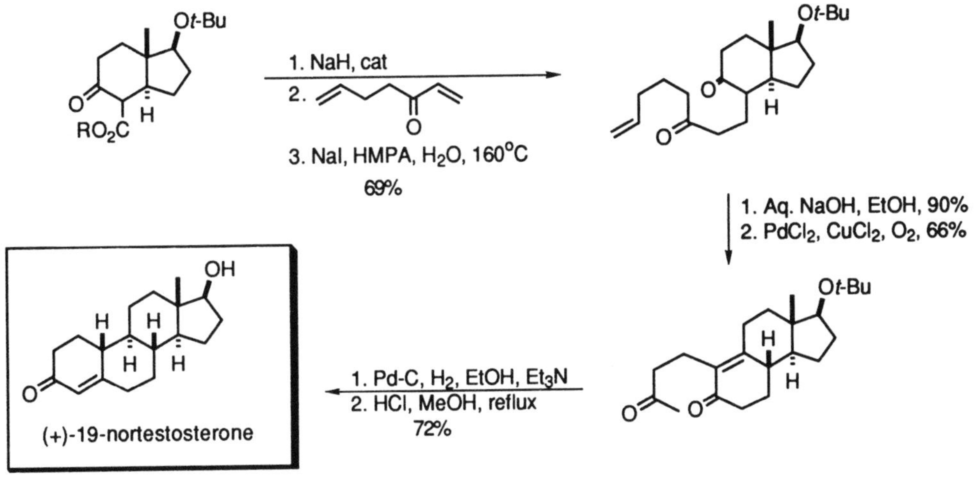

(+)-19-nortestosterone

The addition of 2-bromomalonates to alkenals and alkenones provides a simple synthesis of cyclopropyl derivatives.[111]

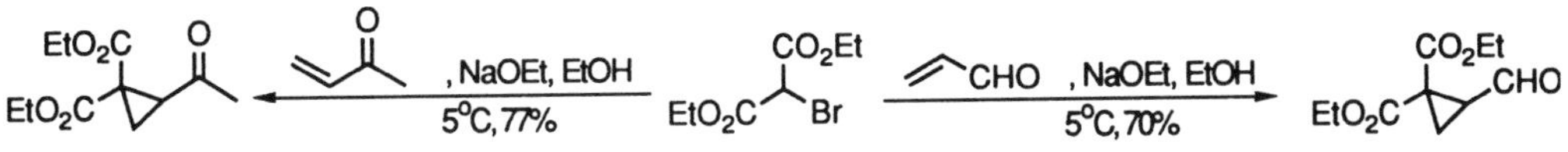

α-Metallated *iso*-cyanides (e.g. from tosylmethyl*iso*-cyanide, "TOSMIC") react with alkenones providing simple access to a variety of heterocycles, including pyrroles.[302]

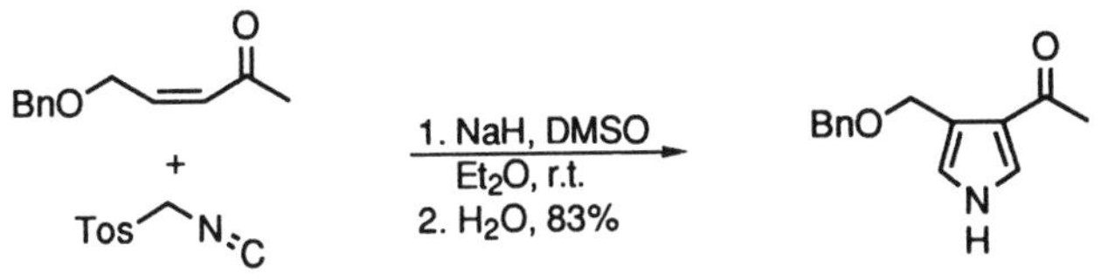

Benzannulated macrocyclic lactones have been prepared by a route which involves a PTC addition of a β-keto nitrile to propenal.[112]

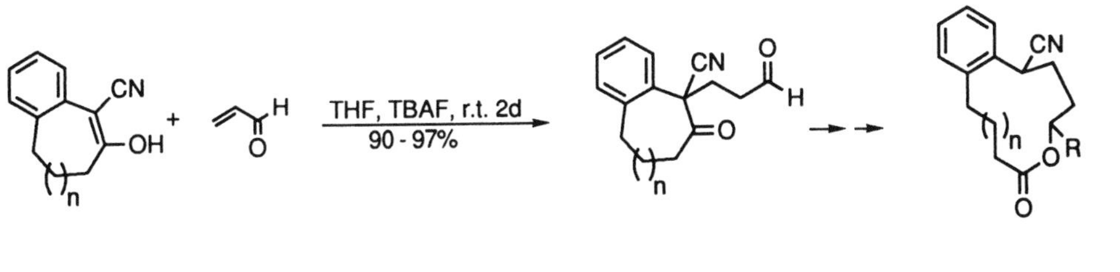

Substituted malonates add efficiently to alkenones. For example, the following addition was used in a short synthesis of (±)-malyngolide, an antibiotic from the marine blue-green alga, *Lyngbya majuscula* Gomont.[113]

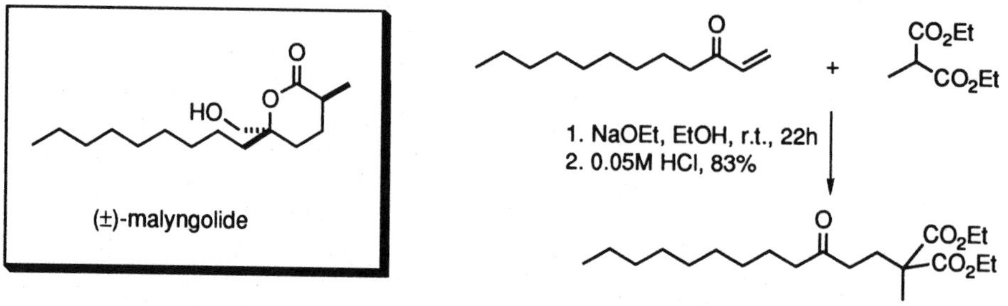

(±)-malyngolide

1. NaOEt, EtOH, r.t., 22h
2. 0.05M HCl, 83%

A synthesis of DL-tryptophan has been developed which incorporates a base-catalyzed conjugate addition of diethyl acetamidomalonate (**14**) to propenal.[114]

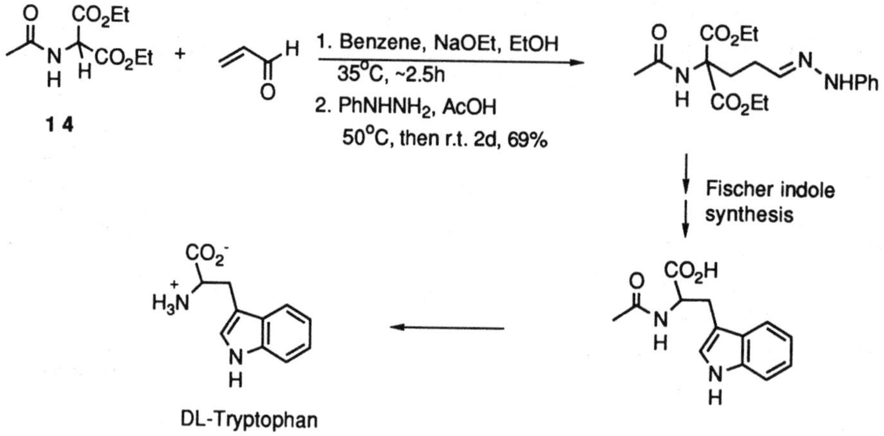

14

1. Benzene, NaOEt, EtOH
   35°C, ~2.5h
2. PhNHNH₂, AcOH
   50°C, then r.t. 2d, 69%

Fischer indole synthesis

DL-Tryptophan

The alkenylimine derived from butenal and benzylamine reacts with both 2,4-pentanedione and ethyl 3-oxobutanoate under acidic conditions.[115] All attempts to react this alkenylimine with a variety of 1,3-dicarbonyl compounds, under basic conditions, failed to yield any isolable products. The products, 1,4-dihydropyridines, are of biological interest as several members of this class are highly active calcium antagonists.[116]

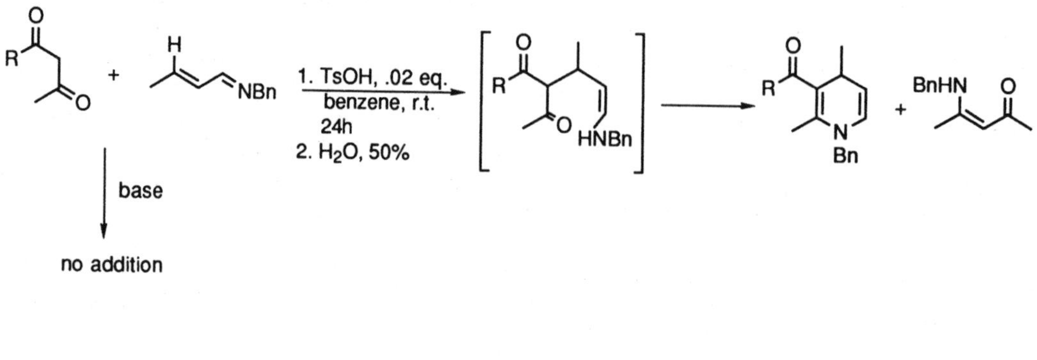

1. TsOH, .02 eq.
   benzene, r.t.
   24h
2. H₂O, 50%

base

no addition

(ii)   *α-Nitroketones*

α-Nitroketones add efficiently to alkenones and alkenals. As mentioned in section 2.1.1.1 above, the nitro group is readily removed reductively.[117]

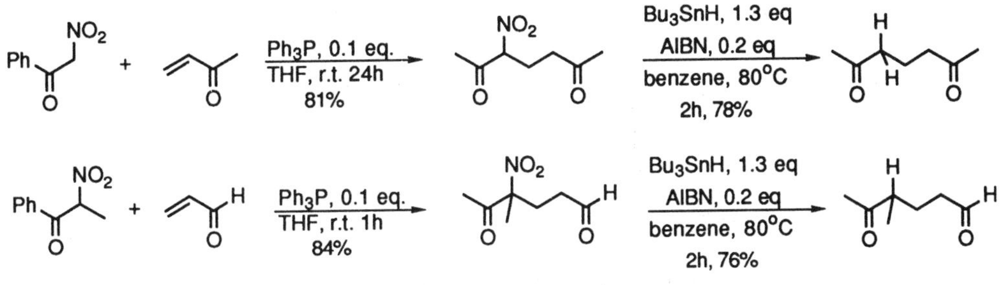

(b)   *Addition of achiral carbanions to chiral alkenones*

As with other classes of nucleophiles, addition of highly-stabilized carbanions to fused ring systems is highly stereoselective.

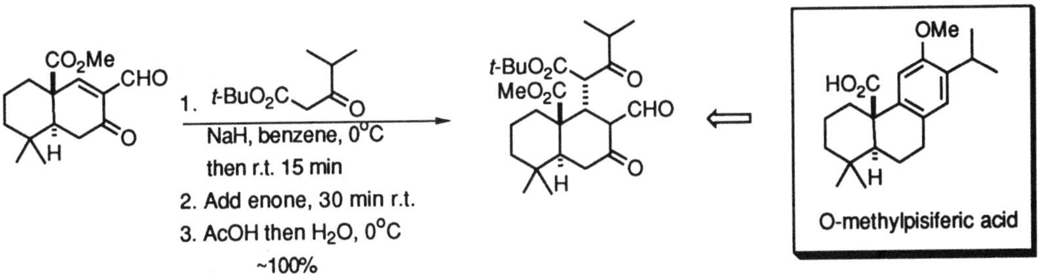

(c)   *Addition of chiral carbanions to achiral alkenones*

(i)   *1,3-Dicarbonyls and related compounds*

Chiral catalysts

The earliest report of attempts to obtain asymmetric induction in the Michael reaction appears to have involved the use of optically active quartz.[118] Subsequent to that, the addition of β-ketoesters to alkenones, with chiral catalysis, has become something of an archetype for the study of asymmetric induction in Michael reactions. The first study of these additions came from Langstrom and Bergson[119], who used the naturally-occurring alkaloid, quinuclidine as the chiral catalyst. However, although they were able to demonstrate that the products contained one enantiomer in excess of the other, neither the configuration of the major enantiomer nor its excess were reported.

Soon afterwards, Wynberg's group reported the first of a series of studies devoted to the development of chiral catalysis in conjugate additions.[120,121] They also used naturally-occurring as well as derivatized (cinchona) alkaloids. They examined several nucleophiles. Of these, the quinine-catalyzed addition of indanone (**15**) to 3-buten-2-one gave the best enantiomeric excess, in favour of the *S* enantiomer. As with many types of chiral transformations, the reaction conditions required considerable optimization.

Solvents of low polarity, such as carbon tetrachloride and toluene gave the best selectivities. Where slightly less acidic β-ketoesters, such as 2-methoxycarbonylcyclohexanone, were used, a combination of quinine methohydroxide and ethanol (1 to 2%) as co-solvent were necessary. This introduction of even a small amount of a protic solvent led to a dramatic decrease in enantioselectivity. Whilst disappointing, this did point to the importance of hydrogen-bonding to the alkaloid hydroxy group. Removing the possibility of hydrogen-bonding by conversion to an acetate also significantly reduced the enantioselection. Subsequently, Wynberg's group[122] and others[123,124] have attempted to use polymer-supported alkaloids as chiral catalysts. However, none of these have achieved the same enantioselectivity obtained using quinine.

The most spectacular results have come from Cram's group which showed that almost complete asymmetric induction could be achieved using a chiral crown ether in combination with an alkoxide base catalyst.[125] Although it is quite likely that these chiral crown ethers provide an asymmetric cavity, into which an enolate can fit, it is still not clear what influences the way in which the enolate fits (or slides) into that cavity. This will determine which enolate face is accessible to an electrophile and hence the enantioselectivity.

This reaction has also been studied using a chiral catalyst based on a complex formed between Co(acac) and a chiral diamine.[126] The chemical yields can be excellent (especially if the chiral ligand is omitted), however the enantiomeric excess is of the order of 66%.

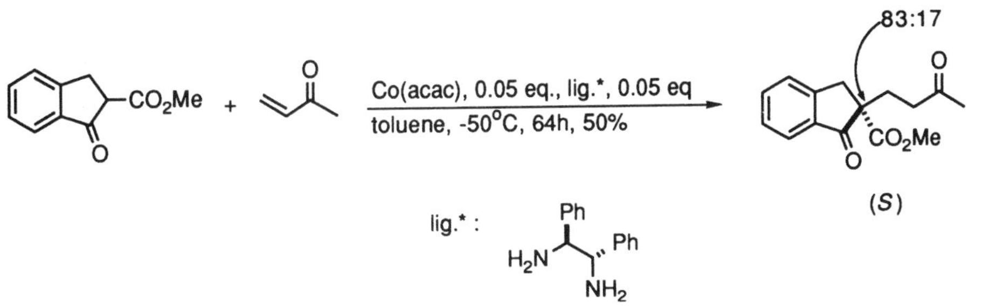

Chiral enolates

In their total synthesis of the trichothecene mycotoxin, anguidine, Brooks' group obtained excellent yields and stereoselectivity in the addition of a β-aldehydolactone to 3-buten-2-one.[127] Only one diastereomer was produced. Cyclization proved to be a problem, requiring a three step sequence as shown below. Attempts to achieve cyclization using equilibrating conditions only led to retro-Michael products. Overall this is another example of a Robinson spiroannulation.

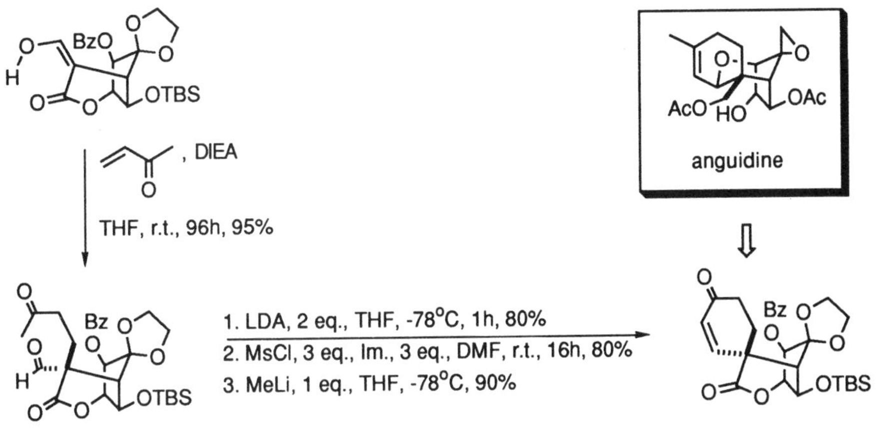

One of the most outstanding examples of stereoselective Michael reactions, so far, has been developed in Taber's group.[128] In their approach, a β-oxo-(1-naphthylbornyl)ester was added to 3-buten-2-one yielding an adduct with a 90% diastereomeric excess. It was found that the use of ethereal solvents, rather than hydrocarbons or halogenated solvents, gave the highest diastereomeric excess. Unfortunately, the major isomer was not that predicted by the accepted solution structure of bornyl esters,[129] i.e. the extended conformation **b** shown on the next page.

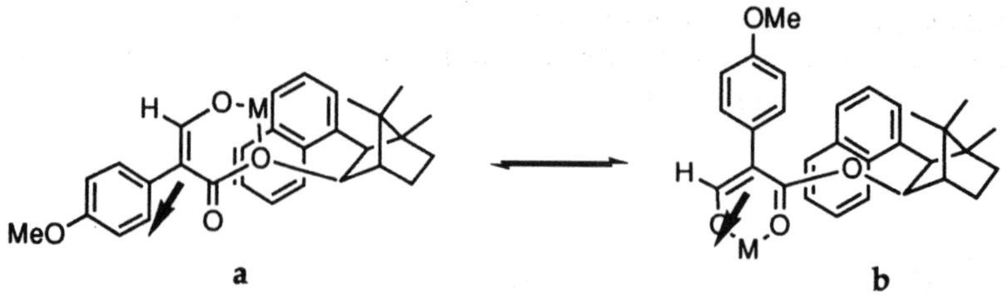

a   b

However, the product was converted into the unnatural enantiomer of the naturally occurring alkaloid, *O*-methyljoubertiamine, in eleven steps. Most of these steps were involved with protection and deprotection while the ester was converted into the dimethylaminoethyl side chain.

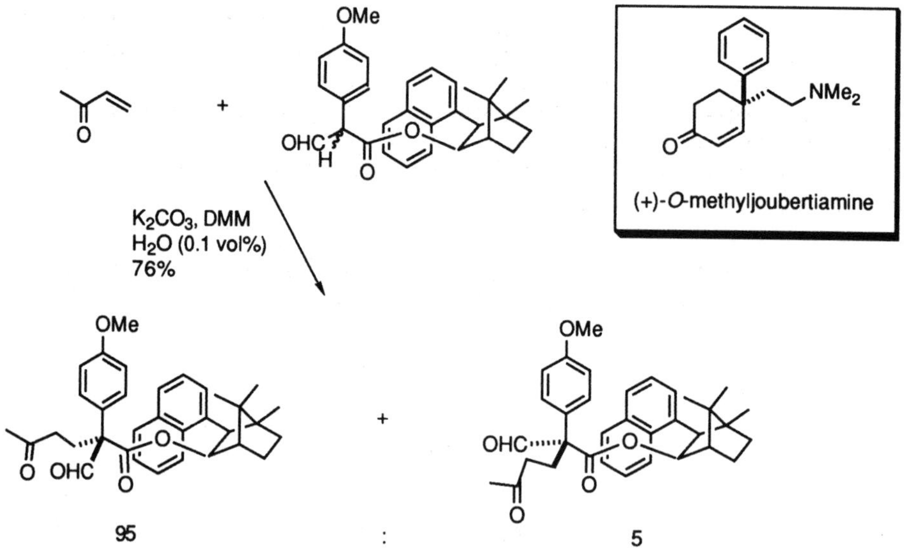

K₂CO₃, DMM
H₂O (0.1 vol%)
76%

(+)-*O*-methyljoubertiamine

95   :   5

Addition of the stabilized lactone enolate, derived from (**16**), to 3-buten-2-one produced an advanced intermediate in a total synthesis of the cytotoxic lignan, (±)-megaphone.[130] Ironically, the newly-formed asymmetric centre is lost in subsequent manipulations and regenerated at a later stage.

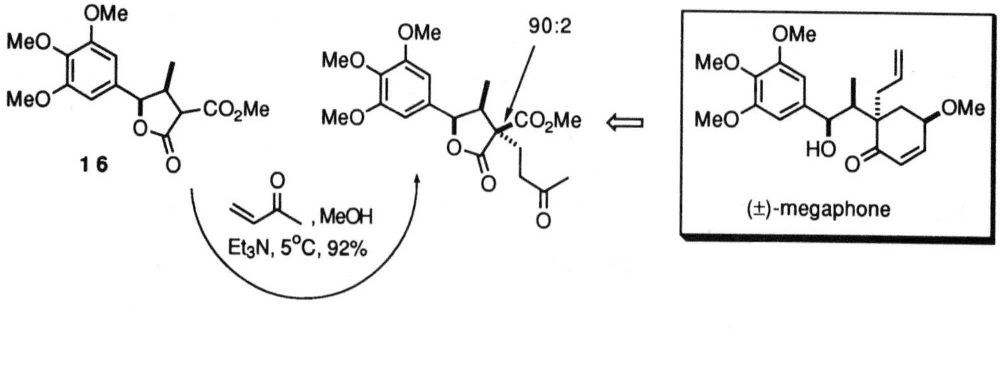

90:2

16

, MeOH
Et₃N, 5°C, 92%

(±)-megaphone

Fodor's group have studied the addition of ascorbic acid to alkenals and alkenones. Many conjugate acceptors have been found to react well and, like the example shown below, two new stereogenic centres are formed with complete stereoselectivity.[131]

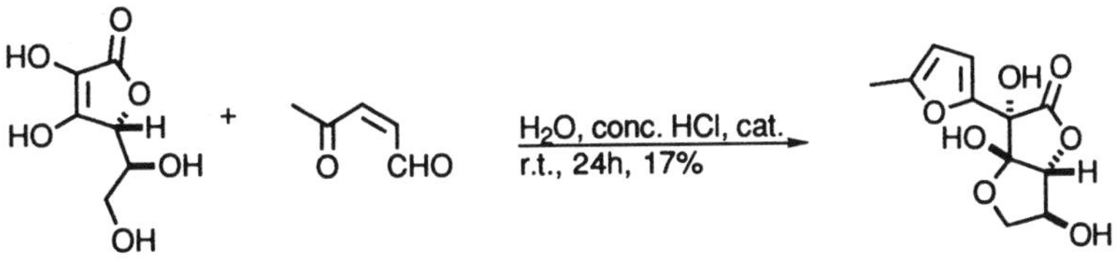

2.1.1.3   Carbanions stabilized by one or more α-heteroatoms

Most carbanions in this class function as masked acyl anions. The most widely used of these have been dithioacetal based reagents.[132] Much effort was put into finding suitable reaction conditions to persuade these, ostensibly hard, nucleophiles to add in a conjugate fashion. Success came with the discovery, by several groups, that the inclusion of one or more equivalents of hexamethylphosphoramide in the reaction solution led to very significant increases in the ratio of 1,4 to 1,2 addition products. Importantly, this held for additions to alkenals, as well as alkenones.[133]

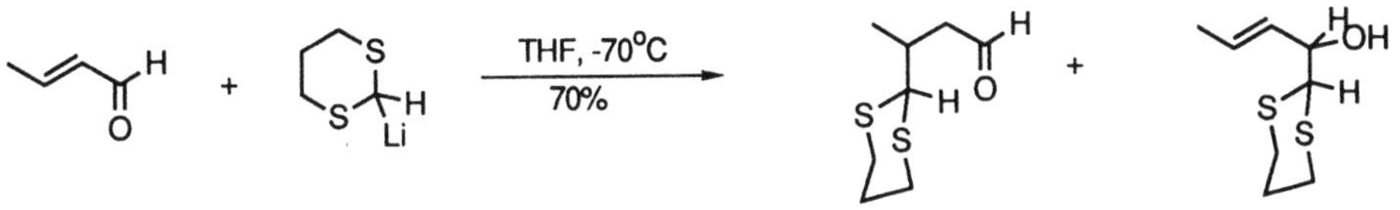

	No HMPA -	0	:	100
	With HMPA -	45	:	55

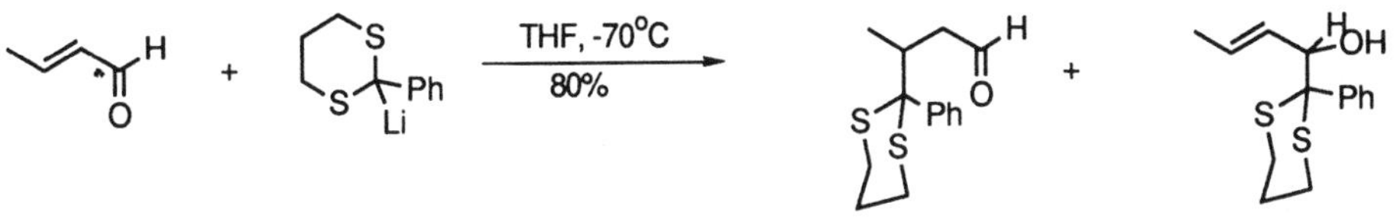

	No HMPA -	?	:	?
	With HMPA -	95	:	5

Although these are very reactive nucleophiles, where there is steric hindrance in both the nucleophile and the β-carbon of the acceptor, the yields drop and 1,2-addition becomes significant as in the following examples.[134,135]

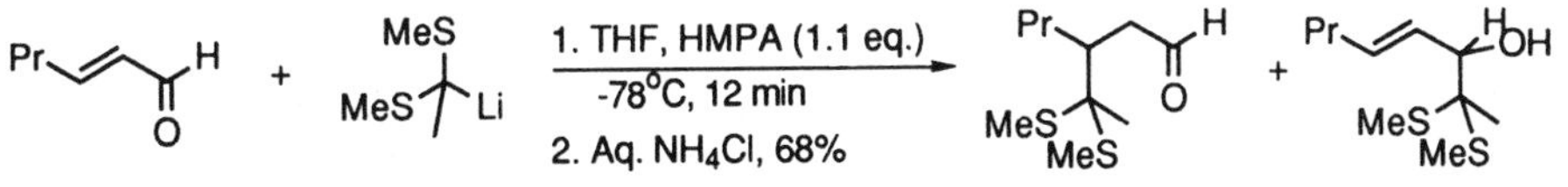

	4.6	:	1

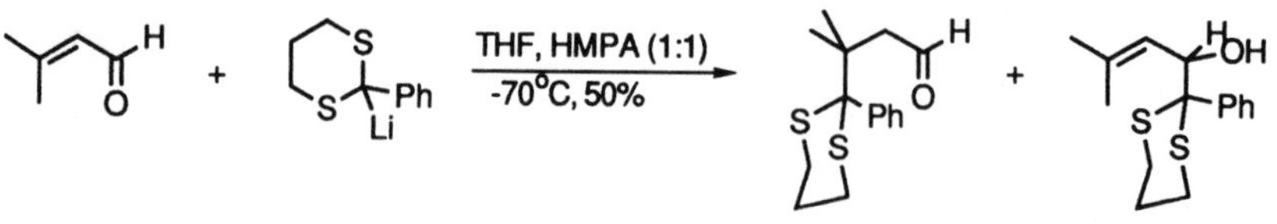

1   :   1.9

For extremely bulky nucleophiles, such as tris(phenylthio)methyllithium, only 1,2 addition is observed (however, addition to *cyclo*alkenones can occur, see section 3.1.1.3).[136]

A nice example of the value of these reagents in synthesis comes from Magnusson's total synthesis of the lignan, (-)-burseran.[137]

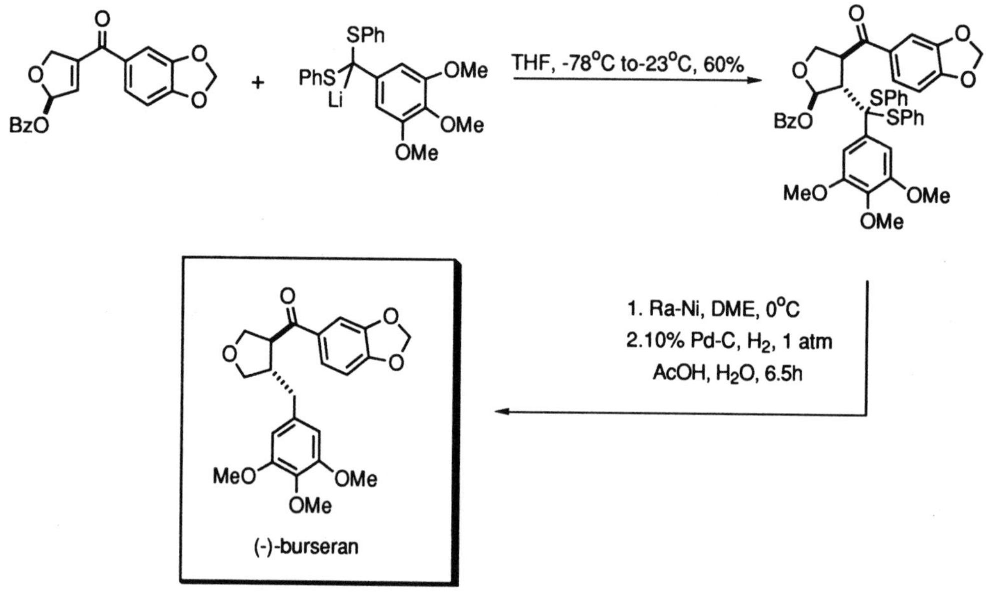

1. Ra-Ni, DME, 0°C
2. 10% Pd-C, H₂, 1 atm
   AcOH, H₂O, 6.5h

(-)-burseran

The following is a compilation of some of the α-heteroatom stabilized carbanions which have been shown to react successfully with alkenals and/or alkenones. (An asterisk indicates a report of a reaction with an alkenal).

R = H[138]*
R = Ph[137]*

R = Me[139]*

R = alkyl[138]*

R = PhS[140]
R = Hex[141]

2.1.2   Organocopper reagents

2.1.2.1   Addition of achiral organocopper reagents to achiral alkenals and alkenones

A wide variety of alkenals and alkenones react smoothly with organocopper reagents. A study by Hallnemo and Ullenius[301] on the addition of LO cuprates to E-4-phenyl-3-buten-2-one, demonstrates the importance of the nature of the solvent for these additions. Thus reaction in hexane, toluene or diethyl ether at 0°C is complete after only one minute. However, in other solvents, especially tetrahydrofuran, only partial reaction had occurred. In highly coordinating solvents, such as DMF or DMSO, no conjugate addition occurred

Solvent	Time, min	Yield, %
Hexane	1	>98
Toluene	1	>98
Et$_2$O	1	>98
THF	1	35
THF	60	90

Figure 2.3   Solvent effect on chemical yield of LO cuprate additions to an alkenone

A selection of examples is included in Table 2.3.[142] Mono, di- and tri-substituted systems are all suitable. It has recently been demonstrated that organocuprates, derived from iodoorganozinc reagents, are tolerated by many functional groups usually incompatible with organolithium or organomagnesium reagents.[143,144] These functional groups include esters, nitriles, alkenoates and imides, e.g. entry 2.

Copper catalyzed additions of organomanganese compounds appears to be a promising alternative to the use of copper catalyzed organomagnesium and organolithium additions.[145,146] The authors claim that yields are better than the more conventional copper catalyzed reactions. The most obvious difference is that the reactions are carried out at 0°C rather than lower temperatures. Even more interesting is the finding that a combination of catalytic quantities of copper (I) chloride and manganese (II) chloride efficiently catalyzes the conjugate addition of butylmagnesium chloride to 4-methyl-3-buten-2-one. The yields are apparently higher than that obtained using copper (I) chloride alone (entry 4).

Table 2.3   Conjugate addition of organocopper reagents to alkenals and alkenones

1[147]	PhS⧸PhS⧸H, R	1. BuLi, 1 eq.,THF, -40 to -30°C 1h; 2. CuI, 0.5 eq., -78°C, 1h; 3. Add (enone), 2h, 80%	PhS, R, PhS (product) R = *n*-hexyl
2[142]	(isopropylidene methyl ketone)	1. EtO₂C(CH₂)₃Cu(CN)ZnI,1.3 eq, BF₃.Et₂O, 3 eq., THF -30°C, 3h; 2.Aq. NH₄Cl, 88%	EtO₂C (product)
3[144]	(enone)	*n*-BuMnCl, CuCl, 0.05 eq, THF, 0°C, 4h, 90%	(product)
4[144]	(enone)	*n*-BuMgCl, CuCl, 0.01 eq, MnCl₂, 0.3 eq, THF, 0°C, 1h, 94%	(product)
5[148]	Ph (enone)	1. Th(2-Py)CuLi, 1.7 eq, Et₂O, 0°C, 10 min; 2. Aq. NH₄OH, NH₄Cl, 83%	Ph, Py (product)
6[149]	Ph (enone)	MeCu, TMSI, 2 eq., Et₂O, 0°C, 4.5h, 69%	Ph, OTMS (product) Z:E (5.1:1)

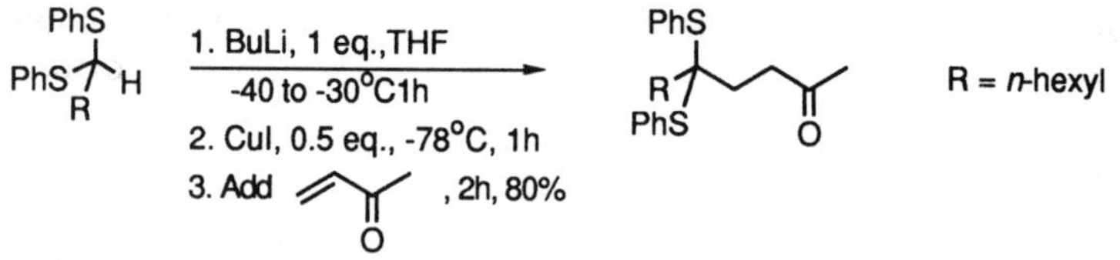

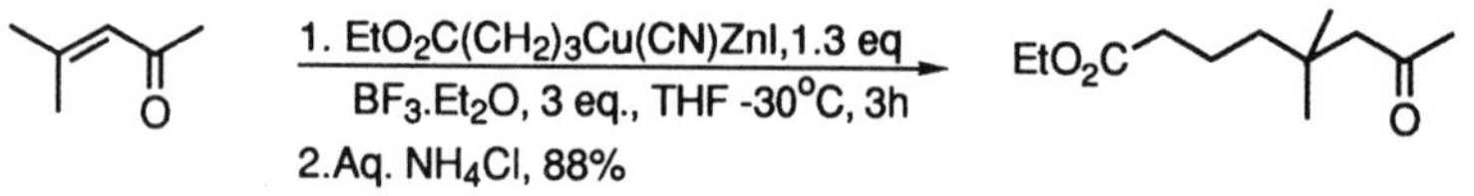

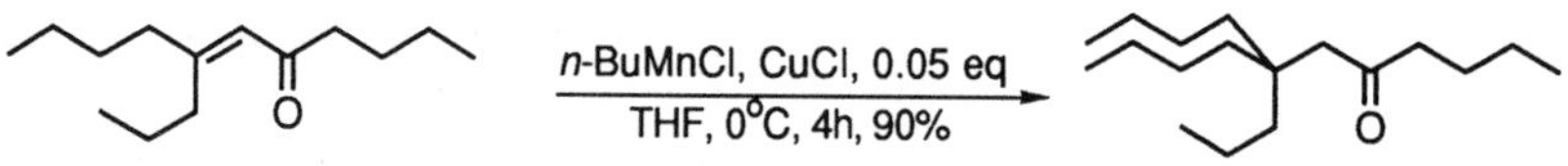

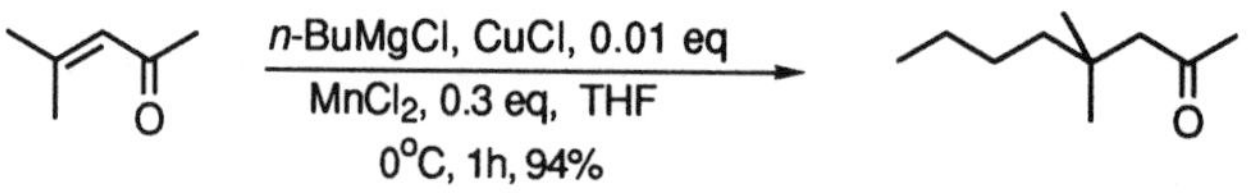

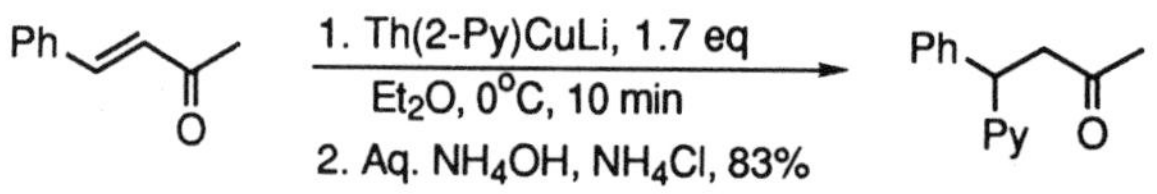

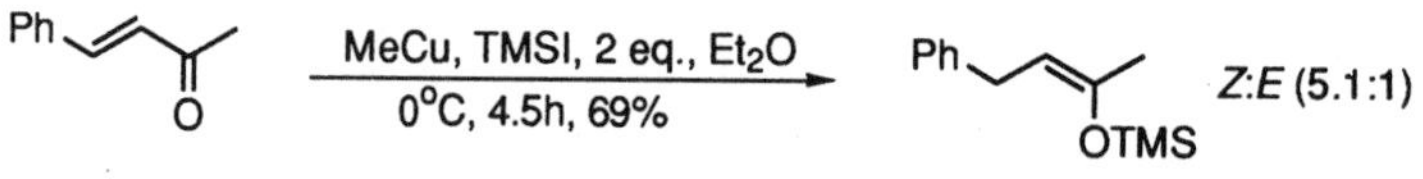

The substitution of β-phenylthioalkenones with HO cuprates has been studied. The reactions give mainly retention of stereochemistry in tetrahydrofuran and inversion in diethyl ether.[150]

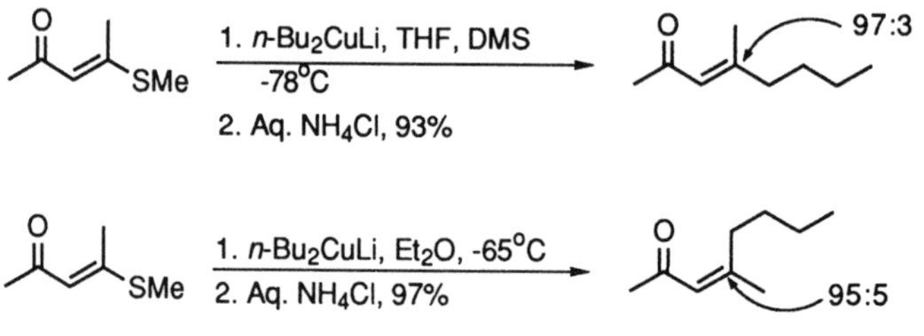

(enone-SMe)
1. *n*-Bu₂CuLi, THF, DMS, -78°C
2. Aq. NH₄Cl, 93%
→ (product) 97:3

(enone-SMe)
1. *n*-Bu₂CuLi, Et₂O, -65°C
2. Aq. NH₄Cl, 97%
→ (product) 95:5

A possible alternative method for preparing Robinson annulation products, using conjugate additions with α-cuprated hydrazones, has been reported.[151] For example, conjugate addition of the α-metallated hydrazone (**17**) from 2-methylcyclohexanone to 3-buten-2-one gave a virtually quantitative yield of adduct after oxidative hydrolysis.

17

1. THF, -30°C,
   then warm to 0°C
   over 12h
2. NaIO$_4$
   90 - 95% overall

EtOH, KOH
100%

Particularly significant are the conjugate additions to alken*als*, as most other classes of organometallic reagents suffer from significant (to complete) competing 1,2 addition.[152] This is rarely a problem with organocuprates. The inclusion of chlorotrimethylsilane and hexamethylphosphoramide in the reaction mixture leads to excellent yields of the trimethylsilyl enol ether of the adduct. The stereochemical purity of the adduct is also often very high. Nakamura's group has studied the use of these additives in both copper-catalyzed[153] as well as stoichiometric[154] additions. Some examples of their results are shown in the following equations.

CHO

n-BuMgBr, CuBr.DMS, 0.05 eq
TMSCl, 2 eq., HMPA, 2 eq., THF
-78°C, 3h, 89%

OTMS

96% *E*

CHO

n-Bu$_2$CuLi   TMSCl   HMPA   -78 to -40°C

OTMS

1.2 eq	2 eq	-	88%	(*EZ*) 89:11
0.6 eq	2 eq	2 eq	80%	(*EZ*) 98:2

n-BuCu

1.2 eq	2 eq	2 eq	71%	(*EZ*) 95:5

CHO   Ph$_2$CuLi

Ph   OTMS

0.6 eq	2 eq	2 eq	99%	(*EZ*) 97:3

O

n-Bu$_2$CuLi

OTMS

0.6 eq	2 eq	2 eq	84%	(*EZ*) 28:72

Many examples exist of organocopper additions followed by enolate trapping.[155] A few are shown in Table 2.4.

Table 2.4   Conjugate addition/enolate trapping reactions of organocopper reagents with acyclic alkenones

1[156]	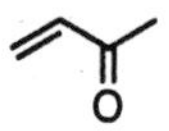	1. n-Bu₂CuLi, 2 eq., Et₂O, 0°C, 30 min 2. MeCHO, 10 eq., ZnCl₂, 2 eq., 0°C, 5 min 3. Aq. NH₄Cl, 92%
2[155]	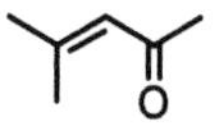	1. Me₂CuLi, 2 eq., Et₂O, 0°C, 30 min 2. MeCHO, 10 eq., ZnCl₂, 2 eq., 0°C, 5 min 3. Aq. NH₄Cl, 96%
3[157]	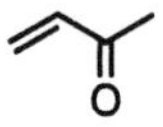	1. Me₂CuLi, THF, -25°C, 1h 2. AcCl, 0°C, 2h 3. Aq. NH₄OH, 30%
4[158]	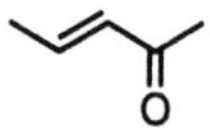	1. Me₂CuLi, 1.5 eq., Et₂O, 0°C, 30 min 2. MeI, DME, 5 min, 46%

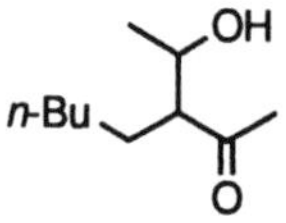

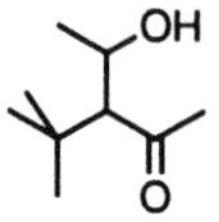

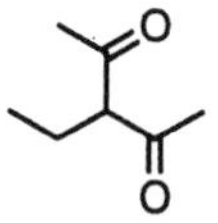

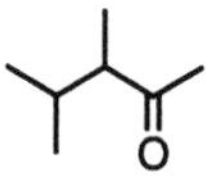

Seyferth has extended his studies on the chemistry of lithium acyls to include "acylcopper" reagents. These are generated by treatment of a lithium acyl with a copper salt in an atmosphere of carbon monoxide.[159,160] LO "acylcuprates" appear to be more stable (at room temperature in one case) than the corresponding HO reagents.[161] By its nature, this reaction provides the most direct method for preparing 1,4-dicarbonyl compounds.

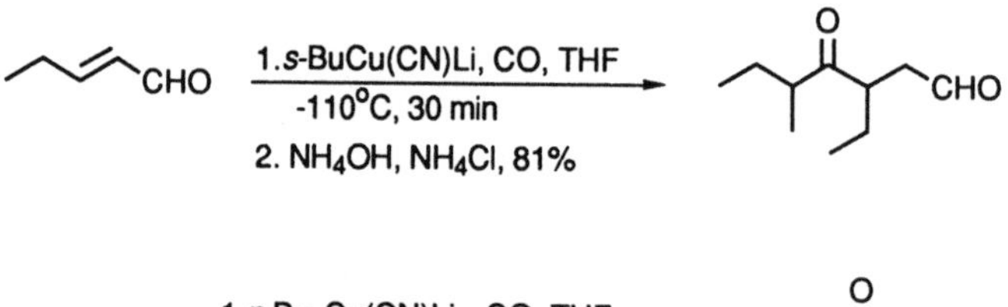

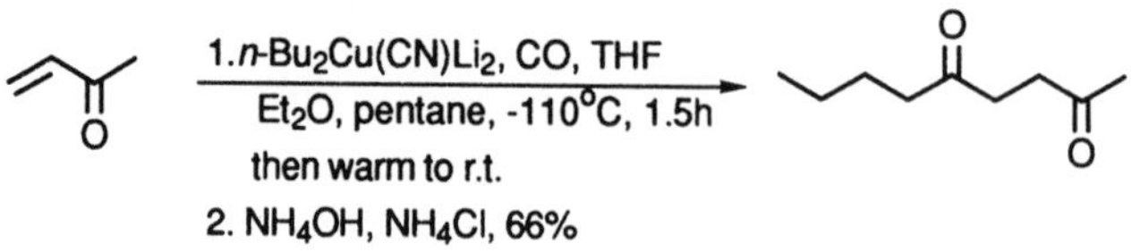

1. s-BuCu(CN)Li, CO, THF
   -110°C, 30 min
2. NH₄OH, NH₄Cl, 81%

1. n-Bu₂Cu(CN)Li₂, CO, THF
   Et₂O, pentane, -110°C, 1.5h
   then warm to r.t.
2. NH₄OH, NH₄Cl, 66%

Alkenylcuprates

Stoichiometric alkenyl cuprates are usually prepared by sequential lithiation and transmetallation of an alkenyl halide.[162] This procedure has found wide use in, especially, prostaglandin synthesis (see section 3.1.2.2). A new method has been developed which obviates the need for proceeding through an alkenyl halide.[163] In this method advantage is taken of the efficiency of the zirconium-catalyzed carboalumination of terminal alkynes.[164] Reaction of the intermediate alkenylalane (**18**) with an HO cuprate leads to transmetallation and the generation of a new, mixed HO cuprate with complete retention of stereochemistry. The reactivity of these HO cuprates is good, as even 3,3-disubstituted alkenones react well. (The use of LO cuprates was not successful). There was also no evidence for interference from the residual ligand in these additions.

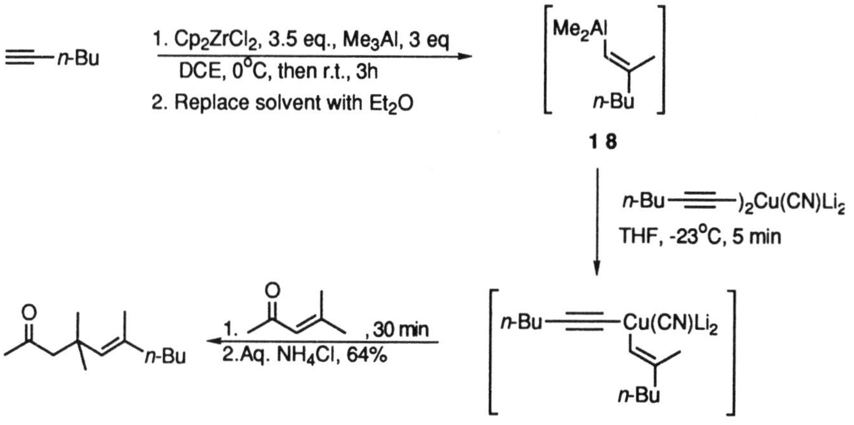

Alkenylcuprates may be prepared from trisylhydrazones.[165] In most cases the heterocuprate added more efficiently than the corresponding homocuprate (hence the use of phenylthiocopper in the example below).

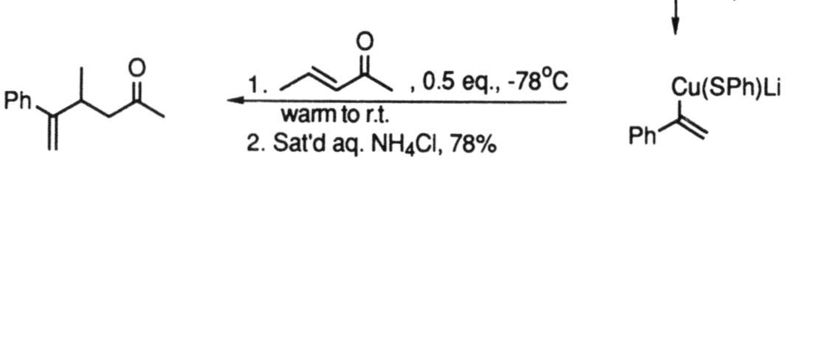

2.1.2.2 Addition of achiral organocopper reagents to chiral alkenals and alkenones

A synthesis of levuglandin E$_2$, a solvent-induced decomposition product of PGH$_2$, has been developed whose stereoselectivity relies upon the conformational bias of a γ,δ–dialkoxyalkenone.[166] As opposed to related additions to γ-alkoxyalkenoates, the alkene stereochemistry in this case does not influence the stereochemical outcome of the addition. Hence, both the E and Z alkenone afforded identical mixtures of diastereomeric products.

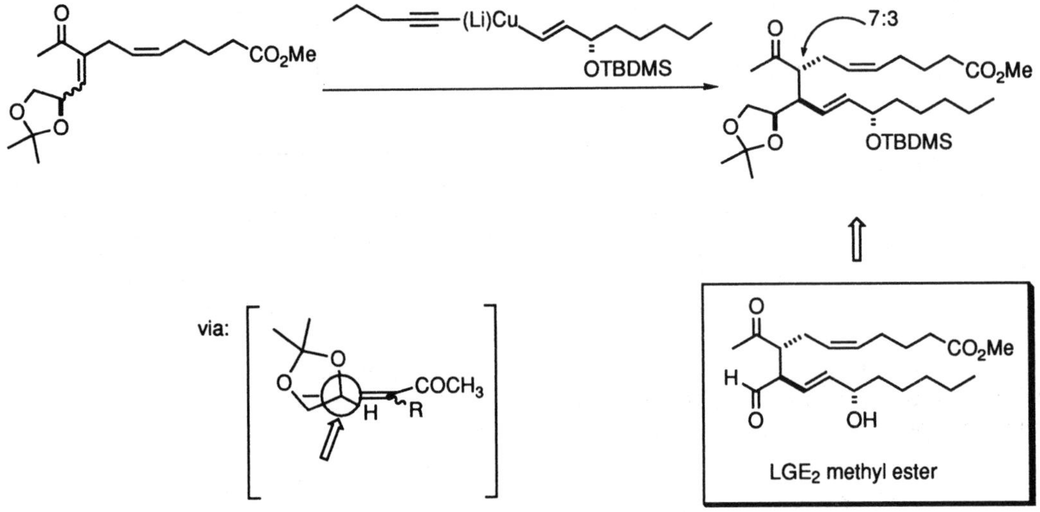

LGE$_2$ methyl ester

Additions to γ-alkoxyalkenones are highly stereoselective, as in the following example.[167]

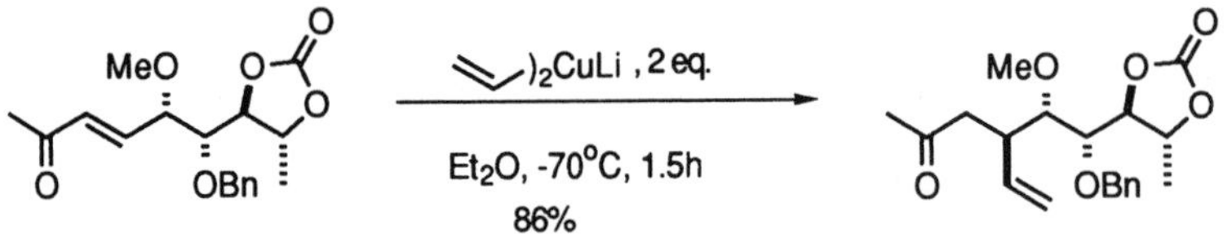

γ-Oxy substituents in rings also provide a high level of stereocontrol.[168]

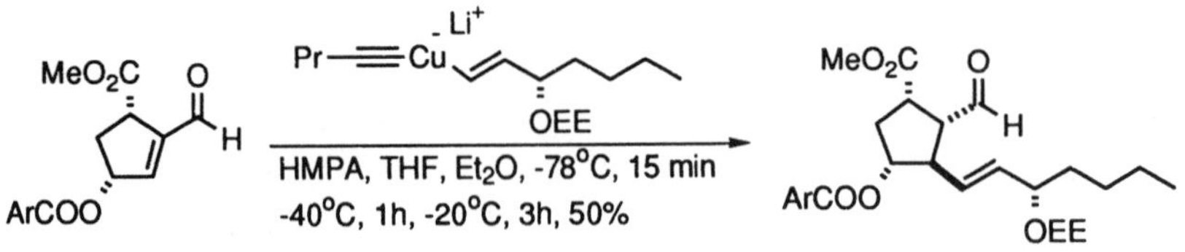

Corey's group has studied the influence of a remote stereogenic centre on cuprate additions to alkenones.[169] Good diastereoselectivity was obtained with the E-isomer (19).

Reactions with the Z-isomer (20) were complicated by isomerization to the E-isomer under the reaction conditions. However, this could be avoided, to some degree, by adding trimethylsilyl triflate prior to addition of the alkenone. The authors proposed that the observed stereoselectivity could be accounted for by assuming that the reactions proceed through the intermediates shown for each case. These assume that a copper(III) intermediate is formed (see section 1.3.3).

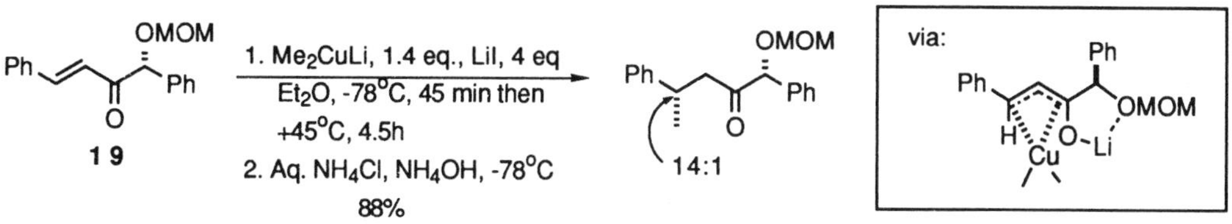

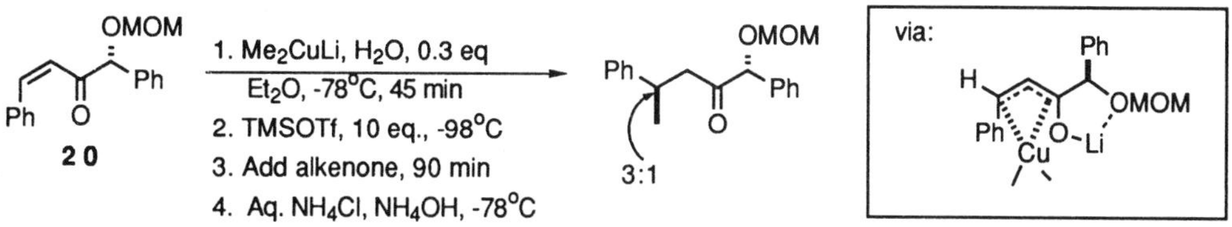

A cyclic aminal, derived from ephedrine, has been used as an auxiliary for additions to alkenals.[170] Conjugate addition to the (+)-ephedrine derivative of E-3-phenylpropenal gives the 3-(R) enantiomer preferentially, after hydrolysis.

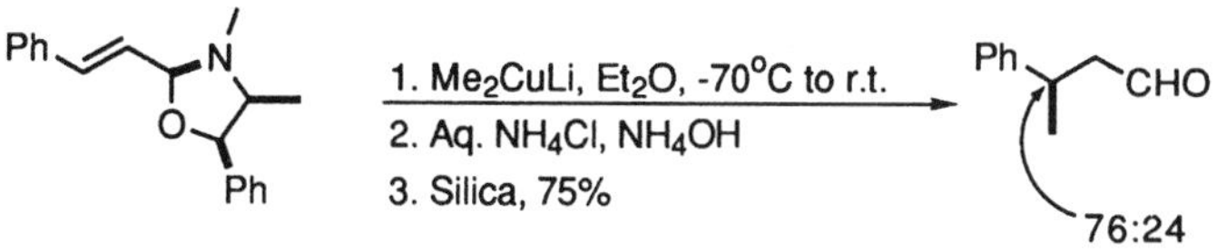

2.1.2.3   Addition of chiral organocopper reagents to achiral alkenals and alkenones

Chiral Transfer Ligands

Nilsson's group has discovered an extremely stereoselective reaction involving the transfer of a group originally designed to be a chiral residual ligand.[171]

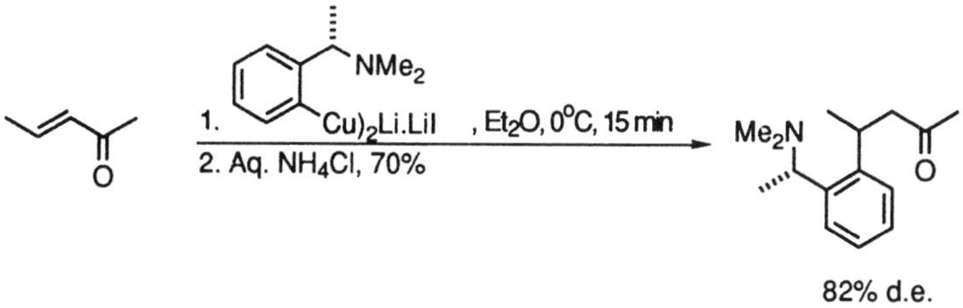

82% d.e.

Chiral Residual ligands

A series of modified, bidentate (*S*)-proline ligands, shown below, has been prepared and employed as chiral residual ligands in heterocuprate additions.[172]

X = OH, OMe SPh, SMe

Enantiomeric excesses as high as 83% were obtained, although the outcome was variable.

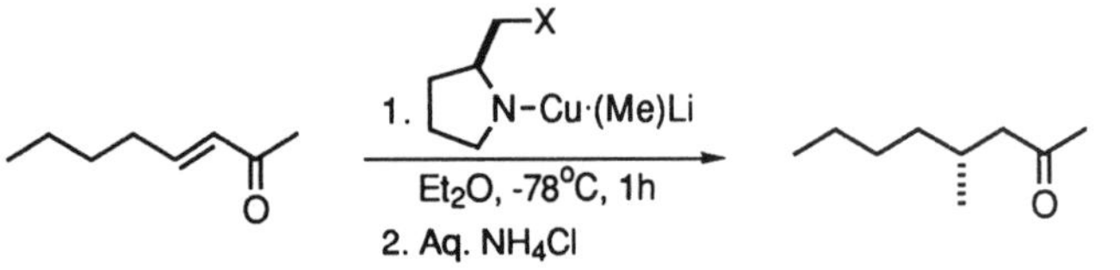

X = OMe	46%	58% e.e.
X = SPh	42%	74% e.e.
X = SPh	78%	83% e.e.

1. [N]–Cu·(Me)Li

Et$_2$O, -78°C, 1h

2. Aq. NH$_4$Cl

2.1.3   Other organometallic reagents

The reaction of hard nucleophiles, such as organolithium and organomagnesium reagents is dominated by 1,2-addition, as might be expected. This is of course one of the reasons why copper catalysis of reactions of these nucleophiles has become so important (see section 2.1.2). As a result only a few classes of organometallic reagents will add reliably to acyclic alkenones.

Triorganozincates[173] add to alkenones (and to cycloalkenones), often with good regioselectivity (see Table 2.5). However, these reagents are very sensitive to steric effects, and do not add to 3,3-disubstituted alkenones. It appears that most additions, especially those of organomagnesium-derived organozincates, are sometimes better carried out at 0°C.[174] A report listing rather poor yields ran all the reactions at much lower temperatures.[175] However, some addition can occur at low temperatures. The nickel-catalyzed addition of dialkylzincs, in the presence of a homochiral aminoalcohol, is a promising method for enantioselective synthesis.[176] Acetonitrile was essential as solvent. Inclusion of an achiral ligand (e.g., 2,2'-bipyridyl) was also essential for obtaining high enantiomeric excesses. Quite a large variety of such ligands were found to be useful.

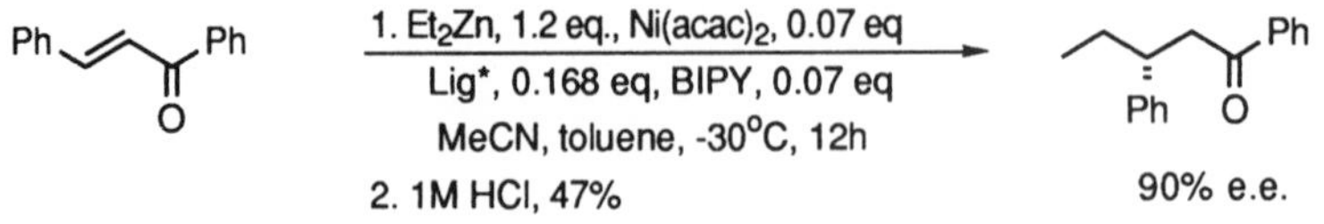

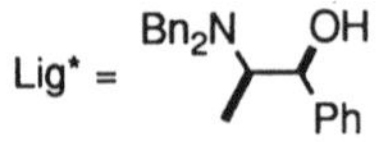

1. Et$_2$Zn, 1.2 eq., Ni(acac)$_2$, 0.07 eq

Lig*, 0.168 eq, BIPY, 0.07 eq

MeCN, toluene, -30°C, 12h

2. 1M HCl, 47%

90% e.e.

Lig* =

Zinc complexes also catalyze conjugate additions of organomagnesium reagents. The yields are comparable to those for stoichiometric organozincates, however, regioselectivity is dependent upon the ligand being transferred (entry 4).

Table 2.5   Conjugate additions of organozincates to alkenones

1[177]		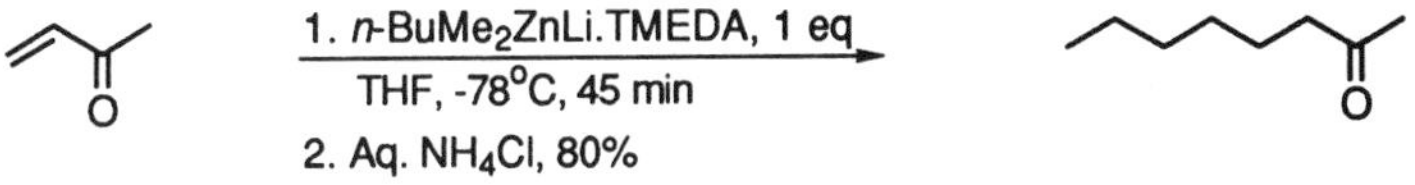 1. *n*-BuMe₂ZnLi.TMEDA, 1 eq THF, -78°C, 45 min 2. Aq. NH₄Cl, 80%	
2[176]	"	1. *n*-BuMe₂ZnLi, 1 eq Et₂O, -78°C, 45 min 2. Aq. NH₄Cl, 46%	"
3[178]		1. *i*-Pr₂(*t*-BuO)ZnMgBr.TMEDA, 1 eq THF, Et₂O, 0°C, 30 min 2. 1NHCl, 95%	

4[179]		1. RMgBr, 1 eq., TMEDA.Zn(O*t*-Bu)Cl, 0.01 eq
		   THF, 0°C, 30 min
		2. Aq. NH₄Cl, R = Et 48% 1,4:1,2 (1:1)
		            R = *i*-Pr 51% 1,4:1,2 (9:1)	

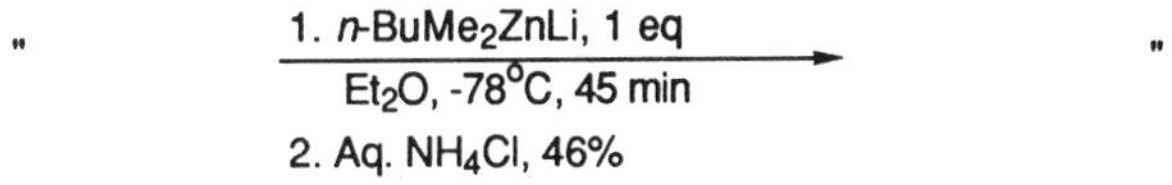

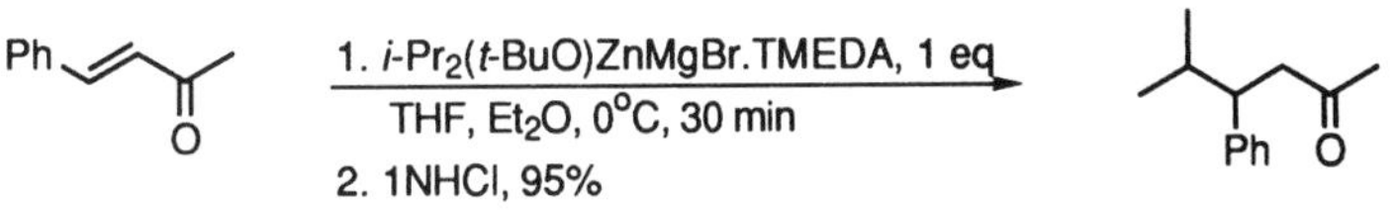

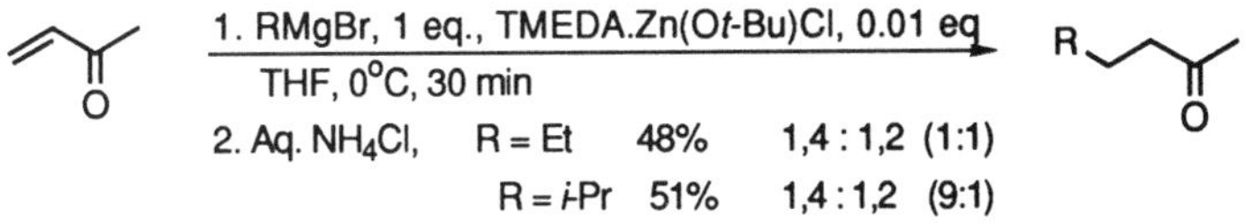

Majetich's group compared the efficacy of fluoride and titanium tetrachloride catalysis in the Sakurai reaction.[180] They found that Lewis acids are the preferred catalyst for additions to alkenones (alkenals fail with both catalysts) whereas fluoride is preferred for additions to alkenoates (see section 4.1.3 and also Table 2.6).

R = H	TBAF, DMF TiCl₄, CH₂Cl₂,	only 1,2 addition decomposition
R = Me	TBAF, DMF TiCl₄, CH₂Cl₂,	50% 1,2 and 25% 1,4 addition 89% (1,4 addition only)
R = *t*-Bu	TBAF, DMF TiCl₄, CH₂Cl₂,	26% 1,2 and 58% 1,4 addition 92% (1,4 addition only)

Table 2.6   Sakurai additions to acyclic alkenones

1 181

+ allyl-TMS

1. TiCl$_4$, 1 eq., CH$_2$Cl$_2$
   -78°C, 1min

2. H$_2$O, 59%

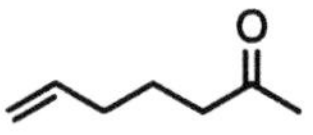

2 179

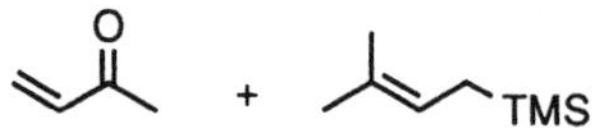

+ TMS

1. TiCl$_4$, 1 eq., CH$_2$Cl$_2$
   -78°C, 3h

2. H$_2$O, 79%

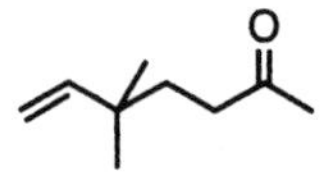

3 180

+ TMS

1. TiCl$_4$, 1 eq., CH$_2$Cl$_2$
   -78°C, 30 min

2. H$_2$O, 81%

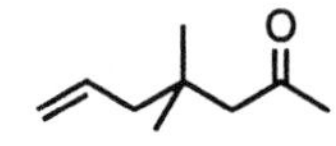

Heathcock's group has examined the stereoselectivity in allylsilane additions to some chiral alkenones.[182,183] Some of their results are summarized in the following series of reactions.

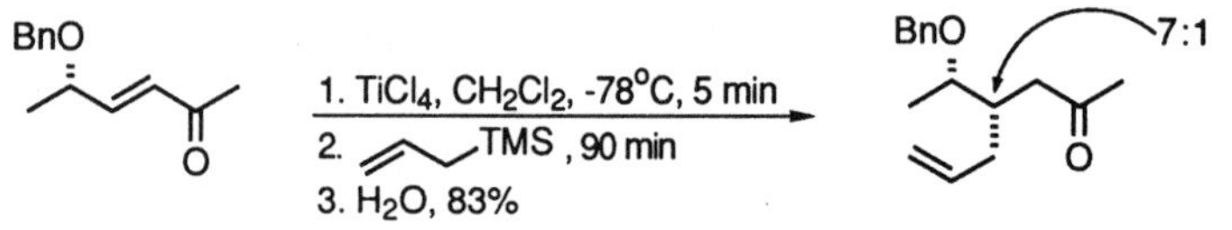

BnO

1. TiCl$_4$, CH$_2$Cl$_2$, -78°C, 5 min
2.   TMS , 90 min
3. H$_2$O, 83%

BnO   7:1

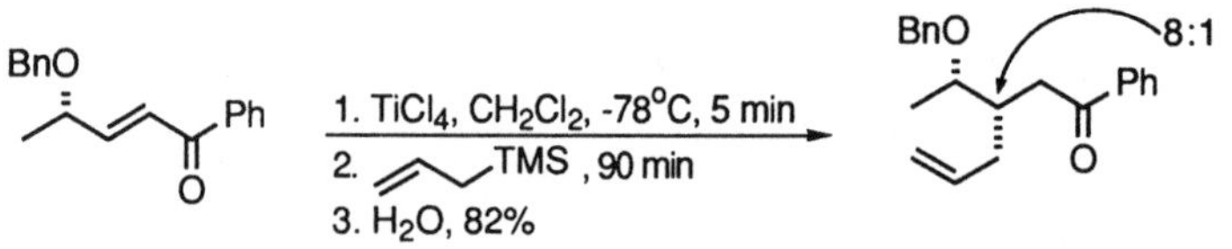

BnO   Ph

1. TiCl$_4$, CH$_2$Cl$_2$, -78°C, 5 min
2.   TMS , 90 min
3. H$_2$O, 82%

BnO   8:1
            Ph

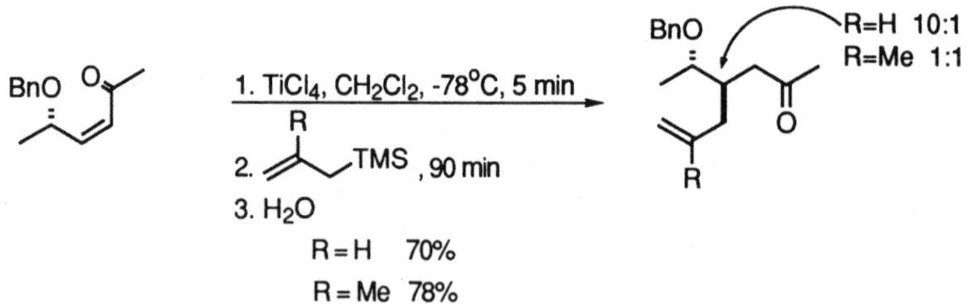

BnO   O

1. TiCl$_4$, CH$_2$Cl$_2$, -78°C, 5 min
   R
2.   TMS , 90 min
3. H$_2$O

   R = H   70%
   R = Me   78%

BnO   R=H 10:1
        R=Me 1:1

R

The reactions are quite selective when R=H. Addition to the *E*-alkenone gives the *syn* isomer whereas the *Z*-alkenone gives the *anti* isomer. Reaction of the 2-methyl derivative gave no stereoselection. The stereochemical outcome of these additions was rationalized using the Felkin model for asymmetric induction. This has the polar substituent orthogonal to the alkene and the smaller substituent in the "inside" position (see section 1.3.7.2, addition to conformer **D**, M = oxygen). Addition to the Z-alkenone was proposed to occur via the complex shown in Figure 2.4.

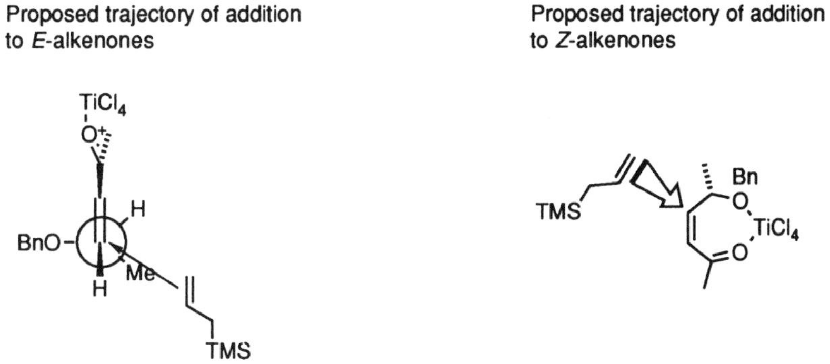

Proposed trajectory of addition
to E-alkenones

Proposed trajectory of addition
to Z-alkenones

Figure 2.4   Proposed trajectories of conjugate addition to chiral E- and Z-alkenones

Allenylstannanes react quite efficiently with alkenones.[184] This is one of the few useful methods for adding a propargyl group in a conjugate fashion.

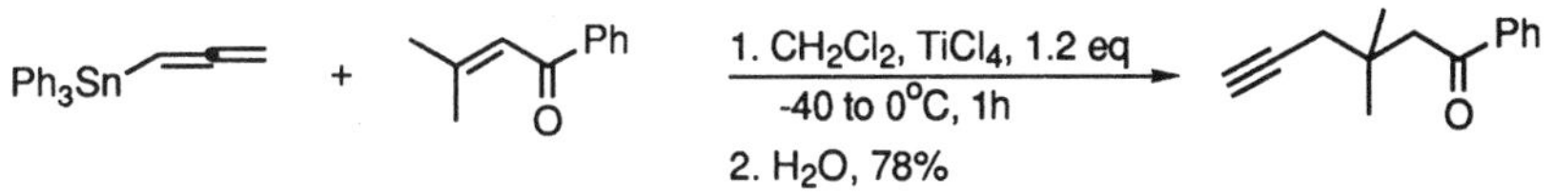

Allenylsilanes, on the other hand, close to produce an annulated product.[185]

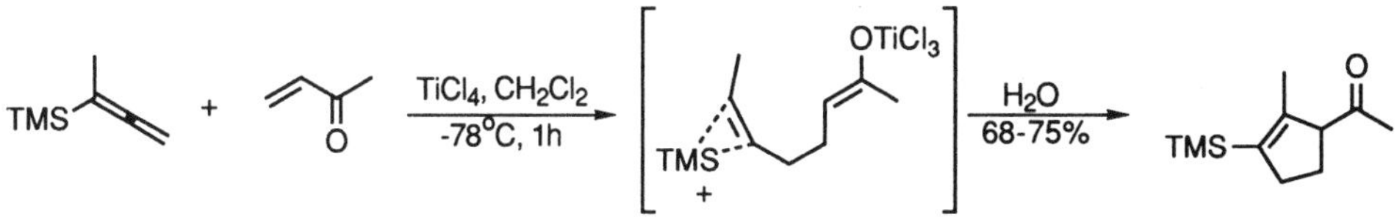

Cooke's group have designed a conjugate acceptor (**21**) with attenuated 1,2 reactivity while maintaining satisfactory 1,4 reactivity (see section 1.3.1).[186]

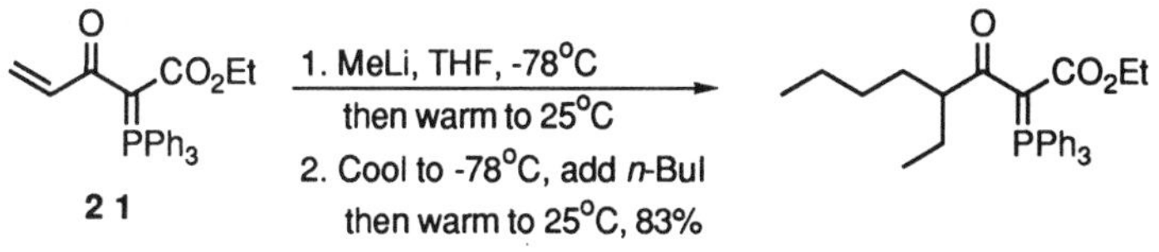

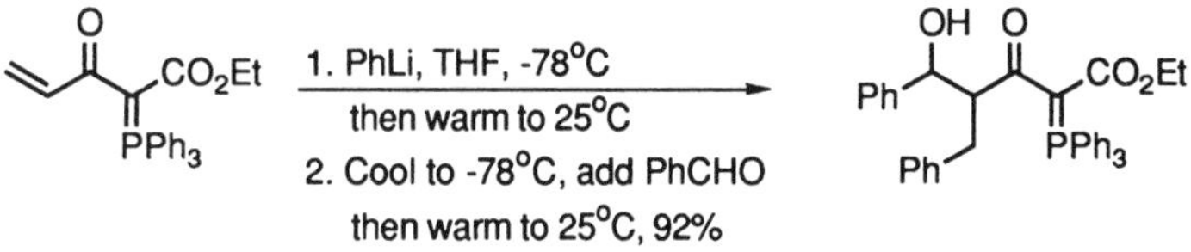

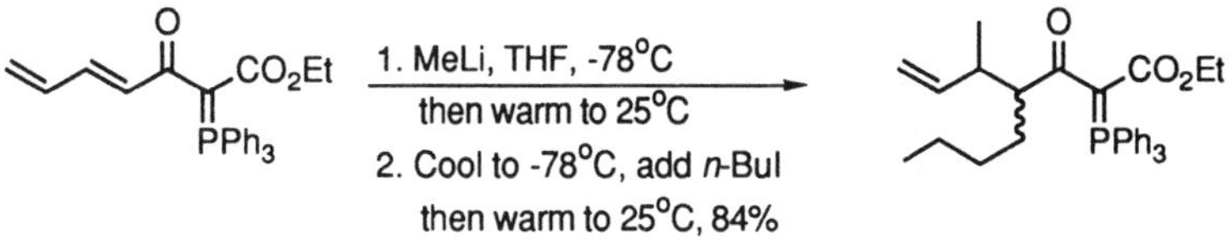

The "attenuator" may be removed by simple acidic methanolysis.

By including a leaving group in the conjugate acceptor, this method has been extended to a synthesis of 5- and 6-membered rings.[187] Conjugate addition to the alkenone is much faster than any other reactions with the ω-chloro group.

The same group has also introduced a method for delivering an acyl anion to alkenones. The mechanism is thought to involve alkylation of the ferrate followed by reaction with the conjugate acceptor.[188]

Hooz and Layton first demonstrated the ability of alkynylalanes to transfer the alkynyl ligand to alkenones in 1971.[189] For example, reaction of diethyl(2-phenylethynyl)alane at 25°C with chalcone proceeded in good yield. They also demonstrated that only alkenones able to adopt an s-*cis* conformation gave 1,4-addition. Thus 1-acetylcyclohexene reacted at C3, whereas cyclohexenone gave only 1,2-addition. (1,4-Addition to E-alkenones is also sometimes possible if a mechanism involving chelation to a hydroxyl group is available).[190,191]

More recently, Schwartz's group have shown that a catalyst, prepared from Ni(acac)$_2$ and DIBAL, successfully catalyzes conjugate addition of alkynyl(dialkyl)alanes to both E- and Z- alkenones.[192,193] However, additions to acyclic alkenones were complicated (in this context) by subsequent aldol reactions (see section 3.1.3 for examples of such additions to cycloalkenones).

Both alkynylboranes[194] and borates deliver their alkynyl ligands in a conjugate fashion to alkenones. Even sterically hindered acceptors, such as 4-methyl-3-buten-2-one, react, although somewhat more slowly (five days!).[195] The authors found that, for these reagents, there was a requirement that the alkenone could adopt an *s-cis* geometry. This implied that a cyclic transition state was operating.

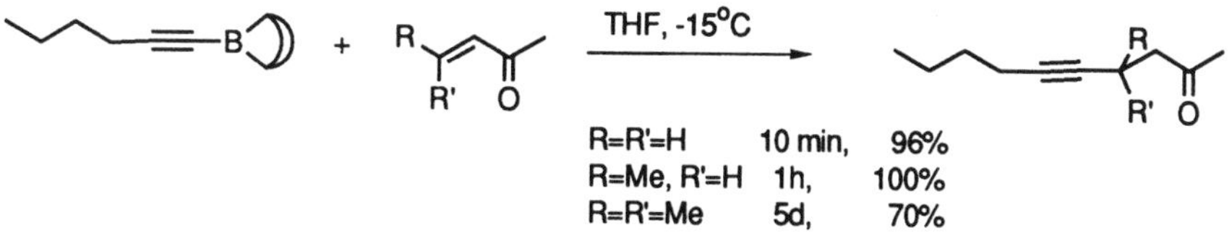

R=R'=H	10 min,	96%
R=Me, R'=H	1h,	100%
R=R'=Me	5d,	70%

If the carbonyl ligand is extremely bulky as in (22) then simple alkynyllithiums add in a conjugate fashion.[196]

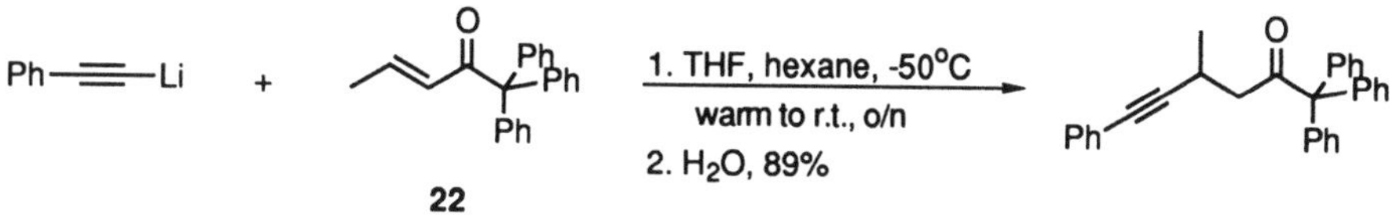

22

The combination of an organolithium and nickeltetracarbonyl serves to introduce an acyl carbanion.[197] Although this method is quite efficient, it inevitably suffers from the hazardous nature of nickeltetracarbonyl.

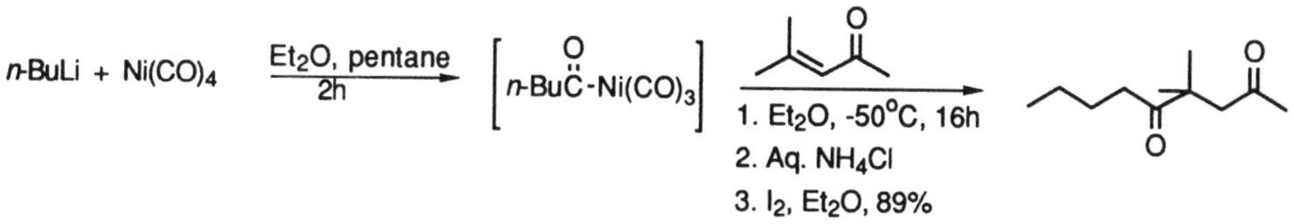

2.1.4   Free radicals[198]

Vitamin B[12a]-catalyzed additions of a variety of radicals give consistently good results with alkenones and alkenals.[199] For example, one such addition was used in a short synthesis of both the *exo-*and *endo-*isomers of brevicomin.[200]

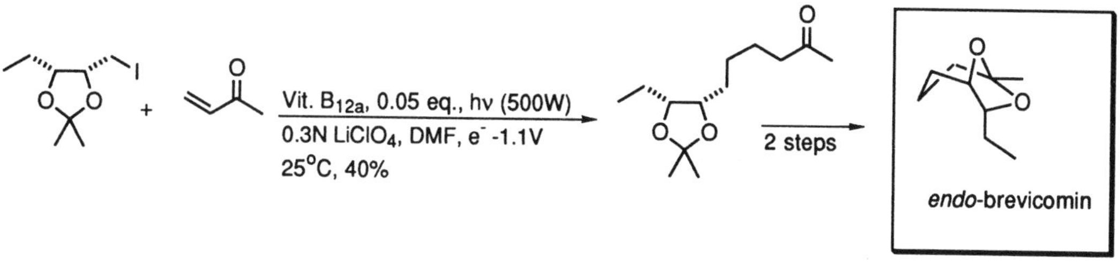

endo-brevicomin

The mechanism is believed to involve the formation and cleavage of a Co-C bond.[201] Using the same catalyst system it is also possible to carry out the equivalent of a nucleophilic acylation of alkenals.[202]

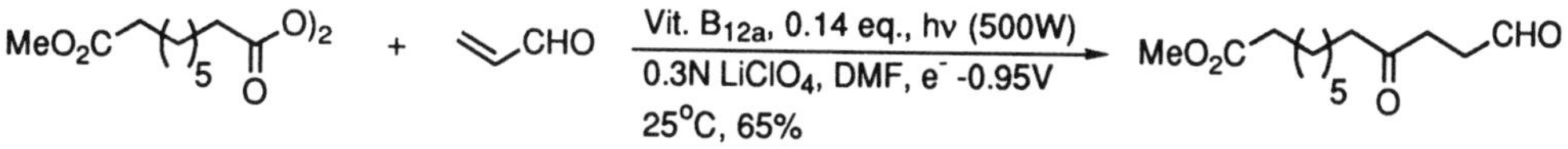

Luche's group has concluded that the conjugate addition of alkylzinc reagents to alkenones, using ultrasonication, proceeds by a free radical mechanism occurring at the metal surface.[203] They have also optimized the reaction conditions for these additions and two examples are given below for additions to alkenals and alkenones.[204]

Only a few examples exist of stereoselective *inter*molecular conjugate additions of free radicals.[205] An innovative use of a chiral auxiliary usually associated with alkenoate chemistry, has been shown to control the stereochemistry of addition to certain alkenones.[206] Thus addition of *n*-hexyl radical to the (S,S)-2,5-dimethylpyrrolidine derivative gave two diastereomeric pairs of regioisomers.

The two diastereomers resulting from conjugate addition to the alkenoic acid amide moiety were formed in a ratio of 3:2 and are not shown in the above diagram (path b). The two diastereomers formed from addition to the alkenone system (path a) were produced in ratios as high as 93:7.[207] To account for this high selectivity, the authors proposed that the approach vector of the incoming nucleophilic radical requires passage over the pyrrolidine. As shown below, one face is shielded by one of the pyrrolidine methyl substituents. Thus addition occurs almost exclusively from the opposite face. Molecular mechanics calculations show that the preferred alkene conformation is that shown below, with the amide nitrogen s-*trans* to the alkene. (The alternative conformation, with the amide nitrogen s-*cis* was calculated to be ~13 kJ mol⁻¹ higher in energy). See section 2.2.4 for application of these ideas to macrocycle synthesis.

Figure 2.5   *Steric control in a free radical addition to a chiral alkenone*

2.1.5   Enamines and enol ethers

2.1.5.1   Enamines

Reaction of enamines with alkenones constitutes an alternative method to the classical Robinson annulation of ketones. For example, in their seminal work on this reaction, Stork's group found that the morpholine enamine of cyclohexanone reacts with

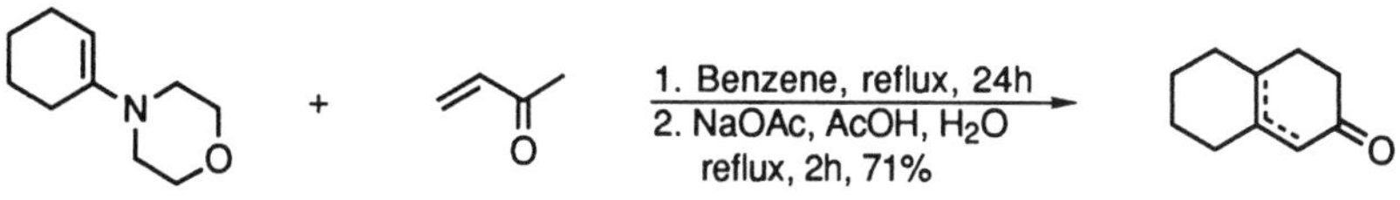

3-buten-2-one to give an adduct which usually wasn't isolated but was cyclized directly to a mixture of octalones in 71% yield.[208,209] A similar addition to ethyl 4-oxopentenoate provided a simple route to the 4-ethoxycarbonyloctalone system.

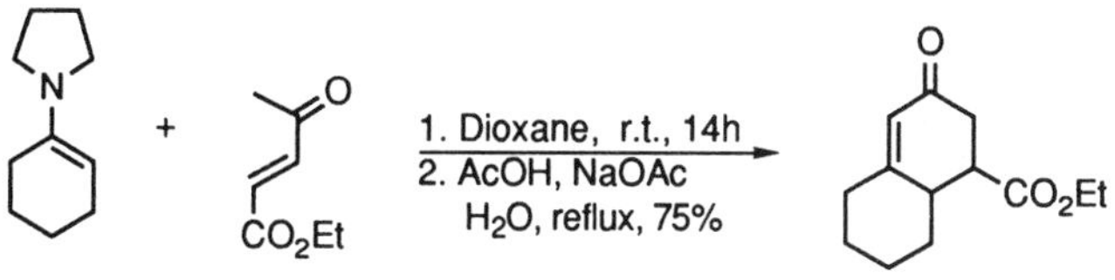

Mixing the reactants together without solvent often allows the reaction to occur at lower temperatures as in the following preparation of 4,4-dimethyl-2-cyclohexenone.[210]

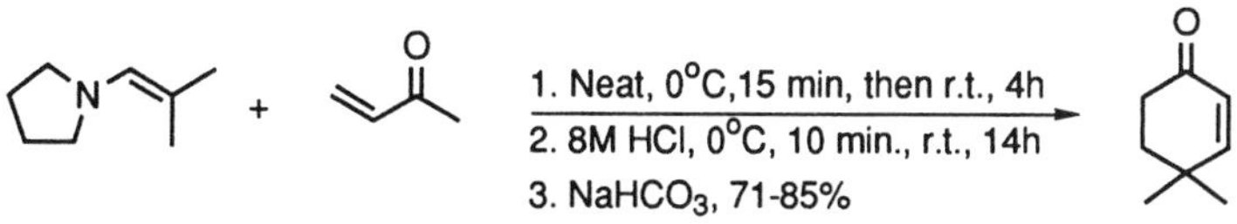

Indoles react with 3-buten-2-one using a clay catalyst.[211] In certain cases cyanide may also be used as catalyst.[212]

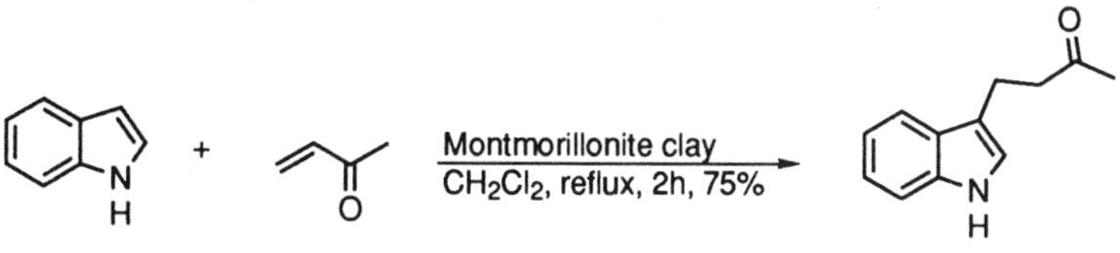

The Hantzsch synthesis of biologically active dihydropyridines is believed to proceed via an enamine intermediate generated *in situ* during the reaction.[213,214] More recently, this reaction has been carried out using a preformed enamine with a variety of conjugate acceptors.[215]

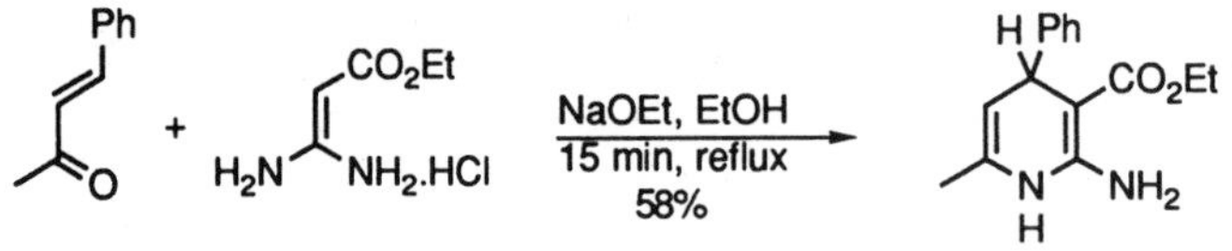

Many fused polycyclic systems may be prepared this way.[216]

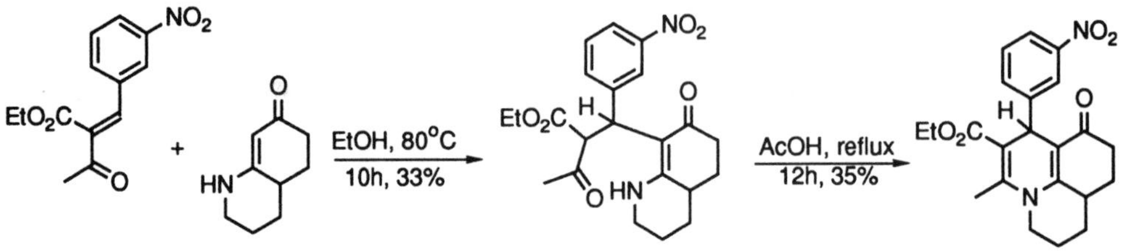

Numerous useful applications of enamine additions to total synthesis have been reported. For example, Stevens' group has developed a general reaction sequence which has enabled them and others to prepare a wide variety of naturally-occurring alkaloids.[217] The two key steps involve

(a)   acid-catalyzed rearrangement of cyclopropylimines to Δ^2-pyrrolines, followed by

(b)   acid-catalyzed[218] annulation of the pyrroline (an enamine) with an alkenone

One example comes from a total synthesis of the alkaloid, (±)-*epi*-elwesine:[219]

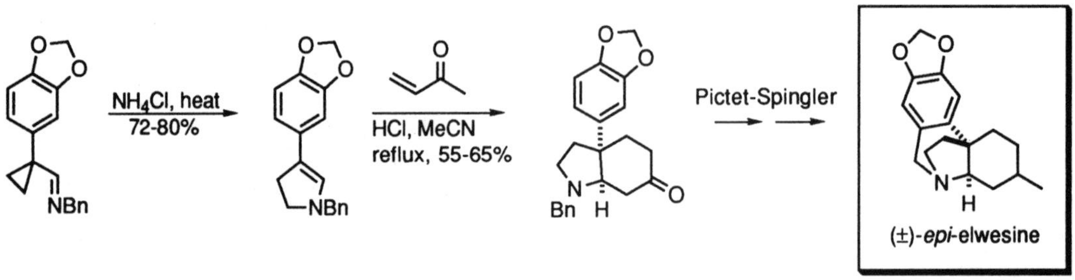

(±)-*epi*-elwesine

Conjugate addition of alkenyliminophosphoranes (available from the Staudinger reaction) to alkenones provides a synthesis of 2,4,6-trisubstituted pyridines.[220] However, the reaction often yields a mixture of pyridines as well as some by-products.

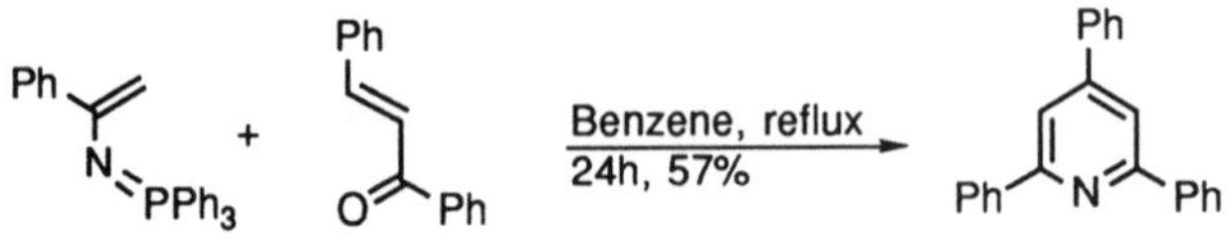

2.1.5.2   *Enol ethers and ketene acetals*

(a)   Addition of achiral enol ethers to achiral alkenones

A viable alternative to the conventional reaction conditions for Michael reactions is the Lewis acid-catalyzed addition of silyl enol ethers and silyl ketene acetals. Many Lewis acids have been tried, including titanium, caesium, tin and trityl salts.

Mukaiyama's group first examined conjugate additions of silyl enol ethers using titanium salts as Lewis acids.[221,222] The reaction is usually carried out in dichloromethane and addition occurs rapidly, even at low temperatures (-78ºC), producing 1,5-dicarbonyl compounds in good yield. Both terminal and internal alkenes react quite well. Where one of the reactants was found to be unstable to TiCl₄, Ti(O*i*-Pr)₄ was included, with good results. As shown in entry 5, additions work equally well with the acetals of alkenones.

Table 2.7 Ti(IV)-promoted conjugate addition of trialkylsilyl enol ethers to alkenones

Entry	Enol ether	Alkenone	Conditions	Product
1[223,224]	OTMS structure	methyl vinyl ketone	1. TiCl₄, 1 eq., CH₂Cl₂, -78°C, 10 min 2. Aq. NaHCO₃, 41%	product
2[221]	Ph OTMS	1. TiCl₄, 1 eq., CH₂Cl₂, -78°C, 2 min 2. Aq. K₂CO₃, 76%	Ph product
3[221]	Ph OTMS	1. TiCl₄, 1 eq., Ti(O*i*-Pr)₄, CH₂Cl₂, -78°C, 30 min 2. Aq. K₂CO₃, 56%	Ph product
4[221]	cyclopentene OTMS	1. TiCl₄, 1 eq., CH₂Cl₂, -78°C, 15 min 2. Aq. K₂CO₃, 66%	product
5[47]	OTMS structure	1. TiCl₄, Ti(O*i*-Pr)₄ (1:1), 1.5 eq CH₂Cl₂, -80°C, 60 min 2. 5% Aq. K₂CO₃, 81%	product

Heathcock's group has examined stereoselection in these additions. Several important observations were made. Firstly, the stereochemical trend in the reaction, i.e. that the major diastereomeric product is *anti*, appears to be independent of both the metal ion (titanium or tin) and the enol ether stereochemistry. Second, the Z-enol ethers of aryl ethyl ketones gave very good stereoselectivity, whereas the corresponding E-isomers gave somewhat poorer results. The authors speculated that the stereoselectivity is due to equilibration, via a retro-Michael reaction, to the more stable isomer.

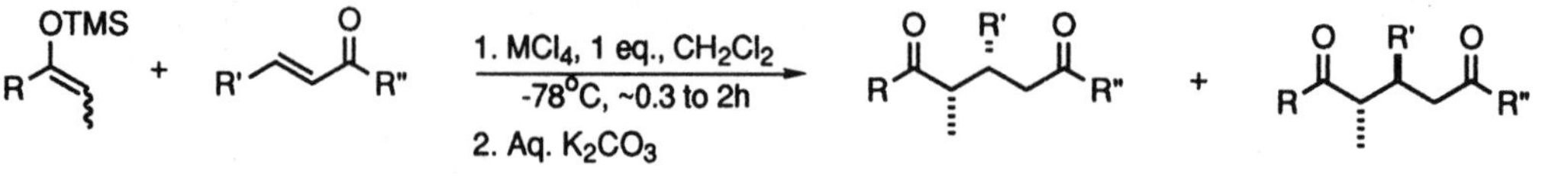

R = Et	Z enol ether	M = Ti	R' = Me	R" = tert-Bu	52%	12	:		88
R = Et	Z enol ether	M = Sn	R' = Me	R" = tert-Bu	57%	11	:		89
R = Et	Z enol ether	M = Sn	R' = Me	R" = i-Pr	52%	24	:		76
R = Et	E enol ether	M = Sn	R' = Me	R" = tert-Bu	59%	13	:		87
R = Ph	Z enol ether	M = Sn	R' = Me	R" = tert-Bu	69%	<5	:		>95
R = Ph	Z enol ether	M = Sn	R' = Me	R" = Ph	75%	5	:		95
R = Ph	E enol ether	M = Sn	R' = Me	R" = tert-Bu	77%	27	:		73

Reaction of 2-(trimethylsilyloxy)butadienes with alkenones provides another alternative to the Diels-Alder reaction. Although the reaction is highly stereoselective, a stepwise, rather than concerted, mechanism appears to be operating.[225] For example, addition of 2-(trimethylsilyloxy)butadiene to E-4-phenyl-3-buten-2-one produces the 3,4-disubstituted cyclohexanone (23) as a single diastereomer.

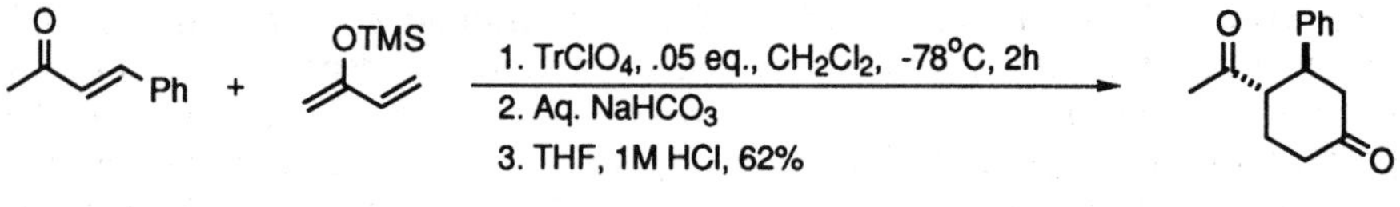

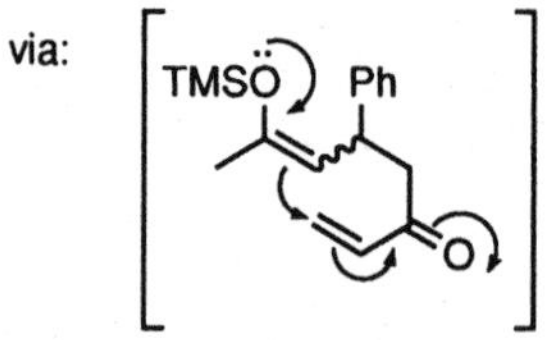

The Lewis acid promoted conjugate addition of ketene acetals has also been studied.[226] Again, the use of Ti(Oi-Pr)4 with TiCl4 proved essential in those cases where the use of TiCl4 alone led to polymerization (compare entries 2 and 3, Table 2.8).

Table 2.8   Lewis acid promoted conjugate addition of silylketene acetals to alkenones[227]

	Silylketene acetal + alkenone	Conditions	Product
1	OTMS / MeO +	1. TiCl$_4$, 1.1 eq., CH$_2$Cl$_2$, -78°C, 3h 2. 5% aq. K$_2$CO$_3$, 72%	MeO
2	OTMS / MeO / Ph +	1. TiCl$_4$, 1.1 eq., CH$_2$Cl$_2$, -78°C, 3h 2. 5% aq. K$_2$CO$_3$	polymer
3	OTMS / MeO / Ph +	1. TiCl$_4$, 1.1 eq., Ti(i-OPr)$_4$, 0.55 eq., CH$_2$Cl$_2$, -78°C, 3h 2. 5% aq. K$_2$CO$_3$, 38%	MeO / Ph
4[228]	OTMS / EtO / OEt / OTMS +	1. ZnCl$_2$, cat., CH$_2$Cl$_2$ r.t., 45 min 2. MeOH, 2N HCl r.t., 5 min, 68%	EtO$_2$C

Heathcock's group has studied some stereochemical aspects of such conjugate additions to pro-chiral alkenones.[229] They first examined simple diastereoselection and found that, in most cases, the *syn* product was favoured irrespective of the ketene acetal stereochemistry. This is the opposite result to that obtained for Lewis acid promoted conjugate additions of silyl enol ethers discussed above. The alkenone carbonyl ligand proved to be important for stereoselectivity as changing from a *tert*-butyl to a methyl substituent led to complete loss of selectivity. For these particular ketene acetals, it was found that slow addition of the ketene acetal to the cooled mixture of alkenone and Lewis acid yielded, after work up, exclusively the *tert*-butyl ester. Rapid addition of the ketene acetal led to mixtures of both the *tert*-butyl and *tert*-butyldimethylsilyl esters.

OTBS / t-BuO + R / R'   1. TiCl$_4$, 1 eq., CH$_2$Cl$_2$ -78°C, ~0.3 to 2h   2. Aq. K$_2$CO$_3$ R"O / R / R' + R"O / R / R'

Z-ketene acetal	R = Me	R' = t-Bu	88%	99 : 1
	R = i-Pr	R' = t-Bu	87%	96 : 4
	R = Me	R' = Me	74%	50 : 50
E-ketene acetal	R = i-Pr	R' = t-Bu	76%	98 : 2

It is probable that these additions are under kinetic control. The authors argued that as the initial ionic adduct is better able to delocalize the charge in the oxonium ion than the starting alkenone-Lewis acid complex, the equilibrium should lie far to the right. Desilylation, which is fast relative to the retro-Michael reaction, renders the process irreversible.

The stereoselectivity can be accounted for by assuming that the reactants adopt open transition states and that the lower energy transition state is **24**, which avoids an unfavourable interaction between the *tert*-butyl and R groups.

More recently, Mukaiyama's group has developed the use of trityl salts as catalysts for these additions. A range of these salts was screened and trityl perchlorate was found to give the best yields.[230] No 1,2-additions were observed with alkenones. (However alkenals gave mixtures of products arising from 1,2- and 1,4-additions and were not examined further). This procedure has a distinct advantage over related methods as the product may be isolated as either the 1,5-dicarbonyl compound (using aqueous sodium bicarbonate in the workup) or the intermediate silyl enol ether (using pyridine in the workup). In fact, the intermediate silyl ketene acetal may be successfully reacted with an

aldehyde by simply adding the aldehyde to the reaction mixture after the conjugate addition is complete.[231] This is made possible by the fact that trityl salts also catalyze the addition of ketene acetals to aldehydes.[232] This tandem conjugate addition-aldol process provides a method for preparing stereochemically pure γ-acyl-δ-lactones. However, the trityl perchlorate catalyzed method has found most use in conjugate additions to cycloalkenones (see section 3.1.2.5.2 and 3.1.3.5.2). The same sorts of levels of simple diastereoselection displayed in additions promoted by titanium and tin salts can be obtained using trityl perchlorate catalysis.[231, 233]

The Lewis acid catalyzed addition of 2-ethylthio-5-trimethylsilyloxyfuran to the highly-functionalized chalcone shown below was used in a total synthesis of mitomycin C.[234] Only one diastereomer was produced.

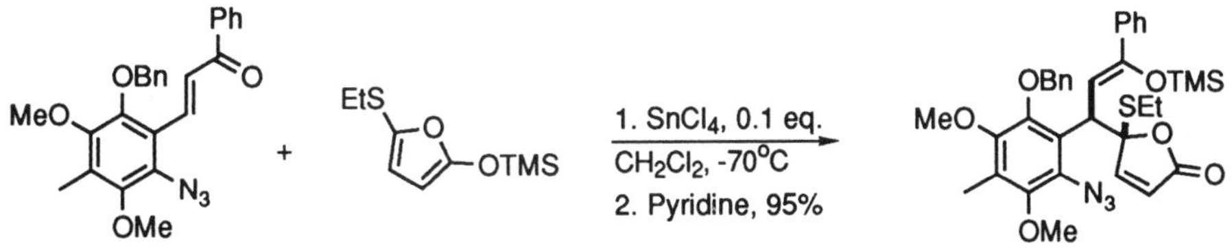

In an early synthesis of (±)-seychellene, Jung's group attempted a Diels Alder-Michael addition sequence with 3-oxo-1,4-pentadiene.[235] The first addition proceeded in almost quantitative yield, giving exclusively the *exo*-adduct. However, the second addition required a Lewis acid and, at best, a very low yield of the required product could be obtained. Apparently under these conditions, a 6-*endo*-trig closure[236] was not favoured compared to a competing retro-conjugate addition. The latter was the major reaction pathway. Attempts to use the corresponding enolate, generated by treatment with methyllithium gave only traces of product. However, more recently it has been demonstrated that reasonable yields are obtainable if diethylaluminium chloride is used instead of titanium tetrachloride.[237]

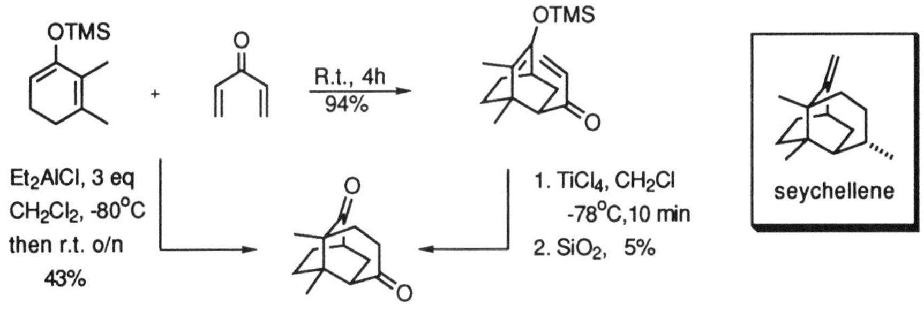

(b)   *Addition of achiral enol ethers to chiral alkenones*

The same group has reported one example of the addition of a silyl ketene acetal to an alkenone containing an asymmetric centre.[238] This gave excellent stereoselectivity, in the same sense as that for the corresponding additions of silyl enol ethers discussed above.

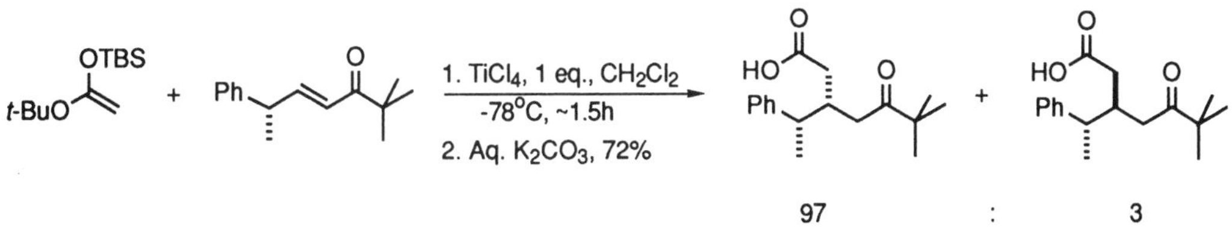

(c) Addition of chiral enol ethers to achiral alkenones

Addition of ketene acetals containing a chiral auxiliary proceeds with reasonable stereoselectivity.[239]

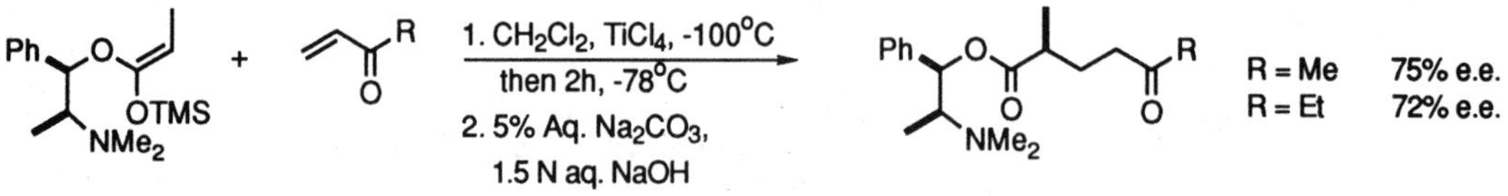

Ph
O
OTMS
NMe2
+
R
O
1. CH2Cl2, TiCl4, -100°C
then 2h, -78°C
2. 5% Aq. Na2CO3,
1.5 N aq. NaOH
Ph
O
O
R
O
NMe2
R = Me 75% e.e.
R = Et 72% e.e.

2.1.6 Conjugate hydrocyanation[240]

Conjugate hydrocyanation of alkenones may be achieved using several different sets of conditions. Steric hindrance, i.e. substitution at the β-position is not usually a problem (see, especially, entry 3 in Table 2.9).

Table 2.9 Conjugate hydrocyanation of alkenones

1[241]	O HCN-AlEt3, THF, r.t. 88% NC O
2[242]	Ph O NaCN, AcOH, H2O, EtOH r.t., 65% NC Ph O
3[243]	O HCN, KCN, 150-160°C 70-75% O CN

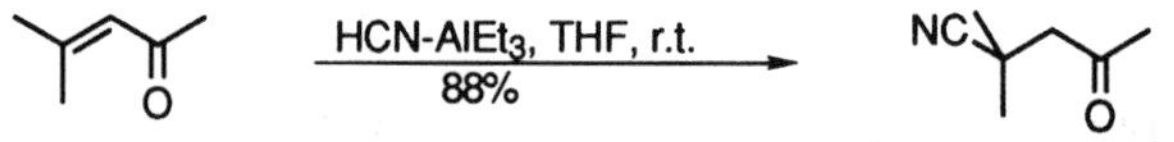

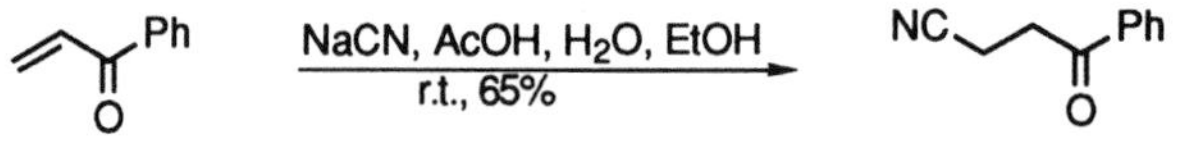

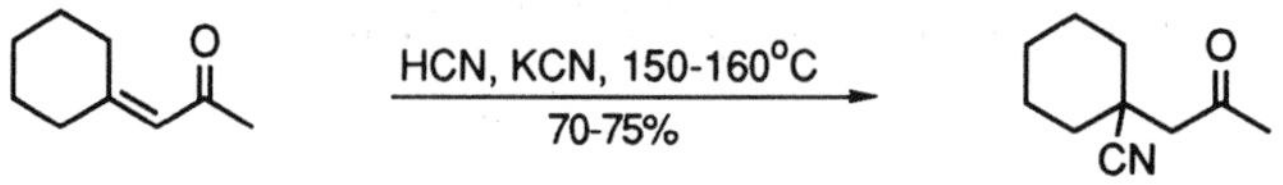

2.1.7 Heteronucleophiles

2.1.7.1 *Nitrogen and phosphorus*

(a) Nitrogen

There are many examples of conjugate additions of amines to alkenones.[244] Imine formation is rarely a problem, even where acid catalysis is used. All classes of alkylamines and most arylamines add (for example, entry 2, Table 2.10). N-haloamines add extremely efficiently yielding overall addition of the halogen and amine (entry 5).

Table 2.10   Conjugate addition of amines to alkenones

1[245]		Anhydrous NH₃ 25 atm., 85°C 1h, 69%	
2[246]	ArNH₂ +	50% Aq. dioxan 0.025M H₂PO₄⁻ buffer r.t., 65-80%	
3[247]		30% Aq. MeNH₂ aq. HCl, 33%	
4[248]		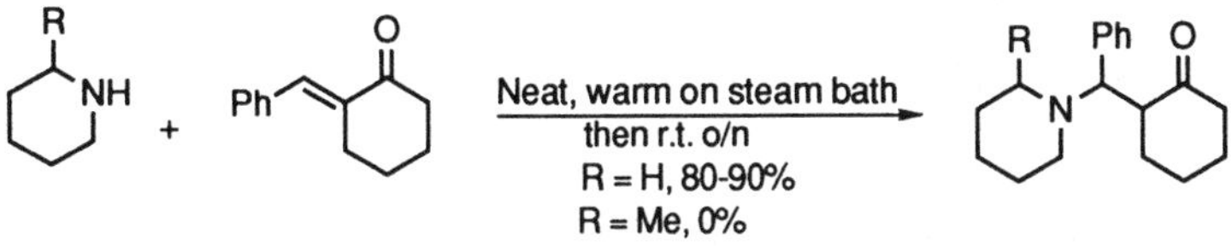 Neat, warm on steam bath then r.t. o/n R = H, 80-90% R = Me, 0%	
5[249]		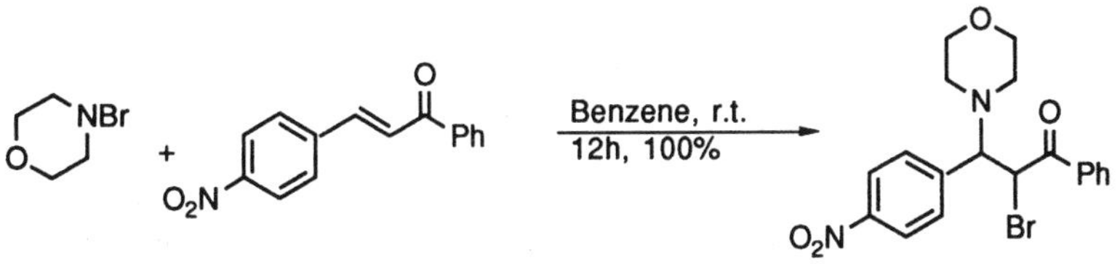 Benzene, r.t. 12h, 100%	
6[250]	Me₂NH +	1. Neat, sealed tube r.t., 2d then 50°C, 1h 2. CHCl₃, HCl, 50%	

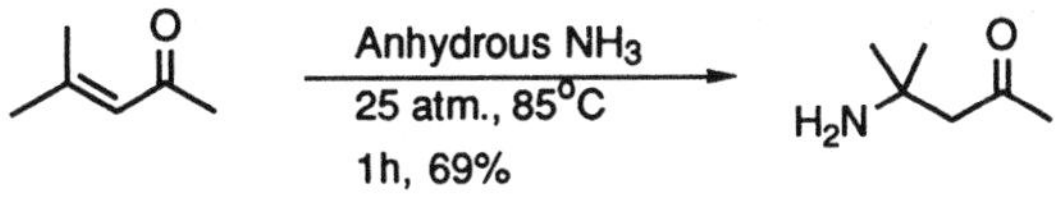

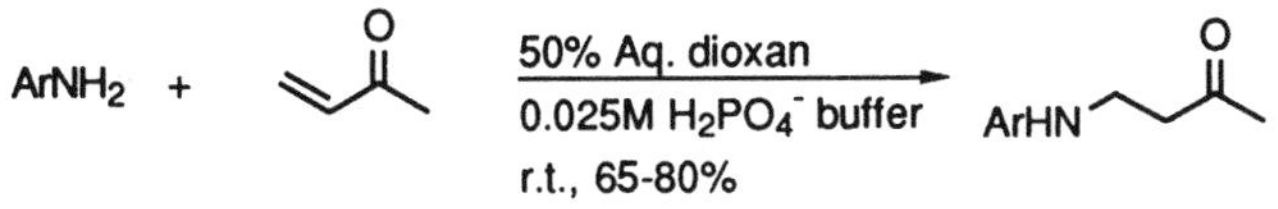

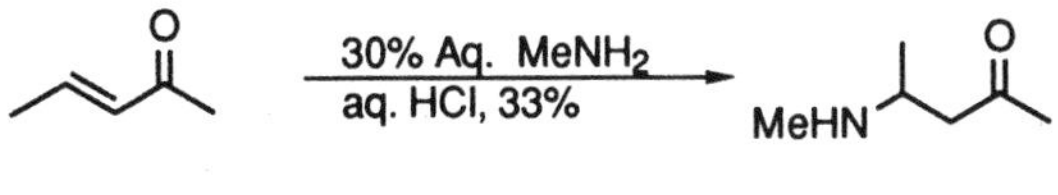

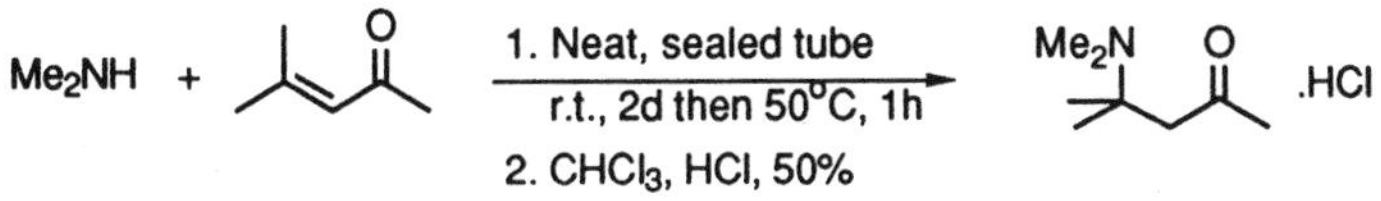

Aziridines may be prepared by standing a benzene or methanol solution of a primary amine and an α-bromoalkenone at room temperature for several hours. [251,252]

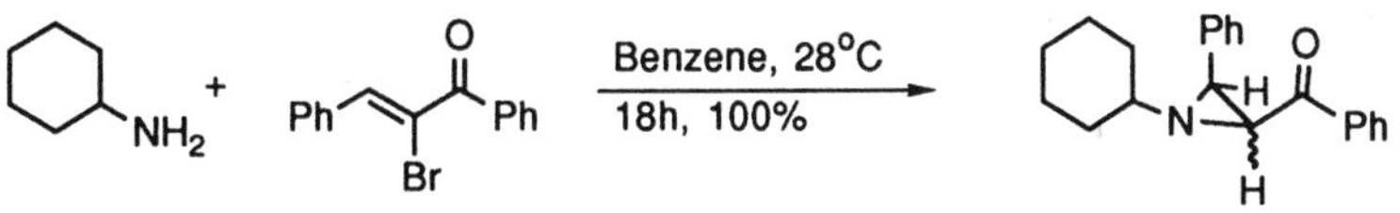

Alternatively, addition to an alkenone may be promoted by iodine.[251] However, the stereoselection for both processes is low.

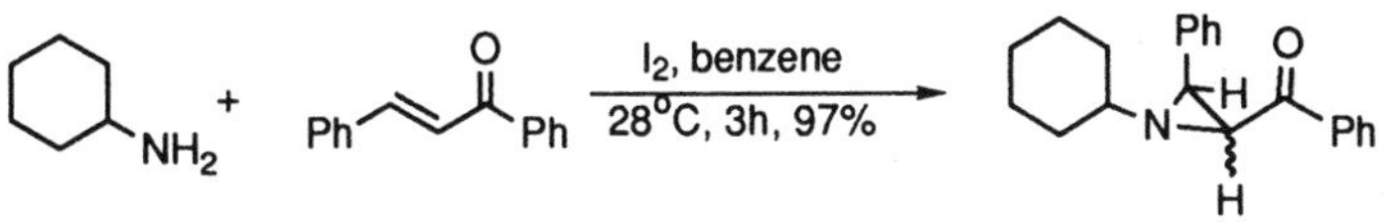

Aziridines also readily add to alkenones at room temperature.[253]

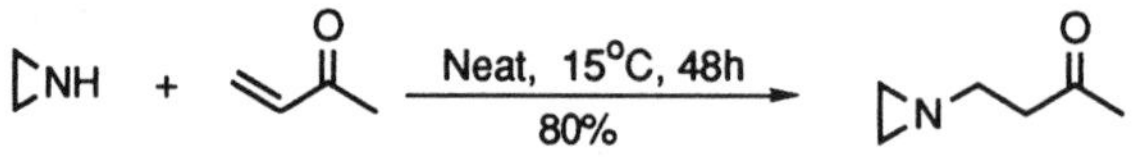

The nitrogen of cyclic thioimidates e.g. (26) adds to Nazarov's reagent[254] (methyl 3-oxo-4-pentenoate) in the presence of mercuric chloride.[255] This reaction was used in a total synthesis of the quinolizidine alkaloid (±)-*epi*-lupinine (as shown below) and is equally applicable to the synthesis of indolizidine alkaloids.

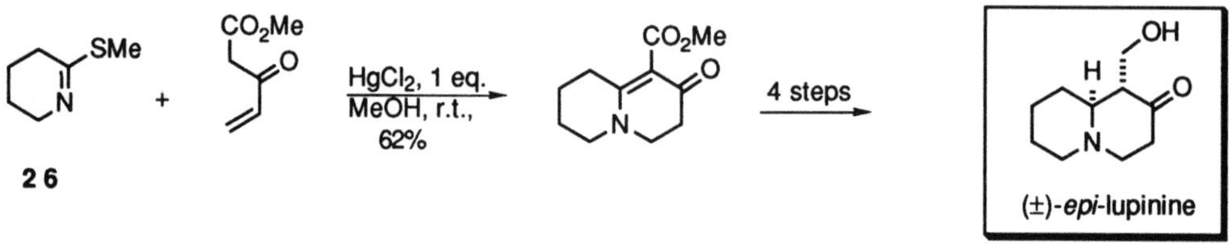

Nitrite adds to both propenal[256] and alkenones[257] in useful yields.

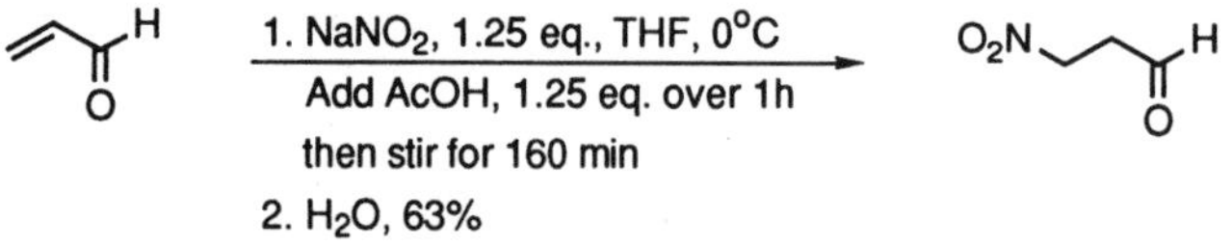

(b)   *Phosphorus*[258]

As mentioned in Chapter One, phosphorus nucleophiles fall into two catagories: (i) neutral and (ii) anionic. Those which have so far been added to alkenals and acyclic alkenones are listed below. In general, neutral reagents have been added to alkenals whereas both neutral and anionic phosphorus nucleophiles have been added to alkenones. Alkenals usually react at room temperature whereas reaction with alkenones often requires heating. As well as this, the types of reactions which are usually carried

Neutral	Anionic
P(OR)3	P(OH)(OR)2
P(OR)2(OSiR'3)	P(OH)2(OR)
P(OR)2(OAc)	P(OH)(OR)R
P(NR3)2(OSiR'3)	P(OH)2R
	P(OH)Ar2

out fall into two groups. The first involves phosphorus addition followed by protonation, e.g.:

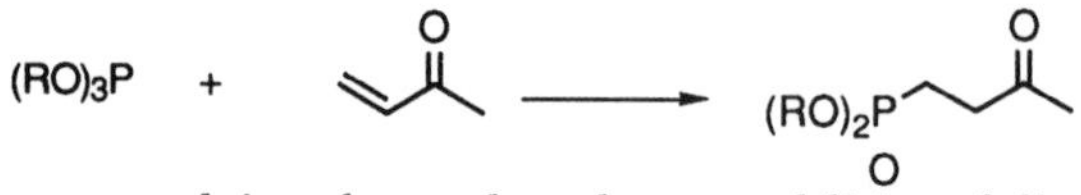

The second involves phosphorus addition followed by enolate trapping with either a dialkyl phosphorochloridate or, more commonly, a silylating reagent. In the latter case, the silyl group can be derived either from an added trialkylchlorosilane or via *intra*molecular silicon transfer. Evans' group have prepared a number of phosphite esters in this manner and have also demonstrated that the transfer of silicon is indeed intramolecular in nature.[259,260]

However, the mechanism of the rearrangement of the putative oxaphospholene intermediate was not investigated. They also prepared an oxaphospholene, which they suggest is an intermediate in the addition of tri*alkyl*phosphites to alkenones, and showed that trimethylchlorosilane reacts exothermically with the oxaphospholene, generating the Z-phosphonate in quantitative yield.

Invariably, phosphorus undergoes an increase in oxidation state from (III) in the starting material to (V) in the product. In the case of phosphite esters, this is achieved by dealkylation of the intermediate, as in the oxaphospholenes just discussed. In the case of phosphinic acids and related acids, this is just a consequence of reaction at phosphorus rather than at oxygen. In general terms, the combination of conjugate addition followed by dealkylation of a phosphonium intermediate may be viewed as a vinylogous Abramov reaction. Kozikowsky's group have taken advantage of the remarkable 1,4-selectivity in the addition of phosphines to alkenals in developing their "P-Si" reaction.[261] The phosphonium intermediate (27) can be converted, in situ, to the ylid and reacted with an aldehyde, providing a useful route to ene/dienes. This method was applied to a model synthesis of intermediates for the synthesis of naturally occurring macrocyclic ketones such as jatrophatrione.[262]

jatrophatrione

2.1.7.2   Oxygen, sulfur and selenium

These nucleophiles add very readily to alkenones and, in the case of selenium, to alkenals as well. Addition/elimination with 3-tosyloxyalkenals is a remarkably efficient process (Table 2.11, entry 3).

Table 2.11   Conjugate addition of oxygen, sulfur and selenium nucleophiles to alkenones

1[263]		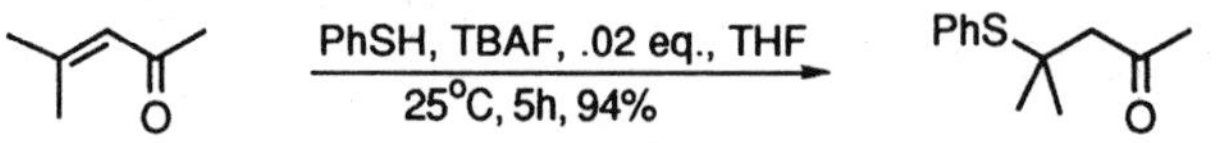 PhSH, TBAF, .02 eq., THF / 25°C, 5h, 94%	
2[262]		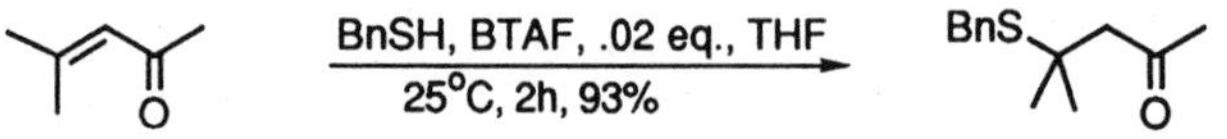 BnSH, BTAF, .02 eq., THF / 25°C, 2h, 93%	
3[264]		1. NaH, THF, 18-c-6, 0.01 eq / 2. TsO—CHO , 2 eq., r.t., 83%	
4[265]		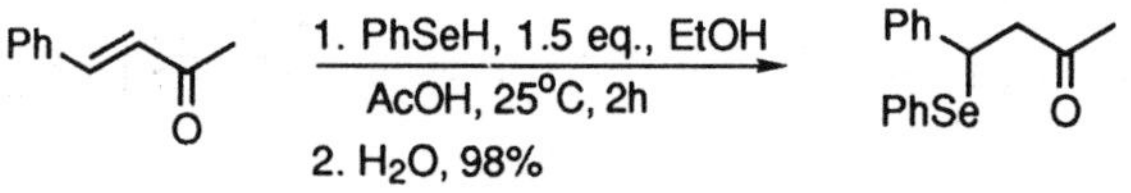 1. PhSeH, 1.5 eq., EtOH / AcOH, 25°C, 2h / 2. H₂O, 98%	
5[265]		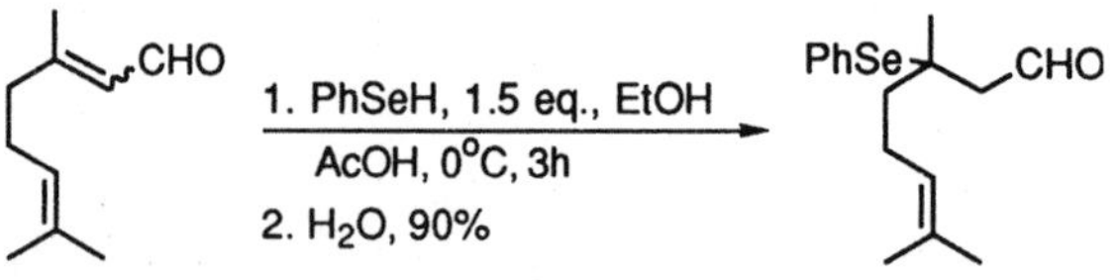 1. PhSeH, 1.5 eq., EtOH / AcOH, 0°C, 3h / 2. H₂O, 90%	

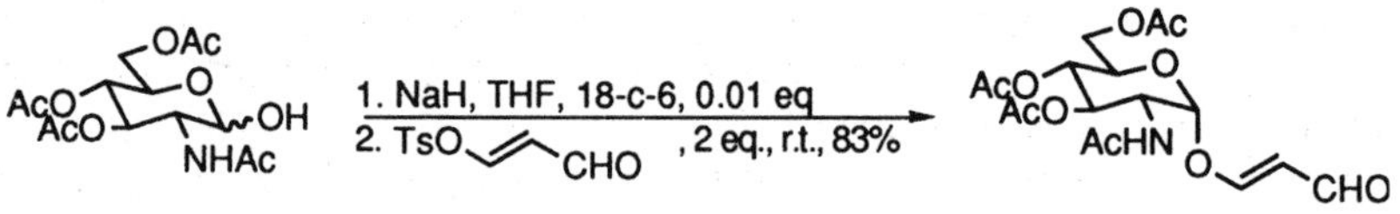

The enolates produced from addition of aluminium thiophenoxides add to alkenones may be efficiently trapped by aldehydes. Subsequent oxidation and elimination provides a very useful route to 2-methylene-3-hydroxyalkanones.[266] (For an application of this reaction to prostaglandin synthesis, see section 3.1.7.2).

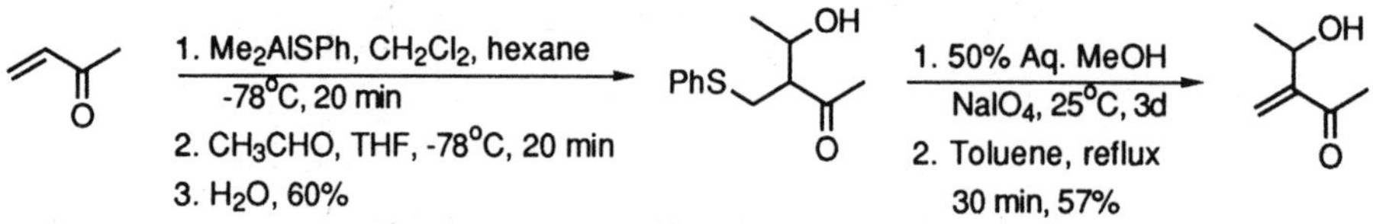

1. Me₂AlSPh, CH₂Cl₂, hexane / -78°C, 20 min
2. CH₃CHO, THF, -78°C, 20 min
3. H₂O, 60%

1. 50% Aq. MeOH / NaIO₄, 25°C, 3d
2. Toluene, reflux / 30 min, 57%

Sometimes phenols add through carbon rather than oxygen.[267]

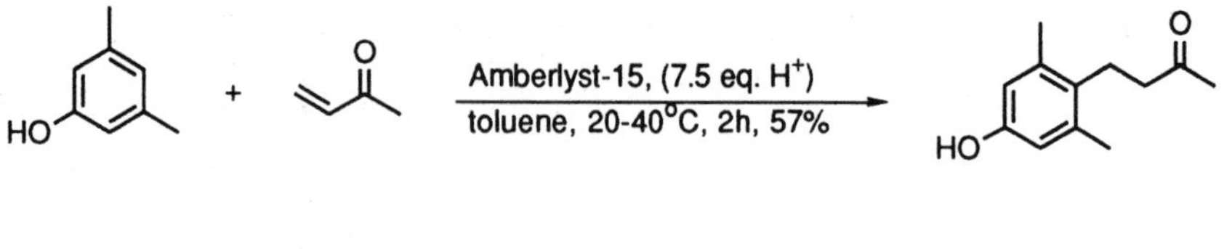

Amberlyst-15, (7.5 eq. H⁺) / toluene, 20-40°C, 2h, 57%

2.1.7.3   *Silicon and tin*

An alternative method for the generation and addition of silicon nucleophiles has been discovered by Seebach's group.[268,269] Addition of the HO cuprate, $Bu_2Cu(CN)Li_2$, to a solution containing an alkenone and a chlorosilane resulted in preferential conjugate addition of the *silyl* rather than the butyl group. This appears to be consistent with Lipshutz's finding that HO cuprates are rapidly modified by chlorosilanes, even at low temperatures (see section 1.4.2.2).

Reaction conditions:
1. $PhMe_2SiCl$
2. Addn to $Bu_2Cu(CN)Li_2$, THF, $-75°C$
3. NH_3, H_2O
4. H_3O^+, 68%

~6   :   1

Rather than using cuprate chemistry, the enolates of β–silylketones may be generated, in situ, using a palladium catalysed addition of disilanes to alkenones followed by treatment with methyllithium.[270] The intermediate in this procedure may be alkylated with the same stereocontrol obtained using Fleming's procedure (see section 4.1.7.3).

$PhCl_2SiSiMe_3$
$Pd(PPh_3)_4$, 0.5 mol%
benzene, 80°C, 1.5h

EtOH, Et_3N
65% overall

MeLi, Et_2O
-70°C

H_3O^+
68% overall

CH_3I, THF
42% overall

Silylorganozincates transfer silicon to alkenones only in moderate yields.[271] Addition of trimethylsilyllithium to cyclohexenone dimethyl-hydrazone followed by quenching with electrophiles of moderate reactivity gives good yields of the *trans* isomer.[272]

1. Me_3SiLi, HMPA, THF, $-78°C$, 1h
2. EtI, $-78°C$ to r.t., 24h
80%

Trialkylstannyllithiums also add well to alkenones.[273,274] Fleming has also used this reaction in a synthesis of cyclopropanes.[273]

1. *n*-Bu_3SnLi, THF, HMPA, $-78°C$, 5 min
2. EtOH, aq. NH_4Cl, $-78°C$ to r.t., 51%

1. MeMgI, Et_2O
0°C, 1h
2. H_2O, 95%

1. BF_3.2AcOH, CH_2Cl_2, 0°C, 15 min
2. 10% Aq. NaOH, 45%

2.1.7.4   Iodine

Iodotrimethylsilane adds to alkenones in a conjugate fashion (Table 2.12, entries 1 to 3). In each case, the initial product is the β-iodotrimethylsilyl enol ether, as in entry 3 below. The corresponding β-iodoketone is easily obtained by treatment of the crude reaction product with aqueous thiosulfate.

Table 2.12   Conjugate addition of iodide to acyclic alkenones

1[275]		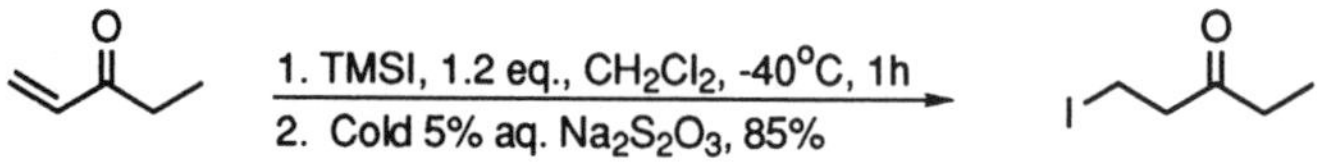 1. TMSI, 1.2 eq., CH$_2$Cl$_2$, -40°C, 1h 2. Cold 5% aq. Na$_2$S$_2$O$_3$, 85%	
2[275]		1. TMSI, 1.2 eq., CH$_2$Cl$_2$, -20°C, 2h then 25°C, 2h 2. Cold 5% aq. Na$_2$S$_2$O$_3$, 93%	
3[275]		TMSI, 1.2 eq., CCl$_4$,-40°C Me ... , excess t-Bu—N—t-Bu	~3:2
4[276]		SiH$_2$I$_2$, <0°C,	

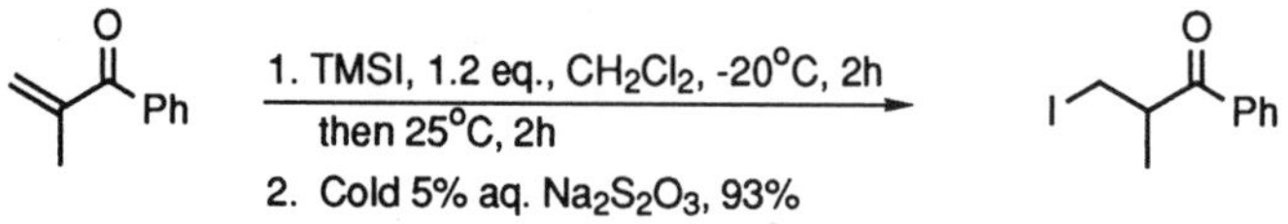

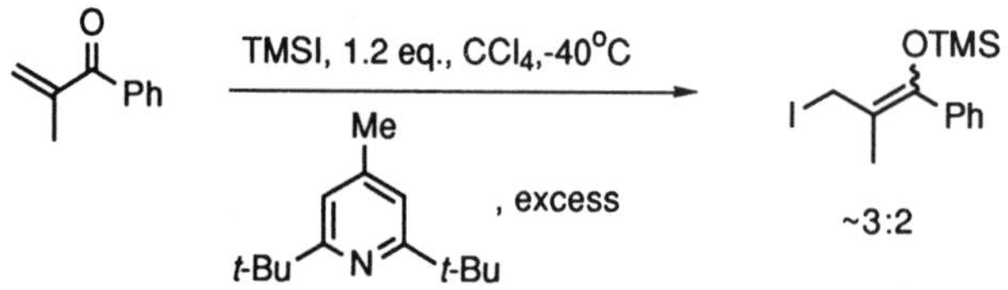

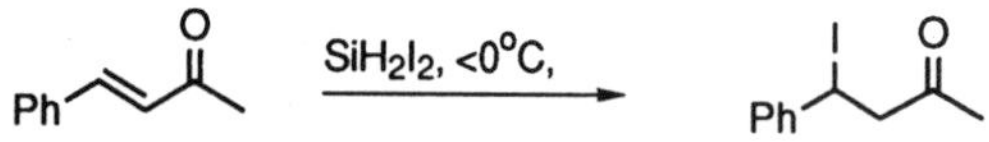

A conjugate addition/Nazarov-type cyclization procedure has been developed using iodotrimethylsilane.[277] (For the preparation of the precursors, see section 7.2). Thus conjugate addition of iodotrimethylsilane to the activated alkenone leads to a dienyl iodide. Ionization to a pentadienyl cation is then followed by electrocyclic closure.

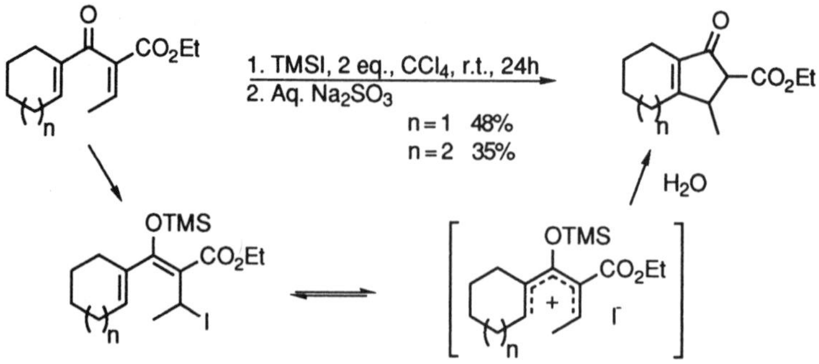

1. TMSI, 2 eq., CCl$_4$, r.t., 24h
 2. Aq. Na$_2$SO$_3$
 n = 1 48%
 n = 2 35%

Figure 2.6   *Proposed mechanism of a tandem conjugate addition/Nazarov cyclization*

2.1.7.5   Conjugate reduction

Semmelhack's group has demonstrated that copper salts modify the reactivity of lithium and sodium hydridoaluminate derivatives so that the reagents deliver hydride in a 1,4 fashion to alkenones (as well as other conjugate acceptors, see sections 3.1.8, 5.1.8 and 8.2.8).[278] They investigated two systems, the first based on lithium trimethoxyaluminium hydride and the second based on sodium bis(2-methoxyethoxy)aluminium hydride ("Vitride"[279] or "Red-Al"[280]). They found that the latter, combined with one molar equivalent of cuprous bromide, to be the more effective reducing agent for acyclic alkenones (entries 1 to 3 in Table 2.13). The addition of excess 2-butanol was found to inhibit polymerization, acting as a weak acid. A phosphine-copper hydride cluster reduces alkenones under very mild conditions (entry 6).

Table 2.13   Conjugate reductions of alkenals and acyclic alkenones

Entry	Substrate	Conditions	Product 1	Product 2 (OH)
1[278]	R⌒⌒C(=O)R	1. Na(MeOCH₂CH₂O)₂AlH₂, 2 mol eq. (4 hydride eq.) CuBr, 1 eq., THF, benzene -78°C, 10 min, -20°C, 1h 2. Sat'd aq NH₄Cl	R=Ph 32 R=t-Bu 98	0 1
2[278]	R⌒⌒C(=O)R	As above, except that 2-butanol (18 eq.) was added before the alkenone	R=Ph 54 R=t-Bu 98	0 0
3[278]	R⌒⌒C(=O)R	1. Li(MeO)₃AlH, 4 mol eq. (4 hydride eq.) CuBr, 0.5 eq., THF -78°C, 10 min, -20°C, 50 min 2. MeOH 3. Sat'd aq NH₄Cl	R=Ph 16 R=t-Bu 80	0 20
4[281]	(structure)—CHO	Ph₂SiH₂, 2 eq., ZnCl₂, 0.12 eq Pd(Ph₃)₄, 0.012 eq CHCl₃, r.t., 1.5h, 96%	(structure)—CHO	
5[282]	Me₅Cp(CO)₂Ru⁺—(structure)	1. NaBH₃CN, CH₂Cl₂, MeOH, 0°C 2. Et₂O, HCl, 68%	(structure)	
6[283,284]	Ph⌒⌒C(=O)(structure)	[(Ph₃P)CuH]₆, 0.24 eq H₂O, benzene, r.t., 20 min 87%	Ph—(structure)	

2.2   Intramolecular additions

2.2.1   Stabilized carbanions

2.2.1.1   Carbanions stabilized by π–conjugation with one heteroatom

Stork's group has demonstrated that ring closure of aldehyde enolates onto alkenones is an effective method for preparing angularly methylated *trans*-hydrindanes.[285,286] They discovered a correlation between the metal counterion's affinity for oxygen and the ratio of *cis* to *trans* product. Thus the higher the counterion's affinity for oxygen the higher the ratio is in favour of the *trans* product, as shown below.

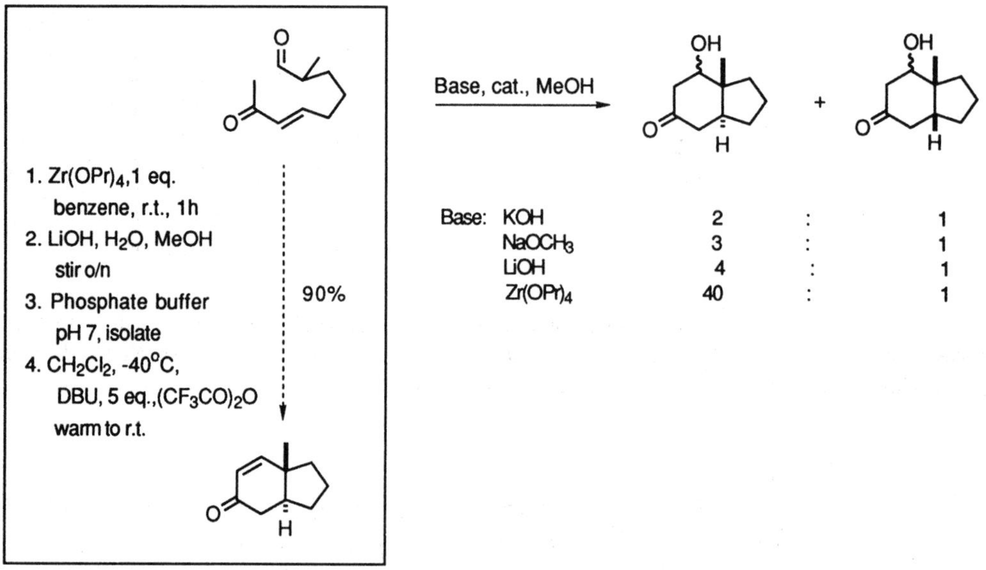

1. Zr(OPr)$_4$,1 eq.
   benzene, r.t., 1h
2. LiOH, H$_2$O, MeOH
   stir o/n
3. Phosphate buffer   90%
   pH 7, isolate
4. CH$_2$Cl$_2$, -40°C,
   DBU, 5 eq.,(CF$_3$CO)$_2$O
   warm to r.t.

Base, cat., MeOH

Base:		
KOH	2	:	1
NaOCH$_3$	3	:	1
LiOH	4	:	1
Zr(OPr)$_4$	40	:	1

A model which is consistent with these findings was also suggested by the authors. The more tight the metal chelate becomes, the closer the two π systems approach each other. The transition state which minimizes their repulsion more effectively should be favoured. In these cases this leads to the formation of the *trans* system.

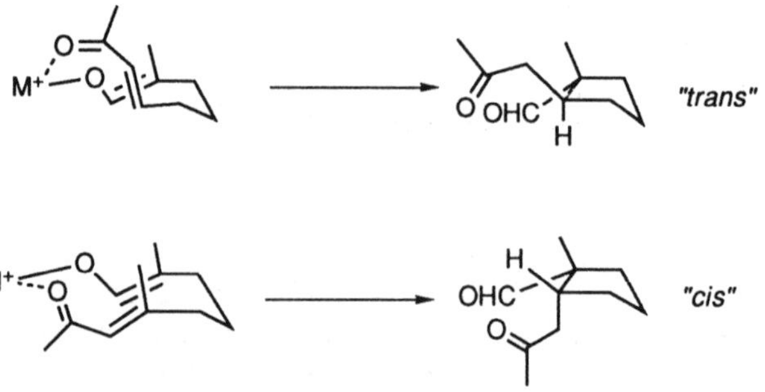

These ideas were applied by Fallis's group to a synthesis of a key intermediate for the total synthesis of retigeranic acid.[287] Their system included the added feature of an *iso-*propyl substituent. Using conditions delineated by Stork's group, but substituting Zr(O-*n*-Pr)$_4$ with Zr(O-*i*-Pr)$_4$, the cyclization proceeded with high stereoselectivity.

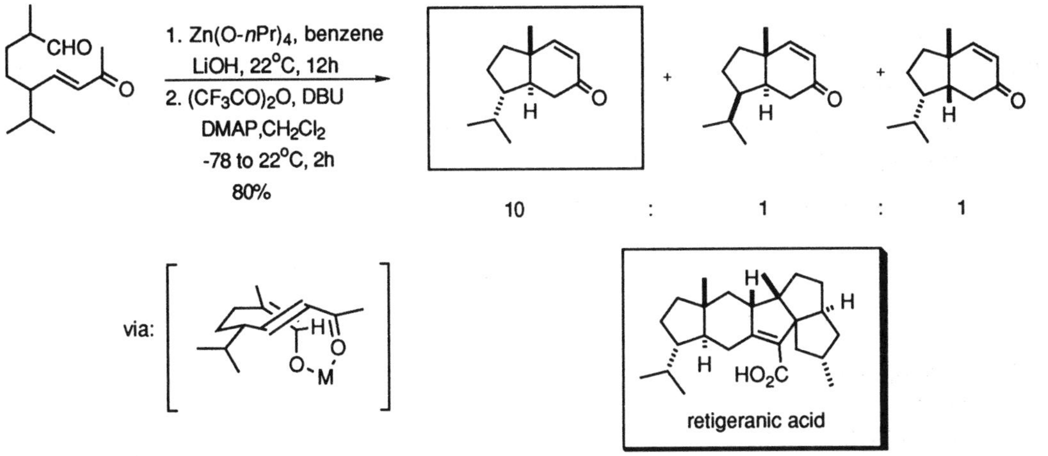

Cooke's alkenone may be closed to a cyclopentane or cyclohexane ring with base catalysis. In contrast to closures initiated by a conjugate addition (see section 2.1.3), this gives variable amounts of both *cis* and *trans* isomers.[187]

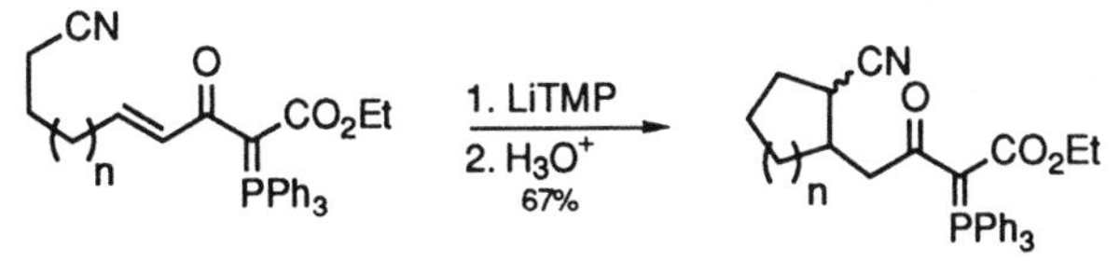

δ-Lactones may be prepared using a ring closure of α-iodoesters promoted by iodotrimethylsilane.[288] (For applications of this reaction to quassinoid synthesis, see section 3.2.3.1).

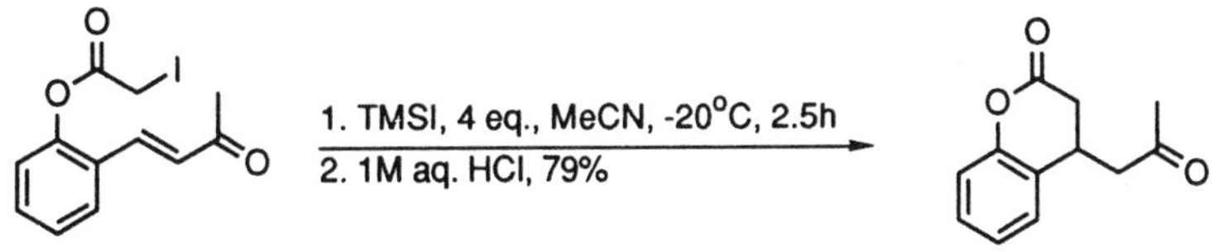

There are several plausible pathways by which this reaction may proceed. One is given below. As it is known that iodotrimethylsilane adds to alkenones in a conjugate fashion, this step may precede cyclization. As well as this the α-iodoester is probably transformed to the mixed ketene acetal. Catalysis by hydrogen iodide may be ruled out as the inclusion of triethylamine in the reaction did not lead to any change in the result.

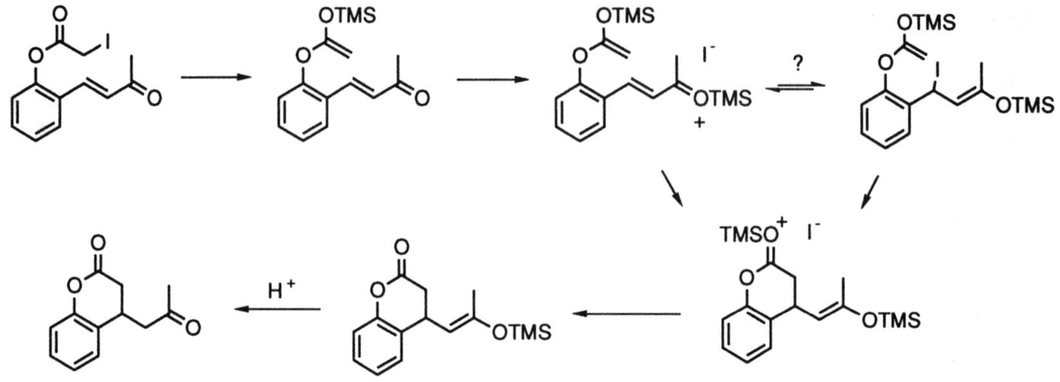
OTMS
OTMS
I⁻
OTMS
+
?
OTMS
I
OTMS
TMSO⁺ I⁻
OTMS
H⁺
OTMS

Fluoride can catalyze the ring closure of 2-trimethylsilyl-1,3-dithianes.[289]

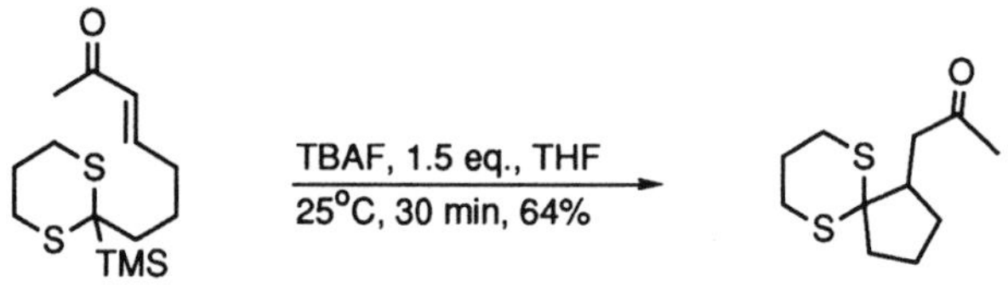
S
S
TMS
TBAF, 1.5 eq., THF
25°C, 30 min, 64%
S
S

2.2.1.2 *Carbanions stabilized by π-conjugation with more than one heteroatom*

Cis-fused bicyclic ketones may be prepared by intramolecular ring closure with β-keto esters.[290] Larger rings may also be produced this way, however the yields are poor and the stereoselectivity eventually decreases.

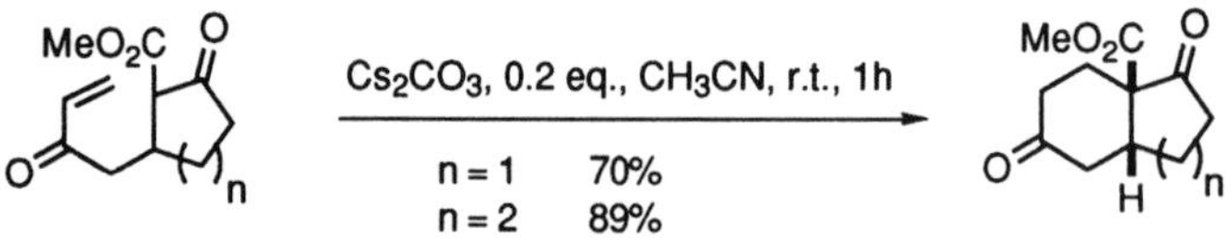
MeO2C
Cs2CO3, 0.2 eq., CH3CN, r.t., 1h
n = 1 70%
n = 2 89%
MeO2C
H

2.2.2 Other organometallic reagents

Cooke's alkenone (see sections 1.3.1 and 2.1.3) can be used to form rings through metallation of an ω-halo precursor.[187]

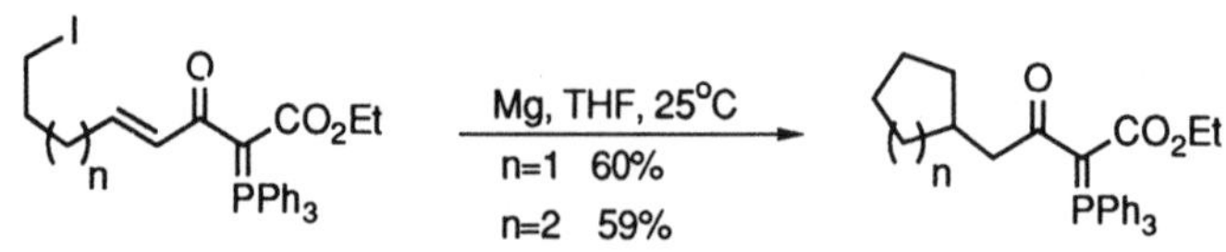
I
CO2Et
PPh3
Mg, THF, 25°C
n=1 60%
n=2 59%
CO2Et
PPh3

2.2.3 Free radicals

By using the combination of a nucleophilic radical and a conjugate acceptor it is possible to close relatively large rings.[291] The value for k_{cyc}, for the closure in entry 1, was calculated to be 1.2 x 10^4 s^{-1} (80°C), which is quite similar to $k_{6\text{-}exo}$ 5.4 x 10^3 s^{-1} (25°C).[292] For smaller rings e.g. 10, H atom transfer predominates (cyclized yield only ~15%). However, larger rings form in good yields. It is also important to minimize transannular steric effects. Thus, closure of a system containing a triple bond is equally efficient (Table 2.14, entry 2).

Table 2.14 Free radical cyclizations of ω-haloalkenones

1[291]

I
O
3-6 mM (iodide), Bu3SnH, 1.1.eq.
AIBN, 0.1 eq., benzene, reflux
argon, 3h, 65%
O
14

2[291]

O
I
3-6 mM (iodide), Bu3SnH, 1.1.eq.
AIBN, 0.1 eq., benzene, reflux
argon, 3h, 65%
O
14

3[293]

O
TMS
N
Bn
hυ, (λ > 320 nm), DCA, N2
15% MeOH-MeCN,>90%
O
N
Bn

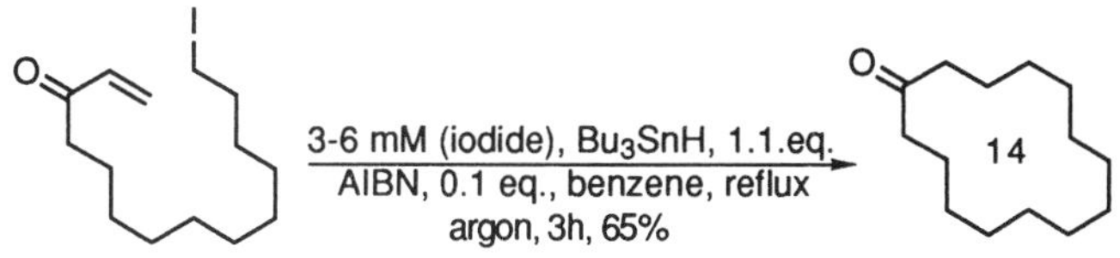

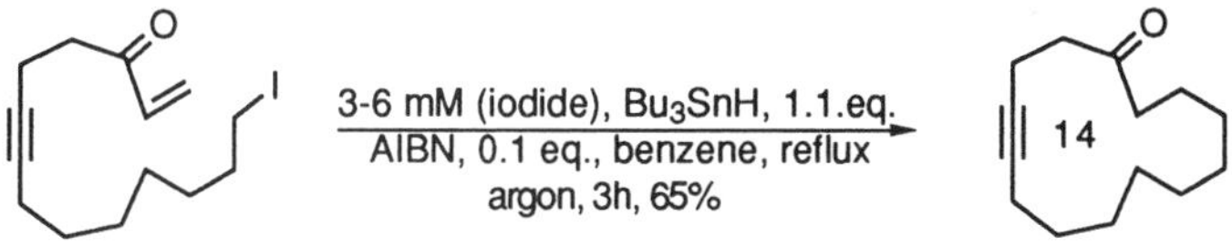

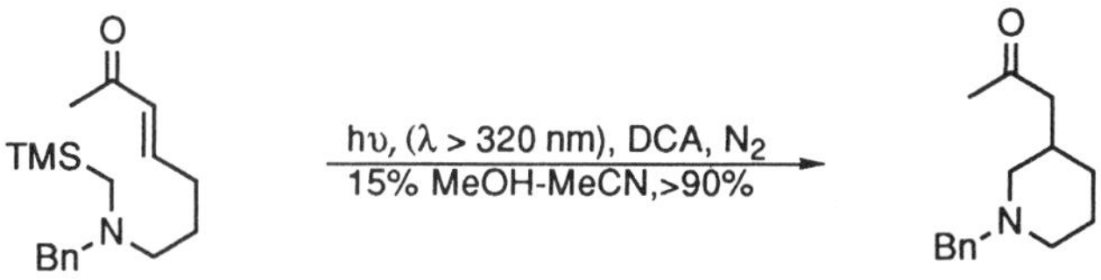

The possibility of tandem cyclizations has also been explored.[294] It was found that the second, transannular, cyclization only occurred if it involved a 5-*exo* mechanism (see section 1.5.1). Even then, only some cases were successful. For example, the two C14 compounds shown below were treated with tributylstannane. Only one of these gave a product resulting from a transannular cyclization.

O
I
3-6 mM (iodide), Bu3SnH, 1.1.eq.
AIBN, 0.1 eq., benzene, reflux
argon, 3h, 70-80%
O
14

O
I
Bu3SnH
as above
O
5-exo
30%
O
H
11
5

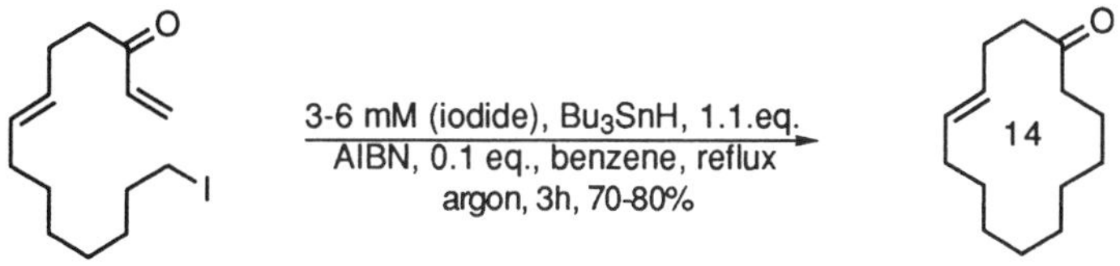

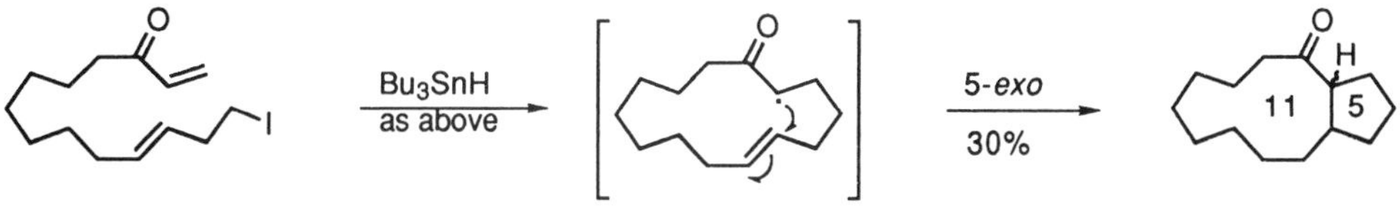

More recently, the same group have developed a regio- and stereoselective free radical macrocyclization procedure.[295] Applying the methodology developed for intermolecular free radical additions (see section 2.1.4), a series of chiral ω-iodo-γ-ketoalkenamides was closed using tribuytlstannane to generate the free radical as before. In general, *endo* selectivity was good. In addition, selectivity for the *R* isomer in each case was also good. This method was used to synthesize (-)-(*R*)-muscone, as shown below.

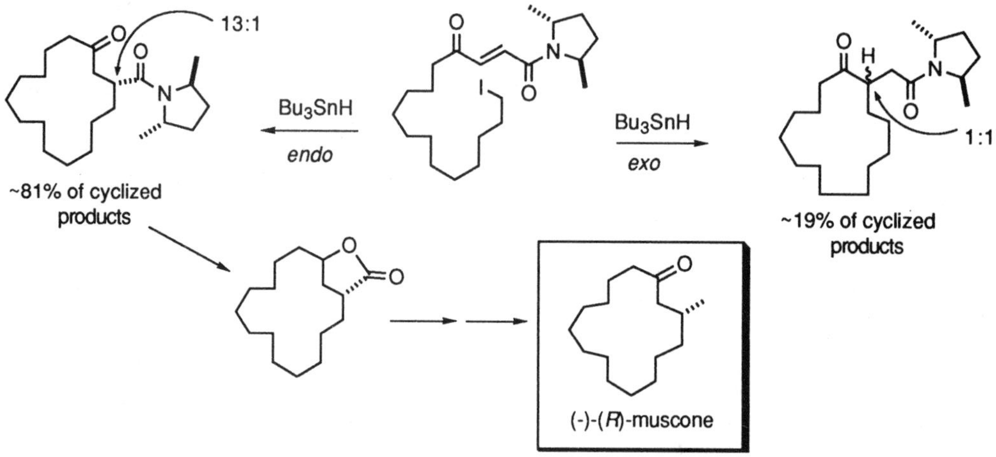
13:1
Bu3SnH
endo
Bu3SnH
exo
1:1
~81% of cyclized products
~19% of cyclized products
(-)-(R)-muscone

2.2.4 Heteronucleophiles

Several approaches to the synthesis of the pseudomonic acids have been developed. One, which has been employed by two groups, employs a sequence of, firstly, a Wittig reaction between acetylmethylenetriphenylphosphorane and a carbohydrate-derived cyclic hemiacetal followed by a stereoselective ring closure of an oxygen nucleophile onto the newly-generated alkenone.[296,297]

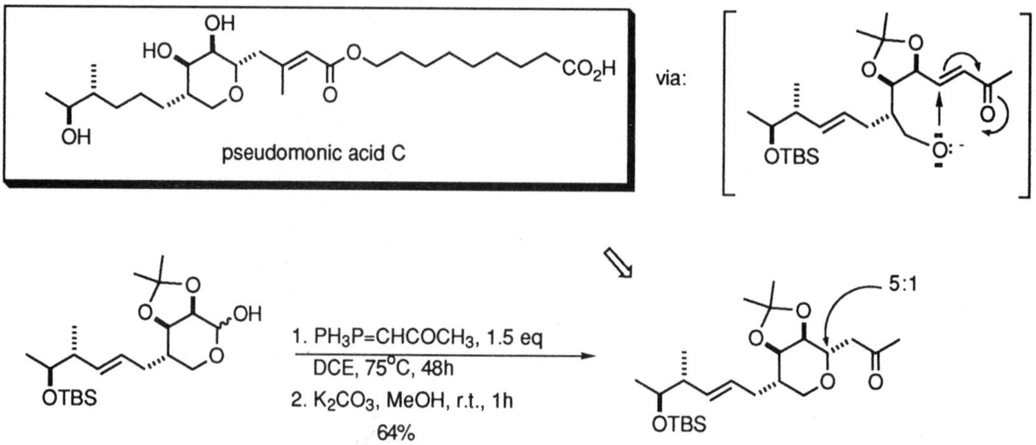
pseudomonic acid C
via:
OTBS
1. PH3P=CHCOCH3, 1.5 eq
DCE, 75°C, 48h
2. K2CO3, MeOH, r.t., 1h
64%
5:1

It has been suggested that the photochemical rearrangement of certain 2,5-disubstituted furans to 2,4-disubstituted 2,3-dihydro-3-oxofurans involves a 5-*endo*-trig addition of an enolic oxygen to an alkenone.[298]

Remarkably large rings may be produced by the intramolecular addition of thiols to alkenones. The precursors are generated as reactive intermediates by an intermolecular conjugate addition/elimination to a trienone bearing a leaving group attached to the substituent at the α-carbon. A variety of different sized rings has been prepared using this method.[299]

References

1.   For a discussion, see D. A. Oare and C. H. Heathcock, in "Topics in Stereochemistry", E. L. Eliel and S. H. Wilen (eds), Wiley, New York, 1989, 227

2.   D. A. Oare and C. H. Heathcock, *J. Org. Chem.*, 1990, **55**, 157

3.   Quite a controversy exists regarding the terms "annulation" and "annelation". Although "annelation" was apparently used in the chemical literature much earlier than "annulation", the latter will be used here as a consideration of the etymology of each word seems to indicate the use of annulation as the more sensible. Both words are derived from Latin. *Anulus*, (one "n"), means "ring", whereas *annelus*, (two "n"s), means "small ring". In addition, the term "annelation" has been used to describe at least two other phenomena. For a discussion of this problem, see A. Nickon and E. F. Silversmith, "Organic Chemistry, The Name Game", Pergamon Press, New York, 1987, 8

4.   Two reviews of the Robinson and related annulations have also appeared: (a) R. E. Gawley, *Synthesis*, 1976, 777 and (b) M. E. Jung, *Tetrahedron*, 1976, **32**, 3. See also, B. P. Mundy, *J. Chem. Ed.*, 1973, **50**, 111. Several chapters in the following book are also devoted to this annulation procedure, T-L. Ho, "Carbocycle Construction in Terpene Synthesis", VCH, New York, 1987.

5.   W. S. Rapson and R. Robinson, *J. Chem. Soc.*, 1935, 1285
6.   E. C. du Feu, F. J. McQuillin and R. Robinson, *J. Chem. Soc.*, 1937, 53
7.   A. J. Birch and R. Robinson, *J. Chem. Soc.*, 1944, 503
8.   C. J. V. Scanio and R. M. Starrett, *J. Amer. Chem. Soc.*, 1971, **93**, 1539
9.   C. D. Ritchie, *J. Amer. Chem. Soc.*, 1969, **91**, 6749
10. Y. Takagi, Y. Nakahara and M. Matsui, *Tetrahedron*, 1978, **34**, 517
11. K. Takaki, M. Ohsugi, M. Okada, M. Yasumura and K. Negoro, *J. Chem. Soc., Perkin Trans. I*, 1984, 741
12. S. A. Monti and Y-L. Yang, *J. Org. Chem.*, 1979, **44**, 897
13. R. B. Woodward, F. Sondheimer, D. Taub, K. Heusler and W. M. McLamore, *J. Amer. Chem. Soc.*, 1952, **74**, 4223
14. Y. Takagi, Y. Nakahara and M. Matsui, *Tetrahedron*, 1978, **34**, 517
15. J. K. Roy, *Chem. Ind.*, 1954, 1393
16. C. H. Heathcock, C. Mahaim, M. F. Schlecht and T. Utawanit, *J. Org. Chem.*, 1984, **49**, 3264
17. I. N. Nazarov and S. I. Zavyalov, *J. Gen. Chem. USSR,*, 1953, **23**, 1793
18. E. Wenkert, A. Afonso, J. B-son Bredenberg, C. Kaneko and A. Tahara, *J. Amer. Chem. Soc.*, 1964, **86**, 2038
19. S. Pelletier, R. I. Chappell and R. S. Prabhakar, *J. Amer. Chem. Soc.*, 1968, **90**, 2889
20. E. Wenkert, *Acc. Chem. Res.*, 1980, **13**, 27
21. K. L. Cook and A. J. Waring, *J. Chem. Soc., Perkin Trans. I*, 1973, 529. For an earlier procedure, upon which this is based, see E. L. Eliel and C. A. Lukach, *J. Amer. Chem. Soc.*, 1957, **79**, 5986
22. S. Ramachandran and M. S. Newman, *Org. Syntheses*, Coll. Vol. V, 1973, 486
23.   See also, S. Danishefsky, P. Cain and A. Nagel, *J. Amer. Chem. Soc.*, 1975, **97**, 380
24. I. N. Nazarov and S. I. Zav'yalov, *Izvest. Akad. Nauk. SSSR Otd. Khim. Nauk*, 300 (1952)
25. K. Balasubramanian, J. P. John and S Swaminathan, *Synthesis*, 1974, 51
26. J. W. Patterson, Jr. and W. Reusch, *Synthesis*, 1971, 155
27. A. M. Chalmers and A. J. Baker, *Synthesis*, 1974, 539
28. N. Harada, T. Sugioka, H. Uda and T. Kuriki, *Synthesis*, 1990, 53
29. U. Eder, G. Sauer and R. Weichert, *Angew. Chem. Int. Ed. Engl.*, 1971, **10**, 496; Z. G. Hajos and D. R. Parrish, *J. Org. Chem.*, 1974, **39**, 1615; J. Gutzwiller, P. Buchschacher and A. Fürst, *Synthesis*, 1977, 167; P. Buchschacher and A. Fürst, *Org. Synth.*, 1986, **63**, 37, the crystallization procedure in this paper apparently requires amendment and will appear in *Org. Synth.*, **Coll. Vol. 7**
30. See also C. Agami, F. Meynier, C. Puchot, J. Guilhem and C. Pascard, *Tetrahedron*, 1984, **40**, 1031
31. For a discussion of the mechanism of the proline catalyzed aldol step, see C. Agami, C. Puchot and H. Sevestre, *Tetrahedron Lett.*, 1986, **27**, 1501
32. R. S. Olsen, Z. A. Fataftah and M. W. Rathke, *Syn. Comm.*, 1986, **16**, 1133
33. A. Mayer and O. Hofer, *J. Amer. Chem. Soc.*, 1980, **102**, 4410
34. G. Stork and B. Ganem, *J. Amer. Chem. Soc.*, 1973, **95**, 6152
35. G. Stork and J. Singh, *J. Amer. Chem. Soc.*, 1974, **96**, 6182
36. R. K. Boekman, Jr., *J. Amer. Chem. Soc.*, 1974, **96**, 6179
37. R. V. Bornert and P. R. Jenkins, *J. Chem. Soc., Chem. Commun.*, 1987, 6
38. D. A. Peak and R. Robinson, *J. Chem. Soc.*, 1958, 1194

39.   P. Wieland and K. Miescher, *Helv. Chim. Acta*, 1950, **33**, 2215

40.   A. Rosan and M. Rosenblum, *J. Org. Chem.*, 1975, **40**, 3621

41.   M. Pesaro and J-P. Bachmann, *J. Chem. Soc., Chem. Commun.*, 1978, 203

42.   C. D. DeBoer, *J. Org. Chem.*, 1974, **39**, 2426

43.   E. J. Corey, M. A. Tius and J. Das, *J. Amer. Chem. Soc.*, 1980, **102**, 1742

44.   C. H. Heathcock, J. E. Ellis, J. E. McMurry and A. Coppolino, *Tetrahedron Lett.*, 1971, 4995. The solvent, benzene, was inadvertently omitted from the manuscript - see footnote 29 in ref. 16

45.   See also P. A. Zoretic, B. Branchaud and T. Maestrone, *Tetrahedron Lett.*, 1975, 527 and P. A. Zoretic, B. Bendiksen and B. Branchaud, *J. Org. Chem.*, 1976, **41**, 3555

46.   P. A. Zoretic, J. A. Golen and M. D. Saltzman, *J. Org. Chem.*, 1981, **46**, 3767

47.   J. W. Huffman, S. M. Potnis and A. V. Satish, *J. Org. Chem.*, 1985, **50**, 4266

48.   R. van der Steen, P. L. Biesheuvel, C. Erkelens, R. A. Mathies and J. Lugtenburg, *Rec. Trav.*, 1989, **108**, 83

49.   J. E. Ellis, J. S. Dutcher and C. H. Heathcock, *Syn. Commun.*, 1974, **4**, 71

50.   For a Lewis acid catalyzed addition of acetophenone to chalcone, see J. A. VanAllan and G. A. Reynolds, *J. Org. Chem.*, 1968, **33**, 1102

51.   P. A. Zoretic and J. A. Golen, *J. Org. Chem.*, 1981, **46**, 3554

52.   G. Stork and J. Singh, *J. Amer. Chem. Soc.*, 1974, **96**, 6181

53.   R. K. Boeckman, Jr., *J. Amer. Chem. Soc.*, 1974, **96**, 6179

54.   M. Pesaro and J-P. Bachmann, *J. Chem. Soc., Chem. Commun.*, 1978, 203

55.   E. J. Corey, M. A. Tius and J. Das, *J. Amer. Chem. Soc.*, 1980, **102**, 1744

56.   Taken from D. A. Oare and C. H. Heathcock, in E. L. Eliel and S. H. Wilen (eds), "Topics in Stereochemistry", Wiley, New York, 1989, 227

57.   D. A. Oare and C. H. Heathcock, *J. Org. Chem.*, 1990, **55**, 157

58.   S. Berrada and P. Metzner, *Tetrahedron Lett.*, 1987, **28**, 409

59.   M. N. Paddon-Row, N. G. Rondan and K. N. Houk, *J. Amer. Chem. Soc.*, 1982, **104**, 7162

60.   For an early review see H. Stetter, *Angew. Chem. Int. Ed. Engl.*, 1976, **15**, 639

61.   For a more recent review, see J. D. Albright, *Tetrahedron*, 1983, **39**, 3207

62.   R. Breslow, *J. Amer. Chem. Soc.*, 1958, **80**, 3719

63.   N. Tagaki and H. Hara, *J. Chem. Soc., Chem. Commun.*, 1973, 891

64.   H. Stetter and H. Kuhlmann, *Tetrahedron Lett.*, 1974, 4505

65.   H. Stetter and M. Schreckenberg, *Chem. Ber.*, 1974, **107**, 2453

66.   H. Stetter and H. Kuhlmann, *Chem. Ber.*, 1976, **109**, 2890

67.   H. Stetter, *Angew. Chem. Int. Ed. Engl.*, 1976, **15**, 639

68.   H. Ahlbrecht and H-M. Kompter, *Synthesis*, 1983, 645

69.   M. Kato, H. Saito and A. Yoshikoshi, *Chem. Lett.*, 1984, 213

70.   N. Ono, A. Kamimura and A. Kaji, *Synthesis*, 1984, 226; S. Hashimoto, K. Matsumoto and S. Otani, *J. Org. Chem.*, 1984, **49**, 4543; B. Kim, M. Kodomari and S. L. Regen, *ibid.*, 1984, **49**, 3233; H. Arzeno, D. H. R. Barton, R-M. Berge-Lurion, X. Lusinchi and B. M. Pinto, *J. Chem. Soc., Perkin Trans. I*, 1984, 2069; D. H. R. Barton, R-M. Berge-Lurion, X. Lusinchi and B. M. Pinto, *ibid*, 1984, 2077

71.   P. G. Baraldi, A. Barco, S. Benetti, G. P.Pollini E. Polo and D. Simoni, *J. Chem. Soc., Chem. Commun.*, 1986, 757; W. Danikiewicz, T. Jaworski and S. Kwiatkowski, *Monatsh Chem.*, 1986, **117**, 1177; T. Miyakoshi, *Synthesis*, 1986, 766

72.   J. Nogami, T. Sonoda and S. Wakabayashi, *Synthesis*, 1983, 763

73.   T. Miyakoshi and S. Saito, *Yukagaku*, 1983, **32**, 749

74.   I. Belsky, *J. Chem. Soc., Chem. Commun.*, 1977, 237
75.   W. T. Monte, M. M. Baizer and R. D. Little, *J. Org. Chem.*, 1983, **48**, 803
76.   For reviews of the Nef reaction, see S. Pentti, in "Chemistry of the Carbonyl Group", S. Patai (ed), Interscience, New York, 1966, 177 and W. E. Noland, *Chem. Rev.*, 1955, **55**, 137. See also, D. St. C. Black, *Tetrahedron Lett.*, 1972, 1331
77.   For an electrochemical version of the Nef reaction, see T. Miyakoshi, *Synthesis*, 1986, 766; J. Nogami, T. Sonoda and S. Wakabayashi, *ibid.*, 1983, 763 and W. T. Monte, M. M. Baizer and R. D. Little, *J. Org. Chem.*, 1983, **48**, 803
78.   J. E. McMurry and J. Melton, *Org. Synth.*, 1988, **Coll. Vol. 6**, 648
79.   B. S. Furniss, A. J. Hannaford, P. W. G. Smith and A. R. Tatchell, "Vogel's Textbook of Practical Organic Chemistry", Longman, Burnt Mill, 1989, 637
80.   J. H. Clark, D. G. Cork and H. W. Gibbs, *J. Chem. Soc., Perkin Trans. I*, 1983, 2253
81.   R. Ballini, M. Petrini and G. Rosini, *Synthesis*, 1987, 711
82.   N. Ono and A. Kamimura, H. Miyake, I. Hamamoto and A. Koji, *J. Org. Chem.*, 1985, **50**, 3692; A. C. Coda, G. Desimoni, A. G. Invernizzi, P. P. Righetti, P. F. Seneci and G. Tacconi, *Gazz. Chim. Ital.*, 1985, **115**, 111; P. Bisiacchi, A. C. Coda, G. Desimoni, P. P. Righetti, and G. Tacconi, *Gazz. Chim. Ital.*, 1985, **115**, 119
83.   J. J. Tufariello, J. J. Tegeler, S. C. Wong and S. A. Ali, *Tetrahedron Lett.*, 1978, 1733
84.   R. V. Stevens and A. W. M. Lee, *J. Chem. Soc., Chem. Commun.*, 1982, 102
85.   A. R. Battersby, C. J. Dutton and C. J. R. Fookes, *J. Chem. Soc., Perkin Trans. I*, 1988, 1569; C. J. Dutton, C. J. R. Fookes and A. R. Battersby, *J. Chem. Soc.*, 1983, 1237
86.   J. E. Richman, J. L. Herrmann and R. H. Schlessinger, *Tetrahedron Lett.*, 1973, 3271
87.   M. Mikolajczyk and P. Balczewski, *Synthesis*, 1984, 691
88.   R. Lawrence and P. Perlmutter, *J. Org. Chem.*, 1990, submitted for publication
89.   W-Y. Zhang, D. J. Jakiela, A. Maul, C. Knors, J. W. Lauher, P. Helquist and D. Enders, *J. Amer. Chem. Soc.*, 1988, **110**, 4652
90.   T. Yura, N. Iwasawa and T. Mukaiyama, *Chem. Lett.*, 1988, 1021
91.   T. Yura, N. Iwasawa, K. Narasaka and T. Mukaiyama, *Chem. Lett.*, 1988, 1025
92.   D. Desmaële and J. d'Angelo, *Tetrahedron Lett.*, 1989, **30**, 345
93.   See also, M. Pfau, G. Revial, A. Guingant and J. d'Angelo, *J. Amer. Chem. Soc.*, 1985, **107**, 273
94.   L. Velluz, G. Nomine, G. Amiard, V. Torelli and J. Cerede, *Compt. Rend.*, 1963, **257**, 3086
95.   L. Velluz, G. Nomine, J. Mathieu, E. Toromanoff, D. Bertin, M. Vignau, and J. Tessier, *Compt. Rend.*, 1960, **250**, 1510
96.   S. Danishefsky, P. Cain and A. Nagel, *J. Amer. Chem. Soc.*, 1975, **97**, 380
97.   J. H. Nelson, P. N. Howells, G. C. DeLullo, G. L. Landen and R. A. Henry, *J. Org. Chem.*, 1980, **45**, 1246
98.   K. Matsumoto, *Synthesis*, 1980, 1013
99.   For reviews of reactions carried out under high pressure conditions, see J. Jurczak and B. Baranowski, "High Pressure Chemical Synthesis", Elsevier, Amsterdam, 1989 and K. Matsumoto, A. Sera and T. Uchida, *Synthesis*, 1985, 1
100.   J. L. Zeilstra and J. B. F. N. Egberts, *J. Org. Chem.*, 1974, **39**, 3215
101.   H. Rubenstein, *J. Org. Chem.*, 1962, **27**, 3886
102.   J-F. Lavallee and P. Deslongchamps, *Tetrahedron Lett.*, 1988, **29**, 6033

103.	U. Eder, G. Sauer and R. Wiechert, *Angew. Chem. Int. Edn. in Eng.*, 1971, **10**, 496; Z. G. Hajos and D. R. Parrish, *J. Org. Chem.*, 1974, **39**, 1612. See also, L. F. Tietze and T. Eicher, "Reactions and Syntheses in the Organic Chemistry Laboratory", University Science Books, Mill Valley, 1989, p. 169
104.	N. Harada, T. Sugioka, H. Uda and T. Kuriki, *Synthesis*, 1990, 53
105.	M. D. Turnbull, G. Hatter and D. E. Ledgerwood, *Tetrahedron Lett.*, 1984, **25**, 5449
106.	T. Naota, H. Taki, M. Mizuno and S-I. Murahashi, *J. Amer. Chem. Soc.*, 1989, **111**, 5954
107.	For a conventional, base-catalyzed addition of a β-ketoester to an alkenal, see R. C. Anand and H. Ranjan, *Indian J. Chem.*, 1984, **23B**, 1054
108.	K. Matsumoto, *Angew. Chem. Int. Ed. Engl.*, 1981, **20**, 770
109.	For a review see, J. Tsuji, *Pure and Appl. Chem.*, 1981, **53**, 2371
110.	J. Tsuji, I. Shimizu, H. Suzuki and Y. Naito, *J. Amer. Chem. Soc.*, 1979, **101**, 5070
111.	D. T. Warner, *J. Org. Chem.*,1959, **24**, 1536
112.	M. Geier and M. Hesse, *Synthesis*, 1990, 56
113.	J. H. Babler, B. J. Invergo and S. J. Sarussi, *J. Org. Chem.*, 1980, **45**, 4241
114.	B. S. Furniss, A. J. Hannaford, P. W. G. Smith and A. R. Tatchell, "Vogel's Textbook of Practical Organic Chemistry", Longman, Burnt Mill, 1989, 757
115.	J. K. F. Geirsson and A. D. Gudmundsdottir, *Acta Chem. Scand.*, 1989, **43**, 618
116.	F. Bossert, H. Meyer and E. Wehinger, *Angew. Chem. Int. Ed. Engl.*, 1981, **20**, 762
117.	N. Ono, H. Miyake and A. Kaji, *J. Chem. Soc., Chem. Commun.*, 1983, 875
118.	A. P. Terent'ev, E. I. Klabunovskii and E. I. Budovskii, *Chem. Abstr.*, 1959, **49**, 5263b
119.	B. Langstrom and G. Bergson, *Acta Chem. Scand.*, 1973, **27**, 3118
120.	H. Wynberg and R. Helder, *Tetrahedron Lett.*, 1975, 4057
121.	K. Hermann and H. Wynberg, *J. Org. Chem.*, 1979, **44**, 2238
122.	K. Hermann and H. Wynberg, *Helv. Chim. Acta*, 1977, **60**, 2208
123.	N. Kobayashi and K. Iwai, *J. Amer. Chem. Soc.*, 1978, **100**, 7071 and *J. Polym. Sci., Polym. Chem. Ed.*, 1980, **18**, 923
124.	P. Hodge, E. Khoshdel and J. Waterhouse, *J. Chem. Soc., Perkin Trans. I*, 1983, 2205
125.	D. J. Cram and G. D. Y. Sogah, *J. Chem. Soc., Chem. Commun.*, 1981, 625
126.	H. Brunner and B. Hammer, *Angew. Chem. Int. Edn. Engl.*, 1984, **23**, 312
127.	D. W. Brooks, P. G. Grothaus and H. Mazdiyashi, *J. Amer. Chem. Soc.*, 1983, **105**, 4472
128.	D. F. Taber, J. F. Mack, A. L. Rheingold and S. J. Geib, *J. Org. Chem.*, 1989, **54**, 3831
129.	For a synthesis of the homochiral 2-naphthylborneol used in this work, see D. F. Taber, K. Raman and M. D. Gaul, *J. Org. Chem.*, 1987, **52**, 28
130.	T. Matsumoto, S. Imai, S. Usui, A. Suetsugu, S. Kawatsu and T. Yamaguchi, *Chem. Letts.*, 1984, 67
131.	K. Sussangkarn, G. Fodor, D. Strope and C. George, *Heterocycles*, 1989, **28**, 467
132.	For a review, see A. Krief, *Tetrahedron*, 1980, **36**, 2531
133.	See also, L. Wartski and M. El Bouz, *Tetrahedron*, 1982, **38**, 3285
134.	J. Luchetti and A. Krief, *J. Organomet. Chem.*, 1980, **194**, C49
135.	M. El-Bouz and L. Wartski, *Tetrahedron Lett.*, 1980, **21**, 2897
136.	R. A. J. Smith and A. R. Lal, *Aust. J. Chem.*, 1979, **32**, 353
137.	N. Rehnberg and G. Magnusson, *Tetrahedron Lett.*, 1988, **29**, 3599

138.   M. El-Bouz and L. Wartski, *Tetrahedron Lett.*, 1980, **21**, 2897

139.   J. Luchetti and A. Krief, *J. Organomet. Chem.*, 1980, **194**, C49

140.   A. R. B. Manas and R. A. Smith, *J. Chem. Soc., Chem. Commun.*, 1975, 216

141.   Copper-catalyzed - see section 2.1.2

142.   For a comprehensive compilation up to the early 1970s, see G. H. Posner in "Organic Reactions", Wiley, New York, 1972, 1

143.   M. C. P. Yeh, P. Knochel, W. M. Butler and S. C. Berk, *Tetrahedron Lett.*, 1988, **29**, 6693

144.   Y. Tamaru, H. Tanigawa, T. Yamamoto and Z-i. Yoshida, *Angew. Chem. Int. Ed. Engl.*, 1989, **28**, 351

145.   G. Cahiez and M. Alami, *Tetrahedron Lett.*, 1989, **30**, 3541

146.   Additions of organomanganese reagents, without copper catalysis, have also been reported, however they were complicated by competing side reactions. See G. Cahiez and M. Alami, *Tetrahedron Lett.*, 1986, **27**, 569.

147.   T. Mukaiyama, K. Narasaka and M. Furusato, *J. Amer. Chem. Soc.*, 1972, **94**, 8641

148.   H. Malmberg and M. Nilsson, *Tetrahedron*, 1986, **42**, 3931

149.   M. Bergdahl, E-L. Lindstedt, M. Nilsson and T. Olsson, *Tetrahedron*, 1989, **45**, 535

150.   R. K. Dieter and L. A. Silks, *J. Org. Chem.*, 1986, **51**, 4687

151.   E. J. Corey and D. Enders, *Tetrahedron Lett.*, 1976, 11

152.   For 1,2-additions of organocopper reagents see B. H. Lipshutz, E. L. Ellsworth, T. J. Siahaan and A. Shinazi, *Tetrahedron Lett.*, 1988, **29**, 6677

153.   Y. Horiguchi, S. Matsuzawa, E. Nakamura and I. Kuwajima, *Tetrahedron Lett.*, 1986, **27**, 4025

154.   E. Nakamura, S. Matsuzawa, Y. Horiguchi and I. Kuwajima, *Tetrahedron Lett.*, 1986, **27**, 4029

155.   For an extensive compilation and review, see M. J. Chapdelaine and M. Hulce, "Organic Reactions", Wiley, New York, 1990, **38**, 225

156.   K. K. Heng and R. A. K. Smith, *Tetrahedron*, 1979, **35**, 425

157.   S. Bernasconi, P. Gariboldi, G. Jommi and M. Sisti, *Tetrahedron Lett.*, 1980, **21**, 2337

158.   R. M. Coates and L. O. Sandefur, *J. Org. Chem.*, 1974, **39**, 275

159.   D. Seyferth and R. C. Hui, *J. Amer. Chem. Soc.*, 1985, **107**, 4551

160.   See also, J. Schwartz, *Tetrahedron Lett.*, 1972, 2803

161.   D. Seyferth and R. C. Hui, *Tetrahedron Lett.*, 1986, **27**, 1473

162.   G. H. Posner, *Org. React.*, 1972, **19**, 1

163.   R. E. Ireland and P. Wipf, *J. Org. Chem.*, 1990, **55**, 1425

164.   E. Negishi, *Pure Appl. Chem.*, 1981, **53**, 2333

165.   A. S. Kende and L. N. Jungheim, *Tetrahedron Lett.*, 1980, **21**, 3849

166.   R. G. Salomon, D. B. Miller, S. R. Raychaudhuri, K. Avasthi, K. Lal and B. S. Levison, *J. Amer. Chem. Soc.*, 1984, **106**, 8296

167.   W. P. Roush and B. M. Lesur, *Tetrahedron Lett.*, 1983, **24**, 2231

168.   B. M. Trost, J. M. Timks and J. L. Stanton, *J. Chem. Soc., Chem. Commun.*, 1978, 436

169.   E. J. Corey, F. J. Hannon and N. W. Boaz, *Tetrahedron*, 1989, **45** , 545

170.   P. Mangeney, A. Alexakis and J. F. Normant, *Tetrahedron Lett.*, 1983, **24**, 373

171.   H. Malmberg, M. Nilsson and C. Ullenius, *Tetrahedron Lett.*, 1982, **23**, 3823

172.   R. K. Dieter and M. Tokles, *J. Amer. Chem. Soc.*, 1987, **109**, 2040

173. As one group has pointed, out there is only a limited amount of evidence for the existence of "triorganozincates" in these reactions. However, the notion serves to describe the stoichiometry satisfactorily. For a spectroscopic study of the reaction between diethylzinc and 1,1-diphenyl-1-lithio-hexane, see R. Waack and M. A. Doran, *J. Amer. Chem. Soc.*, 1963, **85**, 2861

174. R. A. Kjonaas and R. K. Hoffer, *J. Org. Chem.*, 1988, **53**, 4133

175. W. Tuckmantel, K. Oshima and H. Nozaki, *Chem. Ber.*, 1986, **119**, 1581

176. K. Soai, T. Hayasaka and S. Ugajin, *J. Chem. Soc., Chem. Commun.*, 1989, 516

177. R. A. Watson and R. A. Kjonaas, *Tetrahedron Lett.*, 1986, **27**, 1437

178. J. F. G. A. Jansen and B. L. Feringa, *Tetrahedron Lett.*, 1988, **29**, 3593

179. J. F. G. A. Jansen and B. L. Feringa, *J. Chem. Soc., Chem. Commun.*, 1989, 741

180. G. Majetich, A. M. Casares, D. Chapman and M. Behnke, *Tetrahedron Lett.*, 1983, **24**, 1909. See also A. Hosomi, A Shirhata and H. Sakurai, *Tetrahedron Lett.*, 1978, 3543

181. A. Hosomi and H. Sakurai, *J. Amer. Chem. Soc.*, 1977, **99**, 1673

182. C. H. Heathcock, S-i. Kiyooka and T. A. Blumenkopf, *J. Org. Chem.*, 1984, **49**, 4214

183. T. A. Blumenkopf and C. H. Heathcock, *J. Amer. Chem. Soc.*, 1983, **105**, 2354

184. J-I. Hurata, K. Nishi, S. Matsuda, Y. Tamura and Y. Kita, *J. Chem. Soc., Chem. Commun.*, 1989, 1065

185. R. L. Danheiser, D. J. Carini and A. Basak, *J. Amer. Chem. Soc.*, 1981, **103**, 1604

186. M. P. Cooke, Jr. and R. Goswami, *J. Amer. Chem. Soc.*, 1977, **99**, 642

187. M. P. Cooke, Jr., *Tetrahedron Lett.*, 1979, 2199

188. M. P. Cooke, Jr. and R. M. Parlman, *J. Amer. Chem. Soc.*, 1977, **99**, 5222

189. J. Hooz and R. B. Layton, *J. Amer. Chem. Soc.*, 1971, **93**, 7320

190. P. W. Collins, E. Z. Dajani, M. S. Bruhn, C. H. Brown, J. R. Palmer and R. Pappo, *Tetrahedron Lett.*, 1975, 4217

191. R. Pappo and P. W. Collins, *Tetrahedron Lett.*, 1972, 2627

192. R. T. Hansen, D. B. Carr and J. Schwartz, *J. Amer. Chem. Soc.*, 1978, **100**, 2244

193. J. Schwartz, D. B. Carr, R. T. Hansen and F. M. Dayrit, *J. Org. Chem.*, 1980, **45**, 3053

194. J. A. Sinclair and H. C. Brown, *J. Org. Chem.*, 1976, **41**, 1078

195. J. A. Sinclair, G. A. Molander and H. C. Brown, *J. Amer. Chem. Soc.*, 1977, **99**, 954

196. R. Locker and D. Seebach, *Angew. Chem. Int. Ed. Engl.*, 1981, **20**, 569

197. E. J. Corey and L. S. Hegedus, *J. Amer. Chem. Soc.*, 1969, **91**, 4926

198. For an extensive survey and thorough review of the generation and reactions of carbon radicals, see A. Ghosez, B. Giese and H. Zipse in M. Regitz and B. Giese (eds), "Houben-Weyl. Methoden der organischen Chemie", Georg Thieme, Stuttgart, 1989, Band E19a/Teil 2

199. T. Shono, I. Nishiguchi and H. Ohmizu, *J. Amer. Chem. Soc.*, 1977, **99**, 7396

200. R. Scheffold, *Nachr. Chem. Tech. Lab.*, 1988, **36**, 261 and *Chimia*, 1985, **39**, 203

201. For a discussion of Vitamin B_{12a} and related synthetic catalysts, see ref. 198

202. R. Scheffold and L. R. Orlinski, *J. Amer. Chem. Soc.*, 1983, **105**, 7200

203. J. L. Luche and C. Allavena, *Tetrahedron Lett.*, 1988, **29**, 5373

204. J. L. Luche and C. Allavena, *Tetrahedron Lett.*, 1988, **29**, 5369

205. For similar additions to alkenoate derivatives see section 5.1.4

206. N. A. Porter, D. M. Scott, B. Lacher, B. Giese, H. G. Zeitz and H. J. Lindner, *J. Amer. Chem. Soc.*, 1989, **111**, 8311

207. No chemical yields were provided in this paper (i.e. in ref. 206)

208.   G. Stork, A. Brizzolara, H. Landesman, J. Szmuszkovicz and R. Terrell, *J. Amer. Chem. Soc.*, 1963, **85**, 207

209.   See also, G. Stork and H. Landesman, *J. Amer. Chem. Soc.*, 1956, **78**, 5128

210.   Y. Chan and W. W. Epstein, *Org. Synth.*, 1988, **Coll. Vol. 6**, 496

211.   Z. Iqbal, A. H. Jackson and K. R. N. Rao, *Tetrahedron Lett.*, 1988, **29**, 2577

212.   C. H. Mitch, *Tetrahedron Lett.*, 1988, **29**, 6831

213.   A. Hantzsch, *Justus Liebigs Ann. Chem.*, 1882, **215**, 1

214.   U. Eisner and J. Kuthan, *Chem. Rev.*, 1972, **72**, 1

215.   H. Meyer, F. Bossert and H. Horstmann, *Justus Liebigs Ann. Chem.*, 1977, 1895

216.   H. Meyer, F. Bossert and H. Horstmann, *Justus Liebigs Ann. Chem.*, 1977, 1888

217.   This procedure has also been used to prepare *un*natural alkaloids, e.g. see C. P. Forbes, G. L. Wenteler and A. Wiechers, *Tetrahedron*, 1978, **34**, 487

218.   T. J. Curphey and H. L. Kim, *Tetrahedron Lett.*, 1968, 1441

219.   R. V. Stevens and L. E. Du Pree, Jr., *J. Chem. Soc., Chem. Commun.*, 1970, 1585

220.   T. Kobayashi and M. Nitta, *Chem. Lett.*, 1986, 1549. For a review of the Staudinger reaction, see Y. G. Gololobov, I. N. Zhmurova and L. F. Kasukhin, *Tetrahedron*, 1981, **37**, 437

221.   K. Narasaka, K. Soai and T. Mukaiyama, *Chem. Lett.*, 1974, 1223

222.   K. Narasaka, K. Soai, Y. Aikawa and T. Mukaiyama, *Bull. Soc. Chem. Jap.*, 1976, **49**, 779

223.   I. Fleming and T. W. Newton, *J. Chem. Soc., Perkin Trans. I*, 1984, 119

224.   See also J. W. Huffman, S. M. Potnis and A. V. Satish, *J. Org. Chem.*, 1985, **50**, 4266

225.   T. Mukaiyama, Y. Sagawa and S. Kobayashi, *Chem. Lett.*, 1986, 1821

226.   K. Saigo, M. Osaki and T. Mukaiyama, *Chem. Lett.*, 1976, 163

227.   Entries 1 to 3, see ref. 226

228.   M. T. Reetz, H. Heimbach and K. Schwelhus, *Tetrahedron Lett.*, 1984, **25**, 511

229.   C. H. Heathcock, M. H. Norman and D. E. Uehling, *J. Amer. Chem. Soc.*, 1985, **107**, 2797

230.   S. Kobayashi, M. Murakami and T. Mukaiyama, *Chem. Lett.*, 1985, 953

231.   S. Kobayashi and T. Mukaiyama, *Chem. Lett.*, 1986, 1808

232.   S. Kobayashi, M. Murakami and T. Mukaiyama, *Chem. Lett.*, 1985, 1535

233.   T. Mukaiyama, M. Tamura and S. Kobayashi, *Chem. Lett.*, 1986, 1017

234.   T. Fukuyama and L Yang, *J. Amer. Chem. Soc.*, 1989, **111**, 8303

235.   M. E. Jung and Y-G. Pan, *Tetrahedron Lett.*, 1980, **21**, 3127

236.   This might also be regarded as a 6-[enol-*exo*]-*endo*-trig closure. Baldwin has only examined enolate ring closures for *exo*-trig pathways, see section 1.5.1. The rules for these closures have yet to be defined

237.   H. Hagiwara, A. Okano and H. Uda, *J. Chem. Soc., Chem. Commun.*, 1985, 1047

238.   C. H. Heathcock and D. E. Uehling, *J. Org. Chem.*, 1986, **51**, 279

239.   C. Gennari, L. Colombo, G. Bertolini and G. Schimperna, *J. Org. Chem.*, 1987, **52**, 2754

240.   For a review, see W. Nagata and M. Yoshioka, *Org. Reactions*, 1977, **25**, 255

241.   H. B. Dykstra, U.S. Pat. 2,188,340 (1940) [*Chem. Abstr.* 1940, **34**, 3764]

242.   W. Nagata, M. Yoshioka and M. Murakami, *J. Amer. Chem. Soc.*, 1972, **94**, 4654

243.   N. Mayer and G. E. Gantert, Ger. Pat. 1,085,871 (1960) [*Chem. Abstr.*, 1962, **54**, 4639f]

244.   For a review, see S. I. Suminov and A. N. Kost, *Russ. Chem. Rev.*, 1969, **38**, 884

245.   V. E. Haury, Brit. Pat., 578,635; *Chem. Abstr*, 1947, **41**, 3480

246. Y. Ogata, A. Kawasaki and I. Kishi, *J. Chem. Soc. B*, 1968, 703
247. R. Lukes, J. Kovar and K. Blaha, *Coll. Czech. Chem. Commun.*, 1960, **25**, 2179
248. R. Baltzly, E. Lorz, P. B. Russell and F. M. Smith, *J. Amer. Chem. Soc.*, 1855, **77**, 624
249. P. L. Southwick and R. J. Shozda, *J. Amer. Chem. Soc.*, 1959, **81**, 5435
250. E. M. Cherekasova, *Izvest. Akad. Nauk S.S.S.R., Otdel. Khim. Nauk*, 1959, 753; *Chem. Abstr.*, 1959, **53**, 21772d
251. P. L. Southwick and R. J. Shozda, *J. Amer. Chem. Soc.*, 1960, **82**, 2887
252. N. H. Cromwell, R. P. Cahoy, W. E. Franklin and G. D. Mercer, *J. Amer. Chem. Soc.*, 1957, **79**, 922
253. D. Rosenthal, G. Brandrup, K. Davis and M. Wall, *J. Org. Chem.*, 1965, **30**, 3689
254. See B. M. Trost and R. A. Kunz, *J. Org. Chem.*, 1974, **39**, 2648
255. H. Takahata, K. Yamabe, T. Suzuki and T. Yamazaki, *Heterocycles*, 1986, **24**, 37
256. R. Ohrlein, W. Schwab, R. Ehrler and V. Jager, *Synthesis*, 1986, 535; R. Ohrlein and V. Jager, *Tetrahedron Lett.*, 1988, **29**, 6083
257. T. Miyakoshi, S. Saito and J. Kumatoni, *Chem. Lett.*, 1981, 1677 and 1982, 83
258. For a useful review and compilation of conjugate additions of trivalent phosphorus, where a new C-P bond is formed, see R. Engel, "Synthesis of Carbon-Phosphorus Bonds", CRC Press, Boca Raton, 1988
259. D. A. Evans, K. M. Hurst, L. K. Truesdale and J. M. Takacs, *Tetrahedron Lett.*, 1977, 2495
260. D. A. Evans, K. M. Hurst, and J. M. Takacs, *J. Amer. Chem. Soc.*, 1978, **100**, 3467
261. A. P. Kozikowsky and *J. Org. Chem.*, 1989, **54**, 1111
262. A. P. Kozikowsky and S Hojung, *Tetrahedron Lett.*, 1986, **27**, 3227
263. I. Kuwajima, T. Murobushi and E. Nakamura, *Synthesis*, 1978, 602
264. A. Lubineau, H. Beinaymé and J. LeGallic, *J. Chem. Soc., Chem. Commun.*, 1989, 1918
265. M. Miyashita and A. Yoshikoshi, *Synthesis*, 1980, 664
266. A. Itoh, S. Ozawa, K. Oshima and H. Nozaki, *Bull. Chem. Soc. Jap.*, 1981, **54**, 278
267. R. A. Bunce and H. D. Reeves, *Syn. Commun.*, 1989, **19**, 1109
268. W. Amberg and D. Seebach, *Angew. Chem. Int. Ed. Engl.*, 1988, **27**, 1718
269. For the preparation of enantiomerically pure dioxanones like that given in the example, see D. Seebach and J. Zimmermann, *Helv. Chim. Acta*, 1986, **69**, 1147 and J. Zimmermann, D. Seebach and T-K. Ha, 1988, **71**, 1143
270. T. Hayashi, Y. Matsumoto and Y. Ito, *Tetrahedron Lett.*, 1988, **29**, 4147
271. W. Tuckmantel, K. Oshima and H. Nozaki, *Chem. Ber.*, 1986, **119**, 1581
272. P. F. Hudrlik, A. M. Hudrlik, T. Yimenu, M. A. Waugh and G. Nagendrappa, *Tetrahedron*, 1988, **44**, 3791
273. I. Fleming and C. J. Urch, *J. Organomet. Chem.*, 1985, **285**, 173
274. See also, H. Ahlbrecht and P. Weber, *Synthesis*, 1989, 117
275. R. D. Miller and D. R. McKean, *Tetrahedron Lett.*, 1979, 2305
276. E. Keinan, *Pure and Appl. Chem.*, 1989, **61**, 1737
277. J. P. Marino and R. J. Linderman, *J. Org. Chem.*, 1981, **46**, 3696
278. M. F. Semmelhack, R. D. Stauffer and A. Yamashita, *J. Org. Chem.*, 1977, **42**, 3180
279. Trademark of Eastman Organic Chemicals
280. Trademark of the Aldrich Chemical Company
281. E. Keinan and N. Greenspoon, *J. Amer. Chem. Soc.*, 1986, **108**, 7314
282. R. S. Tonke and R. H. Crabtree, *Tetrahedron Lett.*, 1988, **29**, 6737

283.   W. S. Mahoney, D. M. Brestensky and J. M. Stryker, *J. Amer. Chem. Soc.*, 1988, **110**, 291

284.   For an improved preparation of this reagent, see D. M. Brestensky, D. E. Huseland, C. McGettigan and J. M. Stryker, *Tetrahedron Lett.*, 1988, **29**, 3749

285.   G. Stork, C. S. Shiner and J. D. Winkler, *J. Amer. Chem. Soc.*, 1982, **104**, 310

286.   This method has been applied to a total synthesis of (±)-adrenosterone, see G. Stork, J. D. Winkler and C. S. Shiner, *J. Amer. Chem. Soc.*, 1982, **104**, 3767

287.   S. K. Attah-Poku, F. Chau, V. K. Yadav and A. G. Fallis, *J. Org. Chem.*, 1985, **50**, 3418

288.   A. Demir, R. S. Gross, N. K. Dunlap, A. Bashir-Hashemi and D. S. Watt, *Tetrahedron Lett.*, 1986, **27**, 5567

289.   D. B. Grotjahn and N. H. Anderson, *J. Chem. Soc., Chem. Commun.*, 1981, 306

290.   G. Berthiaume, J-F. Lavallee and P. Deslongchamps, *Tetrahedron Lett.*, 1986, **27**, 5451

291.   N. A. Porter, D. R. Magnin and B. T. Wright, *J. Amer. Chem. Soc.*, 1986, **108**, 2787

292.   A. L. J. Beckwith and C. H. Scheisser, *Tetrahedron*, 1985, **41**, 3925

293.   W. Xu, Y. T. Jeon, E. Hasegawa, U. C. Yoon and P. S. Mariano, *J. Amer. Chem. Soc.*, 1989, **111**, 406

294.   N, A. Porter, V. H-T. Chang, D. R. Magnin and B. T. Wright, *J. Amer. Chem. Soc.*, 1988, **110**, 3554

295.   N, A. Porter, B. Lacher, V. H-T. Chang and D. R. Magnin, *J. Amer. Chem. Soc.*, 1989, **111**, 8309

296.   A. P. Kozikowski, R. J. Schmeising and K. L. Sorgi, *J. Amer. Chem. Soc.*, 1980, **102**, 6578

297.   G. E. Keck, D. F. Kachensky and E. J. Enholm, *J. Org. Chem.*, 1985, **50**, 4317

298.   R. Antonioletti, F. Bonadies, T. Prencipe and A. Scettri, *J. Chem. Soc., Chem. Commun.*, 1988, 850

299.   S. J. Brocchini, M. Eberle and R. G. Lawton, *J. Amer. Chem. Soc.*, 1988, **110**, 5211

300.   For reviews on the preparation and chemistry of Mannich salts, see M. Tramontini, *Synthesis*, 1973, 703; M. Tramontini and L. Angiolini, *Tetrahedron*, 1990, **46**, 1791. See also C. Rochin, O. Babot, J. Dunoques and F. Duboudin, *Synthesis*, 1986, 667

301.   G. Hallnemo and C. Ullenius, *Tetrahedron*, 1983, **39**, 1621

302.   A. Gossauer and K. Suhl, *Helv. Chim. Acta*, 1976, **50**, 1698

Chapter Three   Cycloalkenones

3   Introduction

Additions to cycloalkenones have, probably, been the most thoroughly-studied area in this field in the last twenty years or so. Much of the impetus for this has come from the intense interest in the chemistry (and biology) of the prostaglandins (see, especially, section 3.1.2). Being cyclic conjugate acceptors, often exceptionally high levels of stereoselectivity are possible. Because so many natural products are cyclic, many methods for their synthesis have come to rely upon single or multiple conjugate additions. This has also proved to be a fruitful area for the application of the recent developments in organic free radical chemistry.

3.1   Intermolecular reactions

3.1.1   Stabilized carbanions

3.1.1.1   *Carbanions stabilized by π-conjugation with one heteroatom*

(a)   *Ketone enolates*

Additions of ketone enolates to *cyclo*alkenones are not very common. Indeed, attempted reaction of the lithium enolate of propiophenone with 2-cyclohexenone gives no addition products at all (presumably due to competing proton transfer).[1] Rather, these additions are the domain of acyclic alkenones, e.g. the Robinson annulation (see section 2.1.1.1.).

However, at least one, masterful, example of an enolate addition exists. This was developed for the total synthesis of the antimycoplasmal agent (±)-pleuromutilin, from *Pleurotus mutilus*.[2] Only one diastereomer results and the postulated mechanism is shown below.[3]

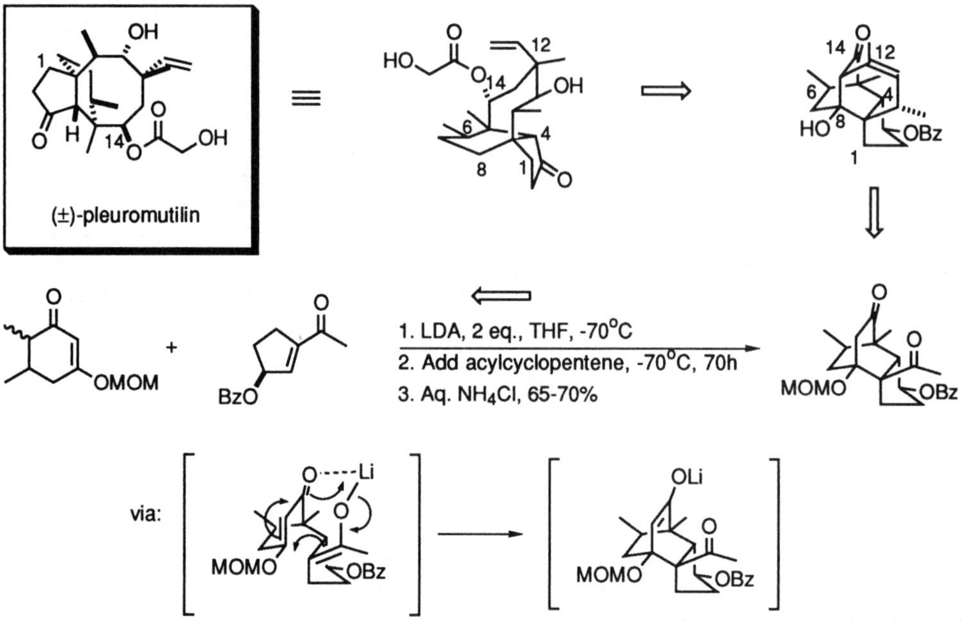

(±)-pleuromutilin

1. LDA, 2 eq., THF, -70°C
2. Add acylcyclopentene, -70°C, 70h
3. Aq. NH₄Cl, 65-70%

via:

(b) *Ester (and related) enolates*

Additions of preformed enolates of amides to cyclohexenones have only recently been reported.[4] The results were disappointing, with 1,2-addition or proton transfer successfully competing with conjugate addition in most of the cases studied. No results were reported for similar reactions with preformed ester enolates.

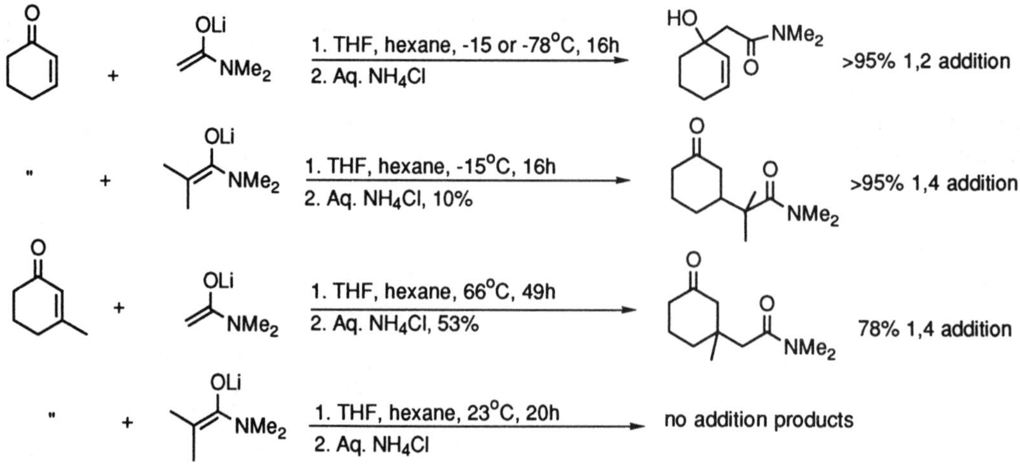

1. THF, hexane, -15 or -78°C, 16h
2. Aq. NH₄Cl

>95% 1,2 addition

1. THF, hexane, -15°C, 16h
2. Aq. NH₄Cl, 10%

>95% 1,4 addition

1. THF, hexane, 66°C, 49h
2. Aq. NH₄Cl, 53%

78% 1,4 addition

1. THF, hexane, 23°C, 20h
2. Aq. NH₄Cl

no addition products

Yamamoto's group has shown that various α-thio esters e.g. (1) add efficiently to cycloalkenones under base catalysis, in the presence of a crown ether.[5] The sulfur substituent is then readily removed, either oxidatively[6] or reductively.[7]

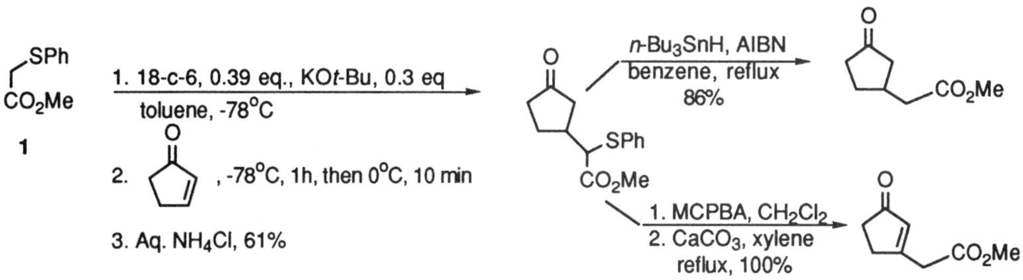

Furthermore, moderate enantioselection (41% e.e.) is obtained if the addition is carried out using the chiral crown ether (derived from an asymmetric cycloaddition of dimenthyl fumarate to anthracene) shown below, instead of 18-crown-6.[8]

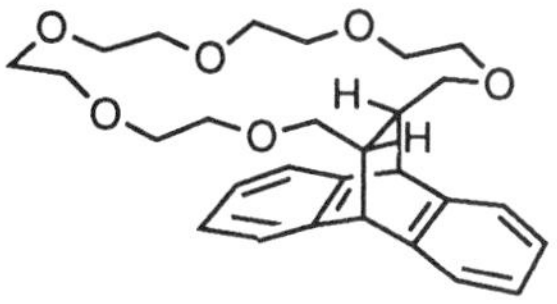

α–Thioacetonitriles, e.g. (2) add to cycloalkenones reversibly. By running the addition initially at low temperature and then allowing the reaction to warm to room temperature, good yields of conjugate adducts may be obtained.[9]

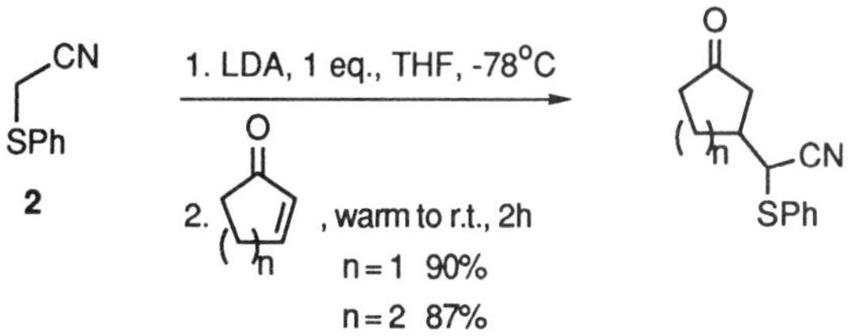

As described in section 2.1.1.1, enethiolates add to 2,3-disubstituted alkenones to provide the *syn* diastereomer as the major product. This stereoselectivity is even greater for similar additions to exo-ethylidenecycloalkenones, e.g. (3).[10]

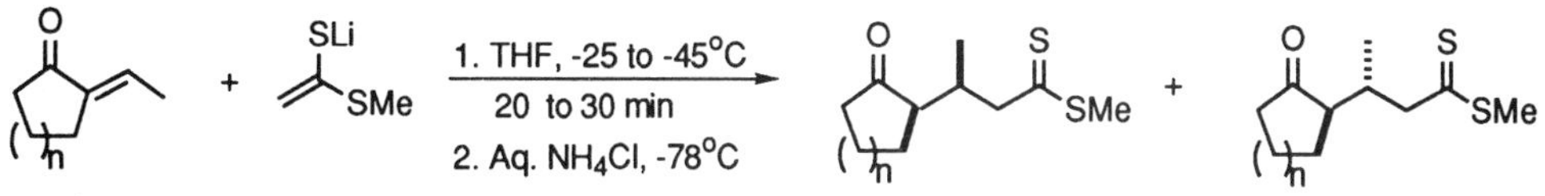

n=1 80%	>95	:	<5
n=2 56%	>95	:	<5
n=3 80%	94	:	6

(c)   *Nitro-stabilized carbanions*

Nitroalkanes add smoothly to cycloalkenones, using a variety of bases, including 1,8-diazabicyclo[5.4.0]undec-7-ene (DBU)[11] and potassium *tert*-butoxide.[12] Phase transfer reagents[13] or basic alumina[14] also efficiently catalyze these additions. One advantage of using a nitro activating group is that it may be chemically manipulated after addition, without disturbing other functionality present in the adduct. For example, Bakuzis et al[15] added ethyl 3-nitropropanoate to 2-cyclopentenone using diisopropylamine in dimethylformamide or, slightly better, potassium *tert*-butoxide in tetrahydrofuran. Under the reaction conditions, elimination occurred, providing, in one pot, a method for introducing a β-acrylate anion equivalent.

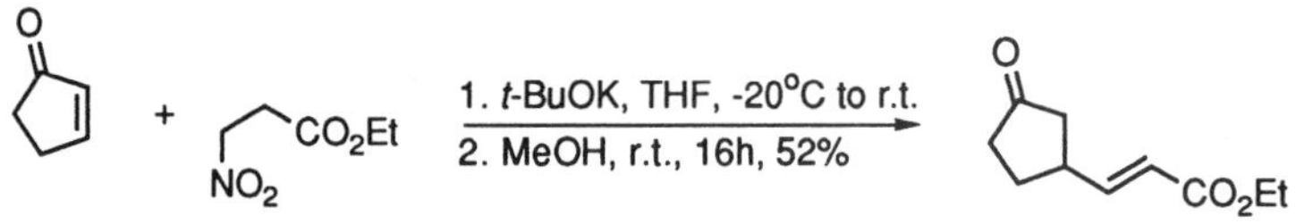

β-Nitroketones do not react in the desired fashion under the above reaction conditions. This problem can be solved by first protecting the ketone as an acetal.[15]

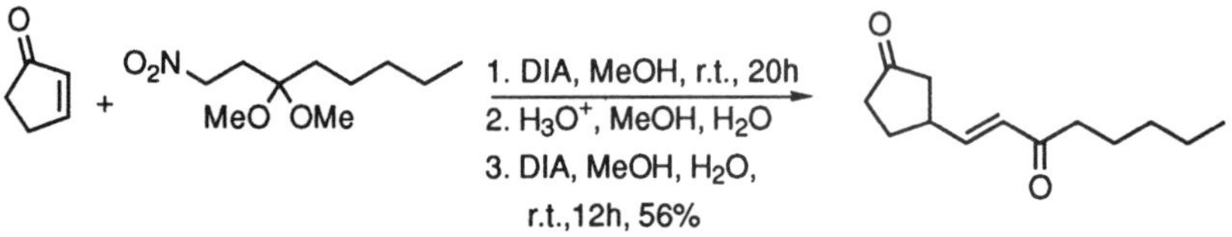

The conjugate addition of nitromethane followed by Nef hydrolysis leads to the overall addition of a formyl anion equivalent.[16]

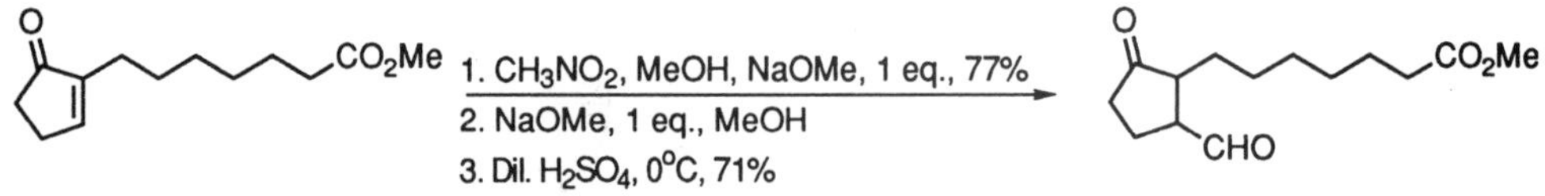

Tandem addition of nitroalkanes to 2-cyclohexenone and a second conjugate acceptor follow a different course to that for other tandem additions. Rather than the second conjugate acceptor quenching the initial enolate, the enolate equilibrates to the nitro-stabilized carbanion which then adds to the second acceptor.[17] For example, PTC conjugate addition of nitroethane to 2-cyclohexenone followed by treatment with methyl propynoate yields, exclusively, the 3-substituted cyclohexanone.

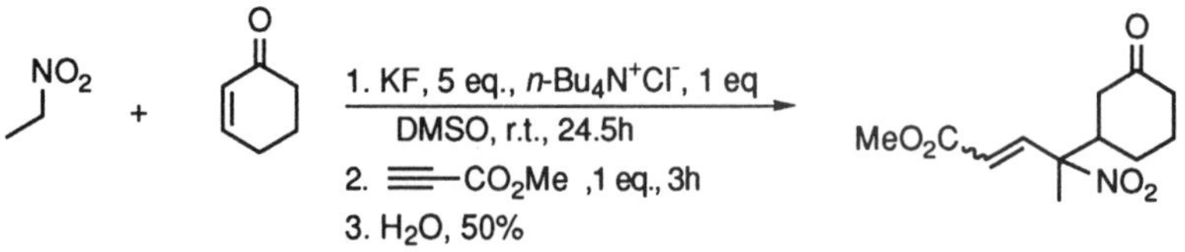

(d)   *Sulfonyl-stabilized carbanions*

Scolastico's group has added the anion of (+)-(4-tolyl) (4-tolyl)thiomethyl sulfoxide to a 2-substituted 2-cyclopentenone. Only 2,3-*trans* products were formed in two pairs, with a ratio of ~11:1.[18] Each pair was isomeric at the carbon adjacent to the sulfoxide group. Removal of this stereocentre by reduction of the sulfoxide produced the homochiral adduct which, in principle, can be used as a prostaglandin synthetic intermediate. Similar addition to 2-cyclopentenone itself resulted in much lower selectivity.

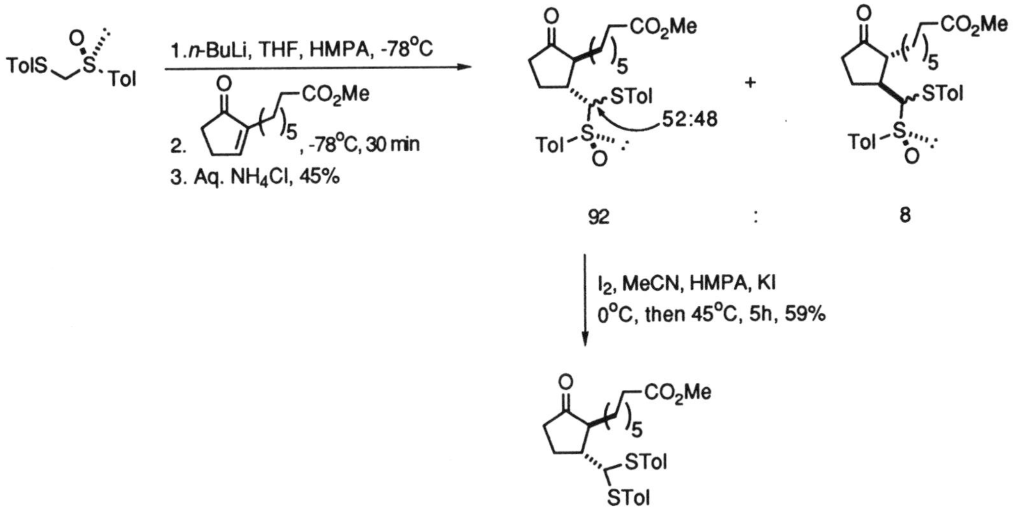

A two-step annulation sequence has been introduced by Ghosez's group.[19] The first step involves the conjugate addition of a sulfonyl-stabilized carbanion to a cycloalkenone followed by trapping with chlorotrimethylsilane. Subsequent treatment with trimethylsilyl trifluoromethanesulfonate yields the cyclized product as a single diastereomer in most cases. The procedure also offers the advantage of providing the two carbonyls in functionally differentiated form.

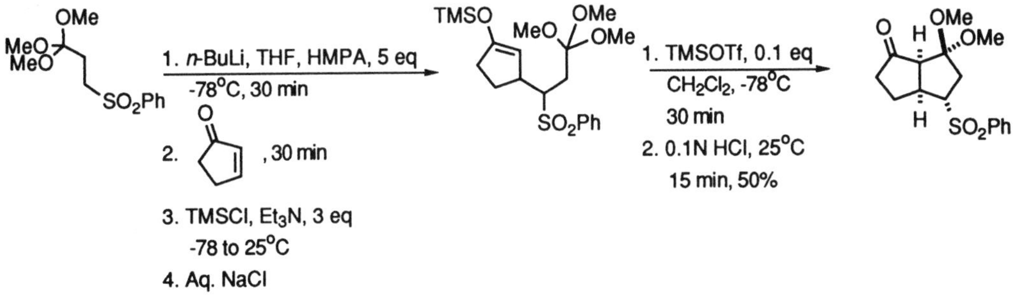

Haynes et al[20] have shown that allyllithium reagents which are also stabilized by sulfoxides[21], phosphine oxides or phosphonate groups add to cycloalkenones. Interestingly, these reagents all add, under kinetically-controlled conditions, at the γ-carbon of the nucleophile. This is to be contrasted with the α-carbon reactivity of allylic sulfones (see discussion below) and allylic sulfides.[22]

As well as being regioselective (1,4-γ), addition of allylic sulfoxides to 4-alkoxy-2-cyclopentenones is highly stereoselective. For example, addition of the lithiated octenyl phenyl sulfoxide shown below gives a single diastereomer.

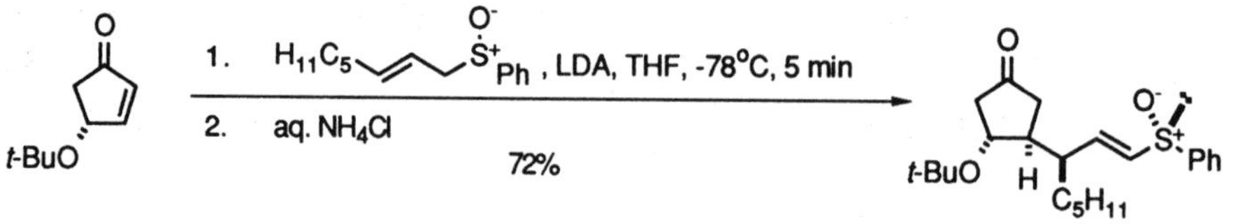

72%

The authors also developed a model which successfully accounted for the stereochemistry of the products by assuming that the lithiated reagent is planar and that the transition state adopts a 10-membered *trans*-fused chair-chair like conformation. The lithium cation, by chelating to the carbonyl oxygen, forms a kind of "anchor" which serves to control the regiochemistry. By assuming a pseudo-equatorial position for the phenyl substituent, this model also correctly predicts the stereochemistry at sulfur.

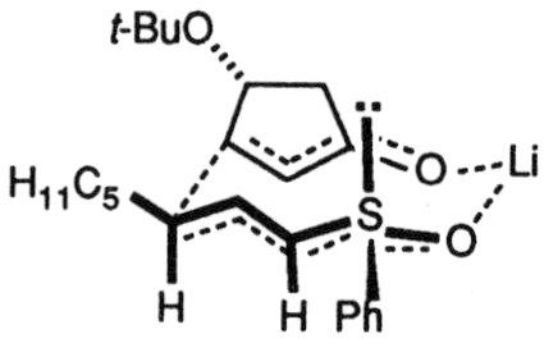

However, such high selectivity is not always the case, as in a similar addition to 2-cyclohexenone.

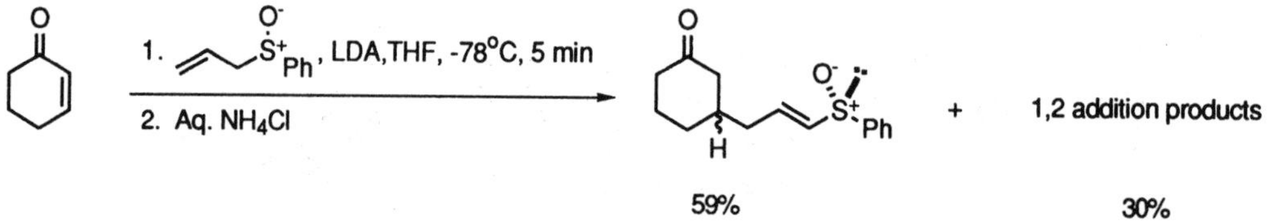

59%   30%

These stabilized allyl carbanions are, of course, ambident in their reactivity. Thus it is possible for the nucleophile to add at the α- or γ-carbon, as well as adding 1,2 or 1,4. Hirama has shown that addition of allylsulfones is exclusively 1,4-α in the presence of hexamethylphosphoramide.[23]

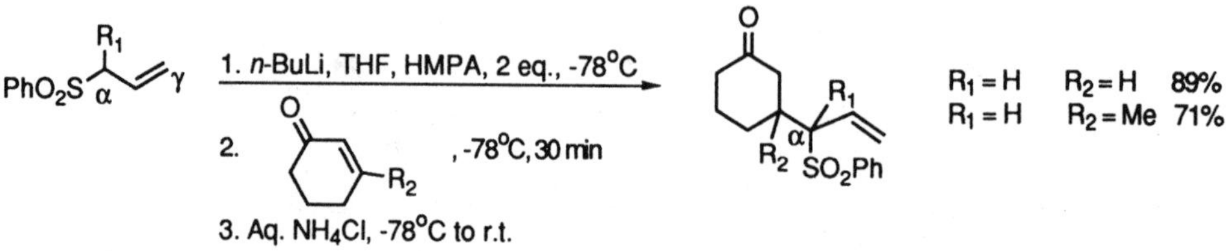

$R_1 = H$	$R_2 = H$	89%
$R_1 = H$	$R_2 = Me$	71%

If the hexamethylphosphoramide is omitted, then 1,2-α addition occurs, followed by rearrangement to the 1,4-γ adduct.

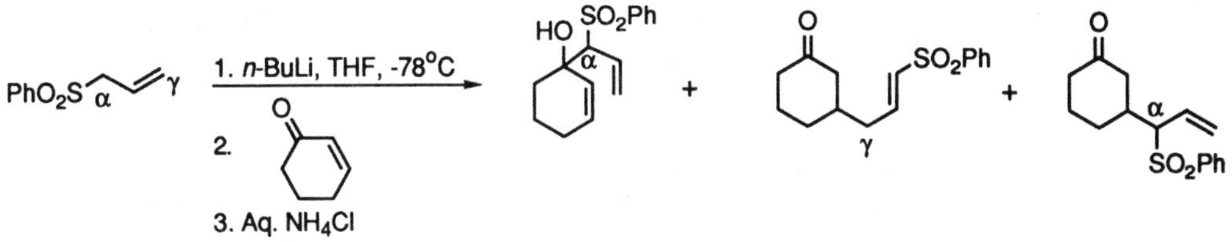

	Yield	1,2-α	1,4-γ	1,4-α
-78°C, 2.5 min	95%	69	22	8
-78°C, 10 min	100%	65	27	8
-78°C, 50 min	89%	43	46	11
1. -78°C,10 min 2. 0°C, 40 min	81%	0	83	17
1. -78°C, 10 min 2. Add HMPA -78°C, 40 min	75%	14	44	42

(e)   *Other stabilized carbanions*

The anion derived from an α-cyanoamine, an acyl anion equivalent, adds smoothly to cycloalkenones.[24]

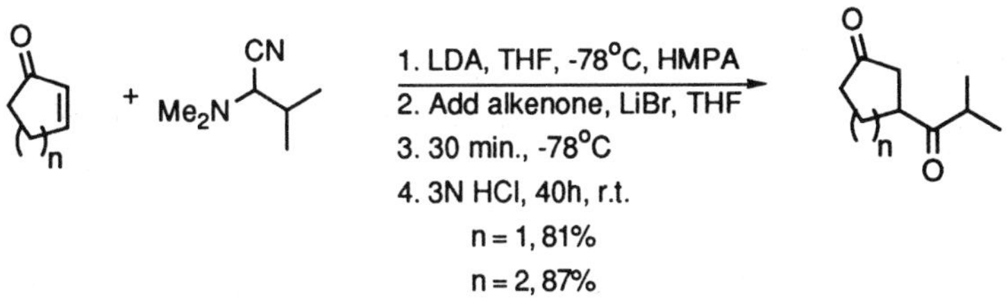

Selected examples of other stabilized carbanions in this class which have been added successfully to cycloalkenones are given below.

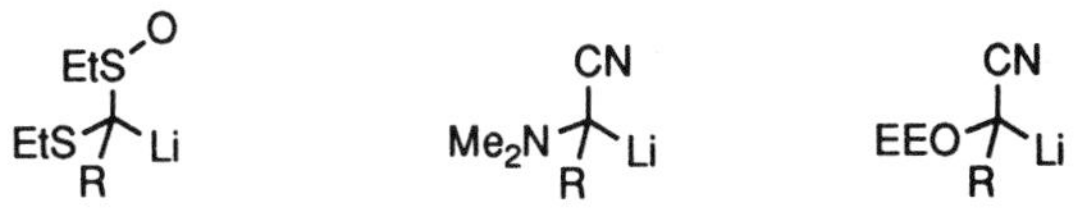

R = Et[25]   R = Ph[26]   R = Ph[27,28]

3.1.1.2   Carbanions stabilized by π-conjugation with more than one heteroatom

(a)   Addition of achiral carbanions to achiral cycloalkenones

The sodium salts (derived from reaction with sodium hydride, a sodium alkoxide or sodium metal) of malonates and their derivatives have proven to be the most popular in Michael additions to cycloalkenones. However, many other bases, such as potassium *tert*-butoxide[29], DABCO[30] and DBU[31] have been used. The addition usually proceeds in good yield. Acyl migration can occur under forcing conditions (Table 3.1, entry 2).

Several tertiary amines have been studied as bases in the Michael addition of ethyl cyanothioacetate to 2-cyclohexenone.[30] DMAP, DABCO, Hünig's base or imidazole all gave yields of around 20%. Pyridine and N,N-dimethylaniline gave no reaction. However, the authors found that addition of *tert*-butyl cyanothioacetate gave good yields when sodium hydride was used as base (see, for example, entry 6 in the Table). These conditions proved successful for Michael additions to a large number of cycloalkenones. The authors also compared yields of these additions with and without chlorotrimethylsilane being present and found that use of the chlorosilane led to a significant decrease in yields.

For more hindered alkenones, where conjugate addition doesn't normally occur, the use of high pressure conditions can often provide acceptable yields of adducts.[32,33]

Addition/elimination reactions with 3-chloro-4-phenylcyclobutenedione are successful, even with rather hindered malonates.[34] Treatment of the product with strong acid removes both ester groups. Overall, this is a rather simple method for introducing a side chain into these cyclobutenediones.

Table 3.1 Conjugate additions of highly stabilized carbanions to cycloalkenones

1[35]	+ CO_2Et / CO_2Et	1. NaOEt, EtOH 2. Add enone, -5°C 3. R.t., 6h 4. AcOH, 90%	

2[36]	+ CN / CO_2Et	1. NaOEt, EtOH
		   Heat, 16h
		2. Cold dil HCl
		   13%	

3[37]	+ CO_2Et / CO_2Et	1. NaOEt, EtOH
		   Heat, 12h
		2. Cold dil HCl
		   85%	

4[38]	+ CO_2Et / CO_2Et	1. NaH, hexane, MeOH 2. Add enone, -5°C, 4h 3. AcOH, 92%	
5[39]	+ CO_2Et	NaH, 0.5 eq benzene, r.t. 64%	
6[40]	+ St-Bu / CN	NaH, 1.5 eq., DME r.t., 17h, 90%	

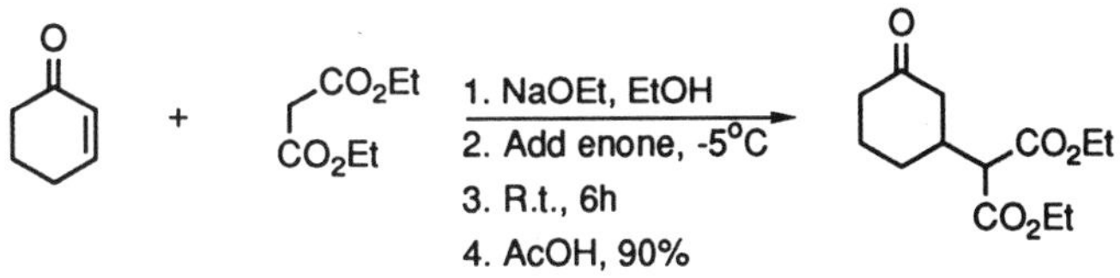

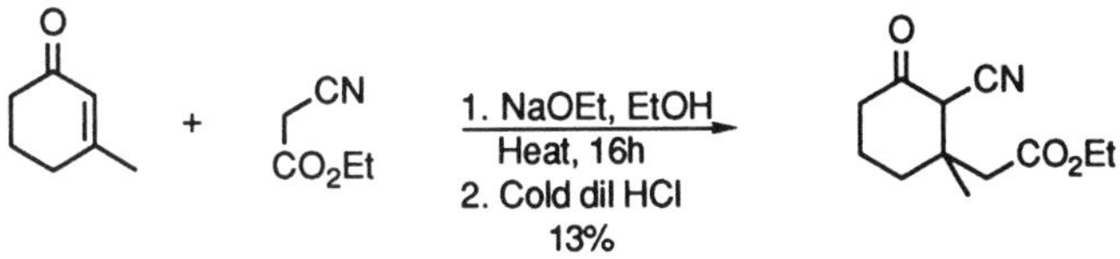

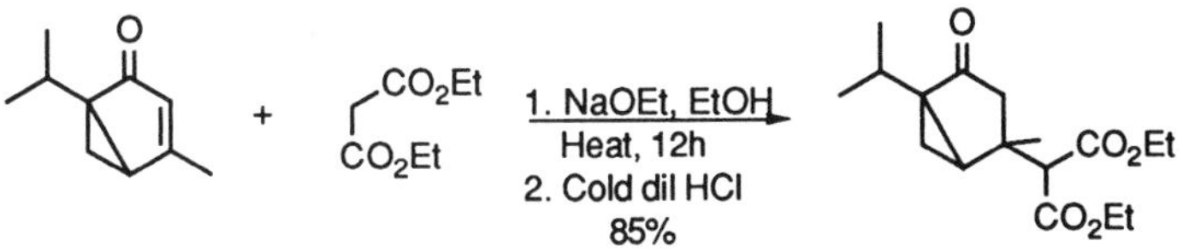

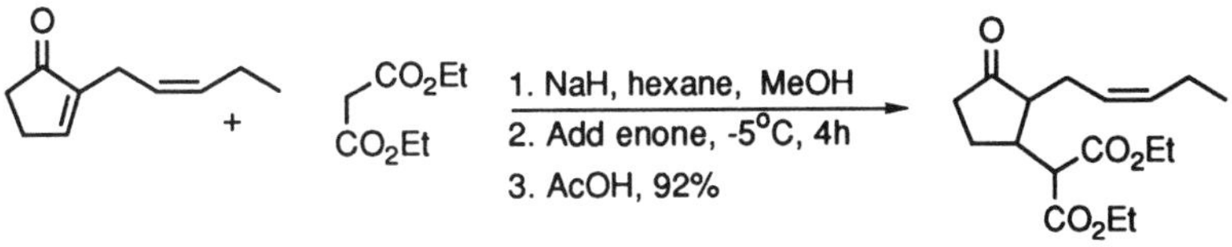

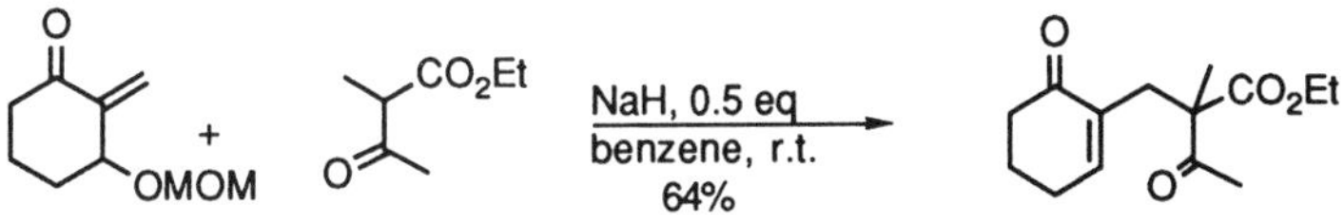

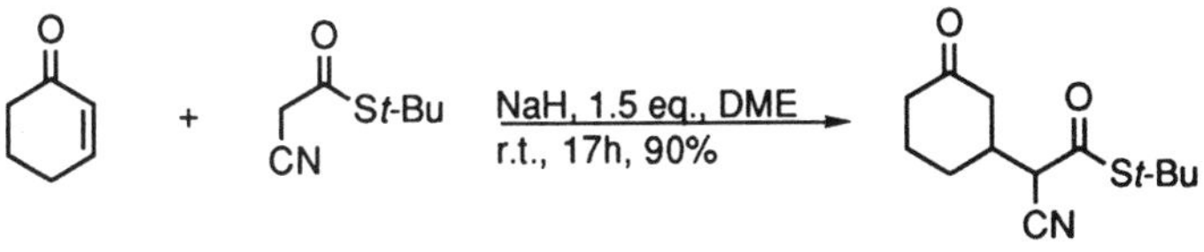

Deslongchamps' group has studied an annulation process based on the addition of the enolate of a typical Nazarov reagent (see section 2.1.1.1) to 2-alkoxycarbonyl-2-cyclohexenones. They concluded that the mechanism of the reaction is probably different in solvents of different polarity. Thus in solvents of low polarity a Diels-Alder mechanism is operating, whereas in more highly polar solvents, sequential double conjugate additions are occurring.[41]

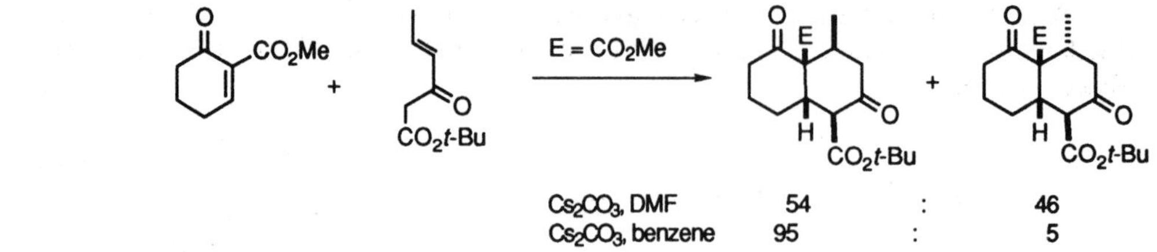

Cs₂CO₃, DMF	54	:	46
Cs₂CO₃, benzene	95	:	5

These studies have culminated in a stereoselective synthesis of the steroid nucleus.[42]

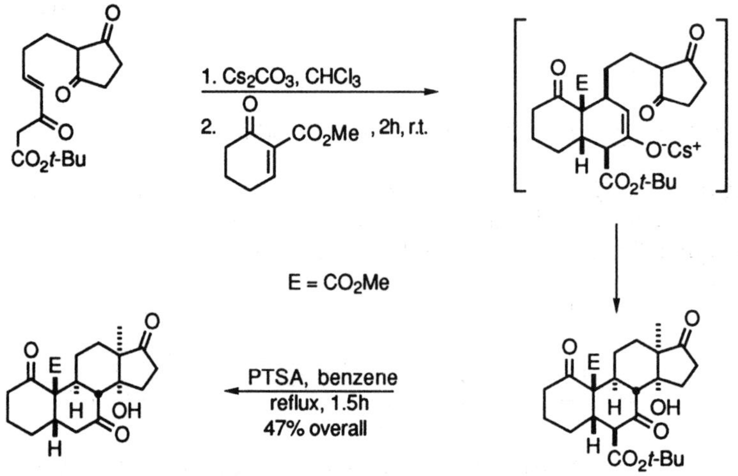

The addition of dimethyl 3-oxoglutarate to 3-methyl-4-methylene-2-cyclohexenone (**4**) provides a simply entry into *cis*-fused decalin-2,7-diones.[43] This probably involves a 1,6-addition followed by an intramolecular 1,4-addition.

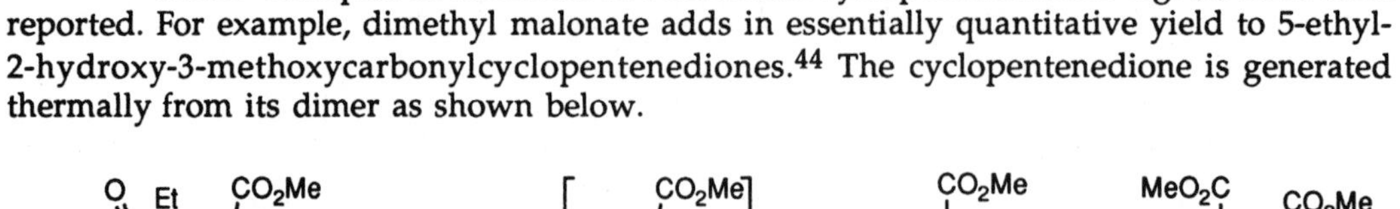

A few examples of additions to substituted cyclopentenediones e.g. (**5**) have been reported. For example, dimethyl malonate adds in essentially quantitative yield to 5-ethyl-2-hydroxy-3-methoxycarbonylcyclopentenediones.[44] The cyclopentenedione is generated thermally from its dimer as shown below.

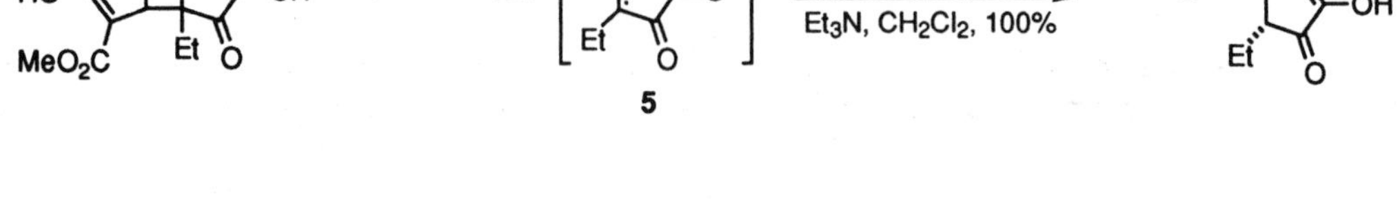

Attachment of a sulfonyl group to acetonitrile assures 1,4 selectivity in additions to cycloalkenones.[45] *In situ* alkylation is also possible. Diastereomeric mixtures are formed in all these reactions.

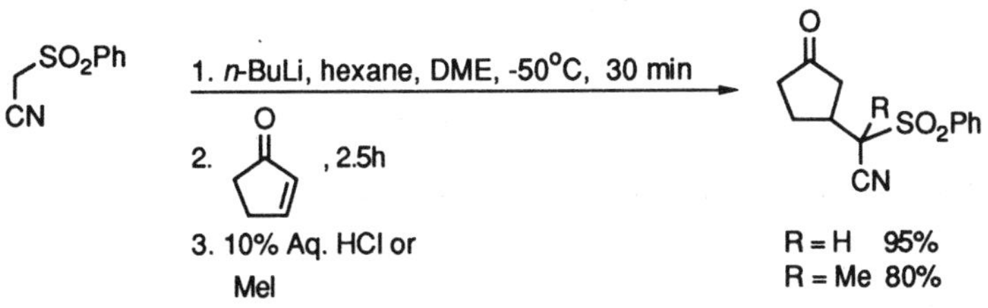
SO2Ph
CN
1. n-BuLi, hexane, DME, -50°C, 30 min
2. O , 2.5h
3. 10% Aq. HCl or
MeI
O
R SO2Ph
CN
R = H 95%
R = Me 80%

(b) Addition of achiral carbanions to chiral cycloalkenones

Conjugate additions to 5- or 6-substituted cyclohexenones are quite stereoselective. For example, addition of the magnesium complex of ethyl hydrogen malonate[46] to 6-methyl-2-cyclohexenone produces a single diastereomer.[47]

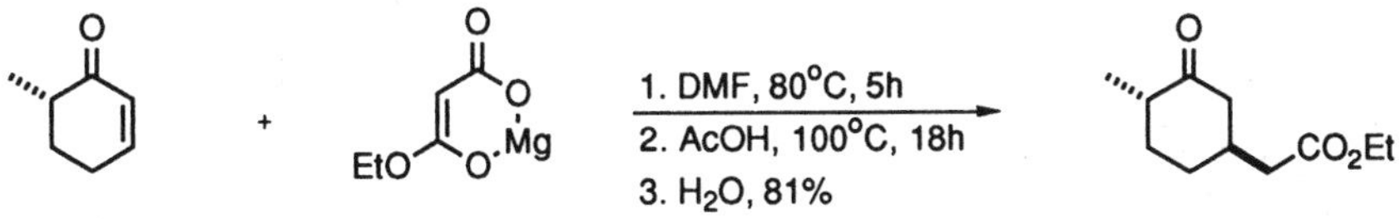
O
+
O
O
EtO
O
Mg
1. DMF, 80°C, 5h
2. AcOH, 100°C, 18h
3. H2O, 81%
O
CO2Et

Addition to 5-trimethylsilyl-2-cyclohexenone is also quite selective.[11]

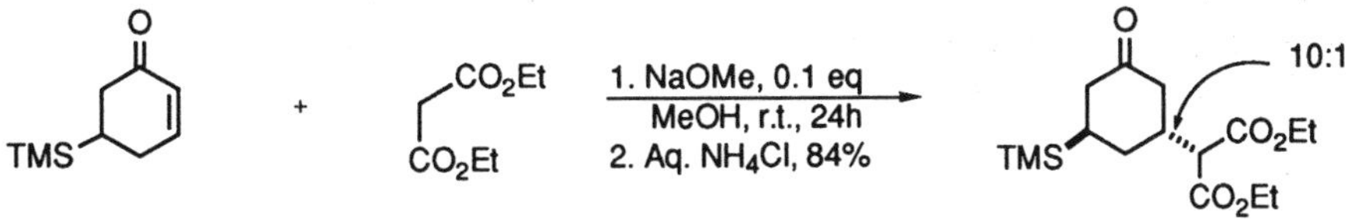
O
TMS
+
CO2Et
CO2Et
1. NaOMe, 0.1 eq
MeOH, r.t., 24h
2. Aq. NH4Cl, 84%
O
TMS
10:1
CO2Et
CO2Et

Addition of diethyl sodiomalonate to 4-methylene-2-cyclohexenone proceeds in an exclusively 1,6 fashion. However, no stereoselectivity was observed at C4.[48]

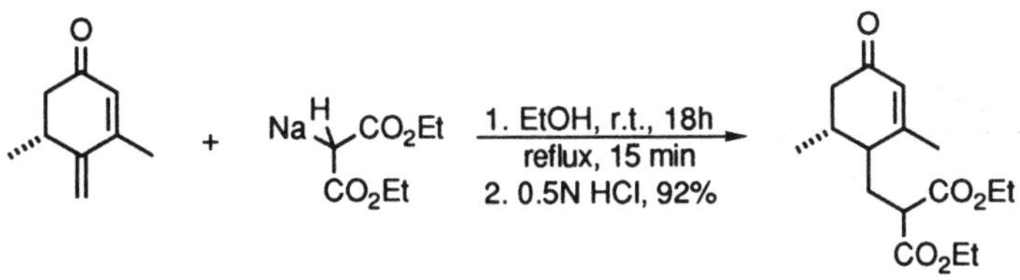
O
+
Na H CO2Et
CO2Et
1. EtOH, r.t., 18h
reflux, 15 min
2. 0.5N HCl, 92%
O
CO2Et
CO2Et

(c) Addition of chiral carbanions to achiral cycloalkenones

As mentioned in 3.1.1.2(a), malonates add to certain cyclopentenediones in high yield. The same group which developed this reaction has extended the study to include the addition of a chiral dioxo-oxazepine (**6**) (prepared from (-)-ephedrine[49]) to a thermally-generated cyclopentenedione.[50] The adduct served as an intermediate in a formal total synthesis of (+)-hirsutene.[51]

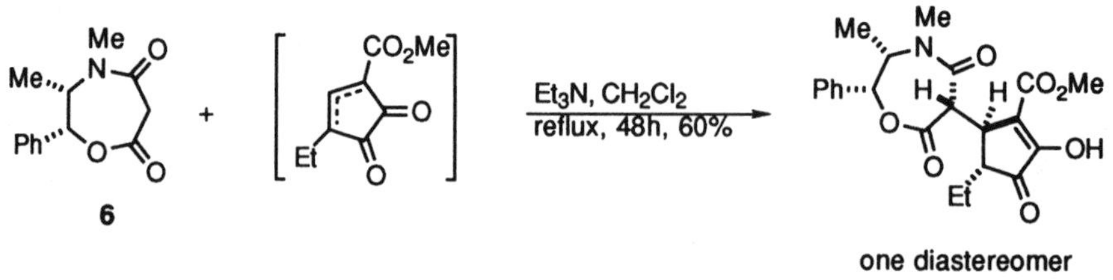

6

one diastereomer

(d)   Addition of chiral carbanions to chiral cycloalkenones

The addition of allyl (4-tolyl) sulfoxide to the fused bicyclic ketone (**7**) shown below gave exclusively *cis*-fused adducts. Some addition-elimination also occurs. This reaction was used in a total synthesis of (±)-pentalenolactone E methyl ester.[52]

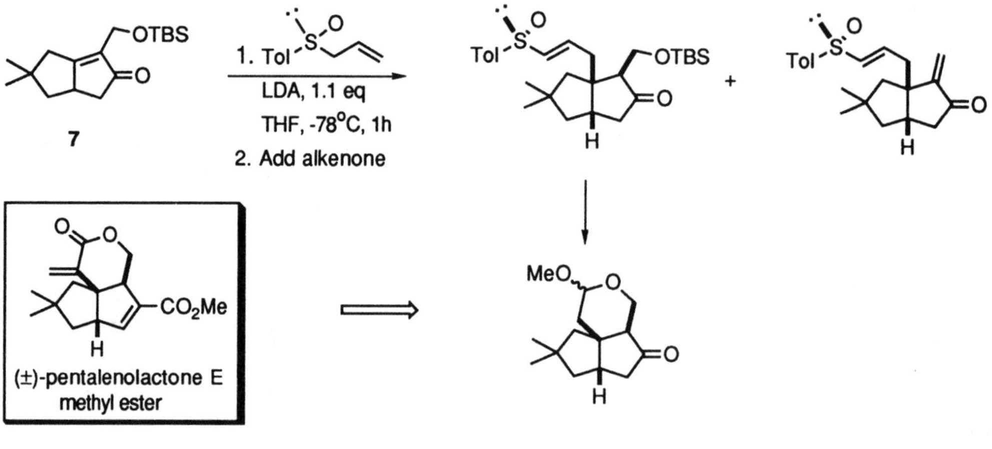

7

1. Tol
   LDA, 1.1 eq
   THF, -78°C, 1h
2. Add alkenone

(±)-pentalenolactone E
methyl ester

3.1.1.3   Carbanions stabilized by   one or more α-heteroatoms

In the late 1970's several papers, seminal to the study of nucleophilic additions of organometallic reagents to cycloalkenones, were published. In an early report on the preparation and reactivity of α-lithio alkyl phenylthio ethers, Dolak and Bryson noted that such anions added in a conjugate fashion to 2-cyclohexenone and 2-methyl-2-cyclopentenone.[53] The conditions they developed for generating these carbanions required the use of hexamethylphosphoramide as co-solvent. The authors commented that these results were somewhat surprising and, in a footnote, suggested that the hexamethylphosphoramide may have been inducing the formation of solvent-separated ion pairs.

Brown and Yamaichi then reported the first study of the (dramatic) influence of hexamethylphosphoramide on regioselection in additions to alkenones.[54] Corey and Crouse had earlier shown that 2-lithio-1,3-dithiane added exclusively to the carbonyl of 2-cyclohexenone.[55] However, if hexamethylphosphoramide were already present in the reaction solution, conjugate addition occurred, almost exclusively. Brown and Yamaichi were also able to demonstrate that these were kinetic additions. That is, if hexamethylphosphoramide were added to the lithium salt of the 1,2-adduct in tetrahydrofuran, no equilibration (to the 1,4-adduct) took place, even after 24 hours.

1. THF, hexane (4:1), -78°C
2. Aq. NH$_4$Cl

n-BuLi, THF
-78 to -20°C, 1h

no HMPA			
n = 1,2 or 3	<1	:	>99
HMPA, 2 eq.,			
n = 0	98	:	2
n = 1	95	:	5
n = 2	74	:	26

The fact that the hexamethylphosphoramide was not equilibrating the 1,2-adduct was significant as Ostrowski and Kane had just shown that similar low temperature additions with 2-phenyl-2-lithio-1,3-dithiane were reversible, simply by warming the reaction mixtures to room temperature.[56] However, the 2-phenyl derivative may be viewed as a π-conjugated carbanion and such highly stabilized carbanions often show reversible behaviour in addition reactions to conjugate acceptors. Tris(phenylthio)methyllithium, a masked carboxylate *carb*anion, adds cleanly in a conjugate fashion to unhindered cycloalkenones. The adducts may be solvolyzed to the corresponding ester or reduced to the methyl derivative.[57,58]

1. (PhS)$_3$CLi, 1 eq.
THF, -78°C
2. H$_2$O, 95%

1. Hg^{2+}, MeOH
2. H$^+$

95%
CO$_2$Me

Ra-Ni

70%

Alternatively, treatment of the initial adduct with a molar equivalent of *sec*-butyllithium generates a dianion (8) which may be quenched with a variety of electrophiles, such as benzaldehyde.[59,60]

(PhS)$_3$CLi, 1 eq
THF, hexane
-78°C, 2.5h

sec-BuLi, 1.05 eq.
-45°C, 5h

1. PhCHO
2. H$_2$O
71% overall

8

Before the value of adding hexamethylphosphoramide was recognized, indirect methods to achieve conjugate addition were developed. For example, addition/elimination to 3-*iso*-butoxy-2-cyclohexenone, followed by reduction and hydrolysis achieves the 1,4 addition of an acyl group.[55]

A much more recent alternative involves equilibrating 1,2-adducts using strong base. However, the yields are low.[61] Some α-heteroatom stabilized carbanions which have been used in conjugate additions to cycloalkenones are listed below.[62]

R = *iso*-Pr[63]   R = H,[64] R = Me[65]   R = H[66]   R = PhS[67]   R[68] = TMS
                     R = Ar,[69] R = metal[70]   R = Me[65]   R = alkyl[71]

3.1.2   Organocopper reagents

3.1.2.1   *Addition of achiral organocopper reagents to achiral conjugate acceptors*

A variety of organocopper reagents add to cycloalkenones. In fact, cyclohexenones are often used as "test cases" for new reagents (see Table 3.2). As with acyclic alkenones (section 2.1.2), cycloalkenones react well with Seyferth's "acylcopper" reagents.[72]

A solution to the problem of transferring an allyl group to a variety of electrophiles has proved one of the more elusive ones in organocopper chemistry. However, Lipshutz's group now believe they have found such a solution. A systematic study led them to the use of a combination of a mono-organocopper in conjunction with chlorotrimethylsilane.[73]

As is evident from the two examples above, chlorotrimethylsilane is an important additive in these additions. Regarding the reactions of LO cuprates, Alexakis has pointed out that 3-alkoxy-2-cyclohexenone is completely unreactive towards dimethylcopperlithium. However, inclusion of chlorotrimethylsilane results in almost quantitative conversion to a 2:1 mixture of 1,2 and 1,4 adducts.[74] Wallace found that addition of LO cuprates to 4-chromones was unsuccessful. However, addition to 2-acyl-4-chromones gave useful yields of adducts.[75]

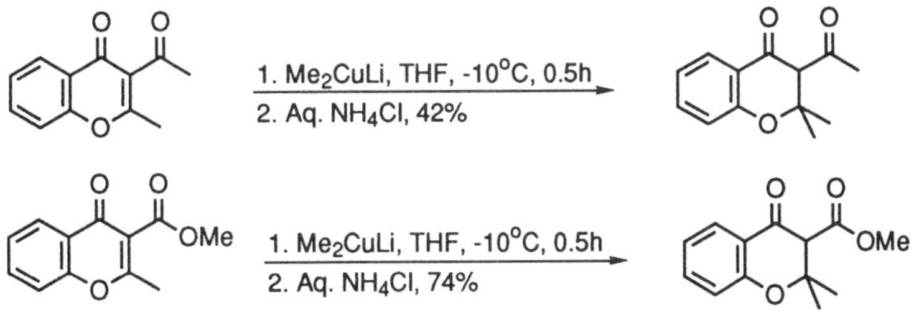

α-Cuprated dialkylhydrazones add to cycloalkenones in high yields.[76] 1,5-Dicarbonyl compounds, as in the example below, can be formed from the conjugate addition of a metallated hydrazone to 2-cyclohexenone in good yield after oxidative hydrolysis.

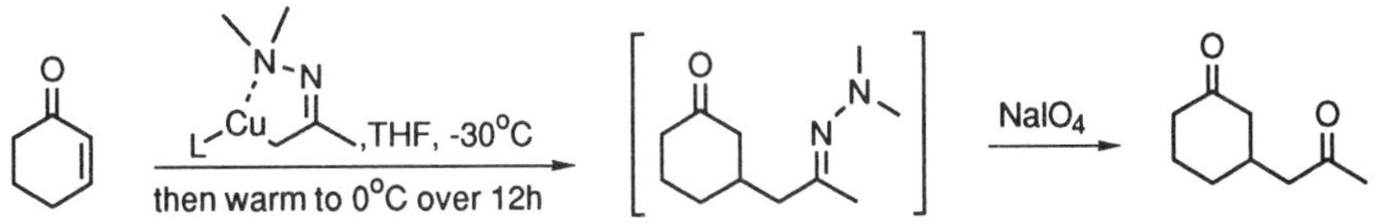

1,6-Addition occurs exclusively in additions to yne-enones.[77]

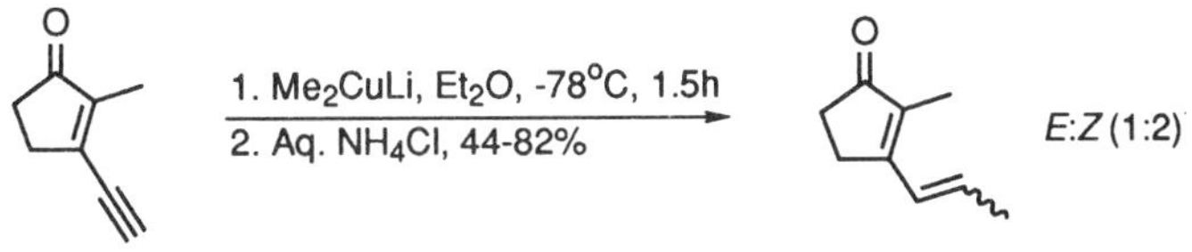

Knochel's group has developed a promising new annulation process which relies upon the use of organocopperzinc reagents.[78] As mentioned above, these reagents are tolerated by many functional groups. In this particular example, an ω-cyanoalkyl ligand is transferred to a 3-substituted 2-cyclohexenone. As before, a Lewis acid was found to be necessary to achieve good yields. The Lewis acid, boron trifluoride, also serves to promote ring closure as shown below. Hydrolysis of the air-stable boron complex (9) then produces the 1,8-dioxodecalin system.

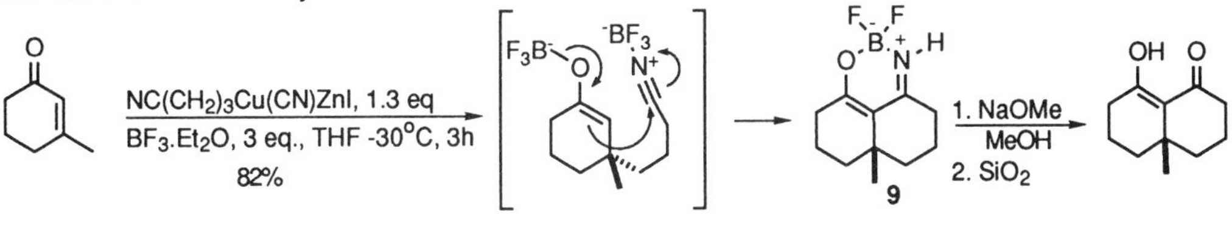

Organozinc reagents are excellent precursors to organocuprates because they are tolerated by a variety of functional groups (Table 3.2, entries 1 to 2).[79] A Lewis acid is generally used in association with the reagent. Copper-catalyzed additions of organomanganese reagents also proceed in excellent yields (entry 4).

Table 3.2   Conjugate addition of organocopper reagents to cycloalkenones

1[78]	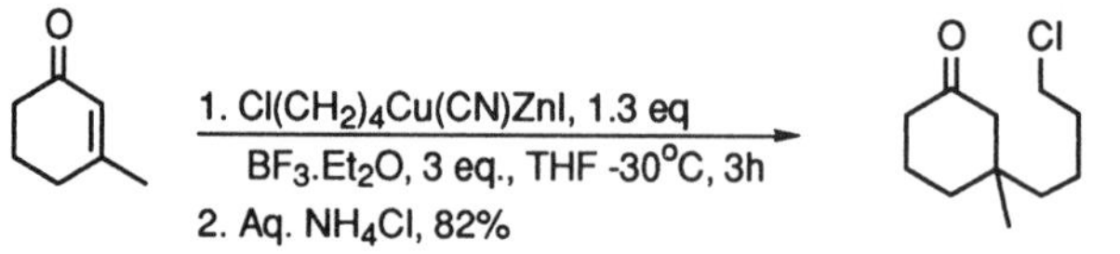
2[78]	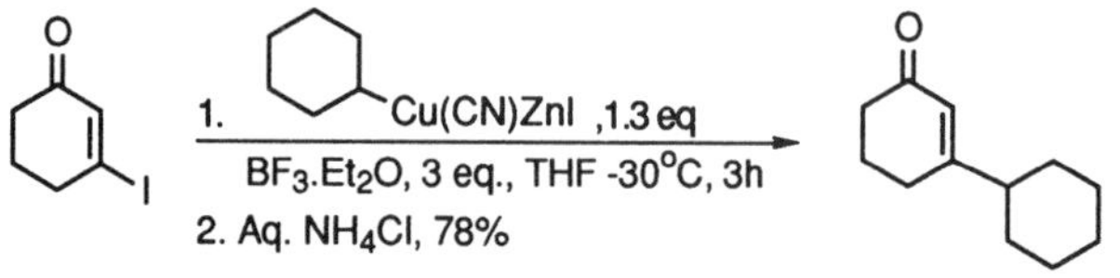
3[80]	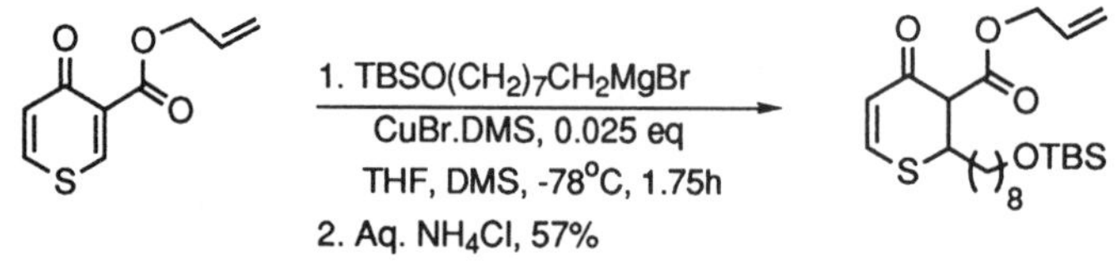
4[81]	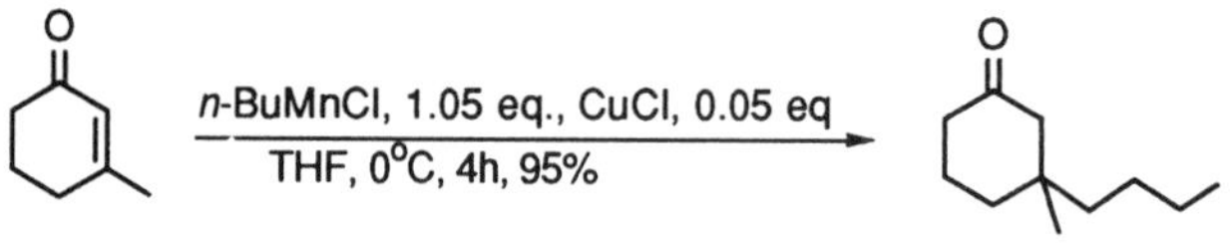
5[82]	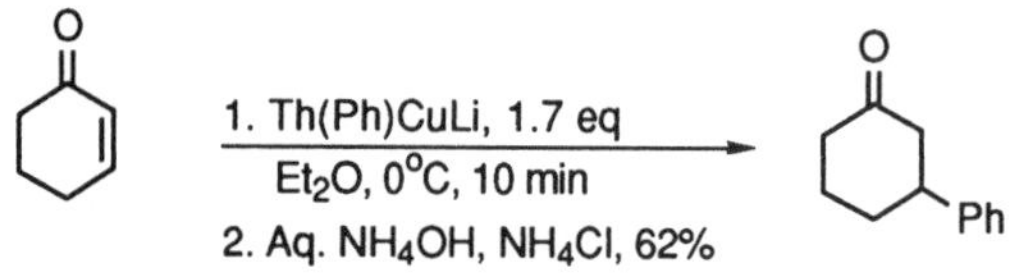
6[83,84]	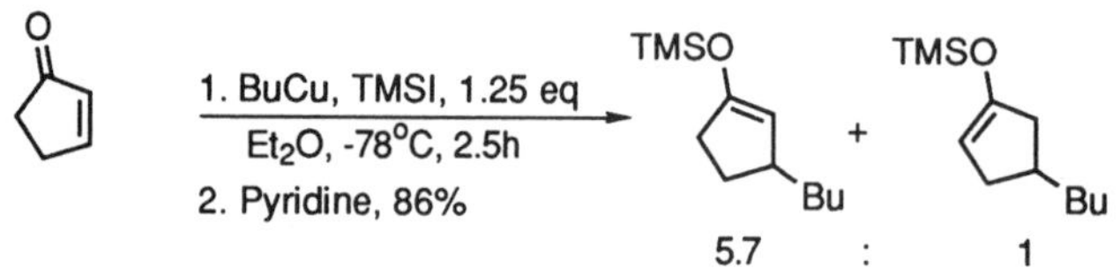

Organocopper additions followed by *in situ* trapping of the enolate intermediate have found many important applications in synthesis.[85] When setting out to use one of these reactions with a new substrate it is advisable to optimize the corresponding conjugate addition/protonation reaction first. Alkylation of the enolate intermediate is very efficient if 1,2-dimethoxyethane is added to the reaction prior to addition of the alkylating agent (Table 3.3, entry 5).

Table 3.3 Tandem organocopper addition/enolate trapping reactions with cycloalkenones

1[86]

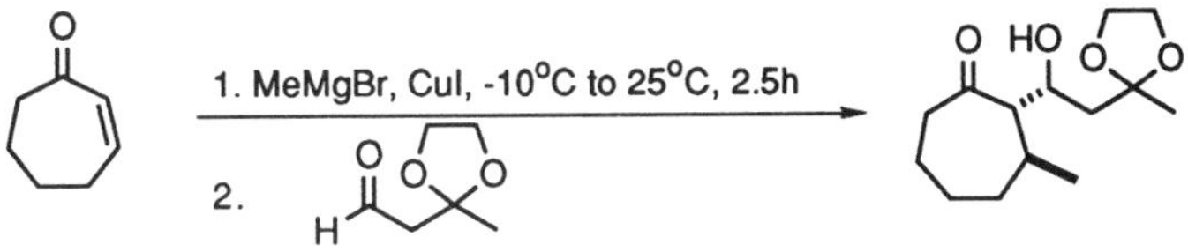
1. MeMgBr, CuI, -10°C to 25°C, 2.5h
2.
HO

2[87,88]

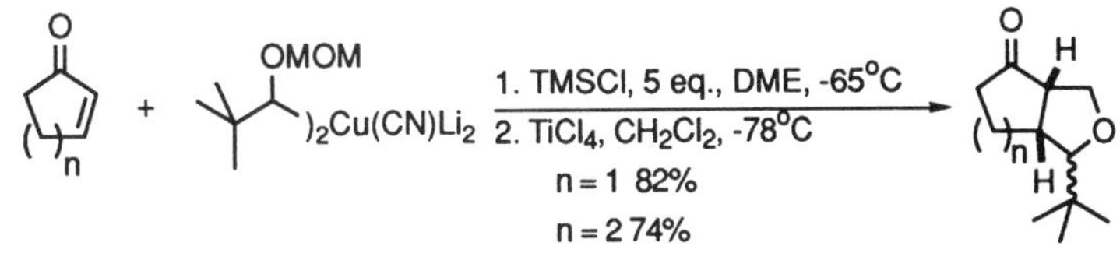
OMOM
)2Cu(CN)Li2
1. TMSCl, 5 eq., DME, -65°C
2. TiCl4, CH2Cl2, -78°C
n = 1 82%
n = 2 74%

3[89]

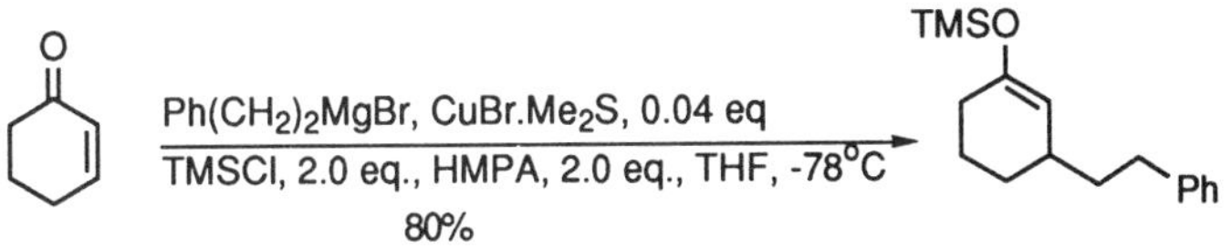
Ph(CH2)2MgBr, CuBr.Me2S, 0.04 eq
TMSCl, 2.0 eq., HMPA, 2.0 eq., THF, -78°C
80%
TMSO
Ph

4[90]

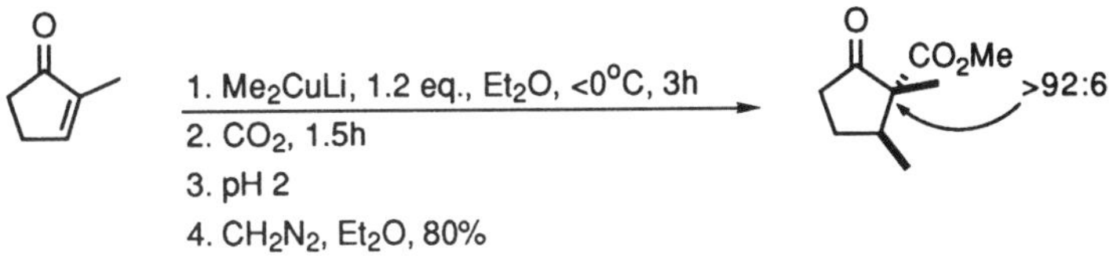
1. Me2CuLi, 1.2 eq., Et2O, <0°C, 3h
2. CO2, 1.5h
3. pH 2
4. CH2N2, Et2O, 80%
CO2Me
>92:6

5[91]

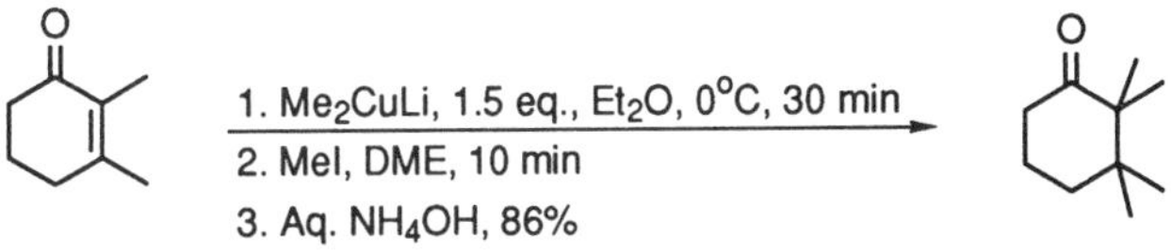
1. Me2CuLi, 1.5 eq., Et2O, 0°C, 30 min
2. MeI, DME, 10 min
3. Aq. NH4OH, 86%

HO (alkoxyalkyl)cuprates react almost quantitatively with cycloalkenones and the adducts may be converted into fused tetrahydrofurans by treatment with a Lewis acid (entry 2). The usefulness of this method is illustrated by a simple synthesis of hop ether.[87]

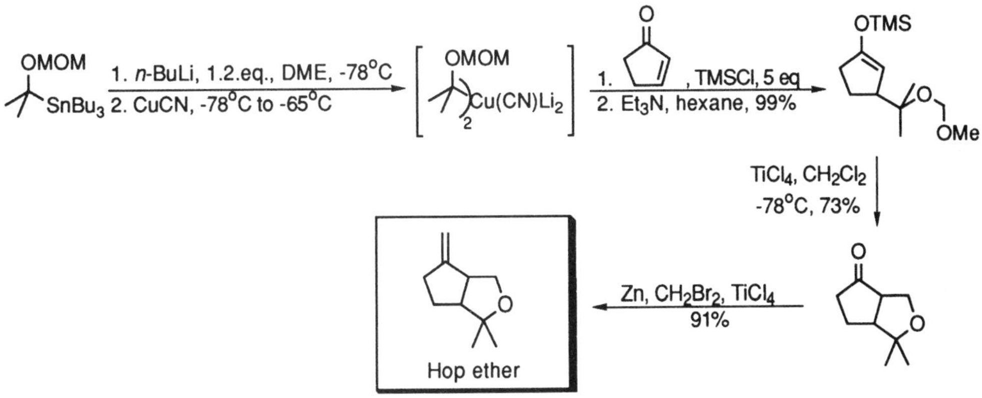
OMOM
SnBu3
1. n-BuLi, 1.2.eq., DME, -78°C
2. CuCN, -78°C to -65°C
OMOM
Cu(CN)Li2
2
1. , TMSCl, 5 eq
2. Et3N, hexane, 99%
OTMS
OMe
TiCl4, CH2Cl2
-78°C, 73%
Zn, CH2Br2, TiCl4
91%
Hop ether

Two procedures, based on LO cuprate addition followed by quenching, have been reported for preparing *cis*-fused bicyclic systems. One involves quenching of the enolate, *in situ* (although a chiral cycloalkenone, it is included here for comparison with the second method).[92] The second regenerates the enolate from an ω-halocycloalkenone (which may be prepared using Knochel's methodology, see Table 3.2, entry 1), which then reacts intramolecularly to close the ring.[93]

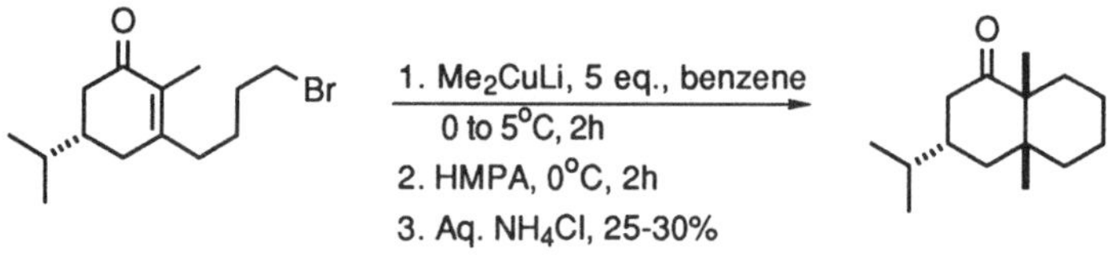

1. Me₂CuLi, 5 eq., benzene
   0 to 5°C, 2h
2. HMPA, 0°C, 2h
3. Aq. NH₄Cl, 25-30%

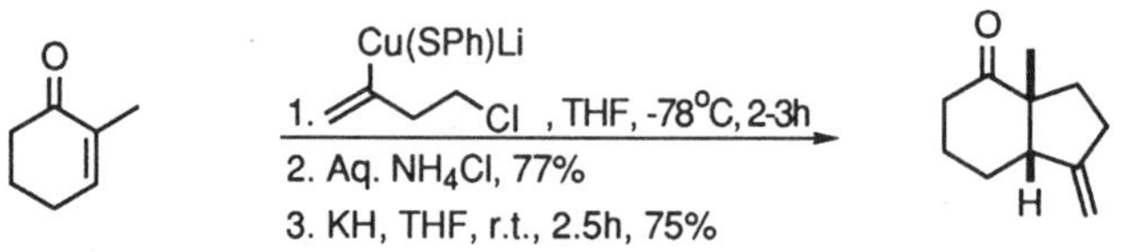

Cu(SPh)Li
1. [structure] Cl , THF, -78°C, 2-3h
2. Aq. NH₄Cl, 77%
3. KH, THF, r.t., 2.5h, 75%

It is often more useful to quench the enolate intermediate with a silylating agent and react the enol ether with an electrophile in a separate step. For example, in Corey's total synthesis of the ginkgolides, one of the spirocyclic moieties was constructed using the following sequence.[94]

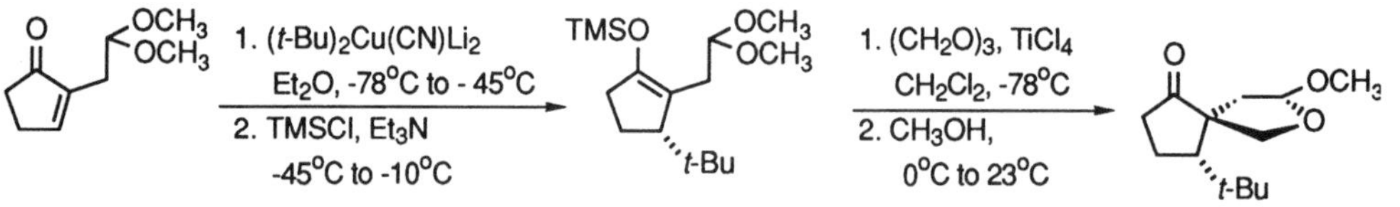

1. (*t*-Bu)₂Cu(CN)Li₂
   Et₂O, -78°C to - 45°C
2. TMSCl, Et₃N
   -45°C to -10°C

1. (CH₂O)₃, TiCl₄
   CH₂Cl₂, -78°C
2. CH₃OH,
   0°C to 23°C

Boeckman and Ganem[95] have devised a procedure for preparing *cis*-4,4a,5,6,7,8-hexahydro-4a,5-dimethyl-2(3H)-naphthalenone (10) an intermediate in syntheses of aristolone[96,97] and fukinone.[98]

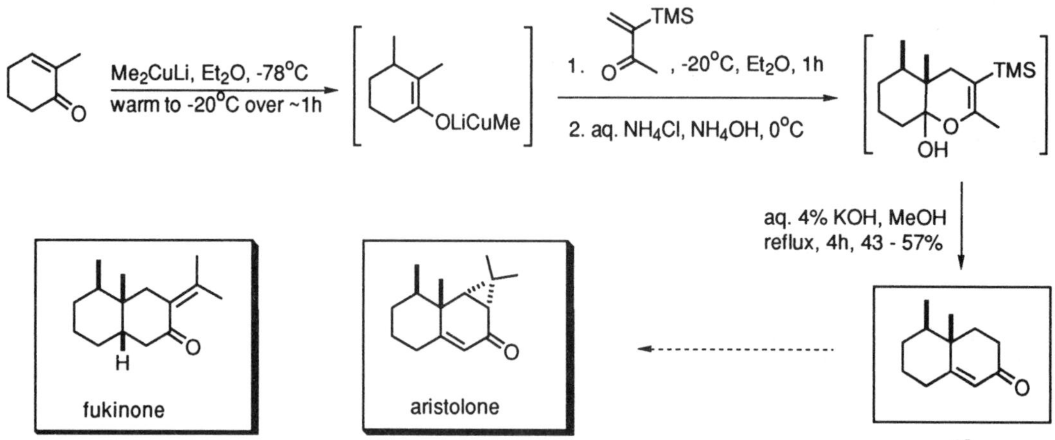

Me₂CuLi, Et₂O, -78°C
warm to -20°C over ~1h

1. [structure] , -20°C, Et₂O, 1h
2. aq. NH₄Cl, NH₄OH, 0°C

aq. 4% KOH, MeOH
reflux, 4h, 43 - 57%

fukinone

aristolone

10

Alkenylcopper reagents

Trisylhydrazones are useful precursors to alkenylcuprates via the Shapiro reaction.[99] The heterocuprate adds efficiently to 2-cyclohexenone (see entry 1 in Table 3.4). Alkenylstannanes undergo a highly efficient transmetallation with the HO cuprate, $Me_2Cu(CN)Li_2$, yielding HO alkenylcuprates (and a tetraalkyltin). These HO alkenylcuprates add to cycloalkenones in excellent yields (e.g. entry 5).[100]

Table 3.4 Conjugate addition of alkenylcopper reagents to achiral cycloalkenones

1[99]	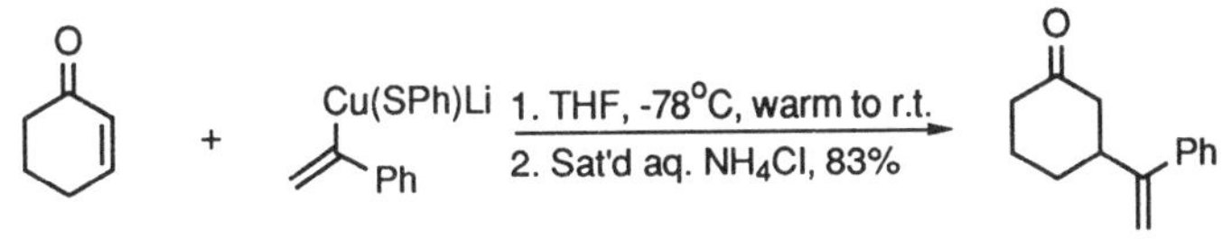
2[74]	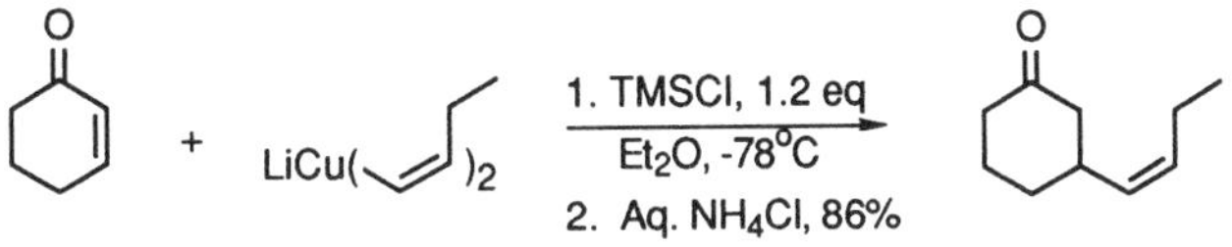
3[101]	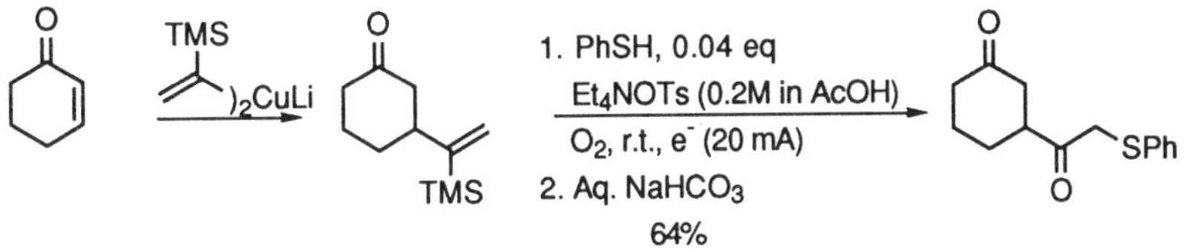
4[102]	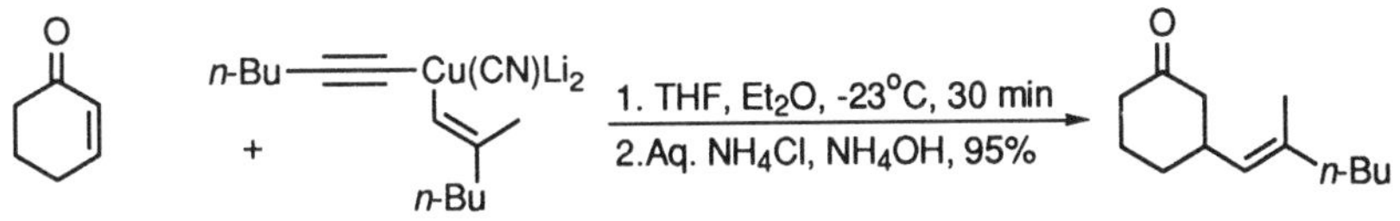
5[100]	

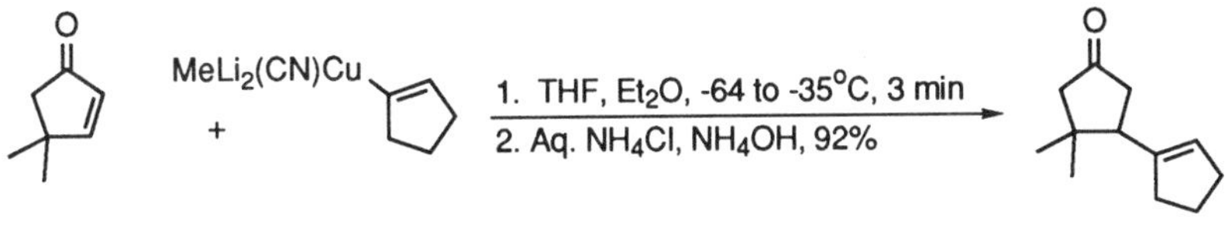

3.1.2.2   Addition of achiral organocopper reagents to chiral cycloalkenones

These additions are dominated by prostaglandin synthesis, which is described below. It is now well-established that high levels of stereocontrol are often possible when a substituent is already present on the ring. Many examples of the use of 4-substituted 2-cyclopentenones exist. Some are given in Table 3.5.

Table 3.5   Stereocontrolled additions of cuprates to 4-substituted 2-cyclopentenones

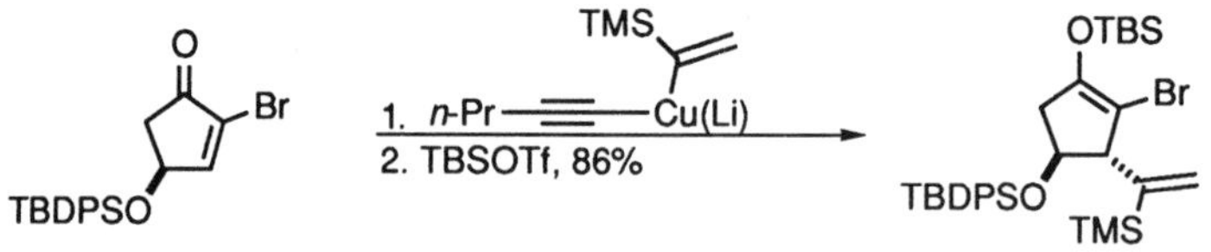

Copper-catalyzed addition of organomagnesium reagents to 3-methyl-5-trimethylsilyl-2-cyclohexenone (**11**) occurs exclusively to the face opposite that of the silyl substituent.[107]

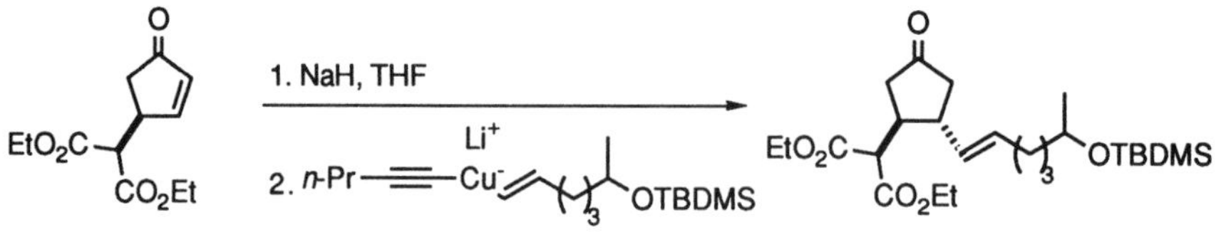

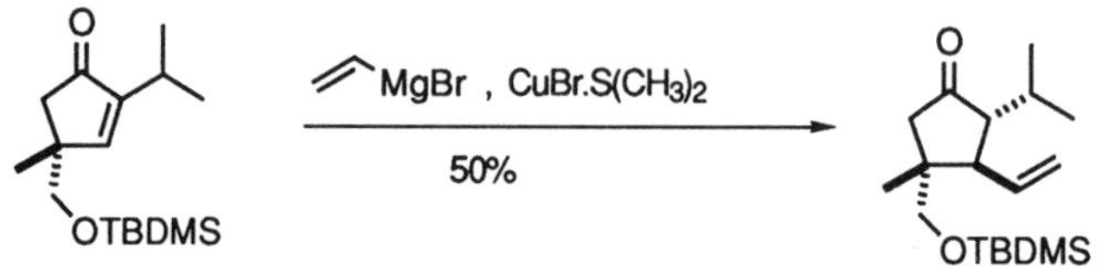

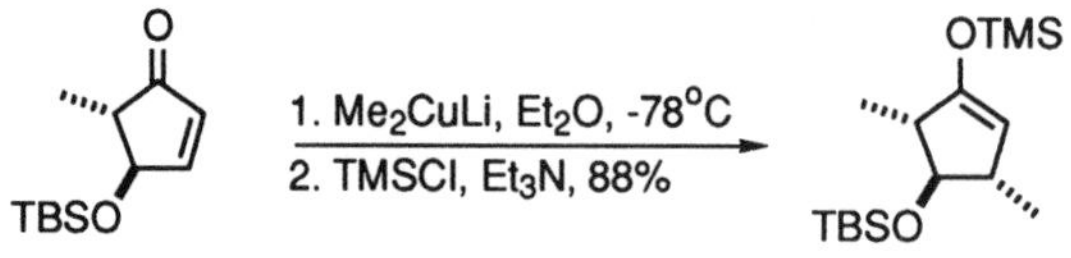

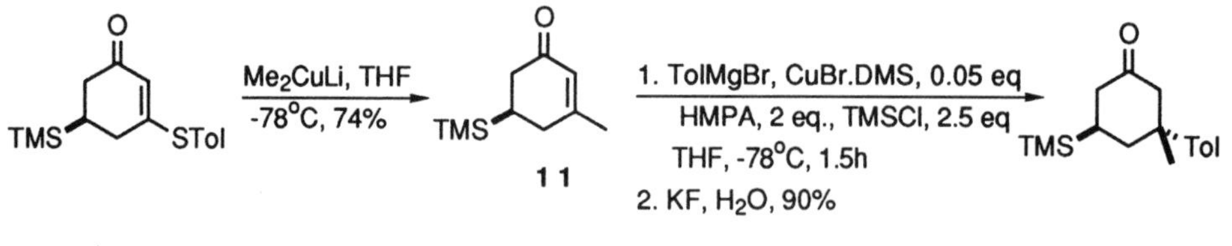

As mentioned in the previous section, LO cuprates add well to 2-acyl-4-chromones. This reaction has been extended to include additions to chiral 3-(toluenesulfinyl)-4-chromones.[108] Some asymmetric induction was obtained.

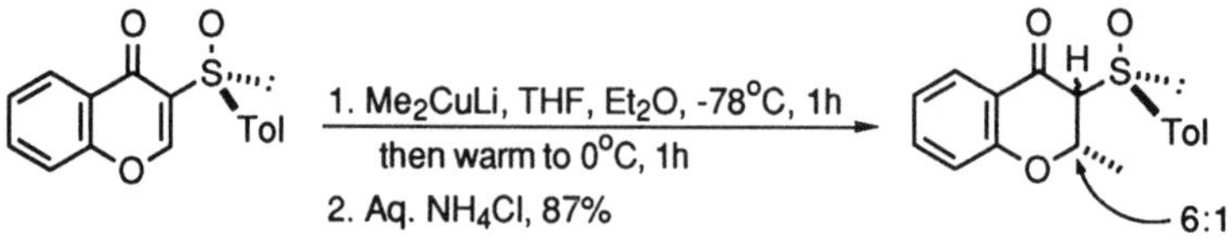

Prostaglandin synthesis

One of the most important applications of conjugate additions of organocuprates to 2-cyclopentenones has been in the area of prostaglandin total synthesis.[109] In particular, the chemistry of *alkenyl*cuprates is dominated by prostanoid synthesis. One of the major synthetic strategies in the total synthesis of prostanoid derivatives involves the conjugate addition of alkenylcuprates to substituted 2-cyclopentenones. Two types of approaches have been developed over the past fifteen or so years:

1. Conjugate addition to 2,4-disubstituted 2-cyclopentenones

2. Conjugate addition to 4-substituted 2-cyclopentenones followed by enolate trapping-the "triply convergent " or "three component coupling" approach

Conjugate addition to 2,4-disubstituted 2-cyclopentenones

This approach was pioneered by Charles Sih's group and began with the preparations of 15-deoxy PGE1[110] and (-)-PGE1[111,112]. The key step in the synthesis of PGE1 involved conjugate addition of a homocuprate to a 2,4-disubstituted 2-cyclopentenone (12).

The 4-alkoxy substituent provided excellent stereocontrol, allowing the incoming nucleophile to approach only from the face opposite to that proximal to the 4-alkoxy substituent. This generates an *anti* relationship between the substituents at the 3- and 4-positions on the ring. Protonation during workup gives the thermodynamically more stable 2,3-*anti*-relationship. The 2,3-*anti*-3,4-*anti* relative stereochemistry is just that found in Nature.

Not surprisingly, it wasn't long before the more economical mixed homocuprates replaced the original homocuprate in these syntheses. Early work seemed to indicate that the Z-alkenylcuprate added more cleanly and efficiently than the E-isomer.[113] However this raised the problem of correction of the geometry at the C13-C14 double bond. An elegant solution to this problem, which exploited Evans's studies on [2,3]-sigmatropic rearrangements of allylic sulfenates,[114] was soon found.[115]

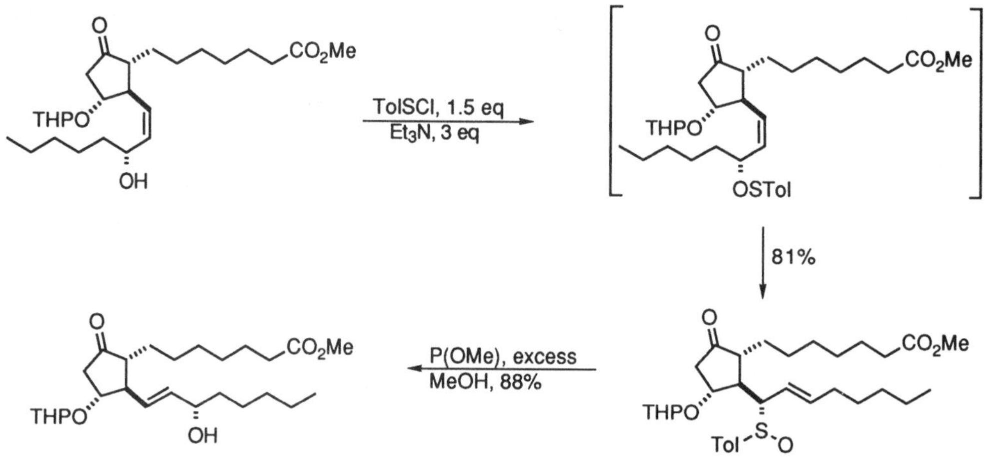

It has since been shown that the *E*-isomer adds well if the required alkenyllithium is generated from the corresponding alkenyltributylstannane.[116]

The "triply convergent " or "three component coupling" approach

A more convergent synthesis of prostaglandins has been developed by Noyori's group and others.[117] This involves the conjugate addition of an alkenylcuprate to a 4-alkoxy-2-cyclopentenone followed by quenching with an appropriate alkylating reagent:

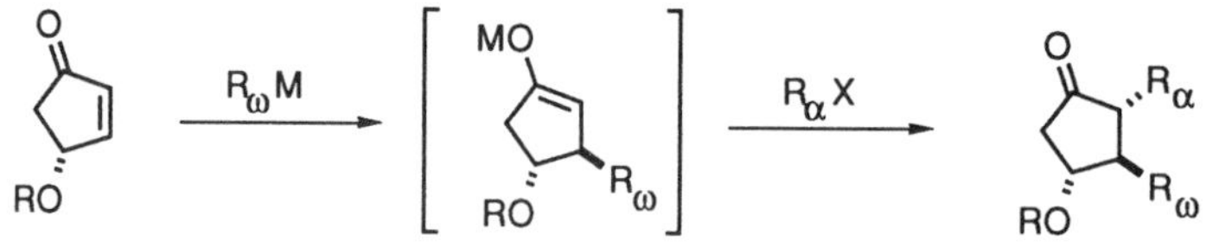

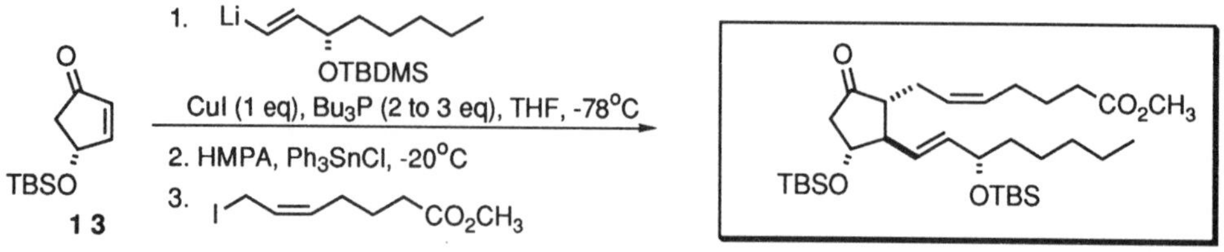

Although this approach appeared to be a potentially efficient and simple one, for several years it was of limited value because the quenching stage (alkylation) of the intermediate enolate suffered from problems of double bond migration, elimination and low yields, often leading to complex product mixtures. These problems have now been overcome using the procedure shown at the bottom of this page. A key element in this modified procedure is the transformation of the intermediate lithium enolate into the corresponding tin enolate in hexamethylphosphoramide. This is sufficiently reactive to effect clean alkylation with either an allylic or propargylic iodide. An added advantage of this method is that enantiomerically pure *R*-4-hydroxy-2-cyclopentenone (the precursor to 4-(*tert*-butyldimethylsilyloxy)-2-cyclopentenone (**13**)) is readily available.[118,119,120,121] This, coupled with the fact that PGE$_2$ may be converted into all of the other prostaglandins, constitutes a flexible, general synthesis of prostaglandins.

An alternative to direct alkylation with R$_\omega$ is quenching of the intermediate enolate with other electrophiles followed by elaboration of the new side chains to give prostaglandins. Examples of electrophiles used include ketene bis(methylthio)acetal monoxide,[122] various aldehydes,[123] carboxylic acid chlorides[124] and chloroformates.[125] An interesting variation to this theme involves a three component coupling using a 4,5-dialkoxy-2-cyclopentenone (14) in place of the monoxy system described above.[126] This system offers the advantage of suppressing any enolate equilibration as well as elimination. Selective deprotection of the of the vicinal diol, followed by selective α deoxygenation, provides the complete prostaglandin system. The dialkoxy-2-cyclopentenone is also available in enantiomerically pure form.

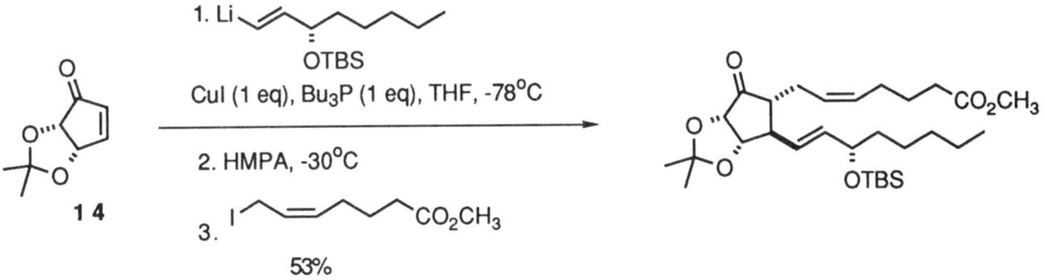

Another recent variation takes advantage of the dual abilities of homochiral pentamethyltricyclo[5.2.1.0^{2,6}]decenone derivatives, such as (15), (i) to completely control the orientation of chemistry at the double bond as well as (ii) to fragment to a 2-cyclopentenone under much milder conditions.[127]

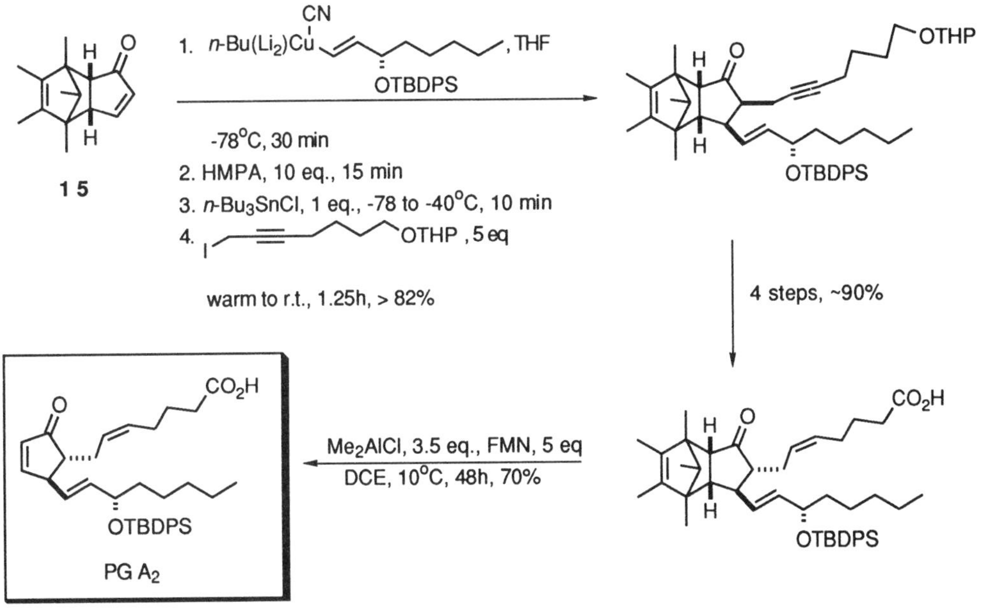

Organocopper reagents add stereoselectively to many other cycloalkenones. Some examples are included in Table 3.6.

Table 3.6   Conjugate additions of organocopper reagents to chiral cycloalkenones

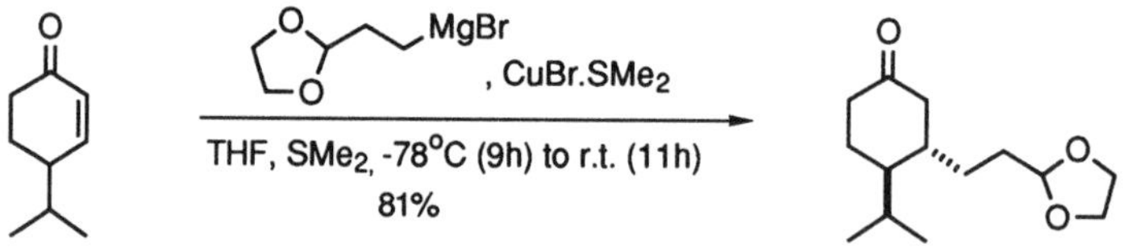

The stereoselectivity in the addition of dimethylcopperlithium to the following spirocycloalkenone is virtually completely reversed if chlorotrimethylsilane is included in the reaction.[132]

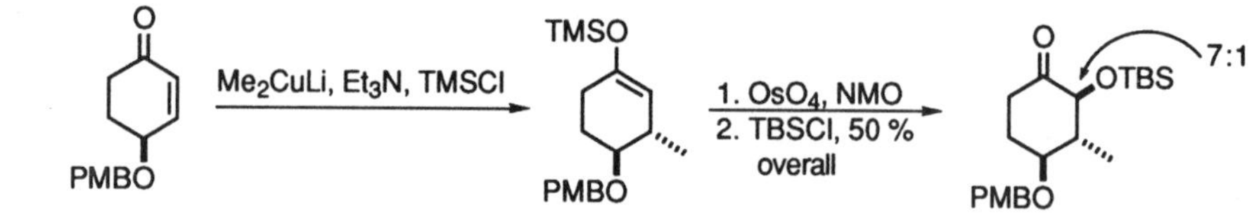

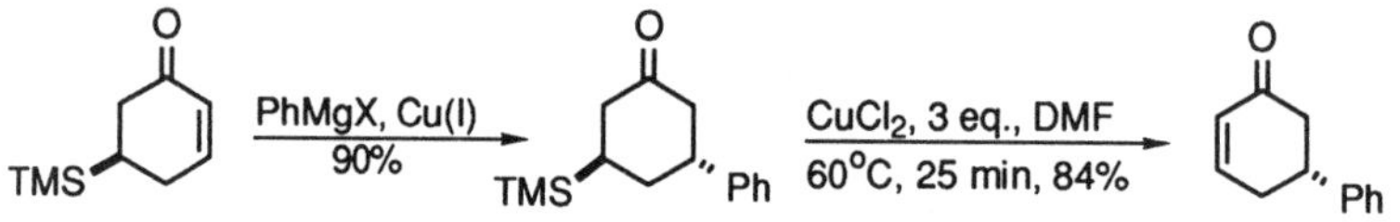

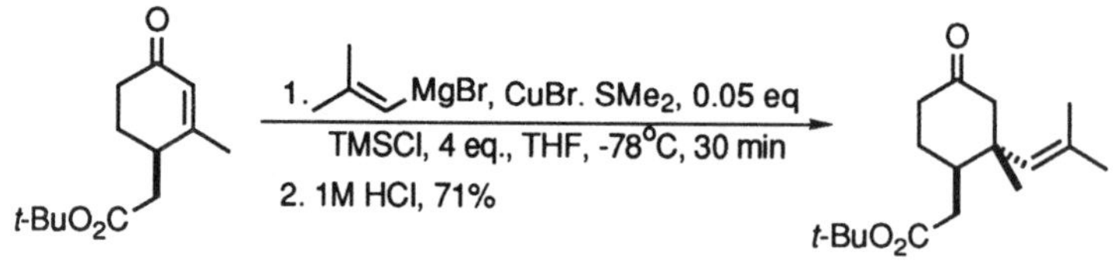

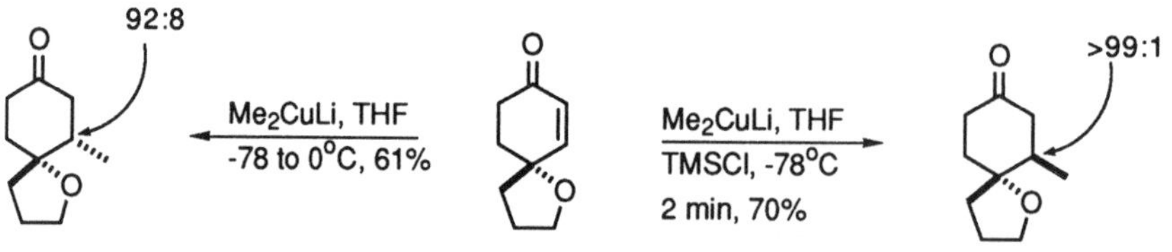

Still and Galynker have examined organocuprate additions to a variety of monosubstituted 8- to 10-membered cycloalkenones[133]. They found that even a single methyl substituent provided a sufficient conformational bias to allow highly stereoselective conjugate additions to be carried out (for a similar study with macrocyclic lactones, see section 5.2).[134] As shown below, addition to the lowest energy conformation(s) of cyclooctenone (determined using MM2 molecular mechanics calculations[135]), accounts for the stereoselection obtained, implying a relatively early transition state. However, in the conjugate addition of lithium dimethylcuprate (in the presence of BF3.Et2O[136]) to either the *E*- or *Z*-9-methyl-2-cyclononenones, the high diastereoselection observed is better accounted for by a relatively late transition state. Finally, conjugate additions to either *E*- or *Z*-10-methylcyclodecenones are also highly diastereoselective. In general, they found that both the yields and the stereoselection were higher if the additions were carried out at -78°C with BF₃ catalysis.

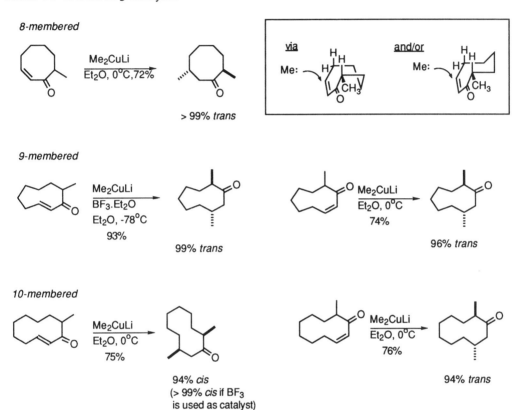

8-membered

Me₂CuLi
Et₂O, 0°C,72%

> 99% trans

via

and/or

9-membered

Me₂CuLi
BF₃.Et₂O
Et₂O, -78°C
93%

99% trans

Me₂CuLi
Et₂O, 0°C
74%

96% trans

10-membered

Me₂CuLi
Et₂O, 0°C
75%

94% cis
(> 99% cis if BF₃
is used as catalyst)

Me₂CuLi
Et₂O, 0°C
76%

94% trans

A synthesis of chiral 3-substituted 2-methylidenecyclohexanones includes as a key element an elimination (really an S$_N$2' type mechanism) of a chiral leaving group.[137]

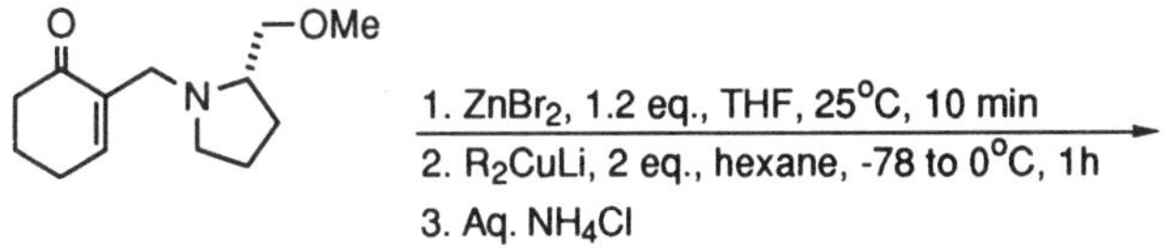

—OMe

1. ZnBr₂, 1.2 eq., THF, 25°C, 10 min
2. R₂CuLi, 2 eq., hexane, -78 to 0°C, 1h
3. Aq. NH₄Cl

R = Me   84%   90% e.e.
R = Et   82%   90% e.e.
R = n-Bu   87%   90% e.e.

The enantioselectivity is consistently around 95% (90% e.e.) and the actual conjugate addition is thought to occur to a metal chelate:

via:

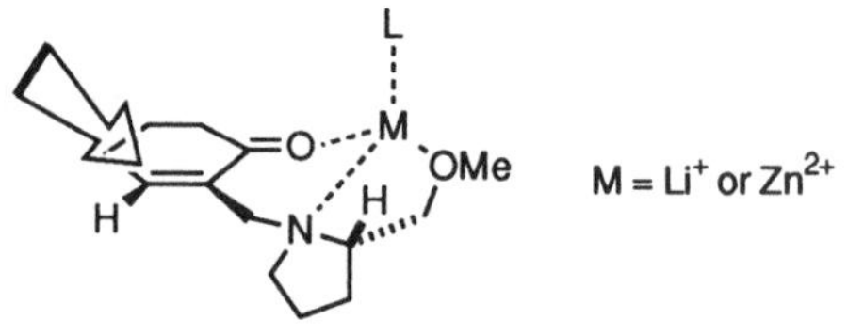

M = Li$^+$ or Zn^{2+}

Polycyclic systems

There are many examples of cuprate additions to polycyclic alkenones where the shape of the system confers a stereoselective bias on the reaction. Some examples are given below. (In the case of the phorbol esters below, it is the protonation of the enolate which generates the new chiral centre).

Forskolin [138]

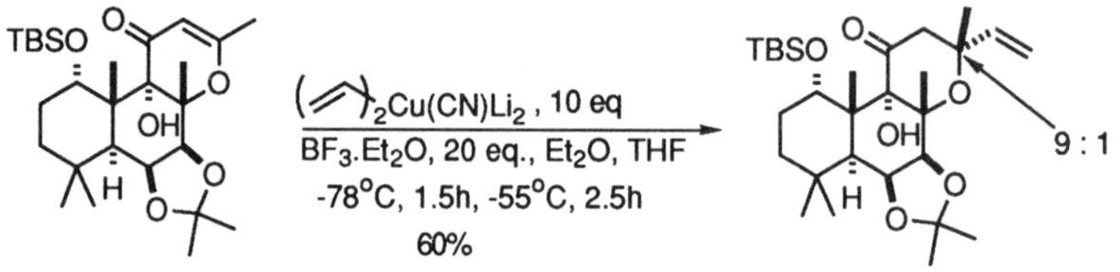

$\left(\diagup\!\diagup\right)_2$Cu(CN)Li$_2$, 10 eq

BF$_3$.Et$_2$O, 20 eq., Et$_2$O, THF

-78°C, 1.5h, -55°C, 2.5h

60%

9 : 1

Phorbol esters[139]

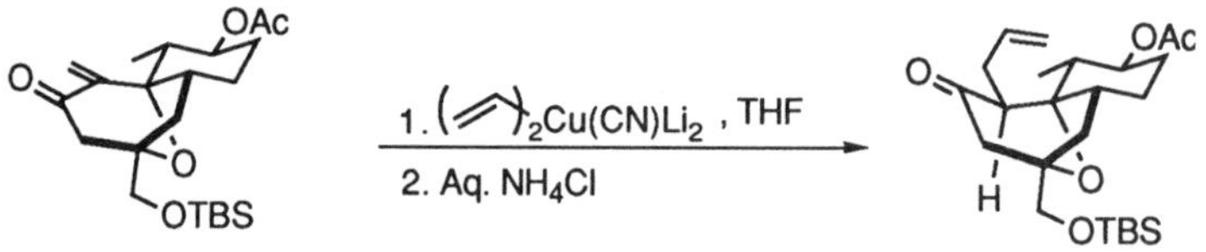

1. $\left(\diagup\!\diagup\right)_2$Cu(CN)Li$_2$, THF

2. Aq. NH$_4$Cl

Erythronolide B seco acid[140]

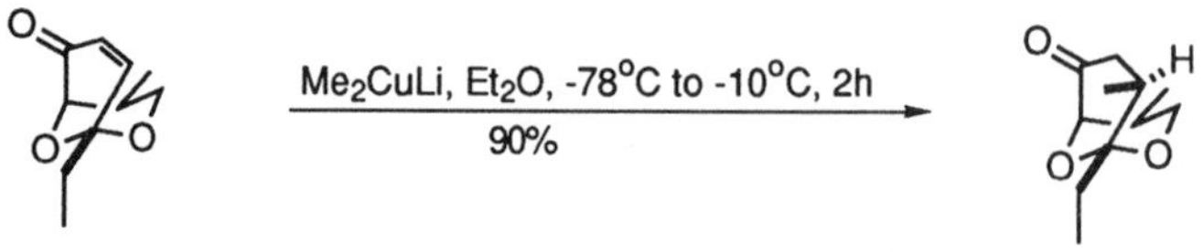

Me$_2$CuLi, Et$_2$O, -78°C to -10°C, 2h

90%

Organocopper reagents often add selectively to bicyclic systems as these systems usually possess a rigid framework which provides steric screening of one alkene face. Entry 1 (Table 3.7) was developed by Heathcock's group as part of a synthesis of vernolepin.

Table 3.7 Addition of organocopper reagents to ring-fused cycloalkenones

1[141]		1.)₂CuLi , 1.5 eq., CuBr.SMe₂, Et₂O THF, DMS, inv addn., -75°C, 45 min 2. HMPA, Et₃N, TMSCl, -40°C 3. 10% HCl, 74%	
2[142]		1. MgBr , CuBr.Me₂S, 0.25 eq DMS, -78°C, 16h, 0°C, 8h 2. Aq. NH₄Cl, pH 8, 72%	
3[143]		1. Me₂CuLi, Et₂O, 1h, 0°C 2. Aq. NH₄Cl, 90%	
4[144]		1. MeMgI, CuBr, cat., Et₂O r.t., 1h then reflux 15 min 2. Dil. HCl, 0°C, 90%	

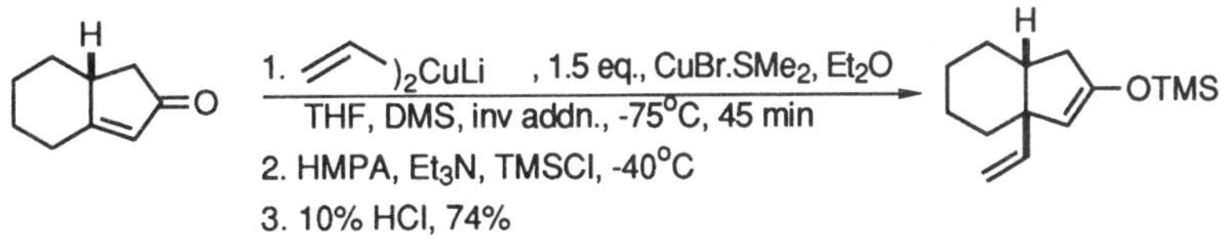

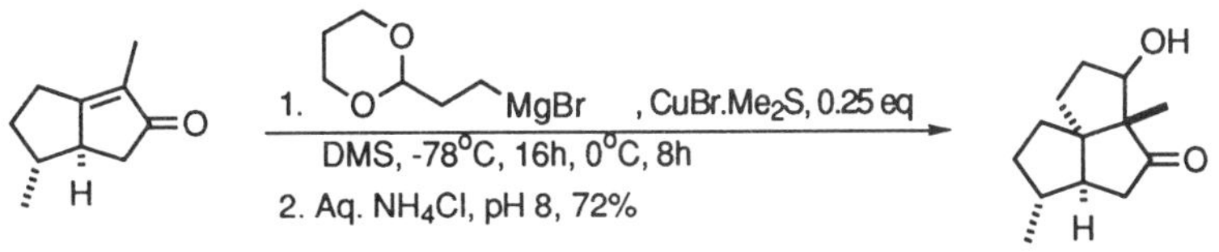

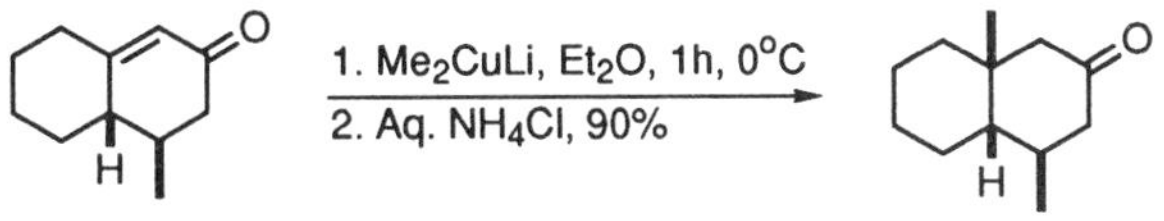

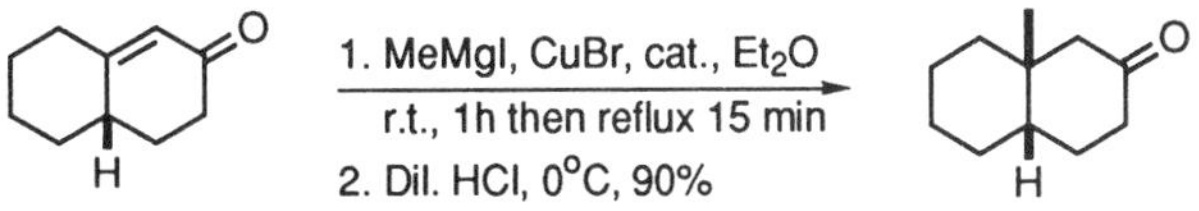

3.1.2.3   Addition of chiral organocopper reagents to achiral conjugate acceptors

Chiral, stoichiometric organocuprates may be formed either by the use of a chiral transfer ligand or a chiral residual ligand.

Chiral residual ligands

Corey's group has designed and studied a chiral ligand based on the commercially available (1R,2S)-(-)-ephedrine.[145]

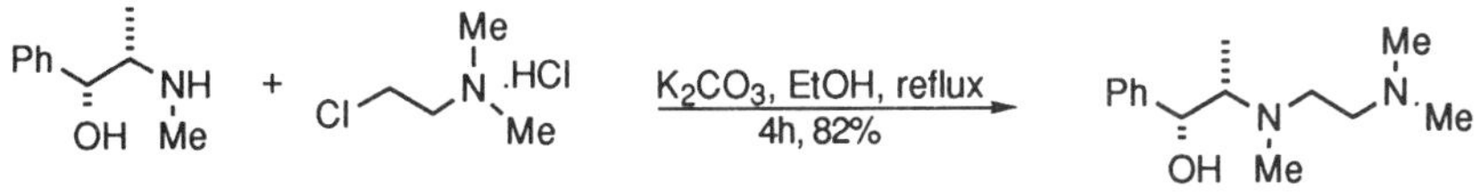

The overall process is that depicted below. The heterocuprate is formed by first deprotonating the ligand with RLi (1 eq.), complexed with cuprous iodide in THF and then treated with an additional equivalent of the organolithium.

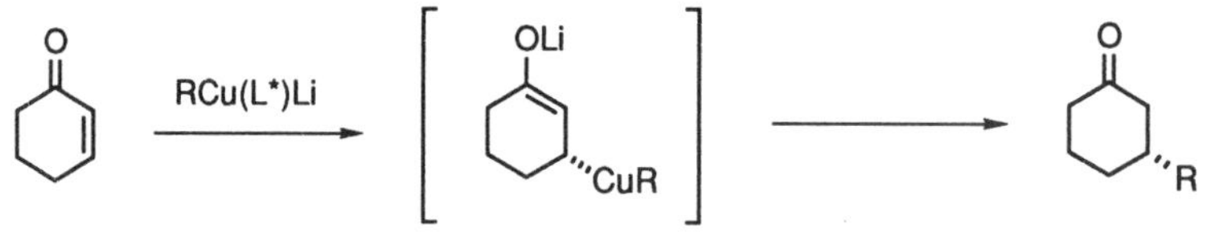

R =	Et	90%	92% e.e.
R =	n-Bu	90%	89% e.e.
R =	t-BuOCH$_2$	73%	85% e.e.

The model developed to account for the good enantioselection in these additions is shown below. The deprotonated diaminoalcohol serves as a tridentate ligand for lithium. This may well be a crucial feature of this system, as this would serve to break up any clusters which might otherwise form, leading to a more homogeneous cuprate species in solution. The authors propose that complexation occurs selectively at the *re*-face of C3 because of the requirement of the carbonyl oxygen to coordinate to a second molecule of "electrophilic" lithium. Complexation at the *si*-face of C3, whilst satisfying this last requirement, would lead to unfavourable steric interactions, although this isn't immediately obvious from the model as drawn. The requirement for this second lithium ion was supported by demonstrating that no reaction took place, at -78°C, until one equivalent of lithium iodide had been added. Conjugate addition then proceeded smoothly.

via:

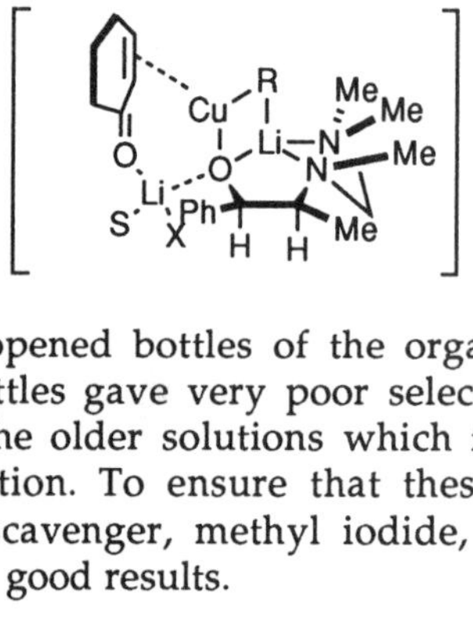

X = I⁻ or THF
S = THF

Only freshly opened bottles of the organolithium reagents could be used as solutions from older bottles gave very poor selectivity. This was thought to be due to alkoxide impurities in the older solutions which formed reactive, non-selective cuprate species during the reaction. To ensure that these alkoxides didn't interfere with the reactions, an alkoxide scavenger, methyl iodide, was added. This also enabled older solutions to be used with good results.

As mentioned in section 2.1.2, a series of modified, bidentate (S)-proline ligands, shown below, has been prepared and employed as chiral residual ligands in heterocuprate additions.[146,241] Enantiomeric excesses as high as 83% were obtained, although the outcome was variable.

X = OH, OMe SPh, SMe

Yamamoto's group has studied the addition of metallated chiral imines prepared from acetone and amino acid-derived amines.[147,148,149] For example, after deprotonation and conversion to the corresponding homocuprate, the acetone imine derived from leucine adds to 2-cyclopentenone with an enantiomeric excess of 75%. Interestingly, similar addition of this cuprate to 2-methyl-2-cyclopentenone is also stereoselective (the reaction proceeds with an enantiomeric excess of 60%) but selects the *opposite* face of the conjugate acceptor!

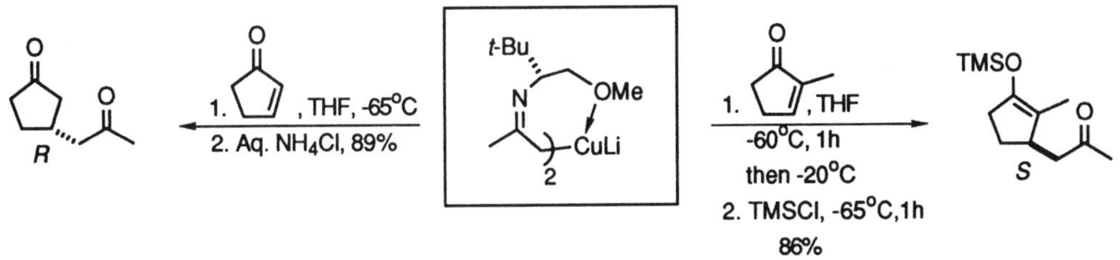

The (S) adduct was then converted into the *trans*-dihydroindanedione in two steps.

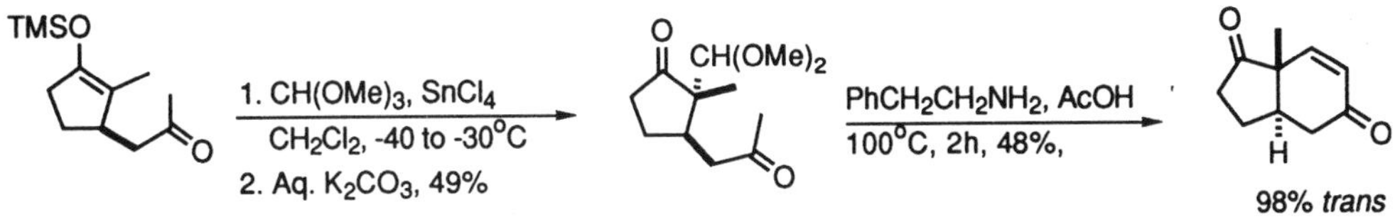

Bertz's group has carried out an extensive survey of chiral amidocuprates prepared from commercially available amines.[150] Best results were obtained using (4S, 5S)-(+)-5-amino-2,2-dimethyl-4-phenyl-1,3-dioxane in diethyl ether at -78°C.

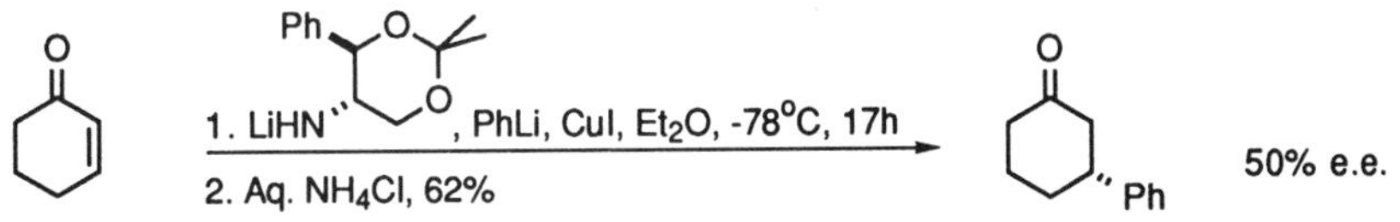

Rossiter has examined several chiral mono and bidentate amine ligands.[151] In one case an excellent stereochemical result was achieved. However, the chemical yield was disappointing.

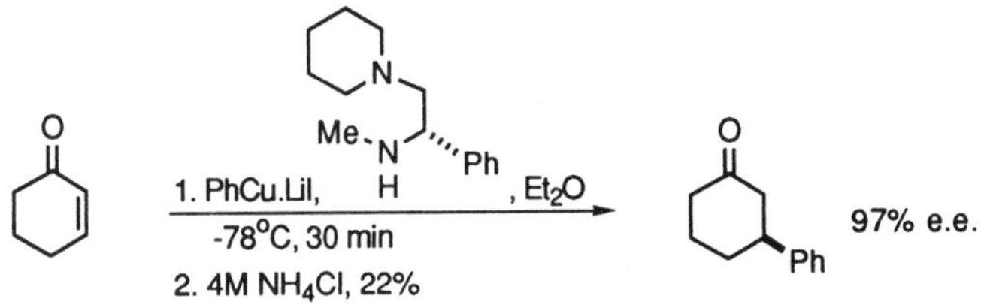

Chiral transfer ligands

Whilst studying the 2-(1-(N,N-dimethylamino)ethyl)phenyl group as a chiral residual ligand, Nilsson's group discovered that it was an excellent chiral *transfer* ligand when the other ligand was an aryl group.[152,153] (In fact, with alkyl transfer ligands it does remain attached to the copper. Unfortunately, these additions show very little enantioselectivity[154]). In each case the new stereogenic centre has the (S) configuration.[155]

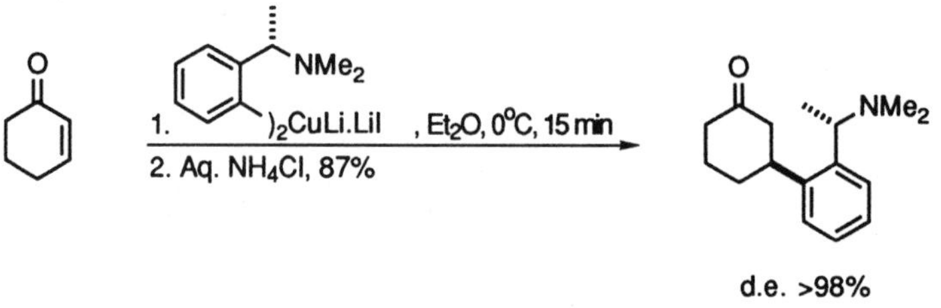
NMe2
1.)2CuLi.LiI , Et2O, 0°C, 15 min
2. Aq. NH4Cl, 87%
NMe2
d.e. >98%

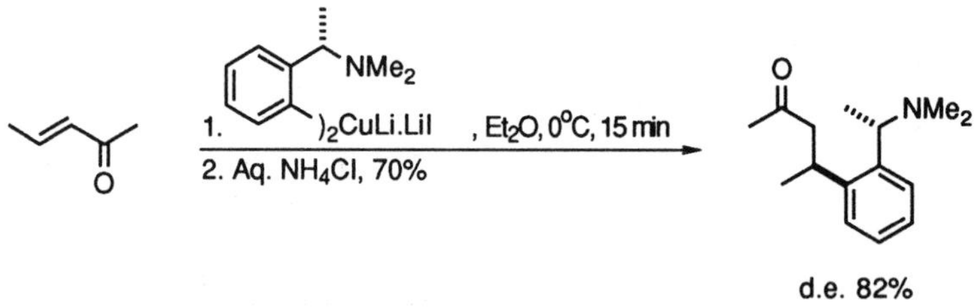
NMe2
1.)2CuLi.LiI , Et2O, 0°C, 15 min
2. Aq. NH4Cl, 70%
NMe2
d.e. 82%

3.1.3 Other organometallic reagents

3.1.3.1 Addition of achiral organometallic reagents to achiral cycloalkenones

Most hard organometallic reagents, such as simple organolithiums and organomagnesiums, add to cycloalkenones in a 1,2 fashion. It is only with copper catalysis that these reagents usually add in a conjugate fashion. One exception to this comes from R. G. Cooke's work on the total synthesis of some naturally-occurring pigments. The synthesis of these compounds, whose structures are based on 9-arylphenalenones, relies upon the addition of arylmagnesium reagents to C9 rather than C3 of the phenalenone. For example, in their total synthesis of haemocorin aglycone, Cooke and Rainbow added phenylmagnesium bromide, without copper catalysis, to 5,6-dimethoxyphenalenone (**16**).[156] No addition occurred at C3.

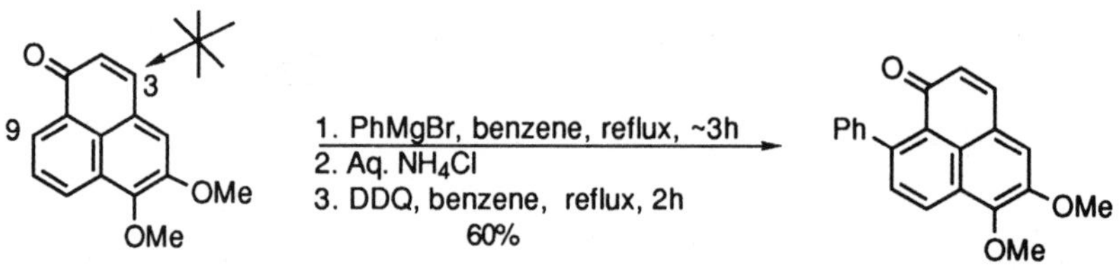
3
9
OMe
OMe
1. PhMgBr, benzene, reflux, ~3h
2. Aq. NH4Cl
3. DDQ, benzene, reflux, 2h
60%
Ph
OMe
OMe

16

Interestingly, when there is a substituent at C2, 1,8 addition can compete successfully with 1,4-addition. In the following example, in a reaction with 2,6-dimethoxyphenalenone (**17**) overall substitution of a methoxy group occurs.[157]

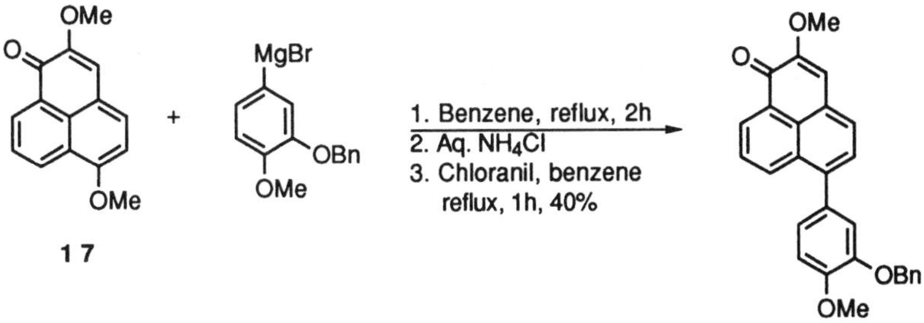

OMe

MgBr

+   OBn
                     OMe

OMe   OMe

17

1. Benzene, reflux, 2h
2. Aq. NH$_4$Cl
3. Chloranil, benzene
   reflux, 1h, 40%

OMe

OBn
OMe

Cycloalkenones do undergo conjugate addition with certain other organometallic reagents. For example, the reaction of triorganozincates with cycloalkenones represent the most successful class of additions for these reagents. Some examples are given in Table 3.8 for both stoichiometric and catalytic reactions.

Table 3.8   Conjugate addition of triorganozincates to cycloalkenones

1[158]

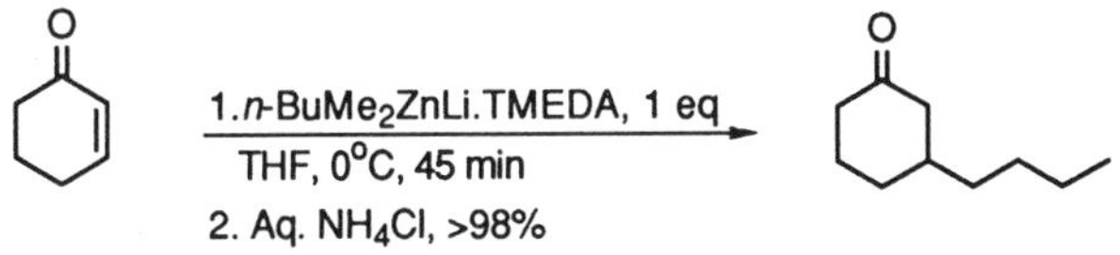

1. *n*-BuMe$_2$ZnLi.TMEDA, 1 eq
   THF, 0°C, 45 min
2. Aq. NH$_4$Cl, >98%

2[158]

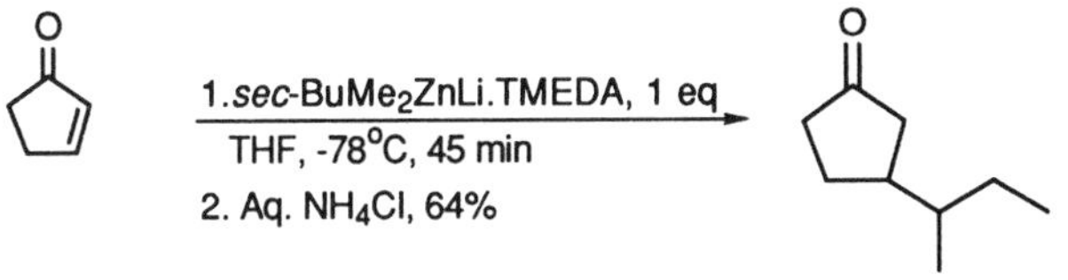

1. *sec*-BuMe$_2$ZnLi.TMEDA, 1 eq
   THF, -78°C, 45 min
2. Aq. NH$_4$Cl, 64%

3[159]

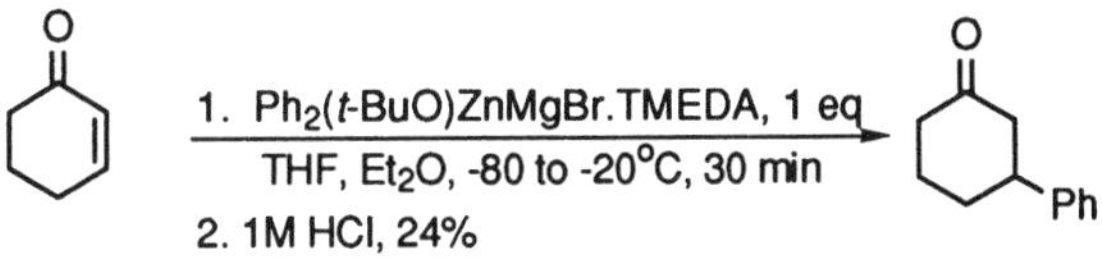

1. Ph$_2$(*t*-BuO)ZnMgBr.TMEDA, 1 eq
   THF, Et$_2$O, -80 to -20°C, 30 min
2. 1M HCl, 24%

Ph

4[160]

1. RMgBr, 1 eq., TMEDA.Zn(O*t*-Bu)Cl, 0.01 eq
   THF, 0°C, 30 min
2. Aq. NH$_4$Cl,   R = Et   50%   1,4 : 1,2 (19:1)
                    R = *i*-Pr 88%   1,4 : 1,2 (1.4:1)

R

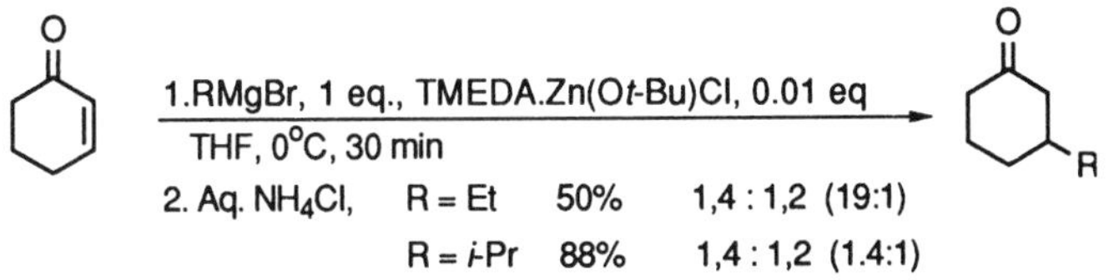

Corey and Hegedus's nickel-based masked acyl anion adds smoothly to cycloalkenones.[161]

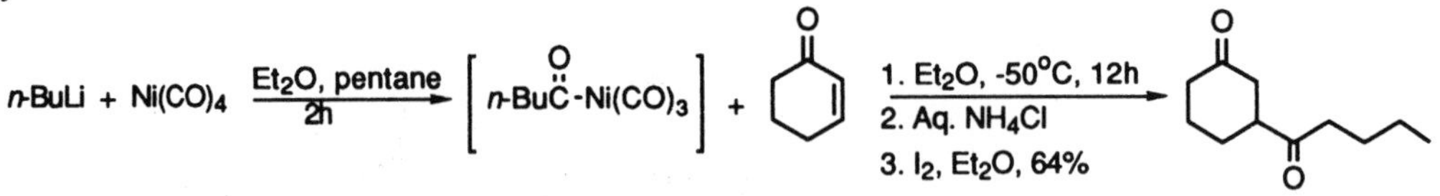

So too does Cooke and Parlman's acyl anion methodology mentioned in section 2.1.3.[162]

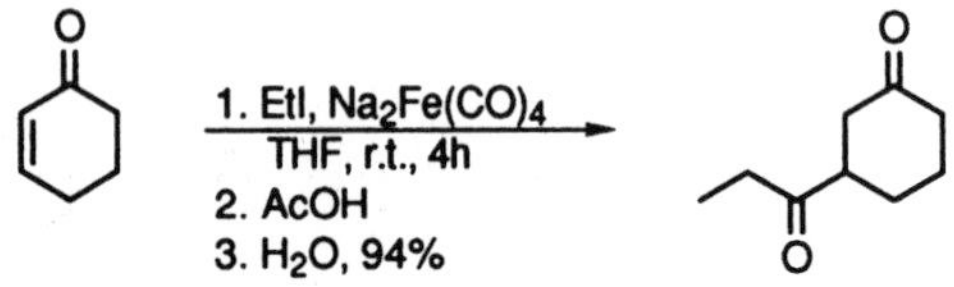

The addition of allyltrialkylsilanes to cycloalkenones (the Sakurai reaction) works best using a powerful Lewis acid, such as titanium tetrachloride. This is one of the most efficient ways of adding an allyl group to a cycloalkenone (see, however, section 3.1.2).

Table 3.9   Sakurai additions to cycloalkenones

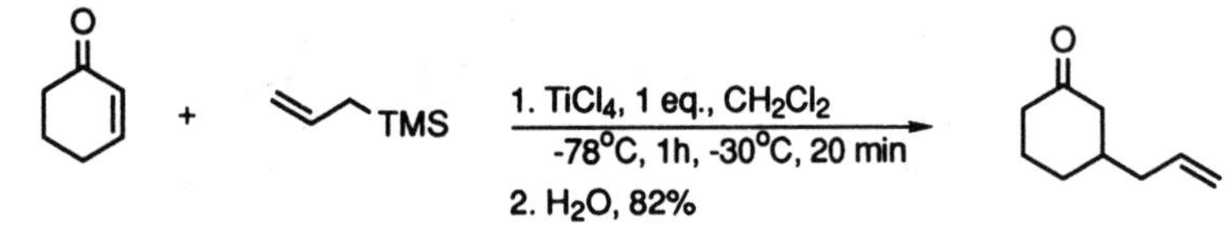

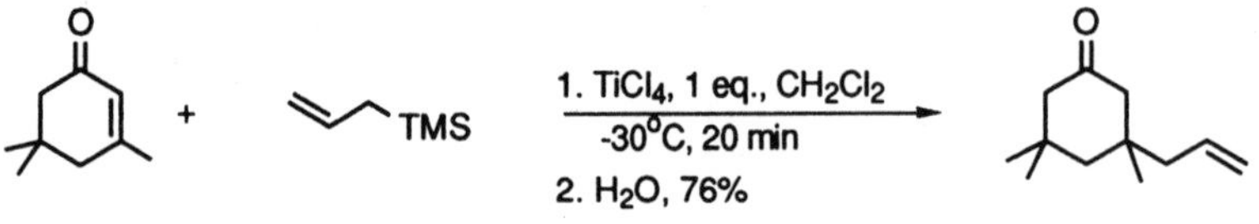

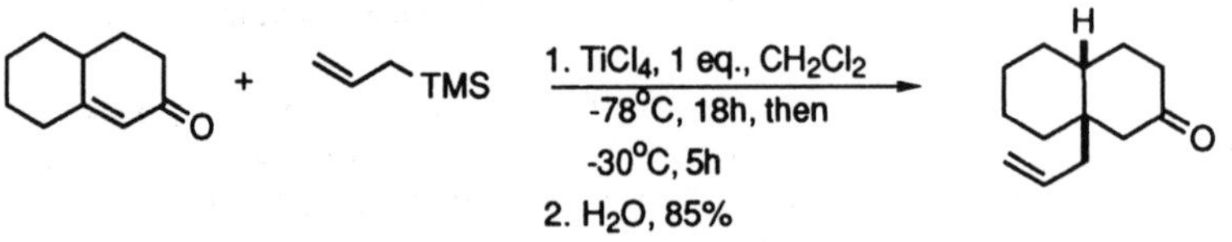

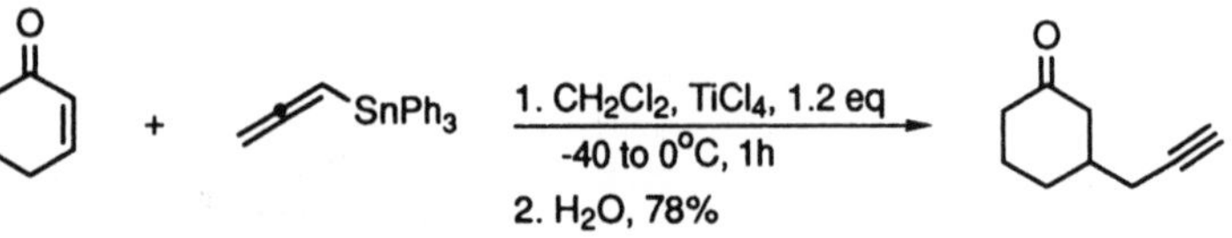

The diastereoselectivity of the titanium tetrachloride promoted Sakurai reaction can be controlled by (i) using either a Z or E alkenyl reagent and (ii) varying the substituents on silicon.[165]

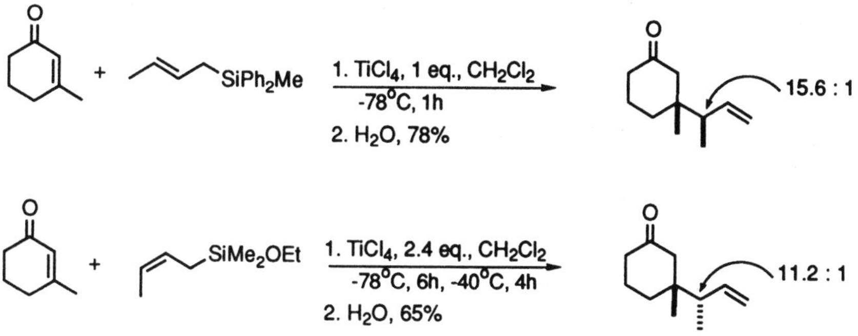

The problem of conjugate addition of alkynyl groups has largely been overcome.[166] As pointed out in section 2.1.3, those reagents that were found to react with alkenones usually failed with Z-alkenones. However, additions to cycloalkenones, which are necessarily Z-alkenones can be achieved if the reaction is carried out in the presence of a catalyst prepared from Ni(acac) and DIBAL.[167]

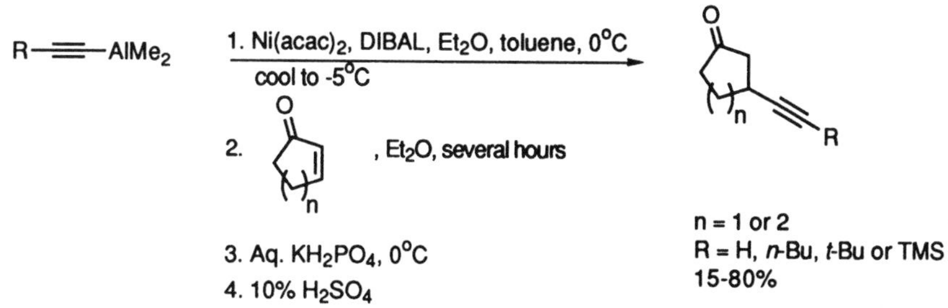

3.1.3.2   Addition of achiral organometallic reagents to chiral cycloalkenones

Organomagnesium reagents sometimes add in a conjugate fashion to cycloalkenones without the need for copper catalysis. For example, addition of 2-propenylmagnesium bromide to pinocarvone (**18**) gives a reasonable yield of the 1,4 adduct.[168,169] (Methyllithium gives exclusively 1,2 addition).

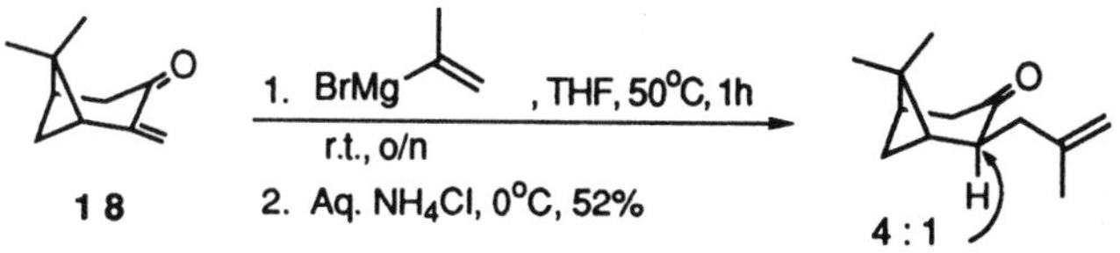

Corey has applied a titanium tetrachloride promoted Sakurai reaction to a total synthesis of (±)-tricyclohexaprenol.[170,171]

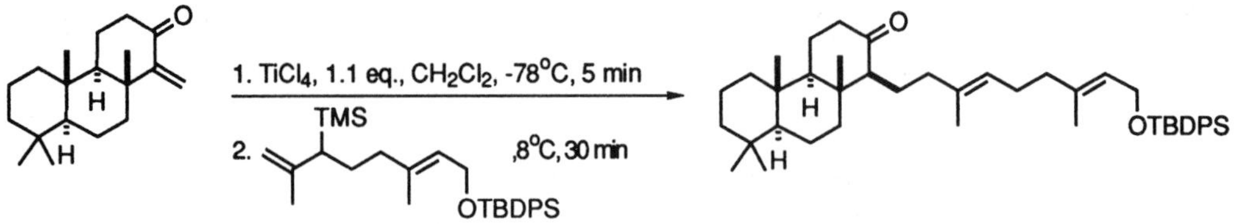

Conjugate additions to chiral 2-sulfinylcycloalkenones, followed by reductive removal of the sulfinyl group, provides an efficient method for preparing homochiral 3-substituted cycloalkenones.[172] Full details of the preparation of 5- and 6-membered 2-sulfinylcycloalkenones, outlined below for (S)-(+)-2-(p-toluenesulfinyl)-2-cyclopentenone (19), have now been published.[173]

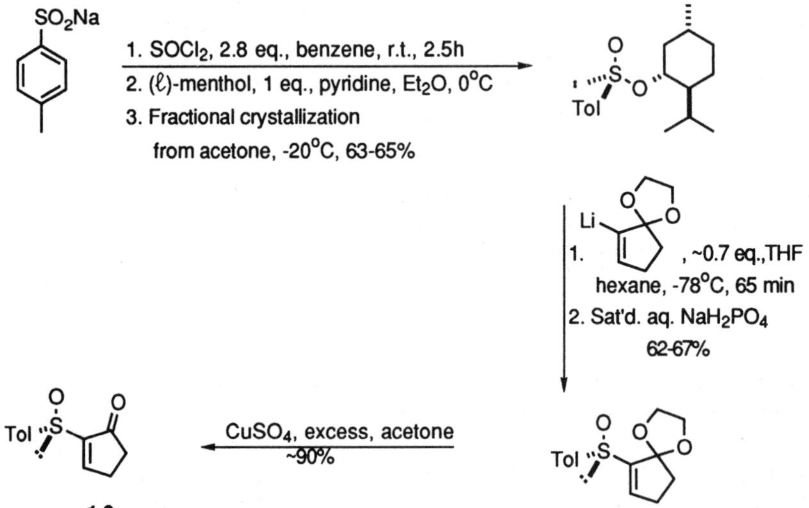

The sense of addition may be controlled by either conformation or chelation control. The addition of tetrahydrofuran solutions of salt-free diorganomagnesium reagents to chiral, 2-sulfinylcycloalkenones and subsequent reductive removal of the auxiliary leads to 3-substituted cycloalkenones in high enantiomeric purity.[174]

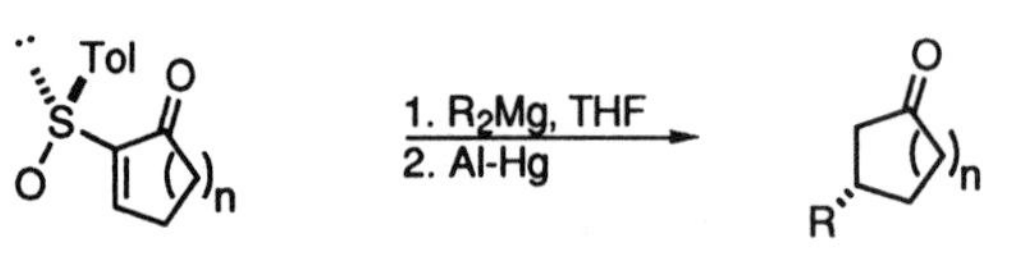

n = 1		
R = Me	60%	97% e.e.
R = Et	81%	81% e.e.
R = Ph	72%	97% e.e.

n = 2		
R = Me	50%	79% e.e.

These results have been interpreted as being the result of addition to a conformation which minimizes electrostatic interactions, with the carbonyl and sulfinyl dipoles *anti* to each other. This necessarily places the tolyl group over one face of the cycloalkenone.[175] This provides a steric screen which directs the incoming nucleophile to the opposite face (the *re*-face at C3).

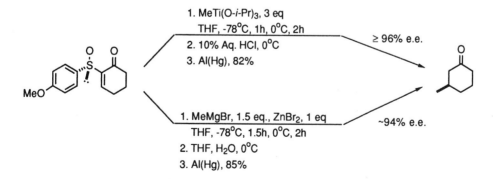

conformation control   chelation control

This stereocontrol may be reversed by introducing a chelating metal ion. A survey of metal halides showed that zinc bromide was the most effective and that extremely high enantiomeric excess are often attainable. The metal-chelate is apparently so electrophilic that even alkyltitanium triisopropoxide reagents add in a conjugate fashion.

RM = MeMgCl	91%	>98% e.e.
RM = EtMgCl, ZnBr$_2$	84%	80% e.e.
RM = EtTi(O-*i*-Pr)$_3$	72%	>98% e.e.
RM = PhMgCl, ZnBr$_2$	70%	92% e.e.
RM = CH$_2$=CHMgBr, ZnBr$_2$	75%	99% e.e.

1. RM, THF
2. Al-Hg

Even without an added chelating metal, titanium reagents add more stereoselectively than the corresponding combination of an organomagnesium reagent with zinc bromide. Changing the aryl substituent, from 4-methylphenyl to 4-methoxyphenyl, as well as reducing the coordinating ability of the solvent, from tetrahydrofuran to 2,5-dimethyltetrahydrofuran, can lead to improved stereoselection in these additions.

1. MeTi(O-*i*-Pr)$_3$, 3 eq
   THF, -78°C, 1h, 0°C, 2h
2. 10% Aq. HCl, 0°C
3. Al(Hg), 82%

≥ 96% e.e.

1. MeMgBr, 1.5 eq., ZnBr$_2$, 1 eq
   THF, -78°C, 1.5h, 0°C, 2h
2. THF, H$_2$O, 0°C
3. Al(Hg), 85%

~94% e.e.

One of the most outstanding examples of this work is the total synthesis of estrone methyl ether (and hence estradiol methyl ether).[176]

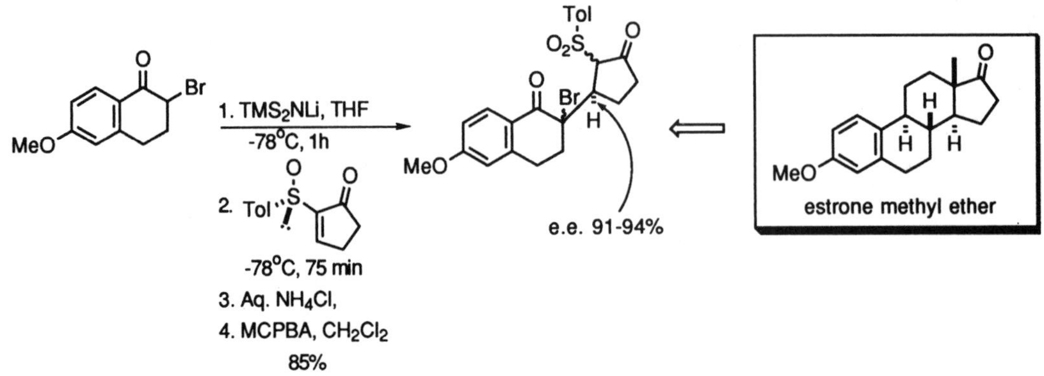

1. TMS$_2$NLi, THF
   -78°C, 1h

2. Tol S (structure)
   -78°C, 75 min
3. Aq. NH$_4$Cl,
4. MCPBA, CH$_2$Cl$_2$
   85%

e.e. 91-94%

estrone methyl ether

3.1.4   Free radicals[177]

Vitamin B$_{12a}$, under reducing conditions, catalyzes the coupling of acyl radicals to many conjugate acceptors, including cycloalkenones.[178] This is one of several direct methods now available for preparing 1,4-dicarbonyl compounds by conjugate addition (see also, section 3.1.1.1 and 3.1.2).

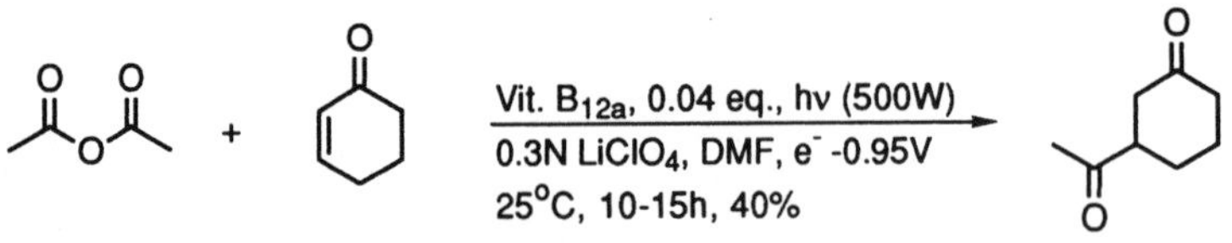

Vit. B$_{12a}$, 0.04 eq., hv (500W)
0.3N LiClO$_4$, DMF, e⁻ -0.95V
25°C, 10-15h, 40%

Alkenyl radicals may also be generated, using this catalyst system, from alkenyl bromides. Their addition to cyclopentenone has obvious relevance to prostaglandin synthesis.[179]

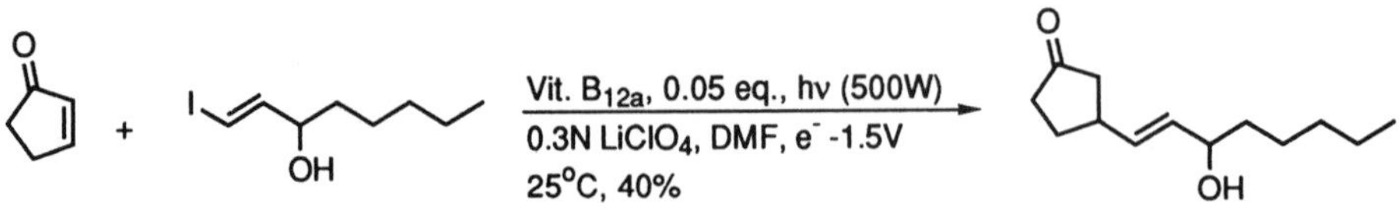

Vit. B$_{12a}$, 0.05 eq., hv (500W)
0.3N LiClO$_4$, DMF, e⁻ -1.5V
25°C, 40%

The use of ultrasound in additions to cycloalkenones (and other acceptors) has recently been re-examined.[180] It was concluded that a free radical reaction was occurring at the metal surface. A set of optimized conditions has also been developed.[181] The most critical parameter was found to be solvent composition. Alcohols were found to be the best organic solvents to use in conjunction with water. Even so the yields of these additions suffer when the β-position is crowded.

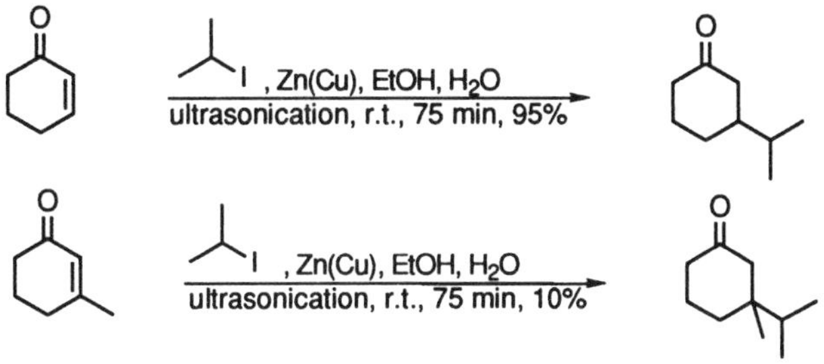

3.1.5   Enol ethers

3.1.5.1   *Addition of achiral enol ethers to achiral cycloalkenones*

As mentioned in section 2.1.5.2, several Lewis acids promote the conjugate addition of trialkylsilyl enol ethers and ketene acetals to acyclic alkenones. This applies equally well to cycloalkenones.

Table 3.10   Lewis acid promoted conjugate addition of trialkylsilyl enol ethers to cycloalkenones

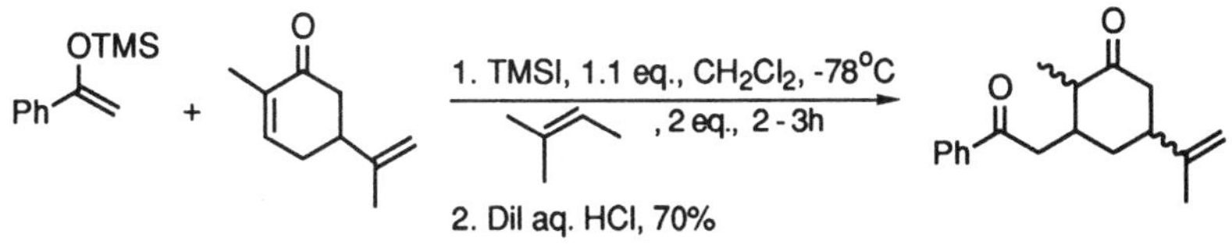

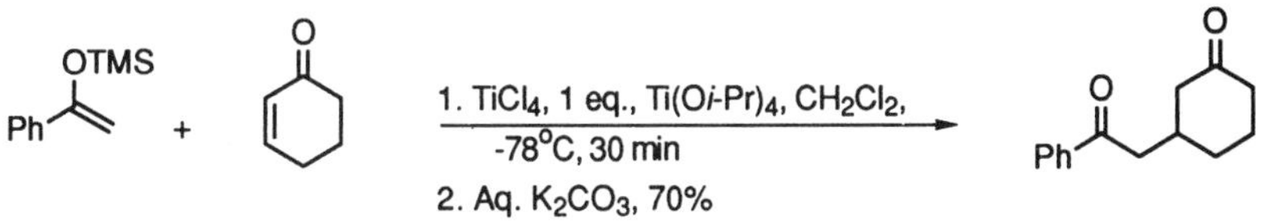

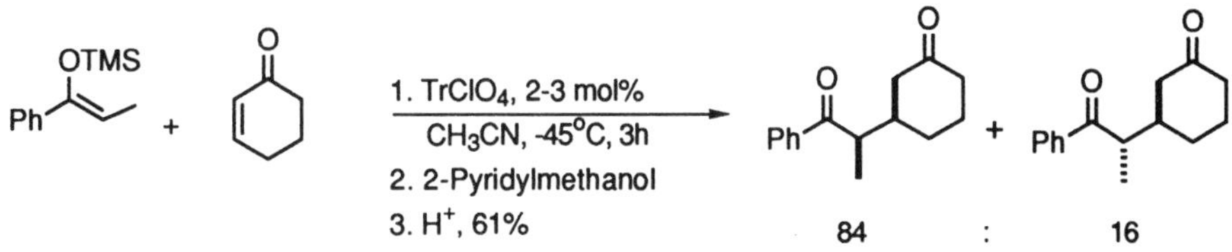

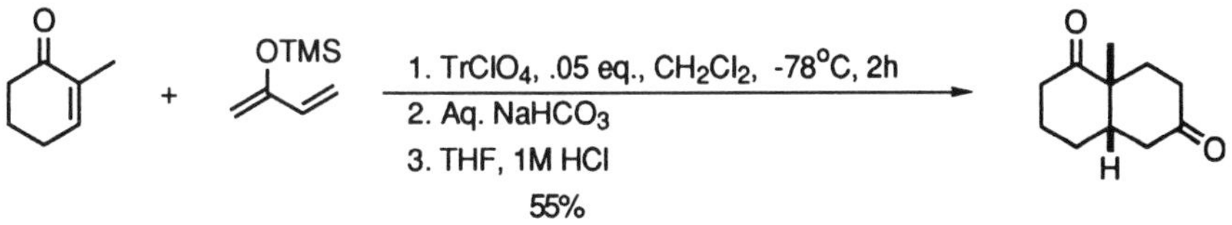

Quite useful levels of stereocontrol may be achieved using trityl perchlorate as catalyst as in entry 3 (Table 3.10, above). Other trityl salts work almost as well, although the use of lower temperatures had no effect on stereocontrol. The authors attribute this stereoselectivity to steric effects in the transition state as shown below.

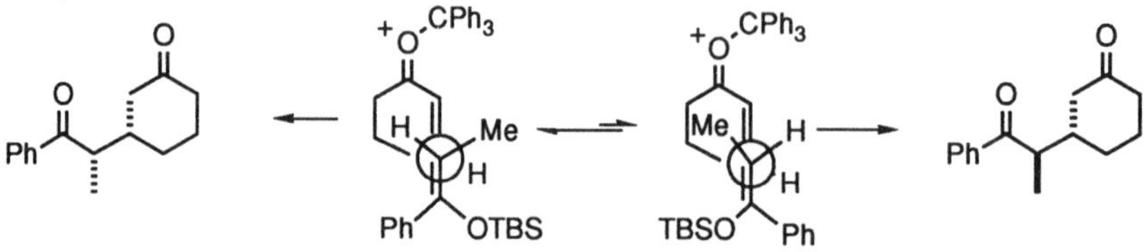

Simple diastereoselection in the addition of ketene thioacetals to 2-cyclopenten-ones is also possible.[186] By systematically varying reaction parameters, it was found that addition of the *Z*-isomer of the dimethylphenylsilyl ketene thioacetal of ethyl thiopropanoate (20) (with trityl hexachloroantimonate as catalyst) gave excellent selectivity for the *syn* product, whereas the trimethylsilyl ketene thioacetal of *tert*-butyl thiopropanoate (21) (with trityl triflate as catalyst) gave the *anti* product with good selectivity. The use of very low temperatures (-100°C) also improved selectivity.

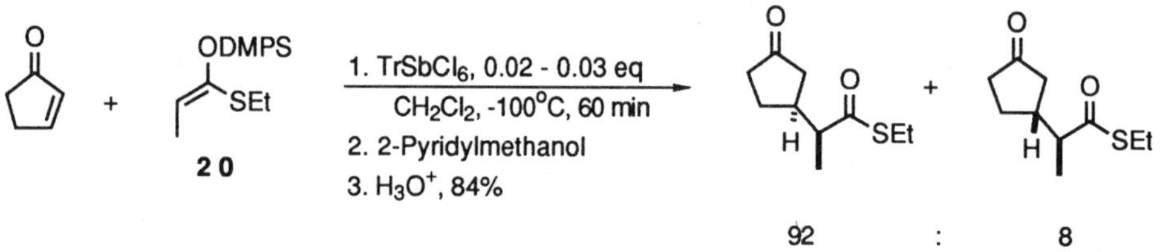

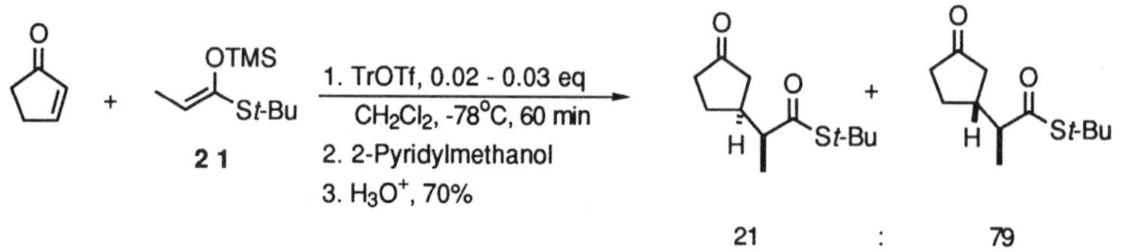

As in the case of acyclic alkenones (see section 2.1.5.2), tandem conjugate addition/aldol reactions with cycloalkenones are highly stereoselective.[187] This methodology was applied to a synthesis of the bicyclo[5.3.0]decane ring system of the pseudoguaianolide, fastigilin-C (see next page).[188] Excellent stereocontrol was achieved in the initial addition and, consistent with the work of Mukaiyama's group, the diastereomeric ratio (which was temperature dependent), at the pseudobenzylic centre, was 6:1 in favour of the desired stereochemistry.

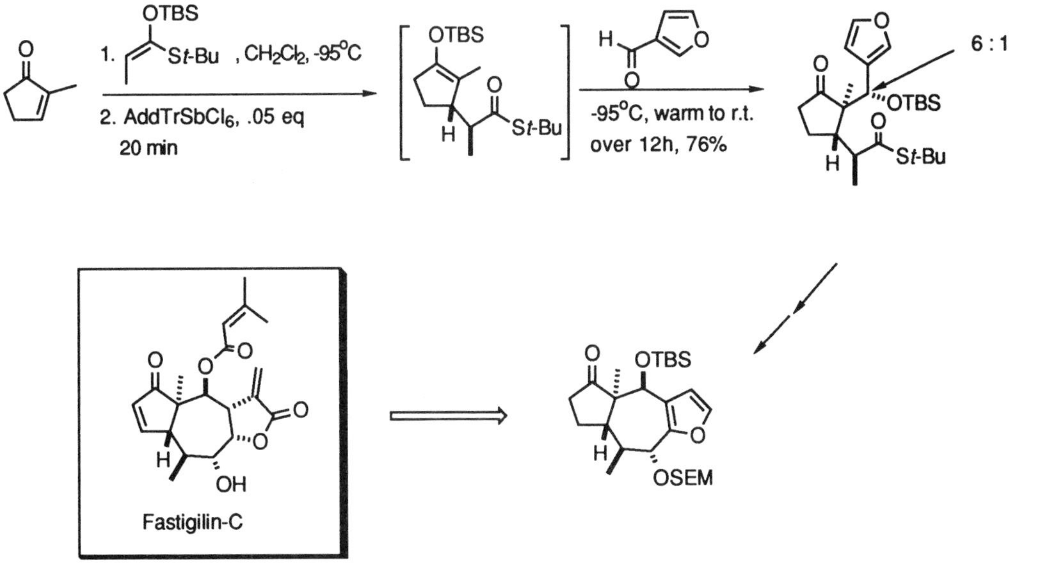

Fastigilin-C

Table 3.11   Lewis acid-promoted conjugate addition of ketene acetals and ketene thioacetals to cycloalkenones

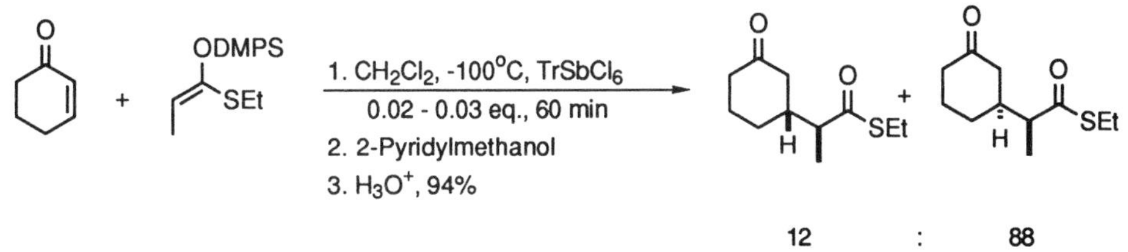

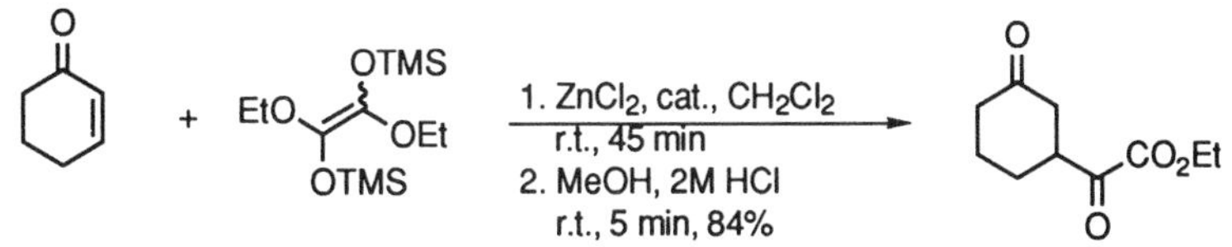

Tandem conjugate addition/aldol reactions followed by acid catalysed lactonization provides an efficient synthesis of fused lactones.[191]

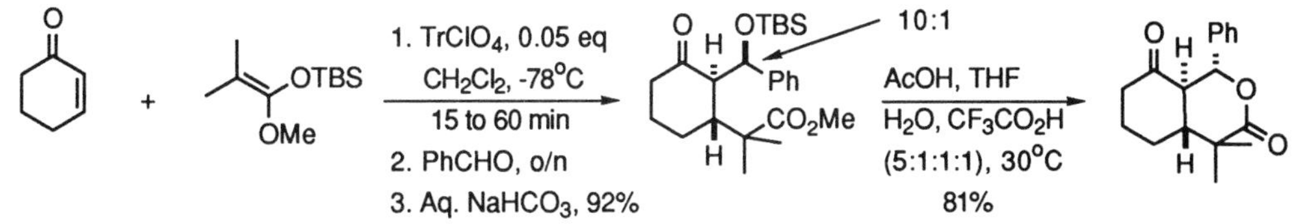

Iodotrimethylsilane catalyzes the addition of furans to alkenones.[182] (2-Methyl-2-butene is included as a trap for any hydrogen iodide generated during the reaction).

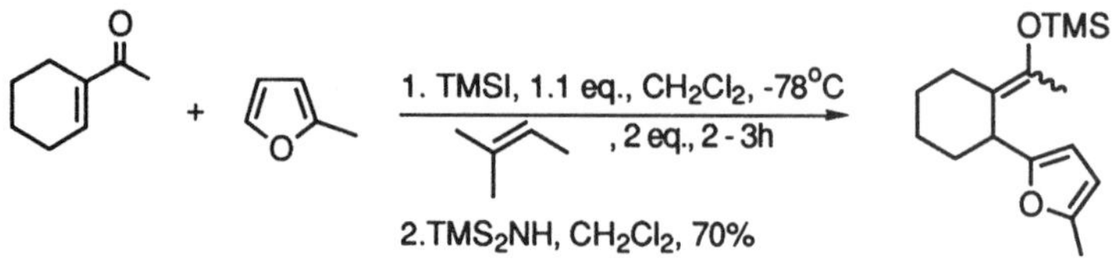

3.1.5.2   *Addition of achiral enol ethers to chiral cycloalkenones*

In their synthesis of the quassinoids, Heathcock's group employed the following reaction.[192] The use of high pressure was found to be essential as the alternative conditions

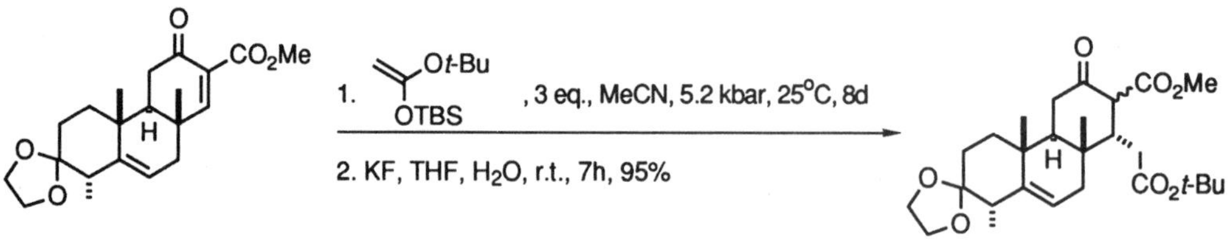

used, heating a mixture of the two reactants[193] in acetonitrile at 155°C for twenty one hours, only gave a 10% yield of the adduct.

However, an iodotrimethylsilane promoted addition of an enol ether to a 2,5-disubsituted cyclohexenone proved to be quite efficient, although without significant stereoselection.[194]

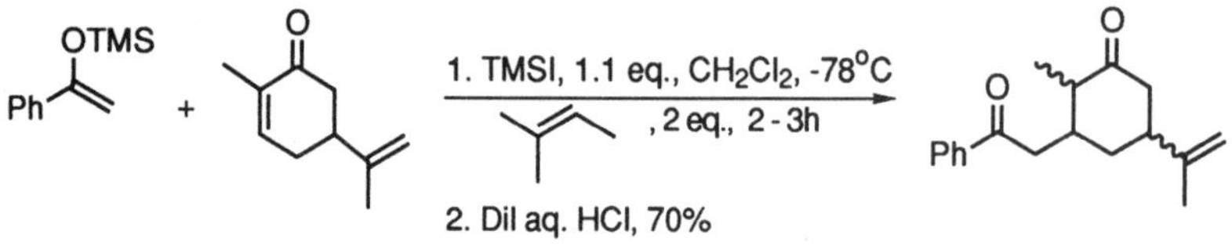

3.1.6   Conjugate hydrocyanation[195]

3.1.6.1   *Conjugate hydrocyanation of achiral cycloalkenones*

Most cycloalkenones undergo hydrocyanation. However, only occasionally is the reaction highly stereoselective. Some examples of these reactions are given in Table 3.12.

Table 3.12   Conjugate hydrocyanation of cycloalkenones

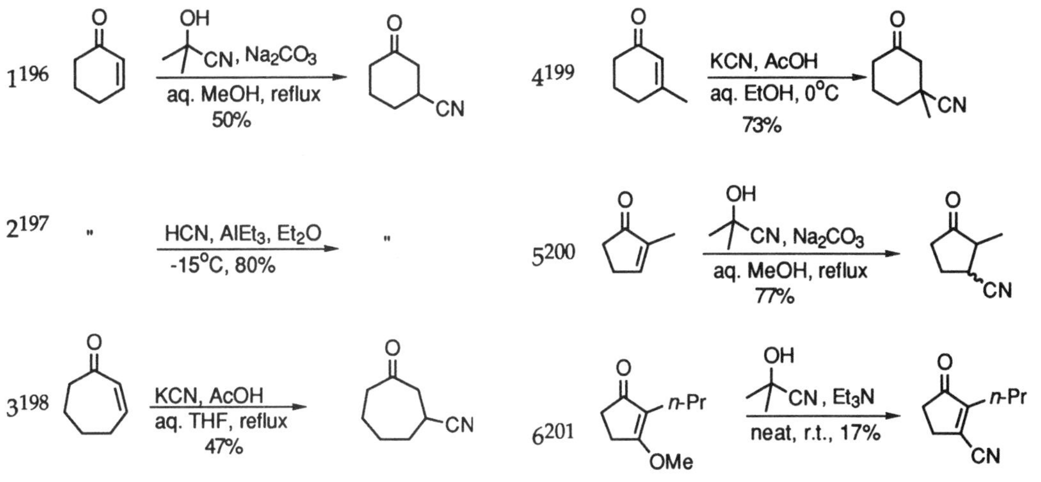

Sometimes side reactions can intervene, for example, in the hydrocyanation of cyclooctenone.[198] As shown in the Table above, this could be avoided by running the reaction in aqueous tetrahydrofuran instead (entry 3).

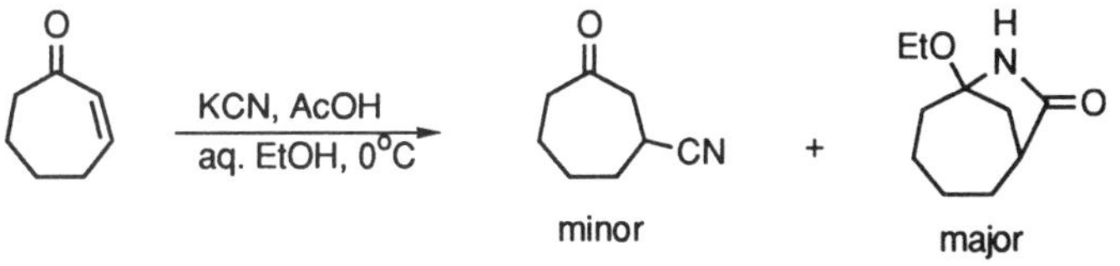

3.1.6.2   Conjugate hydrocyanation of chiral cycloalkenones

As with other nucleophiles, the formation of the kinetic product of cyanide addition to a variety of chiral cycloalkenones, including mono-, fused bi- and poly-cyclic systems, generally proceeds via a chair-like enolate (see section 1.3.5.3).[202]

Hydrocyanation may also be achieved indirectly by reaction with cyanotrimethylsilane. This requires the presence of a Lewis acid.[203] The addition shown in entry 1 in Table 3.13 below is under thermodynamic control. Running the reaction at room temperature with less Lewis acid present leads exclusively to formation of the 1,2 adduct. Heating or longer reaction times transforms this adduct into a mixture of 1,4 adducts, although with better stereoselectivity than is usually observed under alternative hydrocyanation conditions. The following mechanistic scheme is consistent with these results:

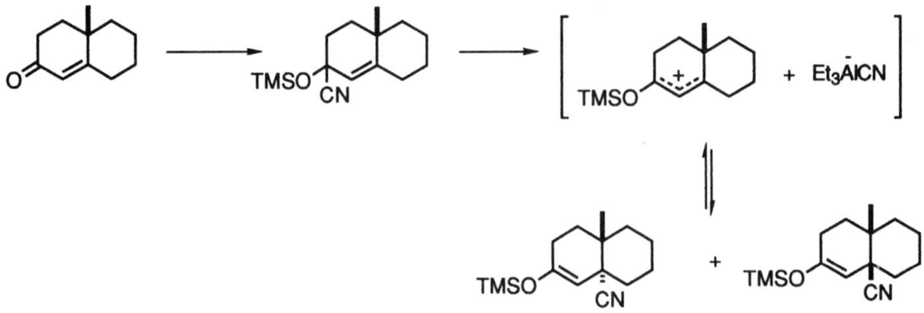

Where no angular methyl group is present, as in entries 2 and 3, only the *trans*-fused system is produced, irrespective of the reaction conditions used. One example of a 1,6 addition was included in the study. Only the *trans* adduct was obtained, in excellent yield.

Table 3.13   Conjugate hydrocyanation of chiral cycloalkenones

1203	1. TMSCN, 2.2 eq., AlEt₃, 2 eq. THF, reflux 20h 2. Aq. NH₄Cl, 100%	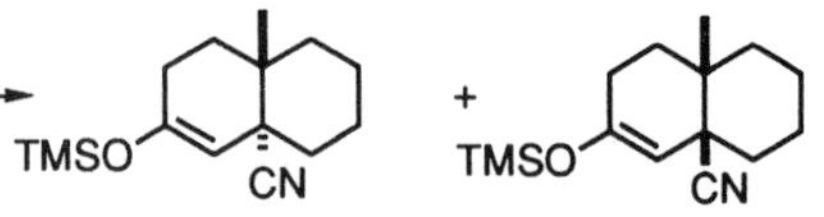 95 : 5
2203	1. TMSCN, 2.2 eq., AlEt₃, 2 eq. toluene, r.t., 20h 2. Aq. NH₄Cl, 80%	
3203	1. TMSCN, 2.2 eq., AlEt₃, 2.eq. THF, reflux 20h 2. Aq. NH₄Cl, 84%	
4203	1. TMSCN, 3 eq., BF₃.Et₂O, 0.1 eq. neat, 70-80°C, 7h 2. Aq. NH₄Cl, 93%	

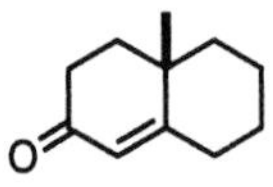

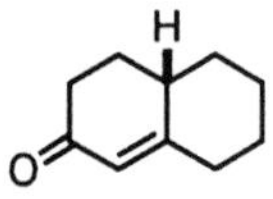

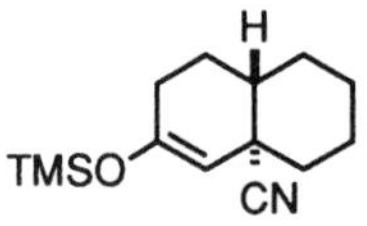

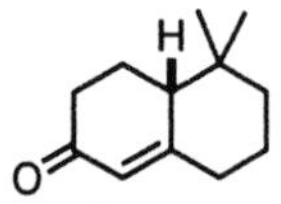

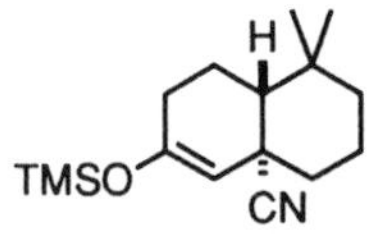

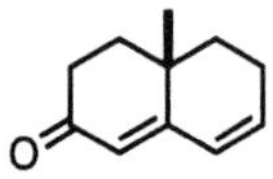

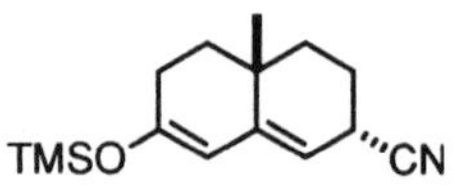

3.1.7   Heteronucleophiles

3.1.7.1   *Nitrogen*

The double addition of (*R*)-(α-methyl)benzylamine to 3-pentyl-2,7-cyclooctadienone (**22**) gave a mixture of diastereomers which could be separated by fractional crystallization. This led to a rather short synthesis of either enantiomer of adaline.[204]

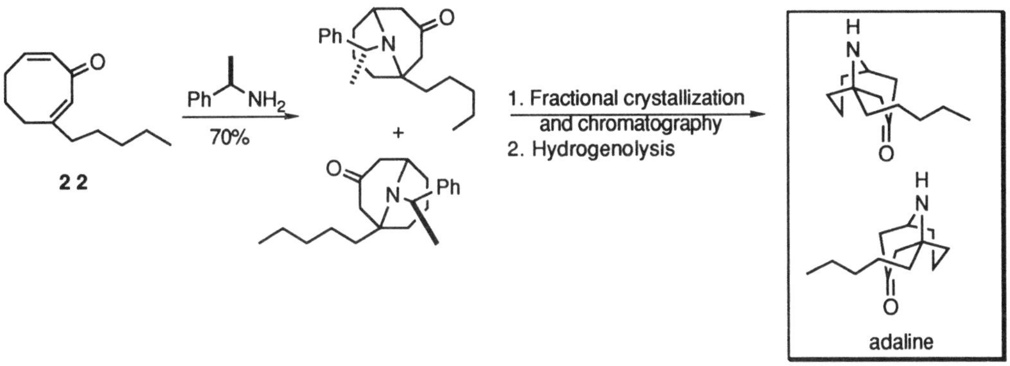

3.1.7.2   Sulfur

Aluminium thiolates add readily to cycloalkenones and the intermediate enolates may be efficiently quenched by aldehydes. Subsequent oxidative elimination regenerates the double bond.[205]

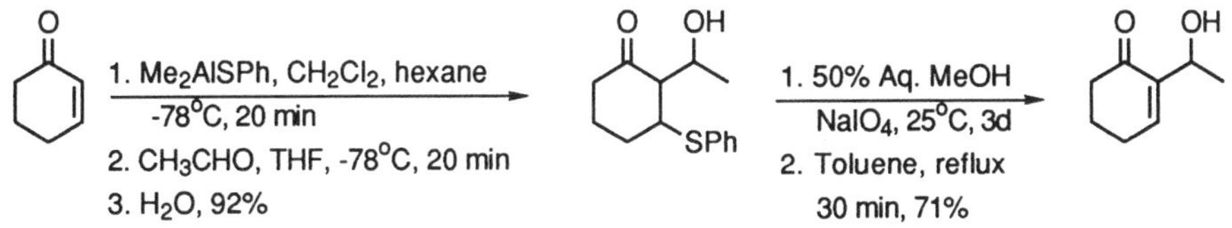

This reaction has found application in total synthesis. Thus addition of thiophenoxide to 4-(*tert*-butyldimethylsilyloxy)-2-cyclopentenone, followed by quenching with an alkynal generated a new intermediate for prostaglandin synthesis.[206]

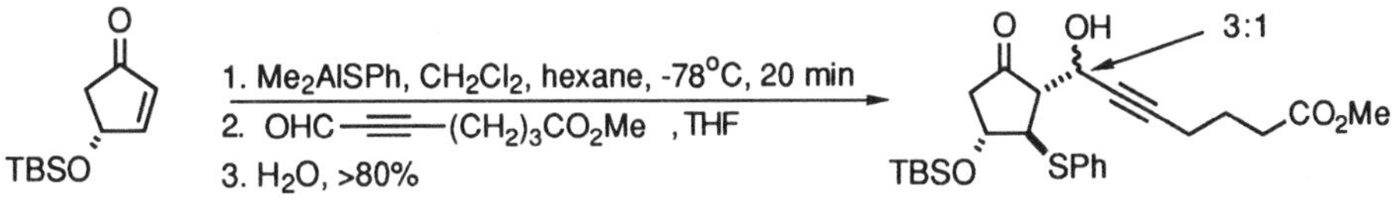

3.1.7.3   Silicon and tin

Addition of trimethylsilyllithium (prepared from hexamethyldisilazane and methyllithium in hexamethylphosphoramide) to 2-cyclohexenone proceeds efficiently, and quenching with methyl iodide, or allyl iodide, produces the trans isomer exclusively.[207] However, reaction with less reactive electrophiles tends to give poor yields due to competing protonation.[208] (See section 4.4 for a solution to this problem).

Table 3.14   Conjugate addition of silicon nucleophiles to cycloalkenones

1207

1. Me₃SiLi , HMPA, Et₂O
   THF, -78°C, ~10 min
2. MeI , warm to 0°C
3. H₂O, 97%

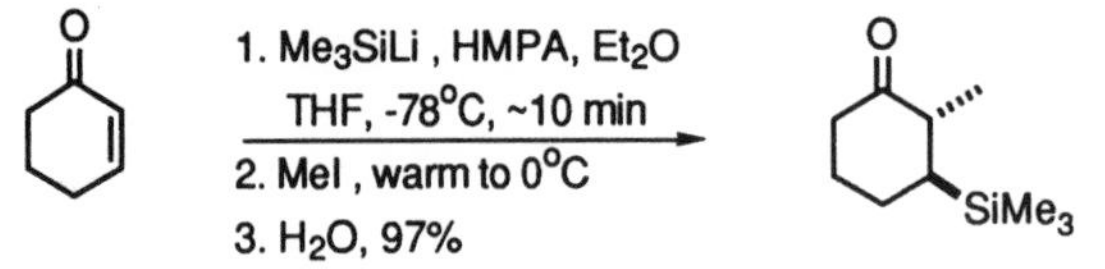

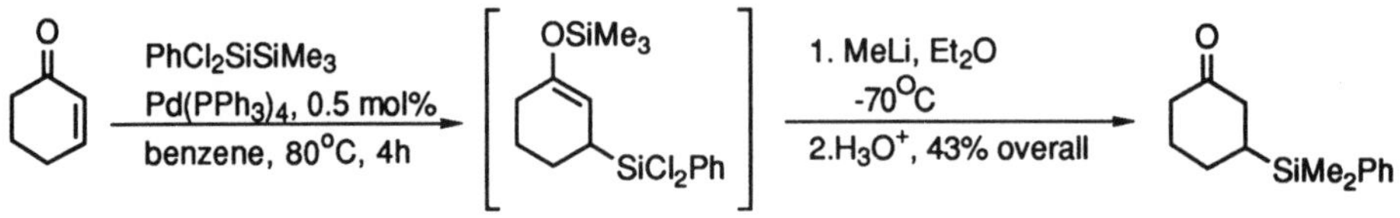

2209

PhCl₂SiSiMe₃
Pd(PPh₃)₄, 0.5 mol%
benzene, 80°C, 4h

1. MeLi, Et₂O
   -70°C
2. H₃O⁺, 43% overall

3210

1. Me₂PhSiZnMe₂Li, 1.2 eq
   THF, -78°C, 1.5h
2. 1N HCl, 67%

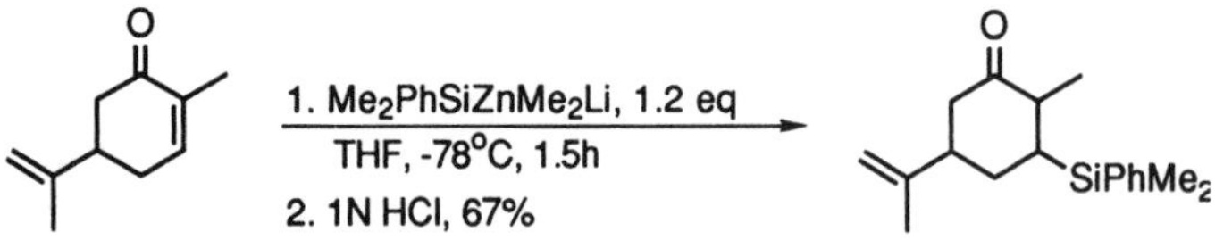

In an extension to this reaction, addition of a trialkylstannyllithium to 2-cyclohexenone has been used as the initial step in a multicomponent annulation procedure.[211,212] The products are suitable for oxidative fragmentation, leading to macrolides.

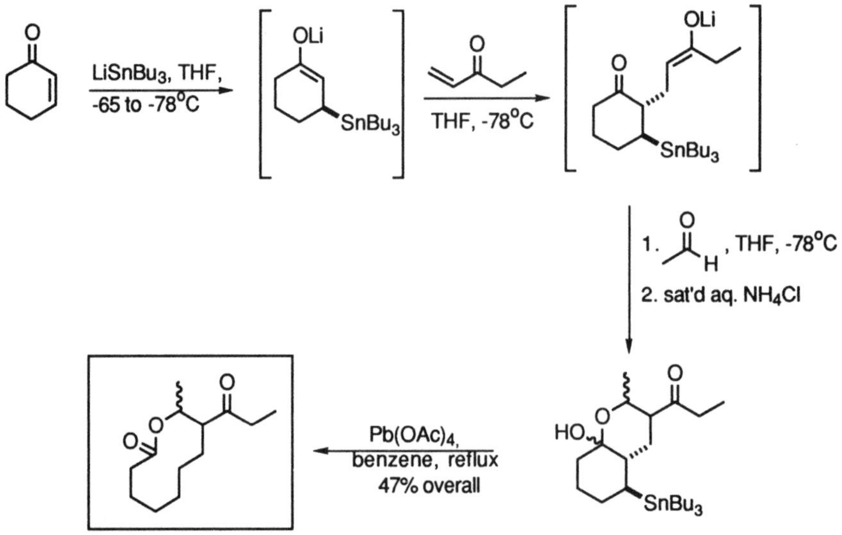

Similar addition, followed by quenching with an alkyl halide, provides access to a series of precursors of cyclononenones and cyclodecenones.[213] The ring expansion, a radical process, is highly dependent upon the stereochemistry of the precursor. Thus *trans*-disubstituted cycloalkenones give ring expanded *trans*-cycloalkenones. However, if the α-substituent is a proton, an intramolecular reduction competes with the ring expansion and the reaction is less useful.

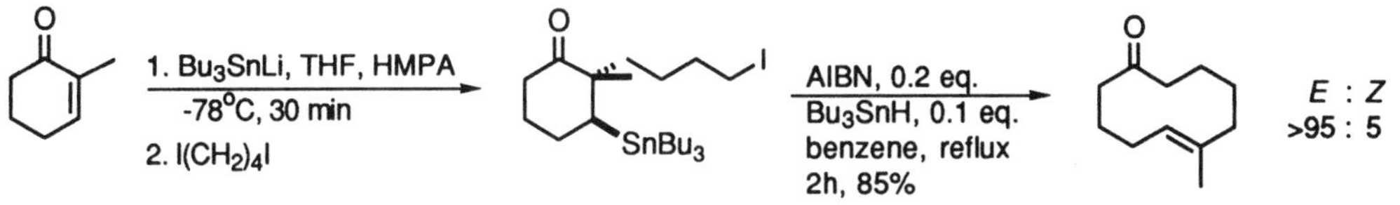

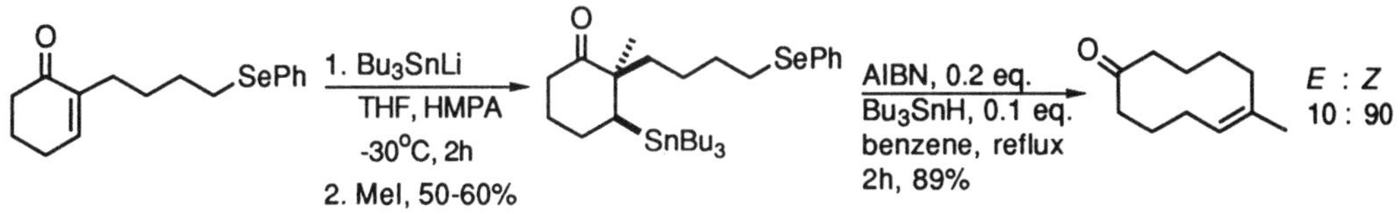

3.1.7.4   Iodine

The addition of iodide to alkenones is easily achieved using iodotrimethylsilane.

Table 3.15   Additions of iodide to cycloalkenones

1[214]		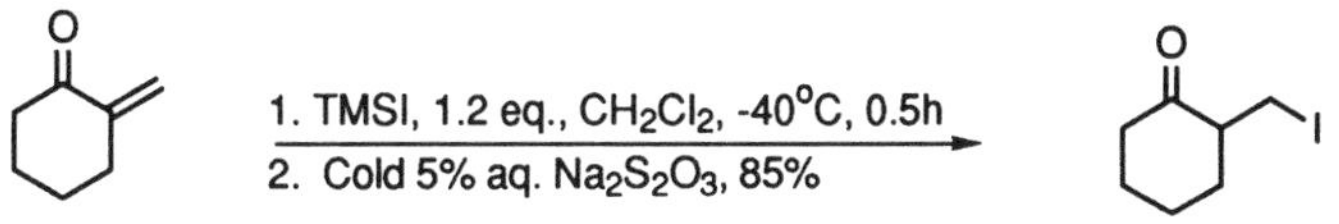 1. TMSI, 1.2 eq., CH$_2$Cl$_2$, -40°C, 0.5h 2. Cold 5% aq. Na$_2$S$_2$O$_3$, 85%
2[214]		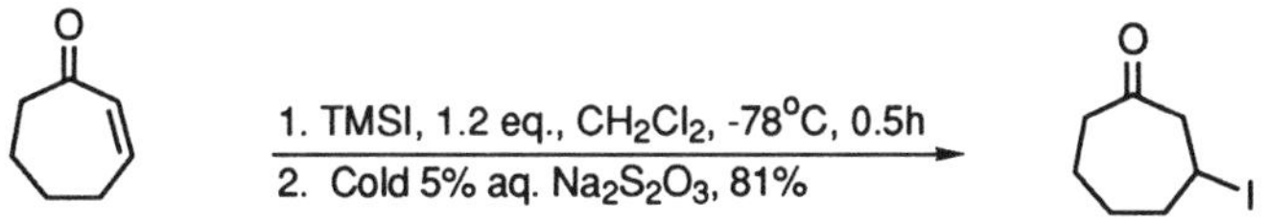 1. TMSI, 1.2 eq., CH$_2$Cl$_2$, -78°C, 0.5h 2. Cold 5% aq. Na$_2$S$_2$O$_3$, 81%
3[214]		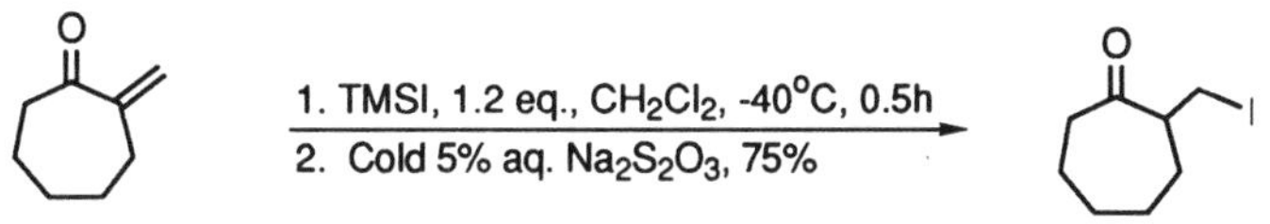 1. TMSI, 1.2 eq., CH$_2$Cl$_2$, -40°C, 0.5h 2. Cold 5% aq. Na$_2$S$_2$O$_3$, 75%

3.1.7.5   Conjugate reduction

As with acyclic alkenones, copper modified sodium and lithium hydride reagents reduce cycloalkenones in a mainly conjugate fashion. However, selectivity is rarely complete.[215]

Table 3.16   Conjugate reduction of cycloalkenones

1[215]	1. Li(OMe)$_3$AlH, 4 mol eq. (4 hydride eq.) CuBr, 1 or 2 eq., THF -78°C, 10 min, -20°C, 50 min 2. MeOH 3. Sat'd aq NH$_4$Cl	R$_1$ = R$_2$ = H 84 3 R$_1$ = R$_2$ = Me 92 6
2[215]	1. Na(MeOCH$_2$CH$_2$O)$_2$AlH$_2$ 2 mol eq. (4 hydride eq.) CuBr, 1.2(R's=Me,4 eq)eq. THF, benzene, -78°C, 10 min (R's=H)0°C, 30 min (R's=Me) -20°C, 1h 2. Sat'd aq NH$_4$Cl	R$_1$ = R$_2$ = H 62 0 R$_1$ = R$_2$ = Me 61 13
3[216]	Ph$_2$SiH$_2$, 1 eq., ZnCl$_2$, 0.36 eq Pd(PPh$_3$)$_4$, 0.018 eq CHCl$_3$, r.t., 1h, 100%	
4[216]	Ph$_2$SiH$_2$, 1.55 eq., ZnCl$_2$, 0.4 eq Pd(PPh$_3$)$_4$, 0.018 eq CHCl$_3$, r.t., 3h, 90%	1:1
5[217,218]	[(Ph$_3$P)CuH]$_6$, 0.24 eq H$_2$O, benzene, r.t., 1h 88%	>100:1

3.2   Intramolecular reactions

3.2.1   Stabilized carbanions

3.2.1.1   *Carbanions stabilized by π-conjugation with one heteroatom*

Probably one of the earliest recorded examples of an intramolecular Michael addition is the base-catalyzed conversion of santonin into santonic acid.[219,220,221] The correct structure for santonic acid was only established in 1948.[222]

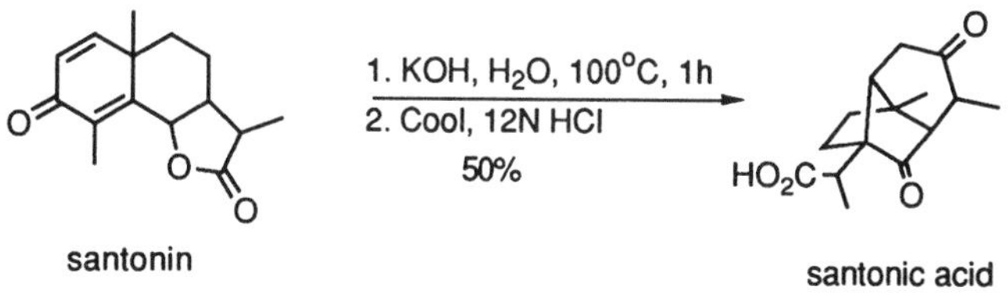

santonin   santonic acid

1. KOH, H₂O, 100°C, 1h
2. Cool, 12N HCl
50%

A mechanism for this rearrangement was later proposed by Corey and is reproduced below.[223]

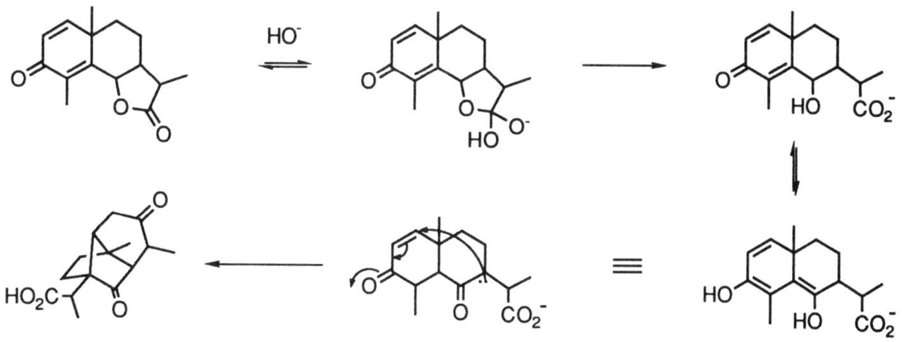

HO⁻

Corey then based his synthetic approach to longifolene on a similar intramolecular Michael addition.[223]

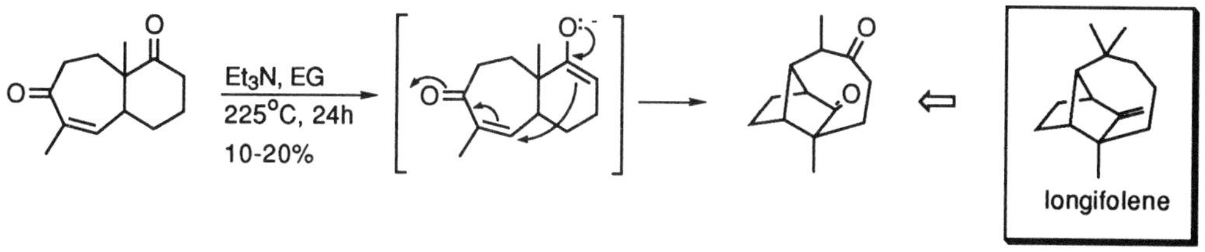

Et₃N, EG
225°C, 24h
10-20%

longifolene

There are only a few examples of the use of retro-conjugate additions in synthesis (for some applications in total synthesis, see section 4.2.4.2). However, they can be useful as in the following retro-Michael reaction which was used to prepare spiro-6,5 compounds.[48]

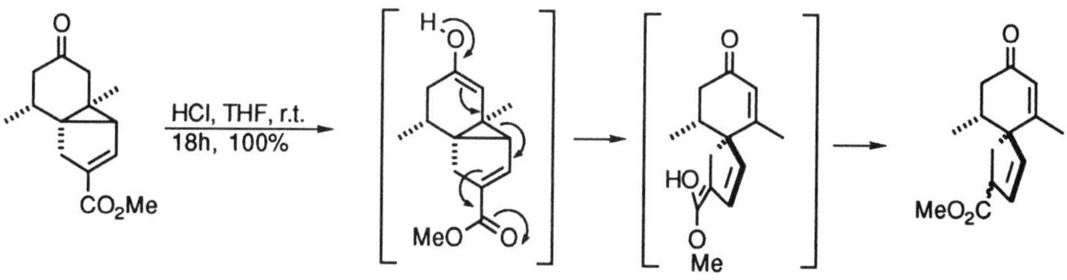

HCl, THF, r.t.
18h, 100%

Iodotrimethylsilane promotes the ring closure of α-iodoesters onto alkenones, producing δ-lactones (see section 3.1.1). This method has been used in the synthesis of quassinoid intermediates.[224] The reaction fails for tri-substituted alkenones, e.g. R =Me in the example below. However this problem can be circumvented by using a free radical ring closure of the corresponding α-bromoacetal (see section 3.2.3).

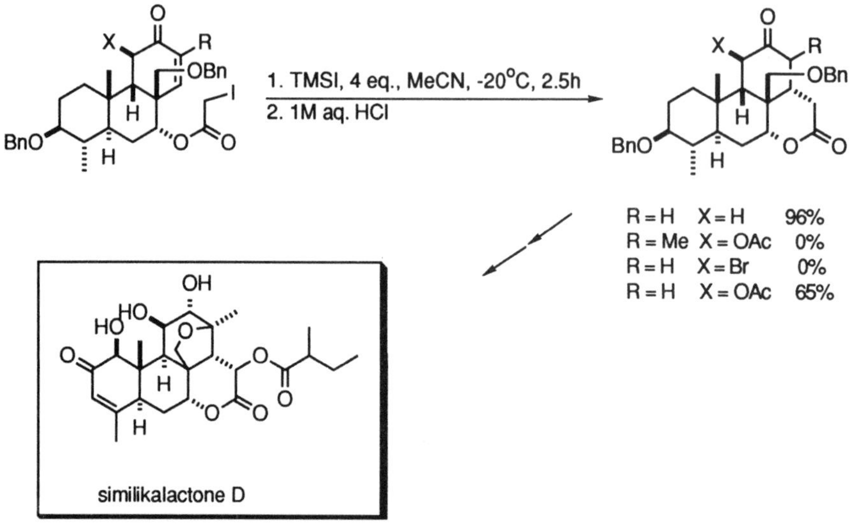

1. TMSI, 4 eq., MeCN, -20°C, 2.5h
2. 1M aq. HCl

R = H	X = H	96%
R = Me	X = OAc	0%
R = H	X = Br	0%
R = H	X = OAc	65%

similikalactone D

Intramolecular double Michael additions to 2-cyclopentenones offer a solution to the problem of synthesizing the skeleton of pentalenene.[225] Using thin layer chromatography, the authors were able to demonstrate that the reaction is not a cycloaddition, but rather it proceeds via tricyclic ketone adducts which are then converted into the silyl enol ethers.

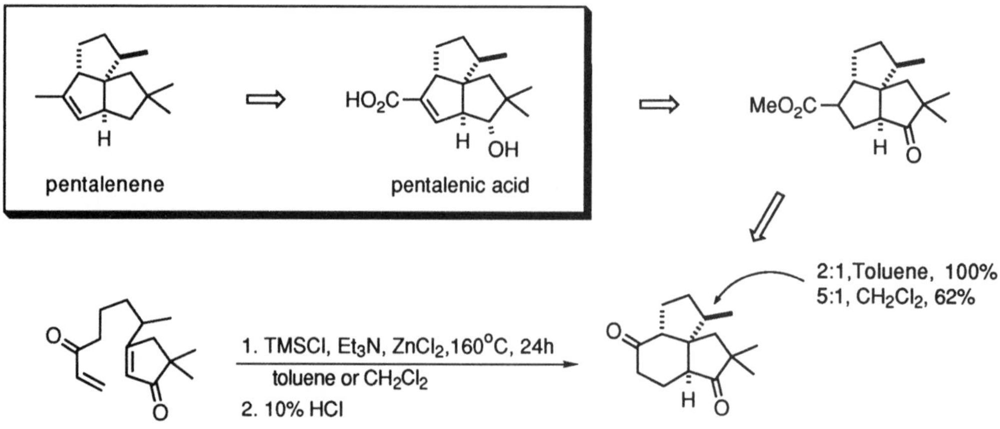

pentalenene

pentalenic acid

2:1,Toluene, 100%
5:1, CH₂Cl₂, 62%

1. TMSCl, Et₃N, ZnCl₂,160°C, 24h
   toluene or CH₂Cl₂
2. 10% HCl

3.2.1.2   *Carbanions stabilized by π-conjugation with more than one heteroatom*

Folded or cup-shaped bicyclic molecules often offer the opportunity for high levels of discrimination between two π-faces. For example, in Barco's synthesis of the sesquiterpene (±)-isoclovene, the closure of ring C proceeded with complete stereoselectivity.[226]

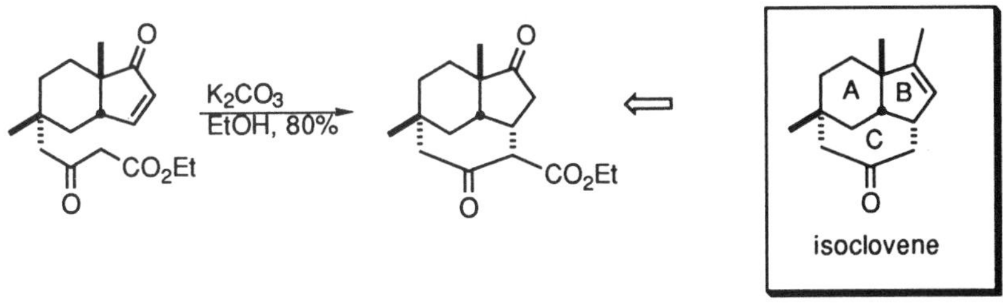

isoclovene

3.2.2   Other organometallic reagents

As with intermolecular additions, intramolecular reactions of allylsilanes have proved to be extremely productive.[227] Possibly the most important generalization to be made from the results so far obtained is that these cyclizations are very reliable for producing *cis*-fused ring systems. There are virtually no exceptions so far. This is, most likely, a consequence of the reactions occurring via a chair-like transition state (see Chapter 1 for a general discussion of intramolecular conjugate additions). This also applies to propargylsilanes.

Two catalysts are used in these reactions, a fluoride catalyst or a Lewis acid, generally titanium tetrachloride or chlorodiethylaluminium. These two systems often provide different stereochemical results. For example, treatment of the following system with each of the catalysts led to a switch in stereoselectivity at C7.[228]

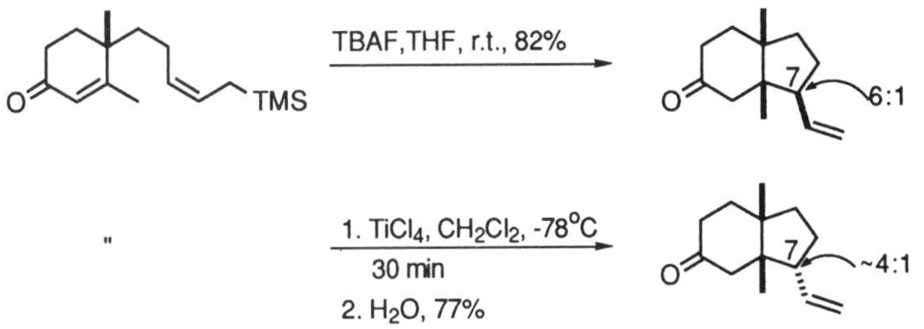

Good yields are often obtainable when the shorter (crotyl) silane is used:

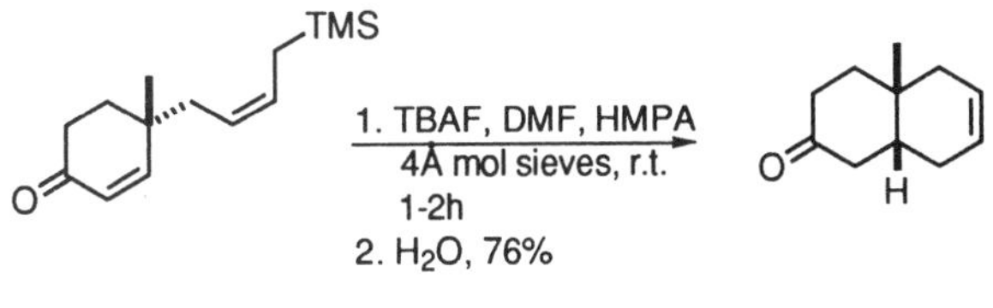

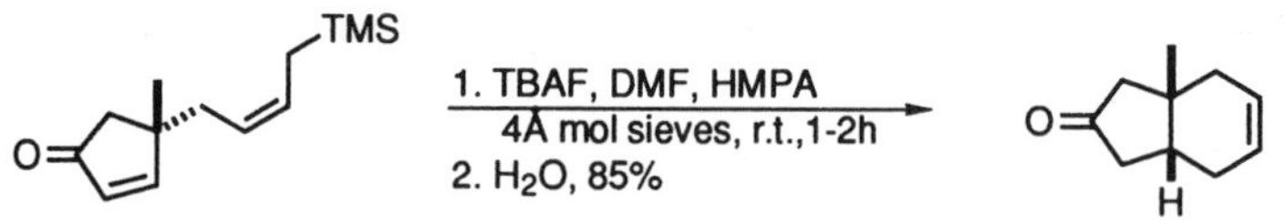

1. TBAF, DMF, HMPA
   4Å mol sieves, r.t.,1-2h
2. H₂O, 85%

1,2-Addition sometimes accompanies these reactions.

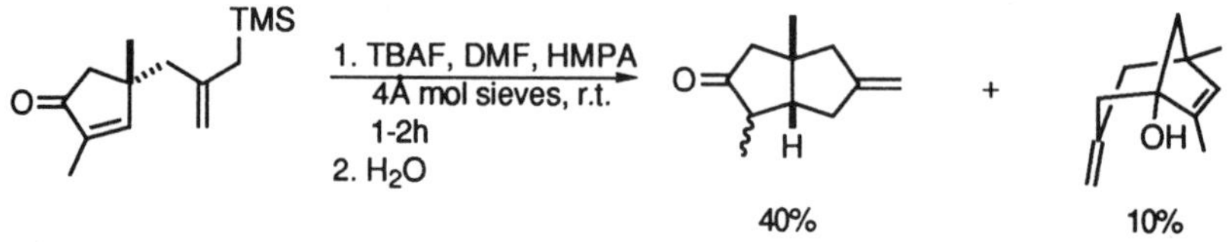

1. TBAF, DMF, HMPA
   4Å mol sieves, r.t.
   1-2h
2. H₂O

40%   10%

However, occasionally reaction occurs at the remote carbon of the allylsilane, leading to a cyclobutyl product. In the following example this pathway was accompanied by 1,2 addition.

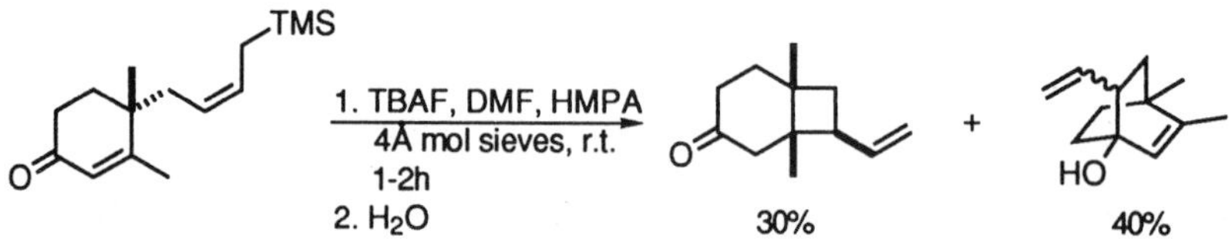

1. TBAF, DMF, HMPA
   4Å mol sieves, r.t.
   1-2h
2. H₂O

30%   40%

This phenomenon can be exploited by including an alkenyl substituent at C3. The divinylcyclobutane intermediate (23) can then undergo an enolate-accelerated Cope rearrangement, providing access to a potential synthetic intermediate for the marine sesquiterpene neolemnane.[229]

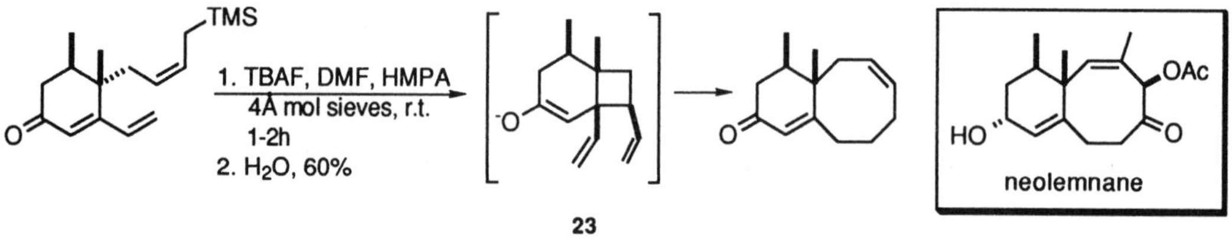

1. TBAF, DMF, HMPA
   4Å mol sieves, r.t.
   1-2h
2. H₂O, 60%

23

neolemnane

3.2.3   Free radicals[230]

Free radical ring closures have been used in the synthesis of quassinoid intermediates. This is particularly valuable as other methods fail for tri-substituted alkenones. Thus the following closure provides a useful entry into the similikalactone D system.[231]

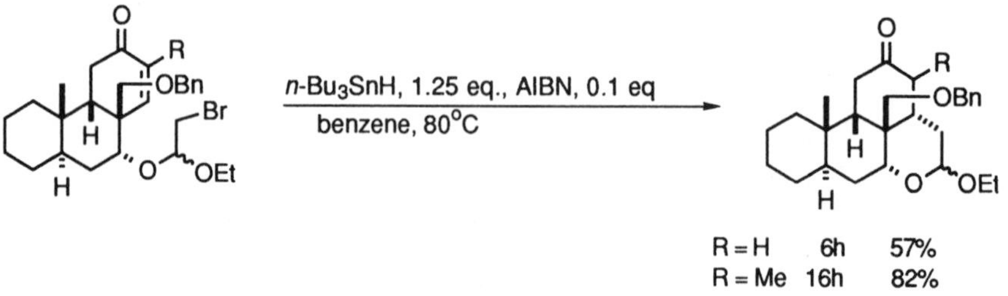

n-Bu₃SnH, 1.25 eq., AIBN, 0.1 eq
benzene, 80°C

R = H   6h   57%
R = Me   16h   82%

The bicyclo[3.2.1]octane system is readily available from the closure of alkenyl free radicals onto cycloalkenones.[232]

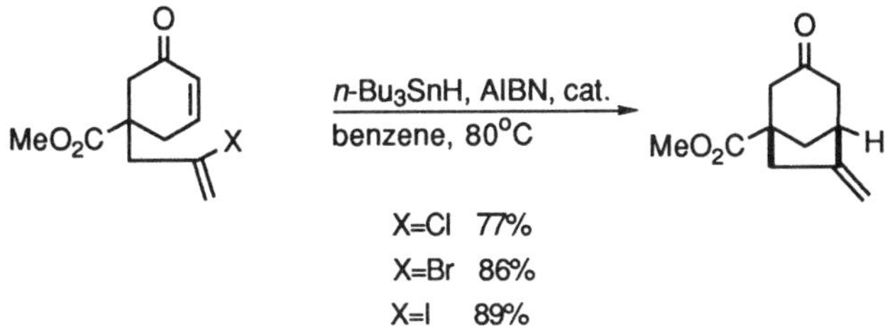
O
MeO2C
X
n-Bu3SnH, AIBN, cat.
benzene, 80°C
X=Cl 77%
X=Br 86%
X=I 89%
O
MeO2C
H

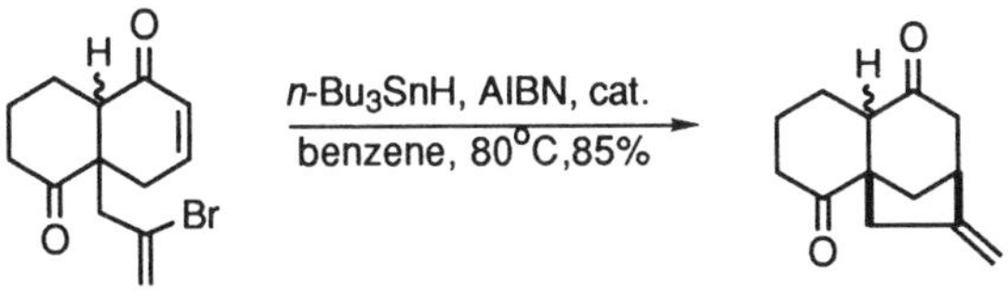
H
O
O
Br
n-Bu3SnH, AIBN, cat.
benzene, 80°C,85%
H
O
O

This also provides one of the few methods available for adding a group to a conjugate acceptor bearing a good leaving group at the β-position, without elimination of the leaving group:

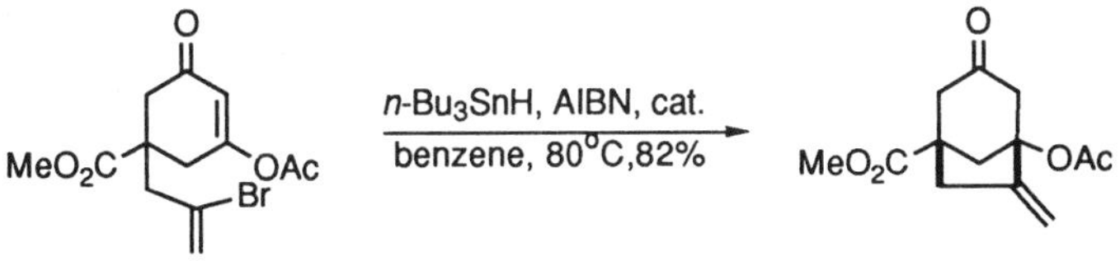
O
MeO2C
OAc
Br
n-Bu3SnH, AIBN, cat.
benzene, 80°C,82%
O
MeO2C
OAc

Other fused ring systems are also possible from this procedure.

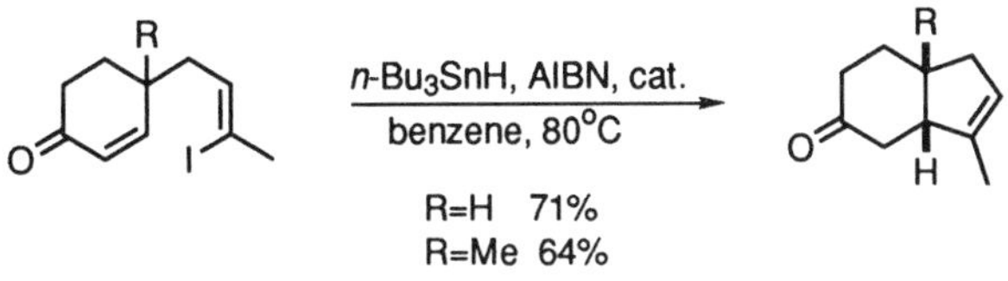
R
O
I
n-Bu3SnH, AIBN, cat.
benzene, 80°C
R=H 71%
R=Me 64%
R
O
H

Certain photoinitiated ring closures are thought to react via a free radical intermediate.[233]

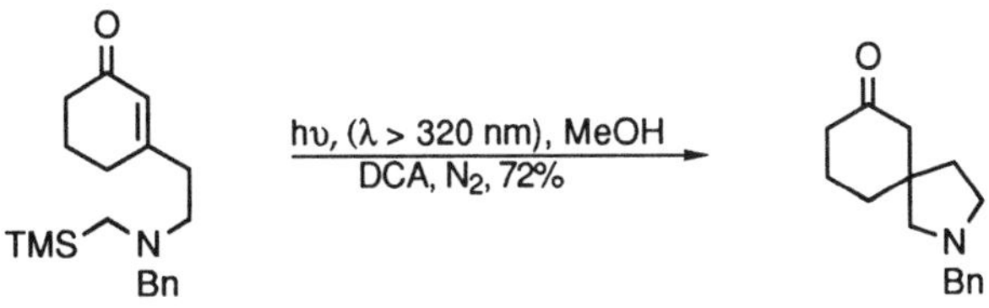
O
TMS
N
Bn
hυ, (λ > 320 nm), MeOH
DCA, N2, 72%
O
N
Bn

3.2.4 Heteronucleophiles

3.2.4.1 *Nitrogen*

Several groups have employed an intramolecular heteroconjugate addition to construct the aza-spirocyclic ring system of perhydrohistrionicotoxin, shown below.

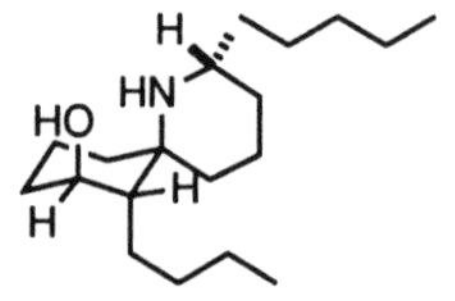
HO
HN
H
H
H

perhydrohistrionicotoxin

In their synthesis of perhydrohistrionicotoxin, Godleski and Heacock used iodotrimethylsilane to catalyze an intramolecular addition of an amine to a cycloalkenone.[234,235]

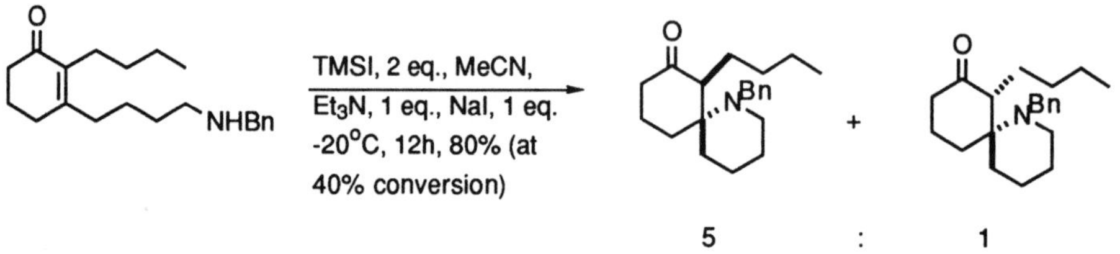
O
NHBn
TMSI, 2 eq., MeCN,
Et_3N, 1 eq., NaI, 1 eq.
-20°C, 12h, 80% (at
40% conversion)
O
NBn
+
O
NBn
5
:
1

Although this appears to be a conjugate addition of the amino group, it is more likely that the iodotrimethylsilane adds first generating an allylic iodide which is then displaced by the amine. As shown in the diagram below internal return of the proton attached to nitrogen would lead to the major diastereomer.

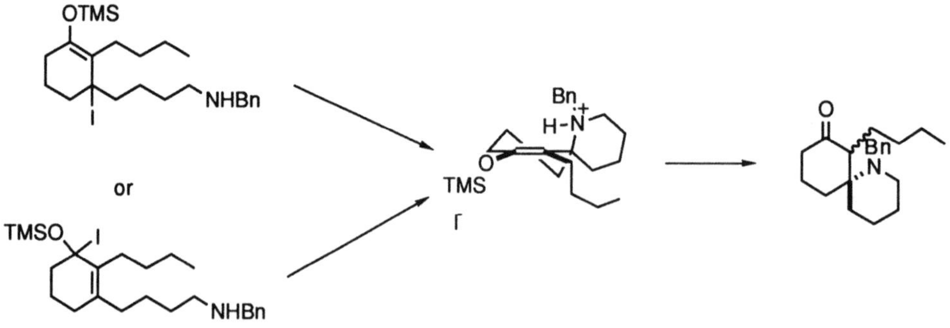
OTMS
NHBn
I
or
TMSO
I
NHBn
Bn
H-N+
O
TMS
I⁻
O
Bn
N

The product ratio of 5:1 could be improved to 13:1 using base-catalyzed epimerization of the initial product mixture. The use of aqueous hydrogen chloride or hydrogen iodide or a variety of bases did not lead to cyclization. The use of either trimethylsilyl acetate, trimethylsilyl trfluoromethanesulfonate or trimethylsilyl trifluoroacetate led to poorer yields of the cyclized material.

Kishi's group has completed two total syntheses of perhydrohistrionicotoxin, one of which employs the following acid-catalyzed cyclization.[236,237] However, the desired diastereomer was the minor product.

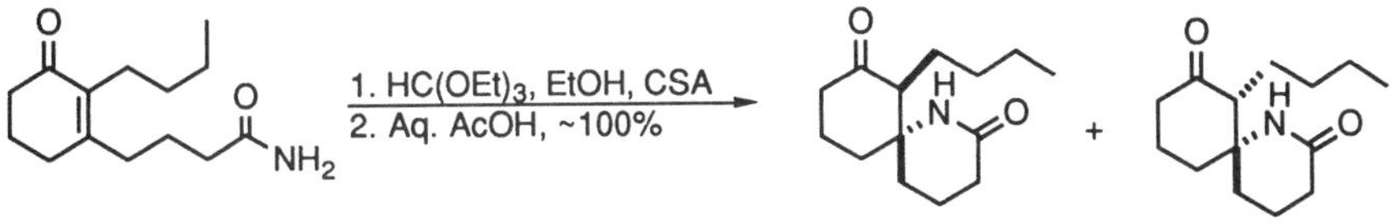

1. HC(OEt)$_3$, EtOH, CSA
2. Aq. AcOH, ~100%

1   :   2

3.2.4.2   Oxygen and sulfur

Heathcock's group attempted to employ a lactonization onto a dienone in their synthetic approach to bruceantin.[192] In their synthesis, which also includes two inter-

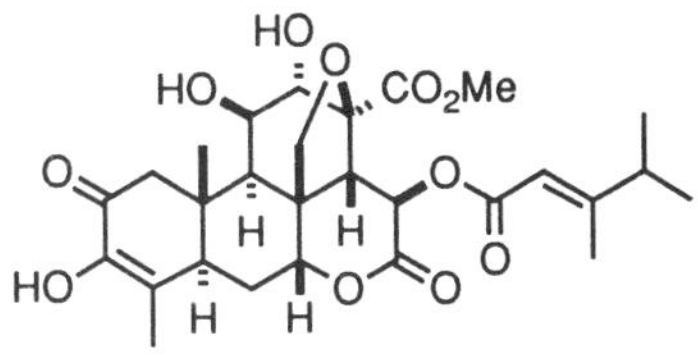

bruceantin

molecular conjugate additions,[238] they arrived at the following intermediate carboxylic acid which they envisaged could undergo intramolecular conjugate addition to yield the lactone. However, all attempts to bring about ring closure failed.

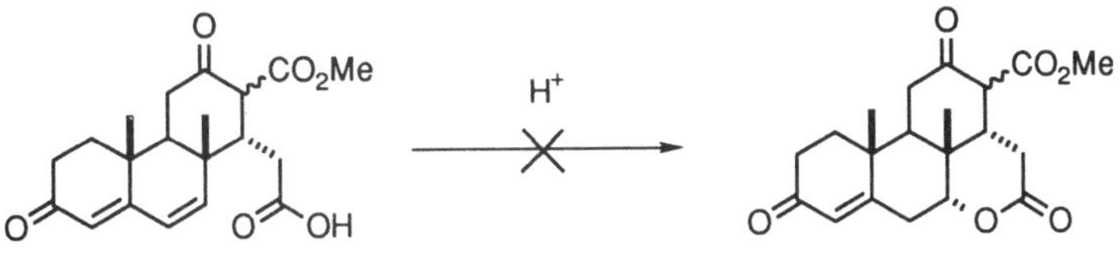

H$^+$

Valenta's group has prepared an advanced intermediate in an effort to synthesize bruceantin. An important step involved the intramolecular addition of a secondary alcohol to the dienone shown below. This is similar to the addition shown above from Heathcock's group. Reaction conditions proved to be crucial. A large excess of the acid as well as 1,3-propanediol were necessary to obtain reasonable yields of the tricyclic product.[239] The authors proposed that protonated 1,3-propanediol exerts a specific protonation/deprotonation action at the interface. Ethylene glycol also works, however the yield is not as good.

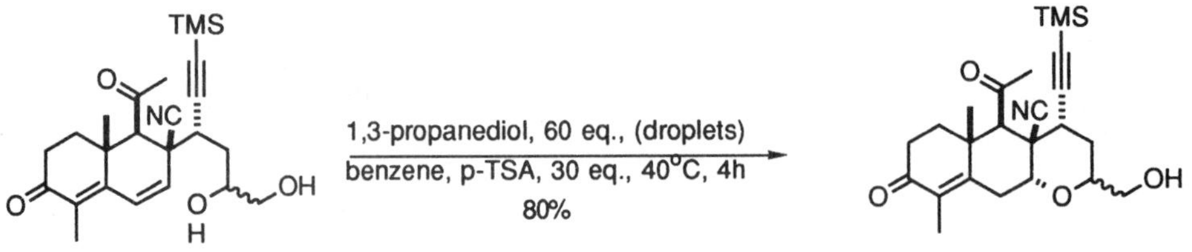

The priming mechanism of the antibiotics, calicheamicin and esperamicin, has been modelled by Danishefsky's group.[240] They found that, even though the conjugate acceptor moiety of a model for these antibiotics contains a bridgehead alkene, no conjugate addition was observed with several different nucleophiles. This prompted the authors to

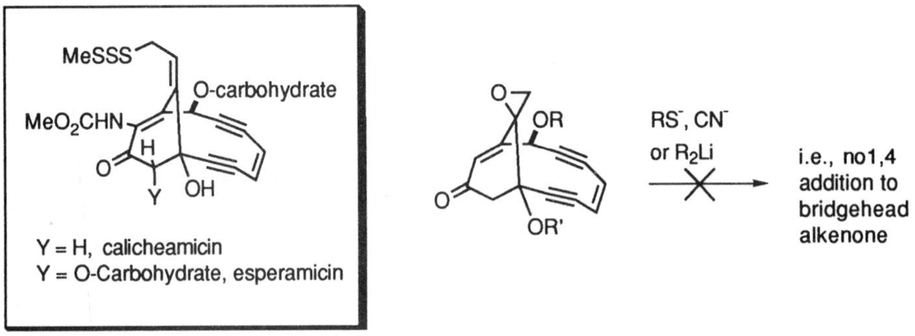

investigate intramolecular additions using a model closer in structure to the antibiotics. Not only did addition occur, but in the presence of 1,4-cyclohexadiene, a quite efficient process occurred, which included isomerization to the aromatic product. Although sulfur is shown below as the nucleophile, oxygen also worked well.

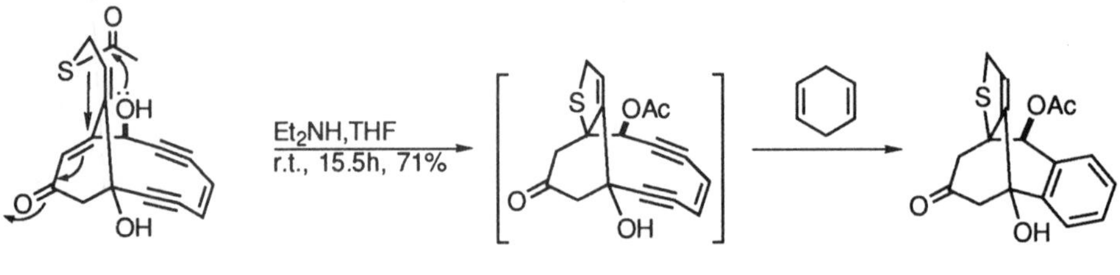

References

1. D. A. Oare and C. H. Heathcock, *J. Org. Chem.*, 1990, **55**, 157
2. E. G. Gibbons, *J. Amer. Chem. Soc.*, 1982, **104**, 1767
3. For a study of this synthetic method see, E. G. Gibbons, *J. Org. Chem.*, 1980, **45**, 1540
4. D. A. Oare and C. H. Heathcock, *J. Org. Chem.*, 1990, **55**, 132
5. M. Takasu, H. Wakabayashi, K. Furuta and H. Yamamoto, *Tetrahedron Lett.*, 1988, **29**, 6943
6. B. M. Trost, T. N. Salzmann and K. Hiroi, *J. Amer. Chem. Soc.*, 1976, **98**, 4887

7.   C. G. Gutierrez and L. R. Summerhays, *J. Org. Chem.*, 1984, **49**, 5206

8.   K. Furuta, H. Iwanaga and H. Yamamoto, *Tetrahedron Lett.*, 1986, **27**, 4507

9.   N-y. Wang, S-s. Su and L-y. Tsai, *Tetrahedron Lett.*, 1979, 1121

10.   S. Berrada and P. Metzner, *Tetrahedron Lett.*, 1987, **28**, 409

11.   M. Asaoka, S. Sonoda, N. Fujii and H. Takei, *Tetrahedron*, 1990, **46**, 1541

12.   S. V. Kessar, Y. P. Gupta, R. K. Mahazan, G. S. Joshi and A. L. Rampal, *Tetrahedron*, 1968, **24**, 899

13.   M. Pohmakotr and S. Popuang, *Tetrahedron Lett.*, 1988, **29**, 4189

14.   M. Mikolajczyk and P. Balczewski, *Tetrahedron*, 1989, **45**, 7023

15.   P. Bakuzis, M. L. Bakuzis and T. F. Weingartner, *Tetrahedron Lett.*, 1978, 2371

16.   J. F. Bagli and T. Bogri, *Tetrahedron Lett.*, 1972, 3815

17.   D. A. Anderson and J. R. Hwu, *J. Org. Chem.*, 1990, **55**, 511

18.   L. Colombo, C. Gennari, G. Resnati and C. Scolastico, *J. Chem. Soc., Perkin Trans.I*, 1981, 1284

19.   S. De Lombaert, I. Nemery, B. Roekens, J. C. Carretero, T. Kimmel and L. Ghosez, *Tetrahedron Lett.*, 1986, **42**, 5099

20.   M. R. Binns, R. K. Haynes, A. G. Katsifis, P. A. Schober and S. C. Vonwiller, *J. Amer. Chem. Soc.*, 1988, **110**, 5411; R. K. Haynes, A. G. Katsifis, S. C. Vonwiller and T. W. Hambley, *J. Amer. Chem. Soc.*, 1988, **110**, 5423

21.   See also, D. H. Hua, S. Venkataraman, R. A. Ostrander, G-Z. Sinai, P. J. McCann, M. J. Coulter and M. R. Xu, *J. Org. Chem.*, 1988, **53**, 507

22.   M. R. Binns, R. K. Haynes, D. E. Lambert and P. A. Schober, *Tetrahedron Lett.*, 1985, **26**, 3385

23.   M. Hirama, *Tetrahedron Lett.*, 1981, **22**, 1905

24.   H. Ahlbrecht and H-M. Kompter, *Synthesis*, 1983, 645

25.   J. E. Richman, J. L. Herrmann and R. H. Schlessinger, *Tetrahedron Lett.*, 1973, 3271

26.   M. Zervos and L. Wartski, *Tetrahedron Lett.*, 1984, **25**, 4641

27.   G. Stork and L. Maldonado, *J. Amer. Chem. Soc.*, 1974, **96**, 5272

28.   N. Seuren, L. Wartski and J. Seyden-Penne, *Tetrahedron Lett.*, 1981, **22**, 2175

29.   E. D. Bergmann and J. Szmuszkovicz, *J. Amer. Chem. Soc.*, 1953, **75**, 3226

30.   H-J. Liu and H. Wynn, *Can. J. Chem.*, 1986, **64**, 649

31.   M. Minamii, K. Tanaka and K. Shioda, *Jpn Kokai Tokkyo Koho* JP 60,224,656 [85, 224, 656], *Chem. Abstr.*, 1986, **104**, 224,622

32.   For a review of organic synthesis under high presssure see, K. Matsumoto, A. Sera and T. Uchida, *Synthesis*, 1985, 1

33.   W. G. Dauben and J. M. Gerdes, *Tetrahedron Lett.*, 1983, **24**, 3841

34.   W. Reid and M. Vogel, *Liebig's Ann. Chem.*, 1982, 355

35.   P. D. Bartlett and G. F. Woods, *J. Amer. Chem. Soc.*, 1940, **62**, 2933

36.   P. R. Shafer, W. E. Loeb and W. S. Johnson, *J. Amer. Chem. Soc.*, 1953, **75**, 5963

37.   R. H. Eastman, *J. Amer. Chem. Soc.*, 1954, **76**, 4115

38.   T. Sato, T. Kawara, K. Sakata and T. Fujisawa, *Bull. Chem. Soc. Japan*, 1981, **54**, 505

39.   T. Takahashi, K. Hori and J. Tsuji, *Tetrahedron Lett.*, 1981, **22**, 119

40.   H-J. Liu and H. Wynn, *Can. J. Chem.*, 1986, **64**, 649

41.   J-F. Lavallée and P. Deslongchamps, *Tetrahedron Lett.*, 1988, **29**, 5117

42.   J-F. Lavallee and P. Deslongchamps, *Tetrahedron Lett.*, 1988, **29**, 6033

43.   H. Irie, Y. Mizuno, T. Taga and K. Osaki, *J. Chem. Soc., Perkin Trans. I*, 1982, 25

44.   W. P. Blackstock, R. T. Brown and M. Wingfield, *Tetrahedron Lett.*, 1984, **25**, 1831

45.   E. Hatzigrigoriou and L. Wartski, *Synth. Commun.*, 1983, **13**, 319

46.   J. E. McMurry, W. A. Andras and J. H. Musser, *Synth. Commun.*, 1978, **8**, 53

47.   F. Bohlmann and H. Suding, *Justus Liebigs Ann. Chem.*, 1985, 160

48.   J. LaFontaine, M. Mongrain, M. Sergent-Guay, L. Ruest and P. Deslongchamps, *Can. J. Chem.*, 1980, **58**, 2460

49.   Y. Hirako, T. Mukaiyama and T. Takeda, *Chem. Lett.*, 1978, 461. For an improved synthesis, see R. T. Brown and M. J. Ford, *Synth. Commun.*, 1988, **18**, 1801

50.   R. T. Brown and M. J. Ford, *Tetrahedron Lett.*, 1990, **31**, 2029

51.   R. T. Brown and M. J. Ford, *Tetrahedron Lett.*, 1990, **31**, 2033

52.   D. H. Hua, M. J. Coulter and I. Badejo, *Tetrahedron Lett.*, 1987, **28**, 5465

53.   T. M. Dolak and T. A. Bryson, *Tetrahedron Lett.*, 1977, 1961

54.   C. A. Brown and A. Yamaichi, *J. Chem. Soc., Chem. Commun.*, 1979, 100

55.   E. J. Corey and D. Crouse, *J. Org. Chem.*, 1968, **33**, 298

56.   P. C. Ostrowski and V. V. Kane, *Tetrahedron Lett.*, 1977, 3549

57.   A. R. B. Manas and R. A. Smith, *J. Chem. Soc., Chem. Commun.*, 1975, 216

58.   See also R. A. Smith and A. R. Lal, *Aust. J. Chem*, 1979, **32**, 353 and T. Cohen and S. M. Nolan, *Tetrahedron Lett.*, 1978, 3533

59.   T. Cohen and L-C. Yu, *J. Org. Chem.*, 1985, **50**, 3265

60.   See also, K. Ramig, M. Bhupathy and T. Cohen, *J. Amer. Chem. Soc.*, 1988, **110**, 2678

61.   S. R. Wilson, R. N. Misra and G. M. Georgiadis, *J. Org. Chem.*, 1980, **45**, 2460

62.   For an extensive tabulation of masked acyl carbanions, see A. Krief, *Tetrahedron*, 1984, **36**, 2531

63.   L. Wartski, M. El-Bouz, J. Seyden-Penne, W. Dumont and A. Krief, *Tetrahedron Lett.*, 1979, 1543

64.   C. A. Brown and A. Yamaichi, *J. Chem. Soc., Chem. Commun.*, 1979, 100

65.   J. Luchetti, W. Dumont and A. Krief, *Tetrahedron Lett.*, 1979, 2695

66.   A. Krief, unpublished results mentioned in reference 62

67.   A. R. B. Manas and R. A. Smith, *J. Chem. Soc., Chem. Commun.*, 1975, 216

68.   For a compilation of dithioacetal-based systems, see B-T. Grobel and D. Seebach, *Synthesis*, 1977, 357 (Table 12)

69.   P. C. Ostrowski and V. V. Kane, *Tetrahedron Lett.*, 1977, 3549

70.   D. Seebach and R. Bürstinghaus, *Angew. Chem. Int. Ed. Engl.*, 1975, **14**, 57

71.   Copper-catalyzed - see section 2.1.2

72.   D. Seyferth and R. C. Hui, *J. Amer. Chem. Soc.*, 1985, **107**, 4551

73.   B. H. Lipshutz, E. L. Ellsworth, S. H. Dimock and R. A. J. Smith, *J. Amer. Chem. Soc.*, 1990, **112**, 4404

74.   A. Alexakis, J. Berlan and Y. Besace, *Tetrahedron Lett.*, 1986, **27**, 1047

75.   T. W. Wallace, *Tetrahedron Lett.*, 1984, **25**, 4299

76.   E. J. Corey and D. Enders, *Tetrahedron Lett.*, 1976, 11

77.   M. Hulce, *Tetrahedron Lett.*, 1988, **29**, 5851

78.   M. C. P. Yeh, P. Knochel, W. M. Butler and S. C. Berk, *Tetrahedron Lett.*, 1988, **29**, 6693

79.   See also Y. Tamaru, H. Tanigawa, T. Yamamoto and Z. Yoshida, *Angew. Chem. Int. Ed. Engl.*, 1989, **28**, 351

80.   G. Casy and R. J. K. Taylor, *Tetrahedron*, 1989, **45**, 455

81.   G. Cahiez and M. Alami, *Tetrahedron Lett.*, 1989, **30**, 3541

82.   E-L. Lindstedt and M. Nilsson, *Acta Chem. Scand. B*, 1986, **40**, 466

83.   M. Bergdahl, E-L. Linstedt, M. Nilsson and T. Olsson, *Tetrahedron*, 1989, **45**, 535

84.   Variation of reaction parameters, such as halosilane, reaction temperature and time still gave a regioisomeric mixture of products

85.   For a review, see R. J. K. Taylor, *Synthesis*, 1985, 364

86.   I. W. J. Still and Y. Shi, *Tetrahedron Lett.*, 1987, **28**, 2489

87.   R. J. Linderman and A. Godfrey, *J. Amer. Chem. Soc.*, 1988, **110**, 6249

88.   R. J. Linderman, A. Godfrey and K. Horne, *Tetrahedron*, 1989, **45**, 495

89.   S-K. Zhao, C, Knors and P. Helquist, *J. Amer. Chem. Soc.*, 1989, **111**, 8527

90.   K. Narasaka, T. Sakakura, T. Uchimaru and D. Guedin-Vuong, *J. Amer. Chem. Soc.*, 1984, **106**, 2954

91.   R. M. Coates and L. O. Sandefur, *J. Org. Chem.*, 1974, **39**, 275

92.   G. H. Posner, J. J. Sterling, C. E. Whitten, C. M. Lentz and D. J. Brunelle, *J. Amer. Chem. Soc.*, 1975, **97**, 107

93.   E. Piers and V. Karunaratne, *J. Chem. Soc., Chem. Commun.*, 1983, 935

94.   E. J. Corey, M.-c. Kang, M. C. Desai, A. K. Ghosh and I. N. Houpis, *J. Amer. Chem. Soc.*, 1988, **110**, 649.

95.   R. K. Boeckman, Jr., D. M. Blum and B. Ganem, *Org. Synth.*, 1988, **Coll. Vol. 6**, 666

96.   C. Berger, M. Franck-Neumann and G. Ourisson, *Tetrahedron Lett.*, 1968, 3451

97.   E. Piers, R. W. Britton and W. de Waal, *Can. J. Chem.*, 1969, **47**, 4307

98.   E. Piers and R. D. Smillie, *J. Org. Chem.*, 1970, **35**, 3997

99.   A. S. Kende and L. N. Jungheim, *Tetrahedron Lett.*, 1980, **21**, 3849

100.   J. R. Behling, K. A. Babiak, J. S. Ng, A. L. Campbell, R. Moretti, M. Koerner and B. H. Lipshutz, *J. Amer. Chem. Soc.*, 1988, **110**, 2641

101.   J. Yoshida, S. Nakatani and S. Isoe, *J. Org. Chem.*, 1989, **54**, 5655

102.   R. E. Ireland and P. Wipf, *J. Org. Chem.*, 1990, **55**, 1425

103.   E. J. Corey and R. H. Wollenberg, *Tetrahedron Lett.*, 1976, 4705.

104.   L. A. Paquette and J. Wright, G. J. Drtina and R. A. Roberts, *J. Org. Chem.*, 1987, **52**, 2960

105.   G. Stork, I. M. Paterson and F. K. C. Lee, *J. Amer. Chem. Soc.*, 1982, **104**, 4686

106.   M. Rowley and Y. Kishi, *Tetrahedron Lett.*, 1988, **29**, 4909

107.   M. Asaoka, K. Takenouchi and H. Takei, *Tetrahedron Lett.*, 1988, **29**, 325

108.   S. T. Saengchantara and T. W. Wallace, *J. Chem. Soc., Chem. Commun.*, 1986, 1592

109.   For a review of the total synthesis of prostaglandins, see E. J. Corey and X-M. Cheng, "The Logic of Chemical Synthesis", Wiley, New York, 1989; Chapter 11

110.   C. J. Sih, R. G. Salomon, P. Price, G. Peruzzoti and R. Sood, *J. Chem. Soc., Chem. Commun.*, 1972, 240

111.   C. J. Sih, P. Price, R. Sood, R. G. Salomon, G. Peruzzoti and M. Casey, *J. Amer. Chem. Soc.*, 1972, **94**, 3643

112.   See also, F. S. Alvarez, D. Wren and A. Prince, *J. Amer. Chem. Soc.*, 1972, **94**, 7823

113.   A. F. Kluge, K. G. Untch and J. H. Fried, *J. Amer. Chem. Soc.*, 1972, **94**, 7827 and *J. Amer. Chem. Soc.*, 1972, **94**, 9256

114.   D. A. Evans and G. C. Andrews, *Accounts. Chem. Res.*, 1974, **7**, 147

115.   J. G. Miller, W. Kurz, K. G. Untch and G. Stork, *J. Amer. Chem. Soc.*, 1974, **96**, 6774

116.   S-M. L. Chen, R. E. Schaub and C. V. Grudzinskas, *J. Org. Chem.*, 1978, **43**, 3450

117.   M. Suzuki, A. Yanagisawa and R. Noyori, *J. Amer. Chem. Soc.*, 1985, **107**, 3348; M. Suzuki, A. Yanagisawa and R. Noyori, *J. Amer. Chem. Soc.*, 1988, **110**, 4718. For reviews of the three-component approach, see R. Noyori, *Chem. In. Brit.*, 1989, **25**, 883, A. D. Baxter and S. M. Roberts, *Chem. and Ind.*, 1986, 510 and

118.   K. Ogura, M. Yamashita and G-i. Tsuchihashi, *Tetrahedron Lett.*, 1976, 759

119.   M. Kitamura, I. Kasahara, K. Manabe, R. Noyori and H. Takaya, *J. Org. Chem.*, 1988, **53**, 708

120.   Y. Kitano, S. Okamoto and F. Sato, *Chem. Lett.*, 1989, 2163

121.   E. Mezzina, *J. Chem. Soc., Perkin Trans. I*, 1989, 845

122.   R. Davis and K. G. Untch, *J. Org. Chem.*, 1979, **44**, 3755

123.   M. Suzuki, T. Kawagishi and R. Noyori, *Tetrahedron Lett.*, 1982, **23**, 5563; G. Stork and M. Isobe, *J. Amer. Chem. Soc.*, 1975, **97**, 6260

124.   T. Tanaka, S. Kurozumi, T. Toru, M. Kobayashi, S. Miura and S. Ishimoto, *Tetrahedron Lett.*, 1975, 1535; *ibid, Tetrahedron*, 1977, 1105.

125.   T. Toru, S. Kurozumi, T. Tanaka, S. Miura, M. Kobayashi, and S. Ishimoto, *Tetrahedron Lett.*, 1976, 4087; A. Hazato, T. Tanaka, K. Watanabe, K. Bannai, T. Toru, N. Okamura, K. Manabe, A. Ohtsu, F. Kamimoto and S. Kurozumi, *Chem. Pharm. Bull.*, 1985, **33**, 1815

126.   C. R. Johnson and T. D. Penning, *J. Amer. Chem. Soc.*, 1986, **108**, 5655; *ibid.*, 1988, **110**, 4726.

127.   P. A. Greico and N. Abood, *J. Chem. Soc., Chem. Commun.*, 1990, 410

128.   A. B. Jones, M. Yamaguchi, A. Patten, S. J. Danishefsky, J. A. Ragan, D. B. Smith and S. L. Schreiber, *J. Org. Chem.*, 1989, **54**, 17

129.   M. Asaoka, K. Shima and H. Takei, *J. Chem. Soc., Chem. Commun.*, 1988, 430

130.   Y. Horiguchi, E. Nakamura and I. Kuwajima, *J. Org. Chem.*, 1986, **51**, 4323

131.   A. R. Matlin and W. C. Agosta, *J. Chem. Soc., Perkin Trans. 1*, 1987, 365

132.   E. J. Corey and N. W. Boaz, *Tetrahedron Lett.*, 1985, **26**, 6015

133.   W. C. Still and I. Galynker, *Tetrahedron*, 1981, **37**, 3981

134.   In all cases, peripheral attack is assumed.

135.   N. L. Allinger, *J. Amer. Chem. Soc.*, 1977, **99**, 8127

136.   A. B. Smith and P. J. Jerris, *J. Amer. Chem. Soc.*, 1981, **103**, 194.

137.   R. Tamura, K-i. Watabe, H. Katayama, H. Suzuki and Y. Yamamoto, *J. Org. Chem.*, 1990, **55**, 408

138.   F. E. Zeigler and B. H. Jaynes, *Tetrahedron Lett.*, 1988, **29**, 2031. For a slightly less selective addition, see B. Delpech and R. Lett, *Tetrahedron Lett.*, 1987, **28**, 4061

139.   P. A. Wender, H. Y. Lee, R. S. Wilhelm and P. D. Williams, *J. Amer. Chem. Soc.*, 1989, **111**, 8954

140.   S. F. Martin, G. J. Pacofsky, R. P. Gist and W-C. Lee, *J. Amer. Chem. Soc.*, 1989, **111**, 7634

141.   R. D. Clark and C. H. Heathcock, *Tetrahedron Lett.*, 1974, 1713

142.   L. A. Paquette, R. A. Roberts and G. J. Ortina, *J. Amer. Chem. Soc.*, 1984, **106**, 6690

143.	J. A. Marshall and R. A. Ruden, *J. Org. Chem.*, 1972, **37**, 659
144.	A. J. Birch and R. Robinson, *J. Chem. Soc.*, 1943, 501. See also A. J. Birch and M. Smith, *Proc. Chem. Soc.*, 1962, 356
145.	E. J. Corey, R. Naef and F. J. Hannon, *J. Amer. Chem. Soc.*, 1986, **108**, 7115
146.	R. K. Dieter and M. Tokles, *J. Amer. Chem. Soc.*, 1987, **109**, 2040
147.	K. Yamamoto, M. Kanoh and N. Yamamoto, *Tetrahedron Lett.*, 1987, **28**, 6347
148.	K. Yamamoto, M. Iijima, Y. Ogimura and J. Tsuji, *Tetrahedron Lett.*, 1984, **25**, 2813
149.	K. Yamamoto, M. Iijima and Y. Ogimura, *Tetrahedron Lett.*, 1982, **23**, 3711
150.	S. H. Bertz , G. Dabbagh and G. Sundararajan, *J. Org. Chem.*, 1986, **51** (25), 4953
151.	B. E. Rossiter and M. Eguchi, *Tetrahedron Lett.*, 1990, **31**, 965
152.	H. Malmberg, M. Nilsson and C. Ullenius, *Tetrahedron Lett.*, 1982, **23**, 3823
153.	See also, M. Nilsson, *Chemica Scripta*, 1985, **25**, 79
154.	H. Malmberg, M. Nilsson and C. Ullenius, *Acta Chem. Scand. B*, 1981, **35**, 625
155.	S. Andersson, S. Jagner, M. Nilsson and F. Urso, *J. Organomet. Chem.*, 1986, **301**, 257
156.	R. G. Cooke and I. J. Rainbow, *Aust. J. Chem.*, 1977, **30**, 2241
157.	A. L. Chaffee, R. G. Cooke, I. J. Dagley, P. Perlmutter and R. L. Thomas, *Aust. J. Chem.*, 1981, **34**, 587
158.	R. A. Watson and R. A. Kjonaas, *Tetrahedron Lett.*, 1986, **27**, 1437
159.	J. F. G. A. Jansen and B. L. Feringa, *Tetrahedron Lett.*, 1988, **29**, 3593
160.	J. F. G. A. Jansen and B. L. Feringa, *J. Chem. Soc., Chem. Commun.*, 1989, 741
161.	E. J. Corey and L. S. Hegedus, *J. Amer. Chem. Soc.*, 1969, **91**, 4926
162.	M. P. Cooke, Jr. and R. M. Parlman, *J. Amer. Chem. Soc.*, 1977, **99**, 5222
163.	A. Hosomi and H. Sakurai, *J. Amer. Chem. Soc.*, 1977, **99**, 1673
164.	J-I. Hurata, K. Nishi, S. Matsuda, Y. Tamura and Y. Kita, *J. Chem. Soc., Chem. Commun.*, 1989, 1065
165.	T. Tokoroyama and L-R. Pan, *Tetrahedron Lett.*, 1989, **30**, 197
166.	For an indirect solution to this problem, see E. J. Corey and R. H. Wollenburg, *J. Amer. Chem. Soc.*, 1974, **96**, 5581
167.	J. Schwartz, D. B. Carr, R. T. Hansen and F. M. Dayrit, *J. Org. Chem.*, 1980, **45**, 3053
168.	W. Treibs, *Chem. Ber.*, 1944, **77**, 572
169.	A. Kover and H. M. R. Hoffmann, *Tetrahedron*, 1988, **44**, 6831.
170.	E. J. Corey and R. M. Burk, *Tetrahedron Lett.*, 1987, **28**, 6413
171.	For a study of the Sakurai reaction with cycloheptenones, see T. A. Blumenkopf and C. H. Heathcock, *J. Amer. Chem. Soc.*, 1983, **105**, 2354
172.	For a review of this method and its applications, see G. H. Posner, *Accts. Chem. Res.*, 1987, **20**, 72
173.	M. Hulce, J. P. Mallamo, L. L. Frye, T. P. Kogan and G. H. Posner, *Org. Synth.*, 1985, **64**, 196
174.	G. H. Posner annd M. Hulce, *Tetrahedron Lett.*, 1984, **25**, 379
175.	For an X-ray crystallographic study of these molecules, see G. H. Posner, M. Weitzberg, T. G. Hamill, E. Asirvartham, H. Cun-Heng and J. Clardy, *Tetrahedron*, 1986, **42**, 2919
176.	G. H. Posner and C. Switzer, *J. Amer. Chem. Soc.*, 1986, **108**, 1239

177.   For an extensive survey and thorough review of the generation and reactions of carbon radicals, see A. Ghosez, B. Giese and H. Zipse in M. Regitz and B. Giese (eds), "Houben-Weyl. Methoden der organischen Chemie", Georg Thieme, Stuttgart, 1989, Band E19a/Teil 2

178.   R. Scheffold and L. R. Orlinski, *J. Amer. Chem. Soc.*, 1983, **105**, 7200

179.   R. Scheffold, *Chimia*, 1985, **39**, 203

180.   J. L. Luche and C. Allavena, *Tetrahedron Lett.*, 1988, **29**, 5373

181.   J. L. Luche and C. Allavena, *Tetrahedron Lett.*, 1988, **29**, 5369

182.   G. A. Kraus and P. Gottschalk, *Tetrahedron Lett.*, 1983, **24**, 2727

183.   K. Narasaka, K. Soai, Y. Aikawa and T. Mukaiyama, *Bull. Soc. Chem. Jap.*, 1976, **49**, 779

184.   T. Mukaiyama, Y. Sagawa and S. Kobayashi, *Chem. Lett.*, 1986, 1017

185.   T. Mukaiyama, Y. Sagawa and S. Kobayashi, *Chem. Lett.*, 1986, 1821

186.   T. Mukaiyama, M. Tamura and S. Kobayashi, *Chem. Lett.*, 1986, 1817

187.   S. Kobayashi and T Mukaiyama, *Chem. Lett.*, 1986, 1805

188.   S. P. Tannis, M. C. McMills, T. A. Scahill and D. A. Kloosterman, *Tetrahedron Lett.*, 1990, **31**, 1997

189.   T. Mukaiyama, M. Tamura and S. Kobayashi, *Chem. Lett.*, 1986, 1817

190.   M. T. Reetz, H. Heimbach and K. Schwelhus, *Tetrahedron Lett.*, 1984, **25**, 511

191.   S. Kobayashi and T. Mukaiyama, *Chem. Lett.*, 1986, 1805

192.   C. H. Heathcock, C. Mahaim, M. F. Schlecht and T. Utawanit, *J. Org. Chem.*, 1984, **49**, 3264

193.   The methyl, rather than *tert*-butyl, ketene acetal was used in this experiment

194.   G. A. Kraus and P. Gottschalk, *Tetrahedron Lett.*, 1983, **24**, 2727

195.   For a review, see W. Nagata and M. Yoshioka, *Org. Reactions*, 1977, **25**, 255

196.   M. P. Mertes, A. A. Ramsey, P. E. Hanna and D. D. Miller, *J. Med. Chem.*, 1970, **13**, 789

197.   W. Nagata, M. Yoshioka and M. Murakami, *J. Amer. Chem. Soc.*, 1972, **94**, 4654

198.   H. Newman and T. L. Fields, *Tetrahedron*, 1972, **28**, 4051

199.   N. K. Chakravarty and D. K. Banarjee, *J. Indian Chem. Soc.*, 1946, **23**, 377

200.   I. N. Nazarov and S. I. Zav'yalov, *J. Gen. Chem. USSR (Engl. Transl.)*, 1954, **24**, 475

201.   J. Katsube and M. Matsui, *Agr. Biol. Chem.* (Tokyo), 1971, **35**, 401

202.   For a detailed discussion, see P. Deslongchamps, "Stereoelectronic Effects in Organic Chemistry", Pergamon Press, Oxford, 1983, 221-226

203.   K. Utimoto, Y. Wakabayashi, T. Horiie, M. Inoue, Y. Shishiyama, M. Obayashi and H. Nozaki, *Tetrahedron*, 1983, **39**, 967

204.   R. K. Hill and L. A. Renbaum, *Tetrahedron*, 1982, **38**, 1959

205.   A. Itoh, S. Ozawa, K. Oshima and H. Nozaki, *Bull. Chem. Soc. Jap.*, 1981, **54**, 278

206.   J. I. Levin, *Tetrahedron Lett.*, 1989, **30**, 13

207.   W. C. Still, *J. Org. Chem.*, 1976, **41**, 3063

208.   P. F. Hudrlik, A. M. Hudrlik, T. Yimenu, M. A. Waugh and G. Nagendrappa, *Tetrahedron*, 1988, **44**, 3791

209.   T. Hayashi, Y. Matsumoto and Y. Ito, *Tetrahedron Lett.*, 1988, **29**, 4147

210.   W. Tuckmantel, K. Oshima and H. Nozaki, *Chem. Ber.*, 1986, **119**, 1581

211.   G. H. Posner and E. Asirvatham, *Tetrahedron Lett.*, 1986, **27**, 663

212.   G. H. Posner, E. Asirvatham, K. S. Webb and S-s. Jew, *Tetrahedron Lett.*, 1987, **28**, 5071

213.   J. E. Baldwin, R. M. Adlington and J. Robertson, *J. Chem. Soc., Chem. Commun.*, 1988, 1404

214.   R. D. Miller and D. R. McKean, *Tetrahedron Lett.*, 1979, 2305

215.   M. F. Semmelhack, R. D. Stauffer and A. Yamashita, *J. Org. Chem.*, 1977, **42**, 3180

216.   E. Keinan, *Pure and Appl. Chem.*, 1989, **61**, 1737

217.   W. S. Mahoney, D. M. Brestensky and J. M. Stryker, *J. Amer. Chem. Soc.*, 1988, **110**, 291

218.   For an improved preparation of this reagent, see D. M. Brestensky, D. E. Huseland, C. McGettigan and J. M. Stryker, *Tetrahedron Lett.*, 1988, **29**, 3749

219.   G. Hooslef, *Førhandlingar vid Skandinaviska Naturforskavemøtat*, 1863, 304

220.   S. Cannizzaro and F. Sestine, *Gazz. Chim. Ital.*, 1873, **2**, 241

221.   H. Schiff, *Chem. Ber.*, 1873, **6**, 1201

222.   R. B. Woodward, F. J. Britschy and H. Baer, *J. Amer. Chem. Soc.*, 1948, **70**, 4217

223.   E. J. Corey, M. Ohno, R. B. Mitra and P. A. Vatakencherry, *J. Amer. Chem. Soc.*, 1964, **86**, 478

224.   M. Voyle, N. K. Dunlap, D. S. Watt and O. P. Anderson, *J. Org. Chem.*, 1983, **48**, 3242

225.   M. Ihara, M. Katogi, K. Fukumoto and T. Kametani, *J. Chem. Soc., Perkin Trans. I*, 1988, 2963

226.   P. G. Baraldi, A. Barco, S. Benetti, G. P. Pollini, E. Polo and D. Simonetta, *J. Org. Chem.*, 1985, **50**, 23

227.   For a review, see D. Schinzer, *Synthesis*, 1988, 263

228.   G. Majetich, J. Defauw, K. Hull and T. Shawe, *Tetrahedron Lett.*, 1985, **26**, 4711

229.   G. Majetich and K. Hull, *Tetrahedron Lett.*, 1988, **29**, 2773

230.   For an early example of these reactions, see S. Danishefsky, S. Chackalamannil and B-J. Uang, *J. Org. Chem.*, 1982, **47**, 2231

231.   M. Kim, R. S. Gross, H. Sevestre, N. K. Dunlap and D. S. Watt, *J. Org. Chem.*, 1988, **53**, 93; see also, M. Kim, K. Kawada, R. S. Gross and D. S. Watt, *J. Org. Chem.*, 1990, **55**, 504

232.   N. N. Marinovic and H. Ramanathan, *Tetrahedron Lett.*, 1983, **24**, 1871

233.   W. Xu, Y. T. Jeon, E. Hasegawa, U. C. Yoon and P. S. Mariano, *J. Amer. Chem. Soc.*, 1989, **111**, 406

234.   S. A. Godleski and D. J. Heacock, *J. Org. Chem.*, 1982, **47**, 4820

235.   See also M. Glanzmann, C. Karalai, B. Ostersehlt, U. Schon, C. Frese and E. Winterfeldt, *Tetrahedron*, 1982, **38**, 2805

236.   T. Fukuyama, L. V. Dunkerton, M. Aratani and Y. Kishi, *J. Org. Chem.*, 1975, **40**, 2011

237.   See also, J. J. Venit and P. Magnus, *Tetrahedron Lett.*, 1980, **21**, 4815

238.   See sections 2.1.1.1 and 3.1.5 for these reactions

239.   S. Darvesh, A. S. Grant, D. I. MaGee and Z. Valenta, *Can. J. Chem.*, 1989, **67**, 2237

240.   J. N. Haseltine, S. J. Danishefsky and G. Schulte, *J. Amer. Chem. Soc.*, 1989, **111**, 7638

241.   See also K. Tanaka and H. Suzuki, *J. Chem. Soc., Chem. Commun.*, 1991, 101

---

4   Introduction

Conjugate additions to alkenoic acid derivatives form one of the most intensively studied areas in this field. They have at least one major advantage over many other conjugate acceptors in that they possess the ability to attach a temporary auxiliary, usually as either an ester or amide. These play an important role in controlling the stereoselectivity of additions as well as sometimes being used to "attenuate" the reactivity of the conjugate acceptor.

All classes of nucleophiles covered in this book add in a conjugate fashion to alkenoic acid derivatives. The conjugate addition of organocopper reagents, which was once regarded as somewhat problematic, has now been developed into a series of reliable procedures.

4.1   Intermolecular Additions

4.1.1   Stabilized carbanions

4.1.1.1   *Carbanions stabilized by π-conjugation with one heteroatom*

(a)   *Addition of achiral carbanions to achiral alkenoate derivatives*

Ketone enolates

The reaction of cyclohexanone enolate with a propenoate has not proved to be the most popular way of preparing the corresponding adduct. In fact, in the case of cyclohexanone, enamine chemistry appears to be preferred (see section 4.1.5.1). In a study on perhydroindanones, House used both enolate and enamine chemistry to obtain both regioisomeric adducts, selectively, from 2-methylcyclohexanone.[1] However, these additions are often complicated by regioisomeric product mixtures as well as dialkylation.[2] (Attempts to quench various cyclohexanone enolates, generated from silyl enol ethers, with propenenitrile, were unsuccessful[3]).

Reaction of 2-ethylcyclopentanone with propenenitrile only proceeds with modest selectivity (Table 4.1, entry 1). However, the product may be used to generate the bicyclic imine, which has been used in a total synthesis of (±)-vallesamidine, in almost quantitative yield.[4]

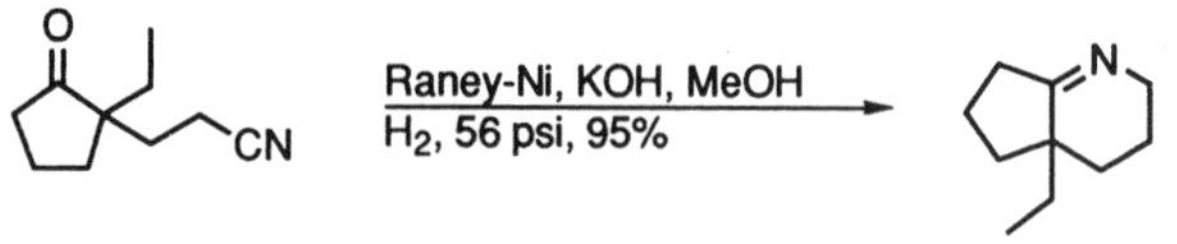

Raney-Ni, KOH, MeOH
H₂, 56 psi, 95%

Table 4.1 Conjugate addition of some ketone enolates to alkenoic acid derivatives

14

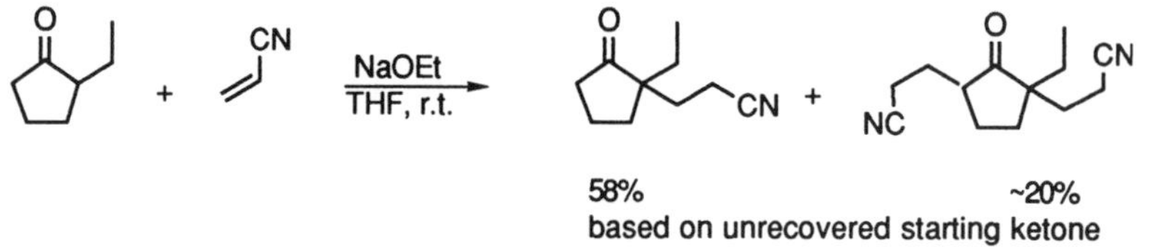

NaOEt
THF, r.t.

58%

~20%
based on unrecovered starting ketone

25

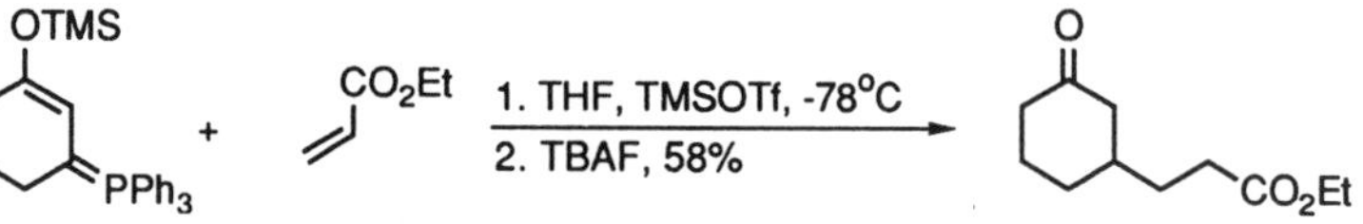

1. THF, TMSOTf, -78°C
2. TBAF, 58%

36

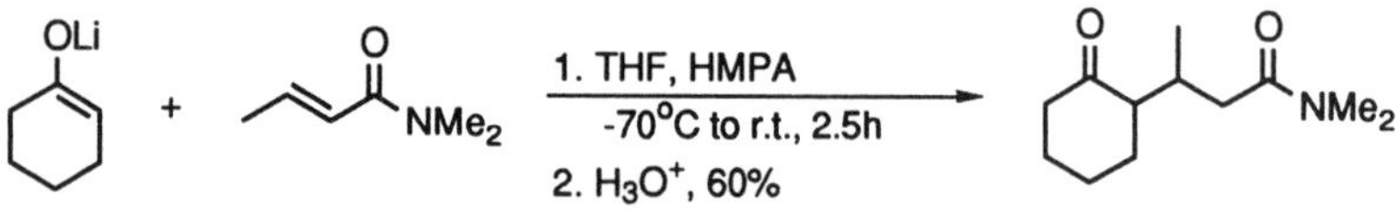

1. THF, HMPA
-70°C to r.t., 2.5h
2. H₃O⁺, 60%

47

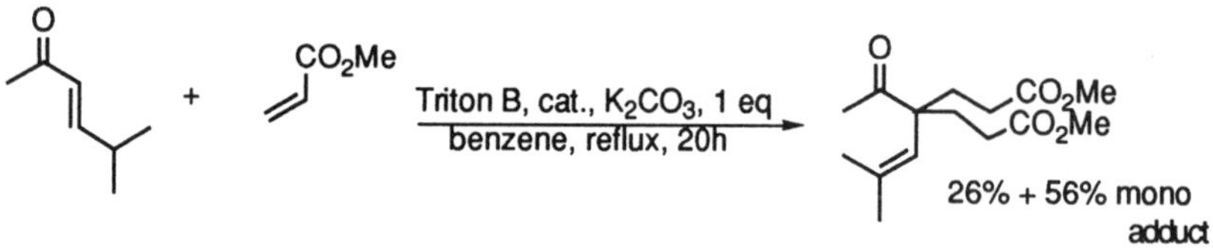

Triton B, cat., K₂CO₃, 1 eq
benzene, reflux, 20h

26% + 56% mono adduct

The minor product shown in entry 4 was used in a total synthesis of the alkaloid, (±)-19-hydroxyaspidofractinine.

(±)-19-hydroxyaspidofractinine

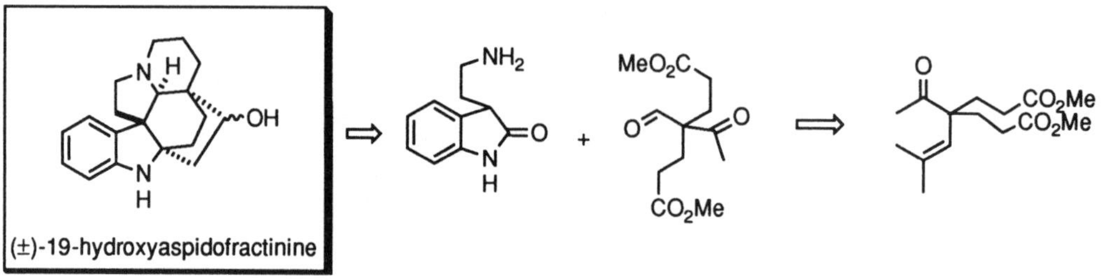

Conjugate addition to crotonamides, followed by quenching with an electrophile, provides a useful method for preparing *syn* adducts.[8]

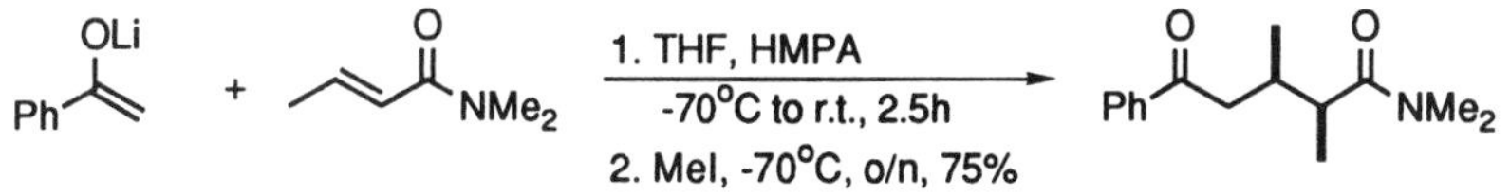

OLi
Ph
+
O
NMe2
1. THF, HMPA
-70°C to r.t., 2.5h
2. MeI, -70°C, o/n, 75%
Ph
O
O
NMe2

One of the many methods in the rapidly evolving area of multi-component reactions is initiated by a ketone enolate conjugate addition to methyl 2-bromopropenoate[9] (an MIMIRC reaction, in Posner's nomenclature[10]).

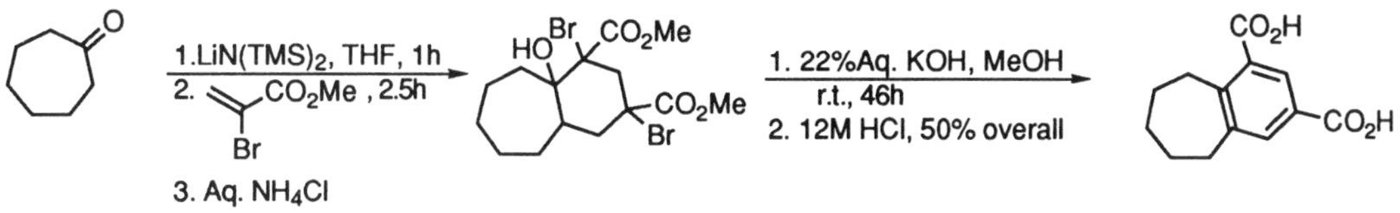

O
1.LiN(TMS)2, THF, 1h
2. CO2Me , 2.5h
Br
3. Aq. NH4Cl
HO Br CO2Me
CO2Me
Br
1. 22%Aq. KOH, MeOH
r.t., 46h
2. 12M HCl, 50% overall
CO2H
CO2H

Lee first demonstrated that the reaction of cyclic dienolates, e.g. (**1**) with alkenoates yields effectively the same products as those obtained from the cycloaddition of the corresponding diene and alkenoate.[11] The reaction is remarkably stereoselective. In each of the cases below a single diastereomer was formed. This process, in both the inter- and intra-(see section 4.2.1.1) molecular form, has been applied to the synthesis of several natural products.[12]

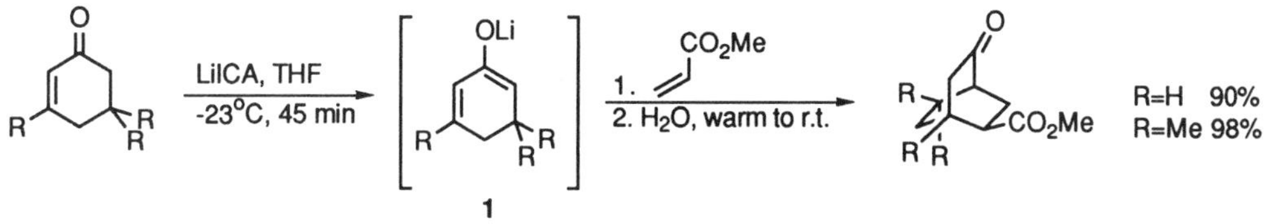

O
R
R
R
LiICA, THF
-23°C, 45 min
OLi
R
R
R
1
CO2Me
1.
2. H2O, warm to r.t.
O
R
R R
CO2Me
R=H 90%
R=Me 98%

If a halogen is attached to C2 of the alkenoate, then a tricyclic ketone is the final product.[13]

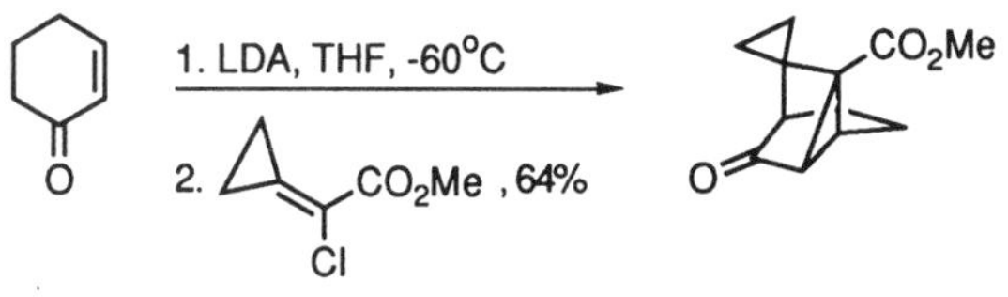

O
1. LDA, THF, -60°C
2.
CO2Me , 64%
Cl
CO2Me
O

The ability of hindered hydrazones, e.g. (**2**) to C-metallate rather than N-metallate has been exploited in conjugate additions to alkenoates.[14] The product can either be isolated and hydrolyzed to the 3-oxo-derivative or reduced *in situ*.

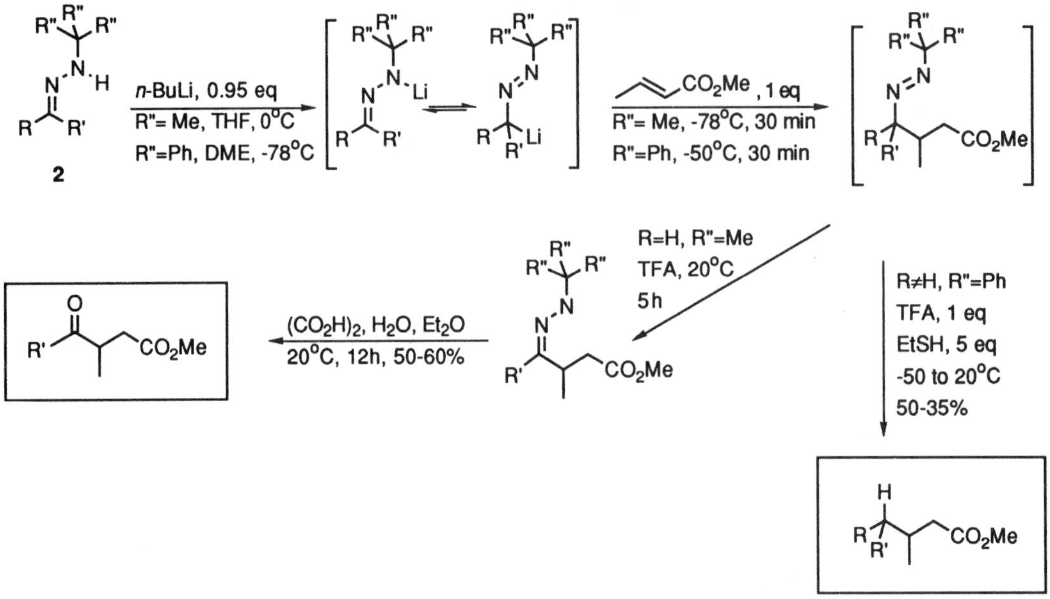

Thermal additions were also successful with the *tert*-butylhydrazones. However, the products could not be reductively cleaved cleanly, so the phenylhydrazones were used instead. These added well, as has already been observed,[15] and reductive cleavage of the azo-ester provides a new route to γ-aminoesters.

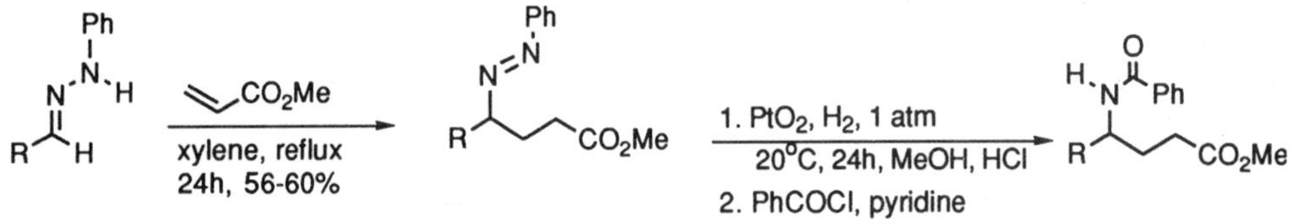

Ester derived enolates

Yamaguchi's group has examined the effect of increasing substitution in both the enolate and the conjugate acceptor on the outcome of their reactions together.[16] As is evident from Table 4.2, steric hindrance plays an important role in determining whether or not 1,4-addition occurs. Reacting the structurally-simplest molecules together, ethanoate enolate and propenoate, only yields polymers. However, changing to an amide enolate provides the conjugate addition product. Quite clearly, it is possible to react most enolates, including tetra-substituted systems with most of the acceptors. The only real exception is the β-disubstituted alkenoate (far right column) where no conjugate addition is observed.

Table 4.2 Reaction of ester and amide enolates with various alkenoates

Entry	Lithium enolate	$CH_2=CH\text{-}CO_2R$	$CH_2=C(CH_3)CO_2R$	$CH_3CH=CH\text{-}CO_2R$	$CH_3CH=C(CH_3)CO_2R$	$(CH_3)_2C=CH\text{-}CO_2R$
1	$CH_2=C(O^-Li^+)OR$	polymer	polymer	1,4	N. R.	N. R.
2	$CH_2=C(O^-Li^+)NR_2$	1,4	1,4	1,4	1,4	N. R.
3	$CH_3CH=C(O^-Li^+)OR$	1,4	1,4	1,4	N. R.	N. R.
4	$CH_3CH=C(O^-Li^+)NR_2$	1,4	1,4	1,4	1,4	N. R.
5	$(CH_3)_2C=C(O^-Li^+)OR$	1,4	1,4	1,4	1,4	N. R.
6	$(CH_3)_2C=C(O^-Li^+)NR_2$	1,4	—	—	—	—

— ..denotes reaction not examined,   N.R. = no reaction

    The same group has also studied simple stereoselection in these additions and found some useful correlations.[17] Thus, the Z-enolate gives *anti* diastereomers with high selectivity. The E-enolate also gives the *anti* diastereomeric product but only if HMPA is added after enolate formation. This suggests a role for the HMPA in promoting *anti* selectivity. If the ester group is *tert*-butoxy, then the E-enolate (the Z-isomer was not included in this study) gives *syn* adducts with excellent stereoselectivity. Thus it is possible, with suitable choice of solvent, enolate isomer and ester group, to synthesize either diastereomer with considerable reliability.

Posner's group has mimicked the termination "back-biting" step in Du Pont's group transfer polymerization process in an example of multicomponent annulation chemistry.[18]

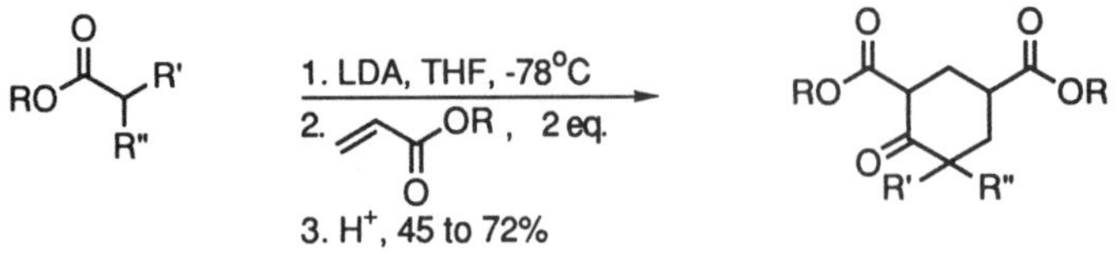

1. LDA, THF, -78°C
2. ⟶OR , 2 eq.
3. H⁺, 45 to 72%

They concluded that the oxyphilic lithium cation is crucial for promoting the intramolecular Claisen ester condensation (rather than polymerization), as illustrated for the proposed intermediate below. However, no evidence, such as metal ion substitution studies, was provided for this contention.

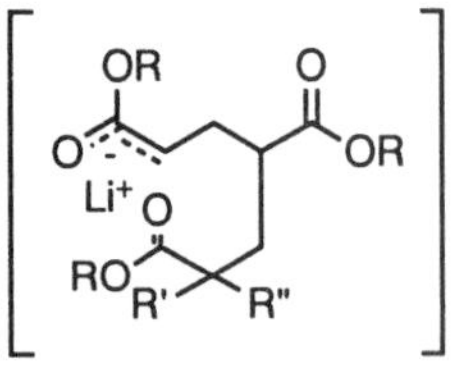

Ahlbrecht's group has shown that α-cyanoamines add to alkenoates, providing an alternative method for preparing γ-oxoalkenoates.[19]

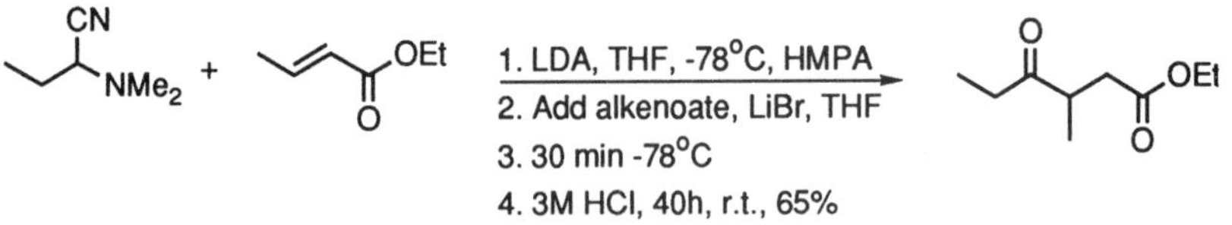

1. LDA, THF, -78°C, HMPA
2. Add alkenoate, LiBr, THF
3. 30 min -78°C
4. 3M HCl, 40h, r.t., 65%

The same group has also used α-cyanoamines in the conjugate acceptor[20] in reactions with nitrile-stabilized carbanions[21]

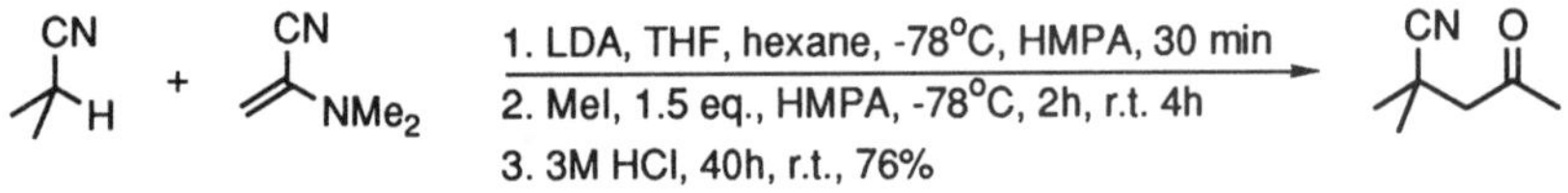

1. LDA, THF, hexane, -78°C, HMPA, 30 min
2. MeI, 1.5 eq., HMPA, -78°C, 2h, r.t. 4h
3. 3M HCl, 40h, r.t., 76%

as well as amide and ester enolates[22] (Also see section 4.1.3).

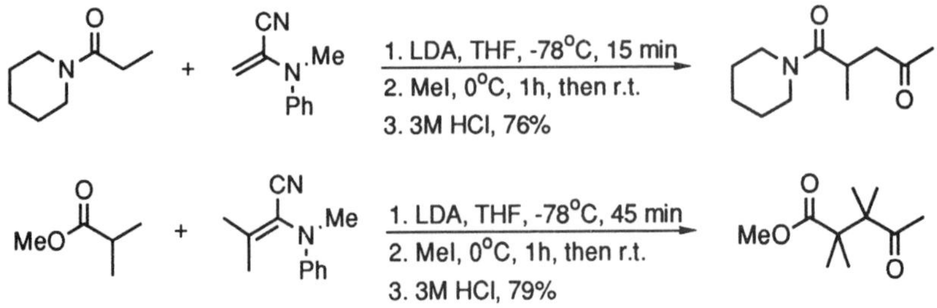

1. LDA, THF, -78°C, 15 min
2. MeI, 0°C, 1h, then r.t.
3. 3M HCl, 76%

1. LDA, THF, -78°C, 45 min
2. MeI, 0°C, 1h, then r.t.
3. 3M HCl, 79%

As with alkenals and alkenones, both cyanide and thiazolium salts catalyze conjugate additions of aldehydes to alkenoate derivatives.[23] Some representative examples are given in Table 4.3.

Table 4.3 Thiazonium-catalyzed addition of aldehydes to alkenoic acid derivatives

1[24]	(acetaldehyde) + (CH₂=CH-CN)	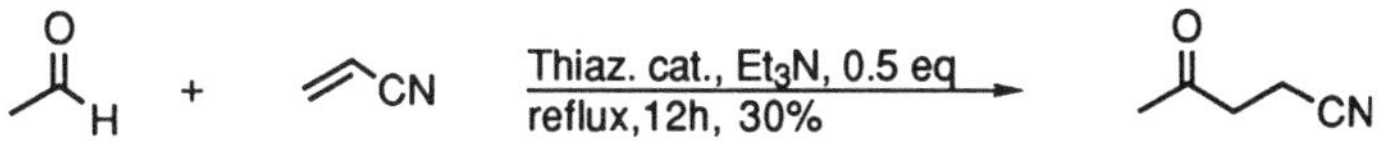 Thiaz. cat., Et₃N, 0.5 eq reflux,12h, 30%	→ (product)
2[25]	(thiophene-2-carbaldehyde) + (crotononitrile)	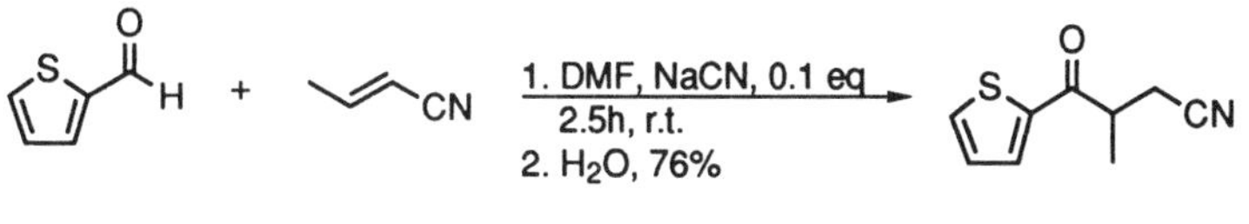 1. DMF, NaCN, 0.1 eq 2.5h, r.t. 2. H₂O, 76%	→ (product)
3[24]	(furfural) + (CH₂=CH-CO₂Et)	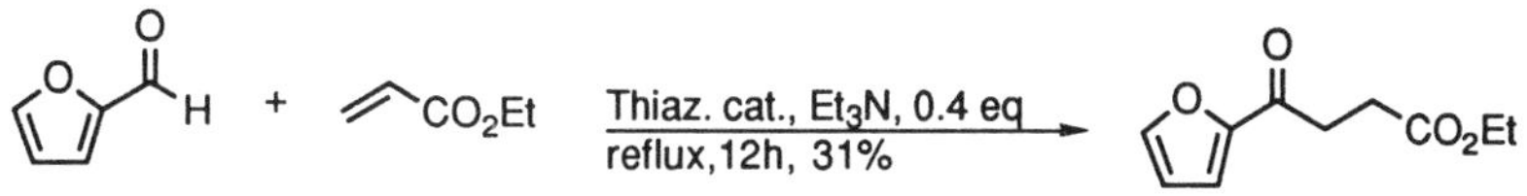 Thiaz. cat., Et₃N, 0.4 eq reflux,12h, 31%	→ (product)
4[25]	(thiophene-2,5-dicarbaldehyde) + 2 (CH₂=CH-CN)	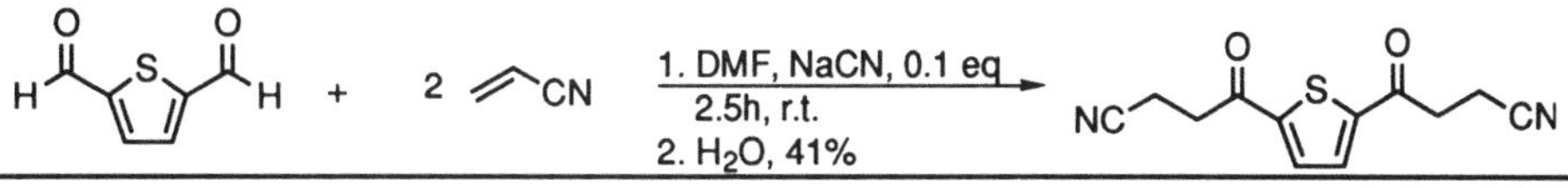 1. DMF, NaCN, 0.1 eq 2.5h, r.t. 2. H₂O, 41%	→ (product)

Nitro-stabilized carbanions

Nitroalkanes also add efficiently to alkenoate derivatives under base catalysis. Bases used include HCO₂Na,[26] TBAB,[27] TBAF[28] and AcOK.[29] Similar reaction conditions to those used for alkenones, although with somewhat longer reaction times, are often employed. The nitro group may then be removed reductively.[30]

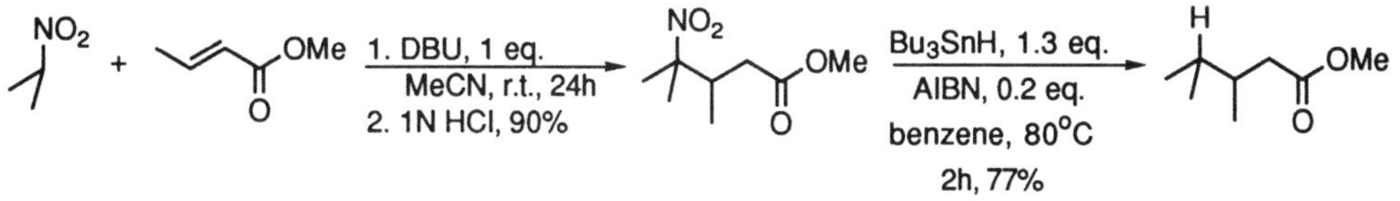

(reaction scheme: nitroalkane + alkenoate, 1. DBU, 1 eq. MeCN, r.t., 24h, 2. 1N HCl, 90% → nitro product; Bu₃SnH, 1.3 eq. AIBN, 0.2 eq. benzene, 80°C 2h, 77% → product)

However, with allylic tertiary nitroalkanes, e.g. (3) this can lead to problems.

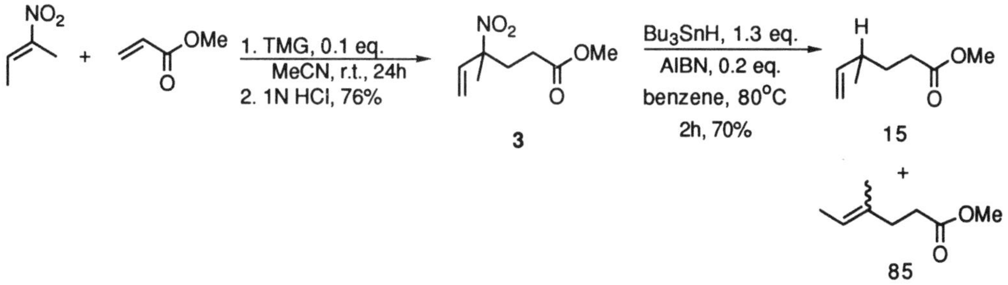

(reaction scheme: allylic nitroalkane + alkenoate, 1. TMG, 0.1 eq. MeCN, r.t., 24h, 2. 1N HCl, 76% → **3**; Bu₃SnH, 1.3 eq. AIBN, 0.2 eq. benzene, 80°C 2h, 70% → **15** + **85**)

Although it is more common to require a single addition, some use has been made of multiple addition products. For example, the addition of nitromethane to three equivalents of propenenitrile was used to prepare the starting material in a synthesis of 1-azoniapropellane (**4**).[31]

CH_3NO_2 $\xrightarrow[\text{CN , 3 eq., 80-85°C, 51%}]{\text{Bu}_4\text{N}^+\text{HSO}_4^-, \text{KOH, H}_2\text{O, dioxane}}$ (structure with NC, CN, CN, NO$_2$) $\xrightarrow{\text{6 steps}}$ (structure **4** Cl⁻ N⁺)

Sulfonyl-stabilized carbanions

(α-Thio)sulfonyl-stabilized carbanions, which are acyl anion equivalents, add well to alkenoates, as in the following examples.[32]

(reaction scheme) EtS, S=O, Li + CO$_2$Me $\xrightarrow[\text{2. H}_2\text{O, 90\%}]{\text{1. THF, -78°C, 2h}}$ EtS, S=O, CO$_2$Et

(reaction scheme) EtS, S=O, Li + SMe, CO$_2$Me $\xrightarrow[\text{2. H}_2\text{O, 93\%}]{\text{1. THF, -78°C, 2h}}$ EtS, S=O, SMe, CO$_2$Et

(reaction scheme) EtS, S=O, Li + CO$_2$Me $\xrightarrow[\text{2. H}_2\text{O, 95\%}]{\text{1. THF, -78°C, 2h}}$ EtS, S=O, CO$_2$Et

(b)   *Addition of achiral carbanions to chiral alkenoate derivatives*

Undoubtedly, one of the most significant advances in stereoselective conjugate additions has been the development of chiral ester auxiliaries, each of which confer a conformational bias on the alkenoate moiety. This, in combination with steric screening provided by the auxiliary, as well as stereoelectronic effects, often leads to significant discrimination of the two π-faces of the conjugate acceptor by the incoming nucleophile.

Several auxiliaries, based on 2,3-disubstituted camphor derivatives, have been prepared. Each is shown on the next page, with the major mode of addition being that indicated.

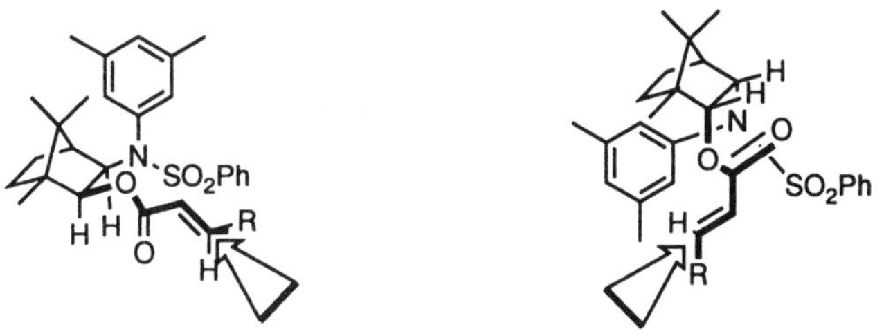

In all cases π-facial selectivity is enforced through steric screening of one face. The alkenoyl conformation, s-*cis* vs s-*trans*, is also a consequence of steric interactions. Oppolzer's group has developed several useful auxiliaries, based on the inexpensive and readily available antipodes of camphorsulfonic acid.[33] The first of these were the enantiomeric dicyclohexylamide derivatives,[34] both of which are now commercially available.[35] The chiral alkenoates apparently adopt an s-*trans* conformation as shown. Nucleophiles then approach from the less hindered side (see below for synthetic applications of these auxiliaries).

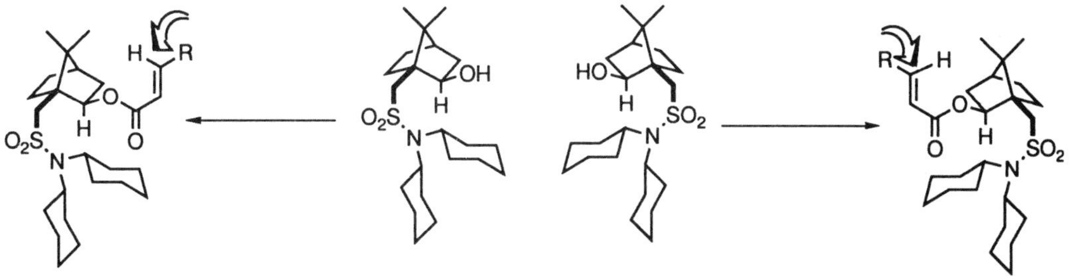

More recently, the same group has prepared a related series of sultams.

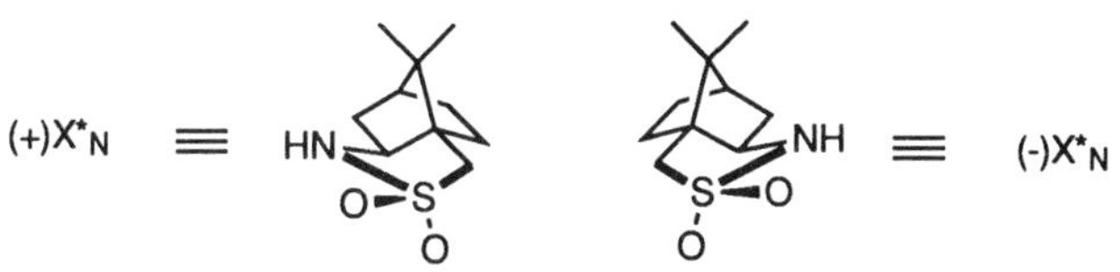

These are readily available from (+)- or (-)-camphor-10-sulfonic acid, as shown on the next page for the (-) antipode.[36,37]

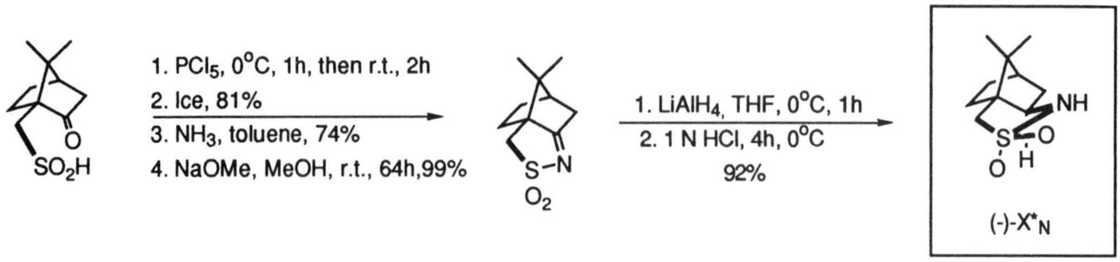

1. PCl$_5$, 0°C, 1h, then r.t., 2h
2. Ice, 81%
3. NH$_3$, toluene, 74%
4. NaOMe, MeOH, r.t., 64h,99%

1. LiAlH$_4$, THF, 0°C, 1h
2. 1 N HCl, 4h, 0°C
92%

(-)-X*$_N$

The corresponding alkenoyl derivatives undergo highly efficient and stereoselective conjugate additions. Analyses of both the crystal and solution structures of these derivatives have shown that their preferred conformation is that which has the CO bond and SO$_2$ group *anti* to each other, with the alkenoyl moiety *s-cis*.

For those reactions where a chelating metal is present, the whole alkenoyl moiety, which remains *s-cis*, flips over so that the carbonyl oxygen can chelate to the metal. As shown below, the choice of either chelation or conformation control provides predictable conjugate addition to either face of the π-system.

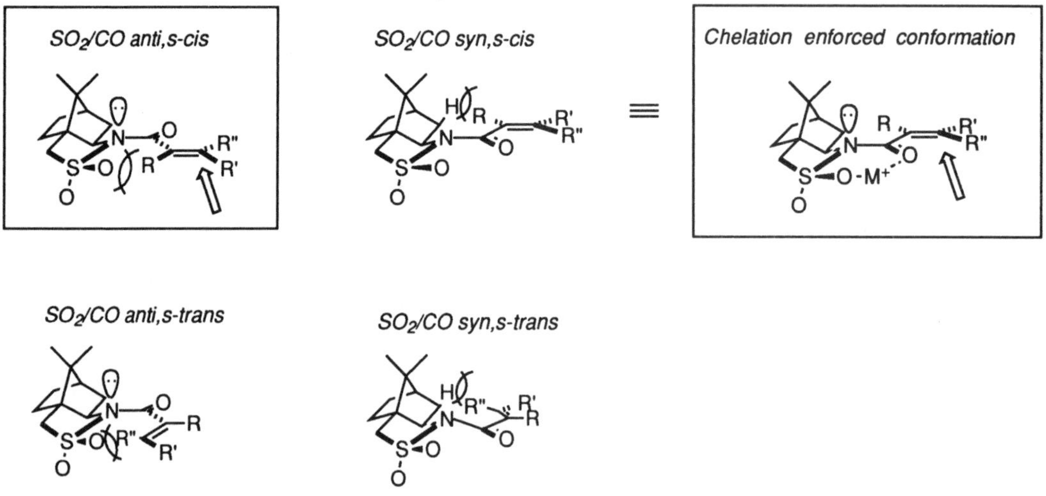

SO$_2$/CO anti,s-cis

SO$_2$/CO syn,s-cis

Chelation enforced conformation

SO$_2$/CO anti,s-trans

SO$_2$/CO syn,s-trans

The reasons for the preferred conformation adopted by these systems appear quite clear. The *anti* disposition of the CO bond and the SO$_2$ group minimizes dipole repulsions, whereas the *s-cis* geometry avoids serious steric interactions between one of the β-substituents (R" in the diagrams above) and the pseudo equatorial SO$_2$ oxygen. The π-facial selectivity appears to be associated with the pyramidal nature of the sulfonamide nitrogen. There is growing evidence that the nitrogen lone pair strongly influences π-facial discrimination in a number of types of reactions. For conjugate additions (see later in this chapter for specific examples) the incoming nucleophile approaches *anti* to the lone pair.[38]

(c)   *Addition of chiral carbanions to achiral alkenoate derivatives*

Ketone and aldehyde derived enolates

Enders' group has demonstrated that lithiated chiral hydrazones from, e.g. (5) prepared from either SAMP or RAMP, add to alkenoates with excellent diasteroselectivity and enantioselectivity.[39,40]

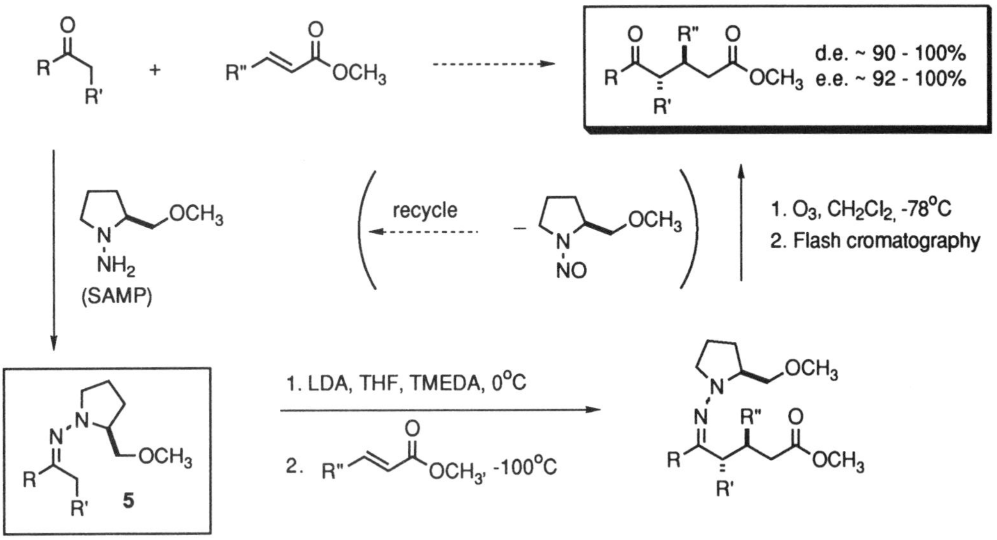

The same group has also developed a remarkably selective nucleophilic acylation using metallated chiral amino cyanides, e.g. from (6).[41]

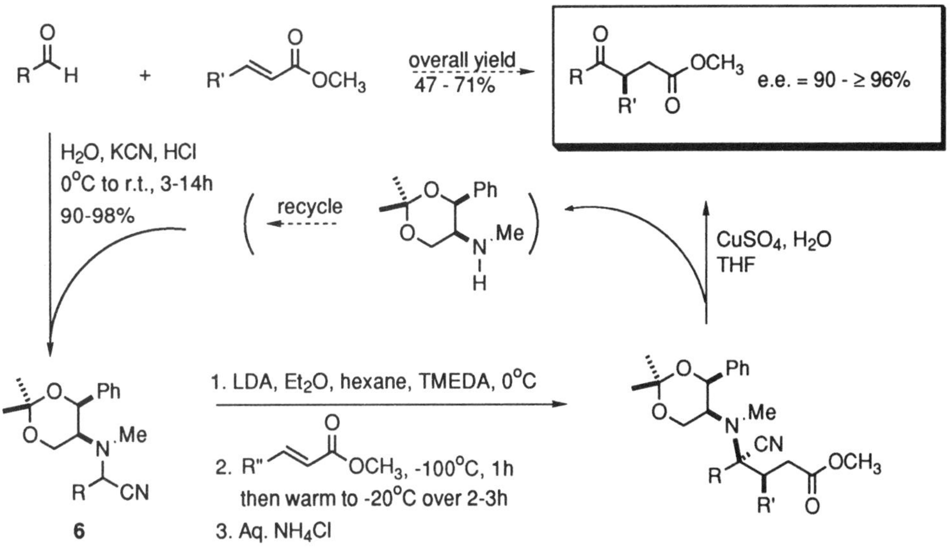

A method for preparing enantiomerically pure spiro-bislactones has been developed which relies on a highly stereoselective addition of the enolate of a chiral imine derived from 2-benzyloxycyclohexanone to methyl propenoate.[42,43]

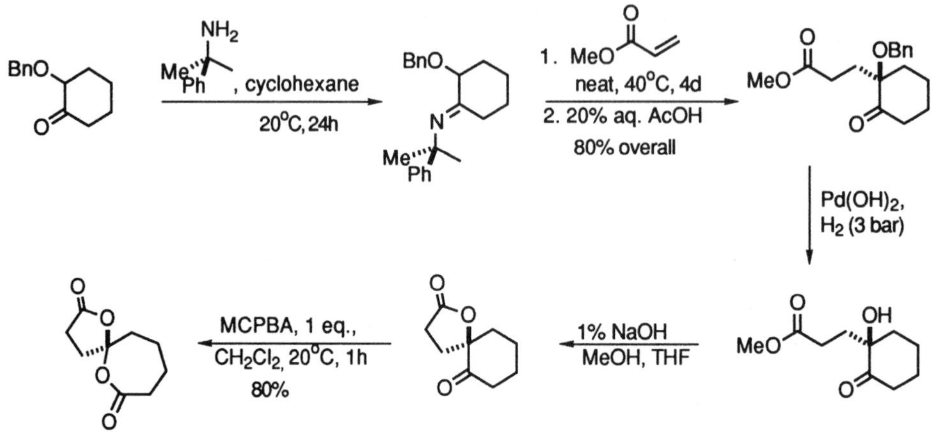

Ester-derived enolates

Corey's group has applied their 8-phenylmenthol auxiliary[44] to chiral enolate chemistry.[45] The major diastereomeric pair were the *syn* isomers as shown below. The *anti* isomers (not shown) were also formed as 10% of the total product.

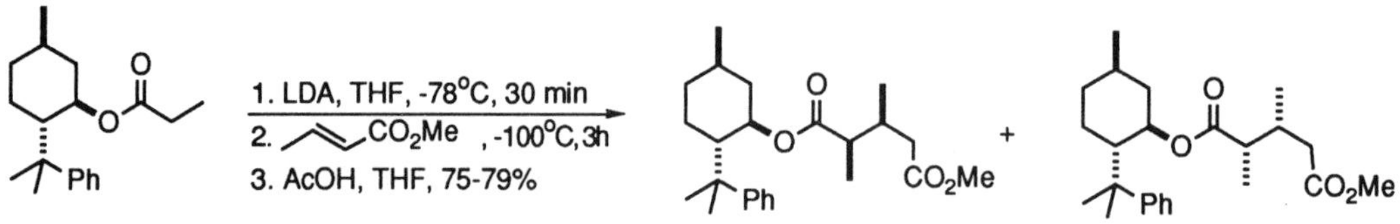

The stereochemical outcome may be rationalized if the major reacting species is the Z-enolate, with the lithium alkoxide *anti* to the benzene ring. The alkenoate is then effectively restricted to approach from the *si* face of the enolate. Chelation of the carbonyl oxygen is also likely.

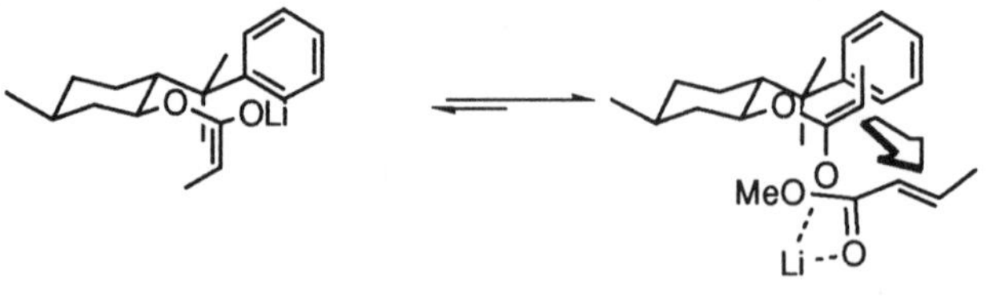

Several studies on the effect of chiral crown ethers on the base-catalyzed addition of methyl phenylacetate to methyl propenoate have been reported. One such study employed an anthracene-derived 18-crown-6 derivative which yielded an adduct with reasonable asymmetric induction (66% e.e.).[46]

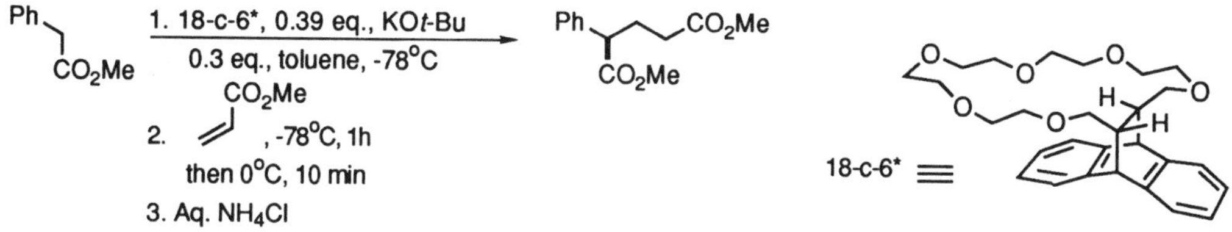

Another group has focussed on carbohydrate-derived chiral crown ethers and several of these have been prepared.[47] Of the many chiral crown ethers prepared the "*bis-lacto*" 18-crown-6 derivatives, e.g. (7) have given the best results so far.[48] The asymmetric induction is comparable to that obtained in the above study.

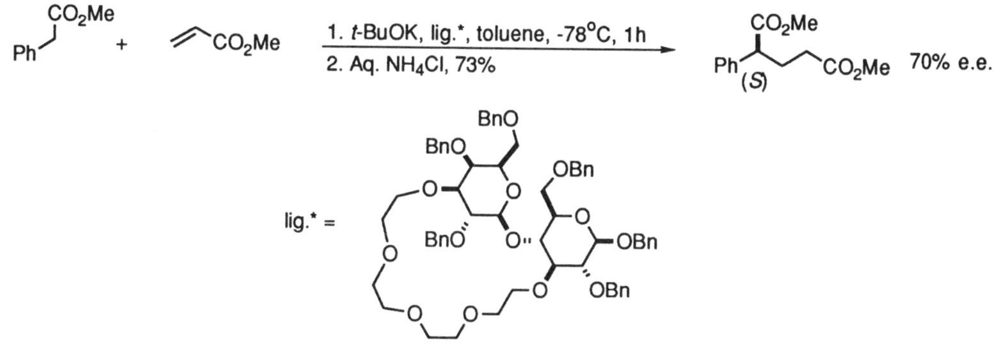

7

The stereoselectivity in the conjugate addition of several enolates of chiral amides has been examined.[49] The pyrrolidine-derived amide enolates gave the best results as shown below.

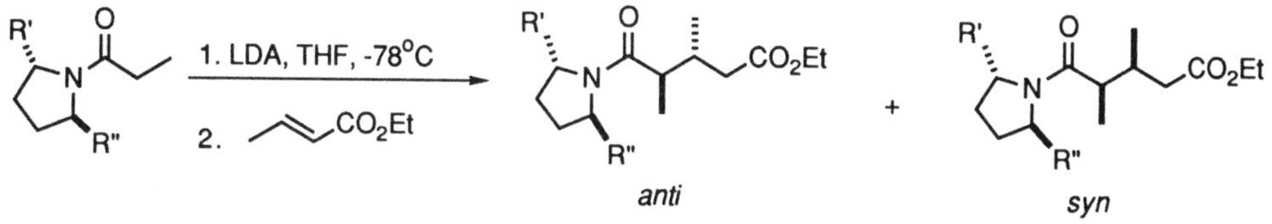

		anti		syn
R' = R" = H	86%	<1	:	>20
R' = CH₂OH R" = H	84%	1	:	20
R' = R" = CH₂OMOM	76%	15	:	1

This reaction was then applied to a synthesis dehydroiridodiols, using a tandem amide enolate conjugate addition/Claisen condensation sequence.

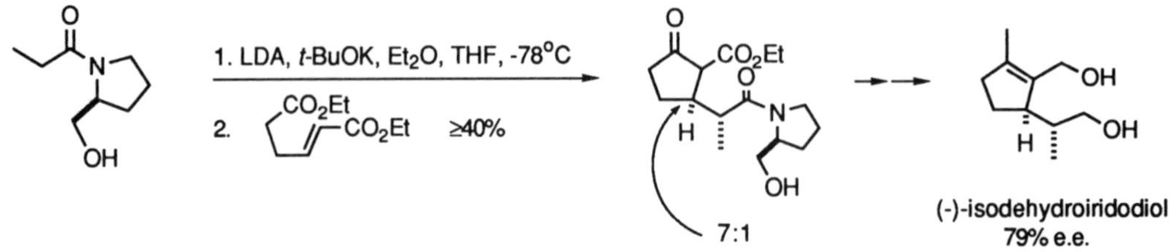

(-)-isodehydroiridodiol
79% e.e.

The addition of lithiated bislactim ethers, e.g. (8) (see section 1.4.1), prepared from cyclo(-L-Val-Ala-),[50] to a variety of alkenoates proceeds with very high selectivity.[51]

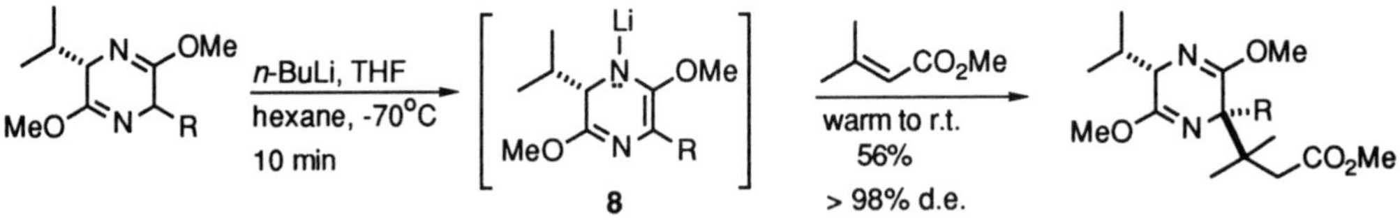

Upon acidic hydrolysis, the adducts may be converted into the corresponding amino acid.

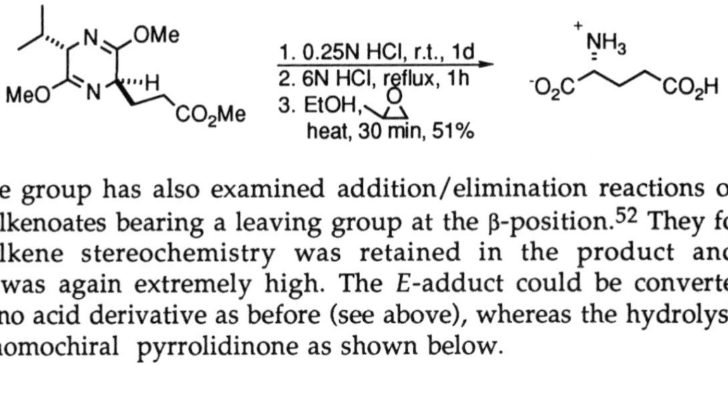

The same group has also examined addition/elimination reactions of this class of nucleophile to alkenoates bearing a leaving group at the β-position.[52] They found, in all cases, that the alkene stereochemistry was retained in the product and that the diastereoselection was again extremely high. The E-adduct could be converted into the corresponding amino acid derivative as before (see above), whereas the hydrolysis of the Z-adduct yields the homochiral pyrrolidinone as shown below.

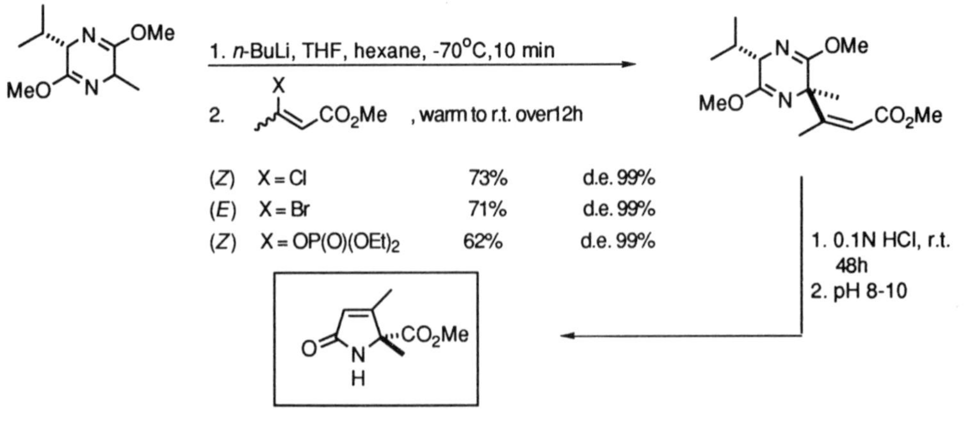

(Z)	X = Cl	73%	d.e. 99%
(E)	X = Br	71%	d.e. 99%
(Z)	X = OP(O)(OEt)$_2$	62%	d.e. 99%

β-Lactone enolates react stereoselectively with dimethyl maleate.[53] Not only is the reaction at the enolate stereoselective (attack occurs at the face opposite that bearing R), but good stereoselectivity is also observed at the β-carbon of the maleate. The authors suggested the lowest energy transition state would be that shown below.

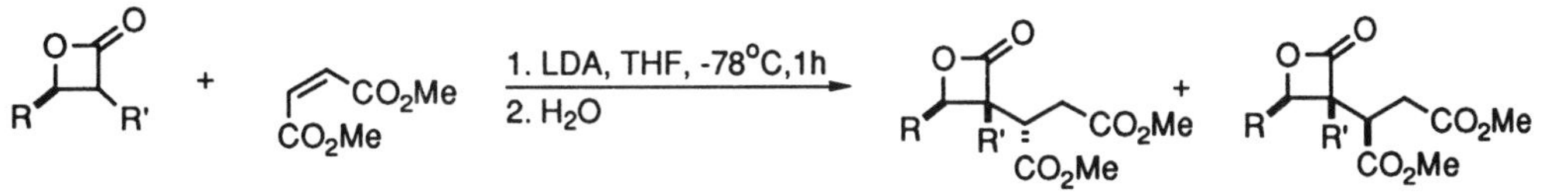

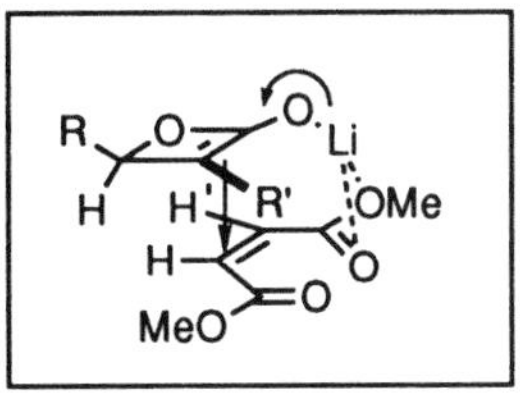

			anti			syn
R = t-Bu	R' = Ph	73%	89	:	11
R = i-Pr	R' = Ph	68%	82	:	18
R = Me	R' = Ph	50%	85	:	15
R = t-Bu	R' = Me	45%	100	:	0

Sulfoxide-stabilized carbanions

Chiral, sulfoxide stabilized carbanions add efficiently to alkenoates. *tert*-Butyl alkyl sulfoxides show excellent stereoselectivity, yielding the *anti, anti,* product with ≥ 10 : 1 diastereomeric ratio.[54] The use of an aryl group in place of the *tert*-butyl group results in almost complete loss of stereoselectivity. However, there are some limitations to this method. 2-Substituted alkenoates, such as methyl tiglate, fail to react cleanly and methyl *tert*-butyl sulfoxides show little stereoselection in their additions.

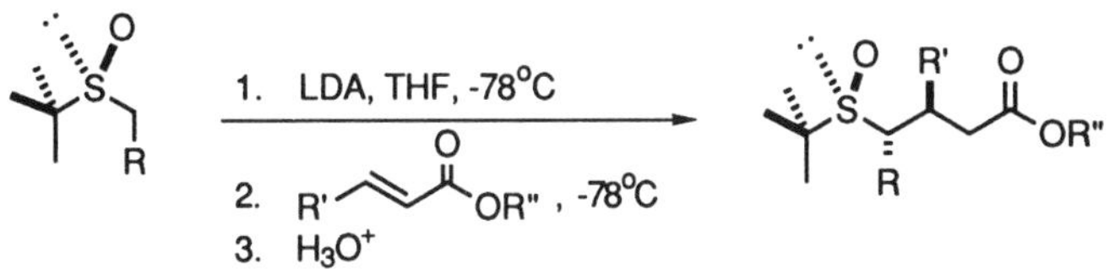

4.1.1.2   *Carbanions stabilized by π-conjugation with more than one heteroatom*

(a)   *Addition of achiral carbanions to achiral alkenoate derivatives*

Malonates add to alkenoates under base catalysis with reasonable efficiency (Table 4.4, entries 1 and 2). Addition/elimination of a sodiomalonate to a β-tosylalkenoate proceeds with complete retention (entry 3).

Table 4.4 Conjugate addition of highly stabilized carbanions to alkenoate derivatives

1[55]	CO$_2$Et / EtO$_2$C + CN	NaOEt, EtOH, 35°C, ~2h 55%	CO$_2$Et / EtO$_2$C—CN
2[56]	CO$_2$Me / MeO$_2$C + CO$_2$Me	MeONa, MeOH reflux, 9h, 76%	MeO$_2$C CO$_2$Me / MeO$_2$C
3[57]	CO$_2$Et / EtO$_2$C—Na + Tos—CO$_2$Me	1. THF, 0°C, 15h 2. Aq. NH$_4$Cl, 70%	EtO$_2$C / EtO$_2$C—CO$_2$Me
4[58]	NO$_2$ / Tos—H + CN	1. Et$_3$N, EtOH, reflux, 5h 2. AcOH, 62%	NO$_2$ / Tos—CN

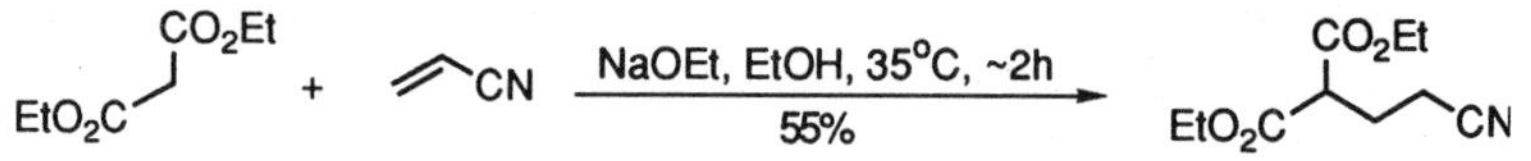

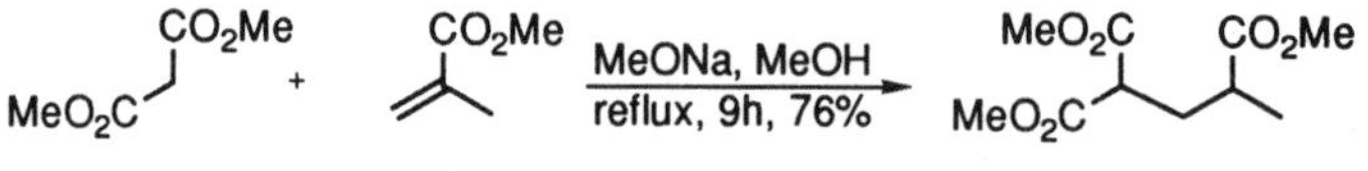

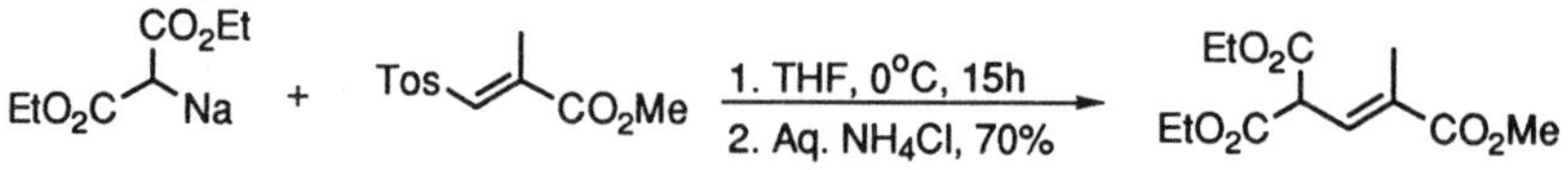

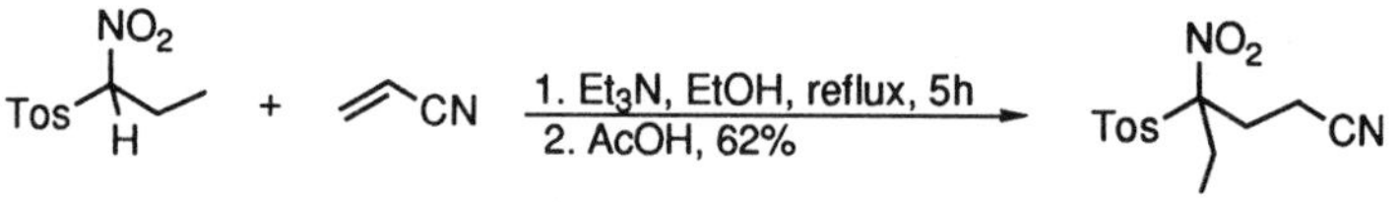

A simple synthesis of 2-substituted glutarates involves the conjugate addition of 2-substituted malonates to propenenitrile as in the following example.[59]

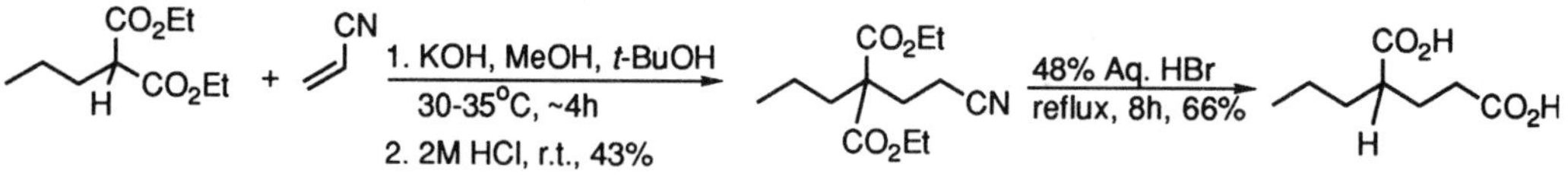

CO$_2$Et / CO$_2$Et—H + CN
1. KOH, MeOH, *t*-BuOH
30-35°C, ~4h
2. 2M HCl, r.t., 43%
→ CO$_2$Et / CN / CO$_2$Et
48% Aq. HBr reflux, 8h, 66%
→ CO$_2$H / CO$_2$H—H

The combination of conjugate additions of malonates to alkenoates with the enantioselective hydrolysis of the derived glutarates has led to a stereoselective synthesis of 4-substituted δ-lactones.[60] The latter, in more elaborate form, are found in the antitumor rhizoxin.[61]

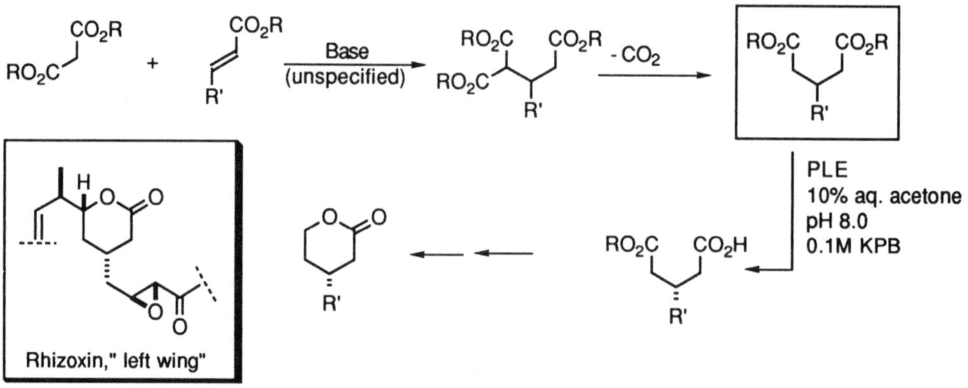

CO$_2$R / RO$_2$C + CO$_2$R / R'
Base (unspecified)
→ RO$_2$C CO$_2$R / RO$_2$C / R'
− CO$_2$
→ RO$_2$C CO$_2$R / R'
PLE
10% aq. acetone
pH 8.0
0.1M KPB
→ RO$_2$C CO$_2$H / R'

Rhizoxin, "left wing"

(b)   *Addition of achiral carbanions to chiral alkenoates*

PTC additions of malonates to 2-[aryl(hydroxy)methyl]-propenoates produce the *syn* diastereomer with excellent selectivity.[62] Under the conditions shown below, addition-eliminations are minimized.

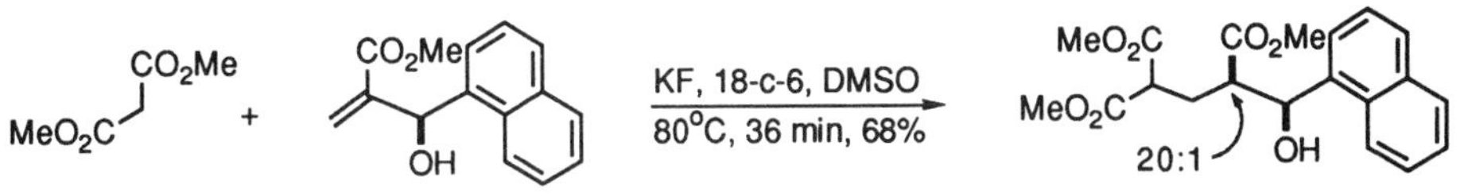

(c)   *Addition of chiral carbanions to achiral   alkenoate derivatives*

Chiral α-sulfinyl ketimine anions, generated from (9) add stereoselectively to alkenoates such as ethyl 1-cyclopentenecarboxylate, generating a chiral tricyclic sulfoxide (accompanied by only 2% of the C6 epimer).[63]

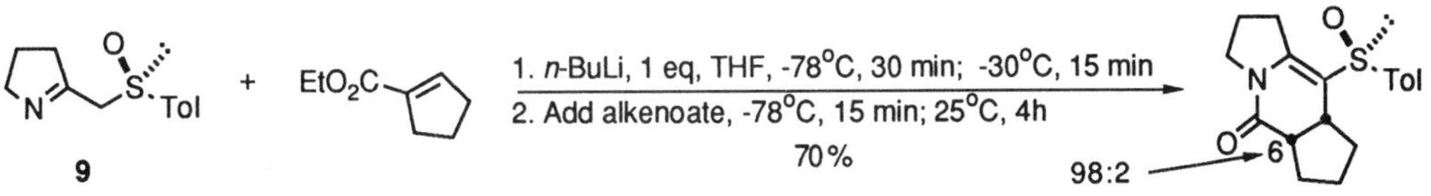

An eight-membered "*trans*-fused" transition state, with the tolyl group occupying a pseudo-equatorial position, was suggested to account for the stereochemical outcome.

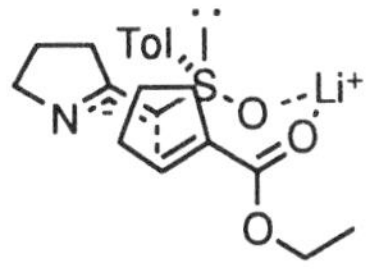

Similar additions to methyl propenoate also offer access to chiral indolizidines as the cyclized adducts undergo highly selective reductions, as in the following synthesis of the naturally occurring octahydroindolo[2,3-a]quinolizine alkaloid.

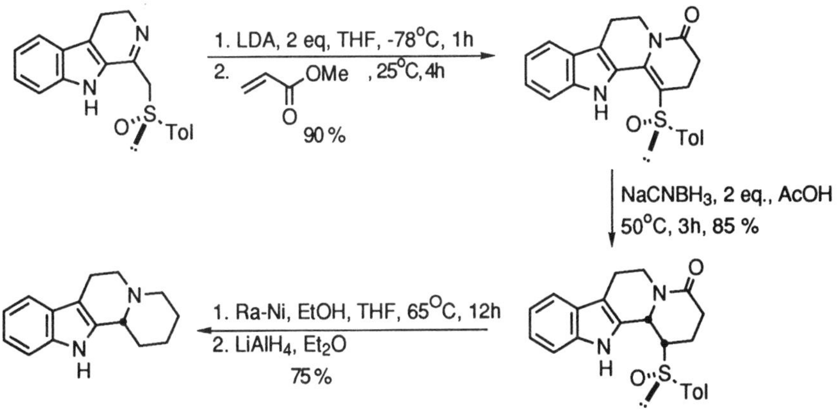

4.1.1.3   Carbanions stabilized by   one or more α-heteroatoms

3-Methyl-1-tetralone, which is not commercially available, was required in a synthesis of the angular anthracyclinones ochromycinone and X-14881C. It was prepared by the addition of 2-lithio-2-phenyl-1,3-dithiane to ethyl E-2-butenoate.[64]

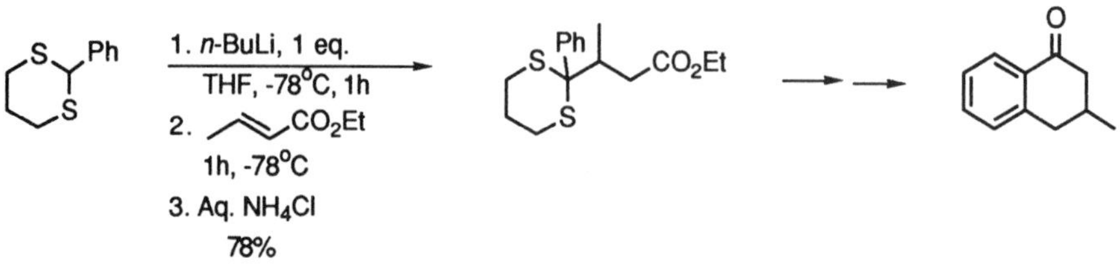

Many masked acyl anions add to alkenoate derivatives. After studying several candidates, Luchetti and Krief concluded that lithio bis(phenylthio)methanes in DME are the best reagents for introducing an acyl group at C3 of alkenoates.[65]

A remarkable, stereoselective cyclopropanation, which involves the addition of an ylid to the E-alkenoate derived from D-glyceraldehyde, has been reported.[66] The first part of the reaction may be viewed as a conjugate addition to a γ-alkoxyalkenoate.

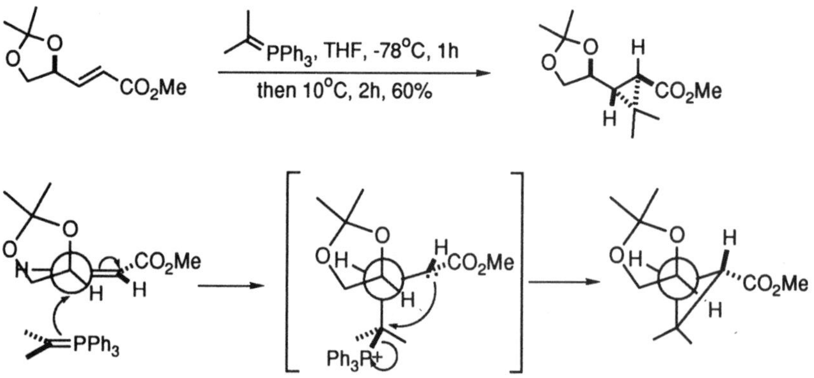

Snieckus's group has demonstrated that dialkylamides of crotonic acid also react well with hard nucleophiles, e.g.:[6]

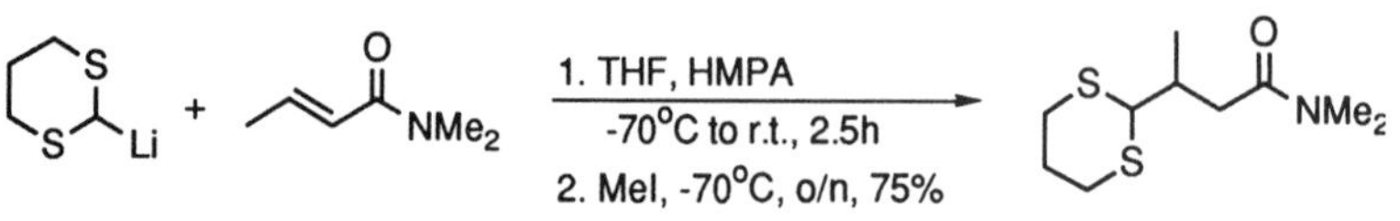

They applied this method to the synthesis of some lignans, including galcatin.[67]

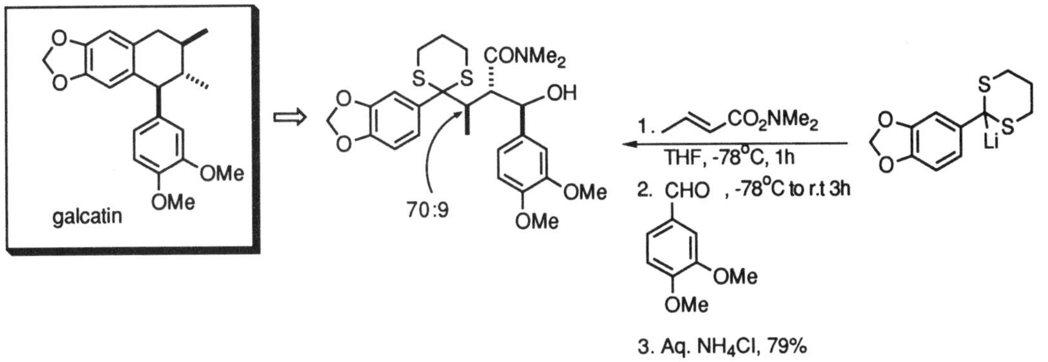

The possibility of asymmetric induction of conjugate additions of dithioacetals, in the presence of a chiral ligand, has been investigated.[68] Although the results were encouraging, the enantiomeric excesses were not consistently high.

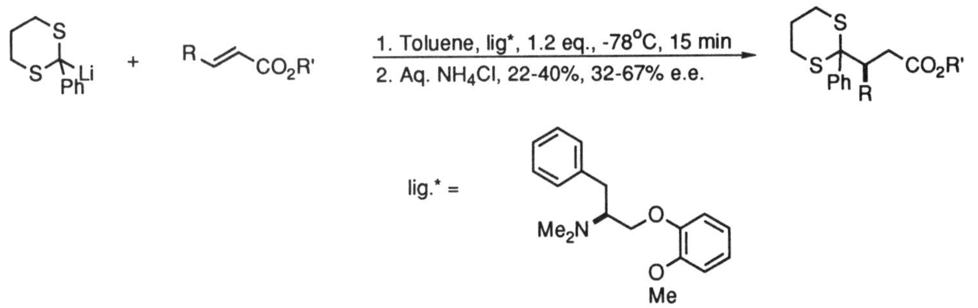

(β-Oxo)alkenyllithiums add well to alkenoates. The adduct cyclizes *in situ* providing a useful cyclopentannulation procedure.[69]

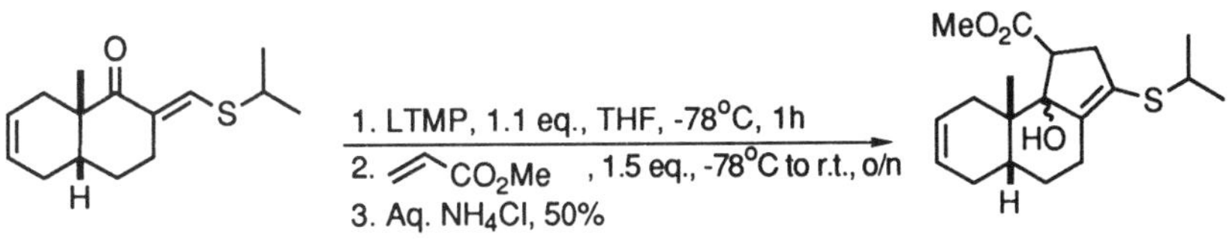

4.1.2   Organocopper reagents

4.1.2.1   Addition of achiral organocopper reagents to achiral alkenoate derivatives

Mixed homocuprates react well with alkenoates. Alkyl(thienyl)cuprates are particularly reactive, adding rapidly at low temperature. Interestingly, in the case of the reagent containing both a phenyl and 2-pyridyl ligand, only the pyridyl is transferred and in good yield (entry 3). A detailed study of the addition of LO cuprates to some *ortho*-substituted cinnamates has produced clear evidence that tetrahydrofuran is the worst for these reactions. Diethyl ether gave acceptable results, however the best solvents were found to be the non-coordinating solvents, toluene and dichloromethane (entry 4).

Table 4.5 Conjugate addition of organocopper reagents to alkenoic acid derivatives

1[70]	CO₂Me structure	1. *n*-Bu(Th)CuLi, 2 eq Et₂O, 0°C, 10 min 2. Aq. NH₄OH, NH₄Cl, 54%	product
2[71]	Ph CO₂Me	1. Me(Th)CuLi, 1.25 eq Et₂O, -50°C 2. TMSCl, 2.5 eq., -50 to 0°C 3. Aq. NH₄OH, NH₄Cl, 75%	product
3[72]	Ph CO₂Et	1. Ph(2-Py)CuLi.LiI, 2 eq Et₂O, r.t., 1h 2. Aq. NH₄OH, NH₄Cl, 85%	product
4[73]	structure OMe	1. Me₂CuLi, CH₂Cl₂, 0°C 60 min 2. Aq. NH₄Cl, 85%	product

Although not as well-studied as in similar reactions with alkenones and cycloalkenones, a number of electrophiles have been used to trap the intermediates in these additions.[74] LO cuprates are not always particularly successful in this type of reaction as in, for example, entry 2 in Table 4.6. Nugent and Hobbs found that, although LO alkenylcuprates worked in their addition/intramolecular Claisen condensation, the yields were not very good (~40%). However, switching to HO cuprates led to a significant improvement in yields (see entry 3).

Table 4.6 Enolate trapping reactions in the addition of organocopper reagents to alkenoic acid derivatives

1[75]	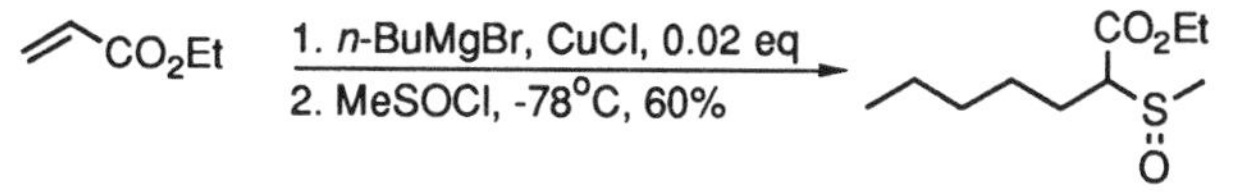 ⌇CO₂Et	1. *n*-BuMgBr, CuCl, 0.02 eq 2. MeSOCl, -78°C, 60%	structure
2[76]	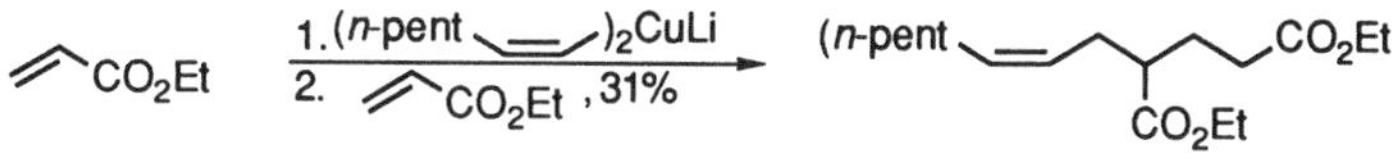 ⌇CO₂Et	1.(*n*-pent⌇)₂CuLi 2. ⌇CO₂Et , 31%	structure
3[77]	MeO₂C ⌇ CO₂Me	1. ⌇)₂CuCNLi₂ , -30°C 2. Aq. NH₄Cl, 77-85%	structure 94:6

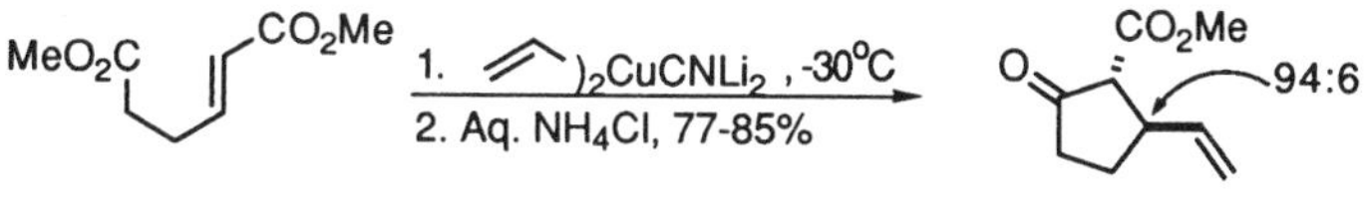

Organocopper reagents participate in addition/elimination reactions with a variety of 3-substituted alkenoates. Additions to enol phosphates of 3-oxoalkanoates is restricted to primary dialkyl cuprates (entry 1). Lower temperatures must be used in additions to enol phosphates of acyclic 3-oxoalkanoates (see synthesis below).

Table 4.7   Addition/elimination reactions of organocopper reagents with 3-substituted alkenoates

1[78]	(EtO)₂P-O structure CO₂Me	1. Me₂CuLi, 1.4 eq Et₂O, -23°C, 3h 2. 5% HCl, 0°C 86-90%	structure CO₂Me
2[79]	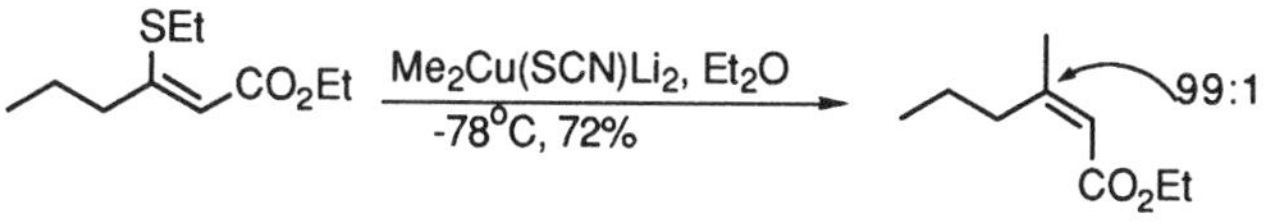 SEt structure CO₂Et	Me₂Cu(SCN)Li₂, Et₂O -78°C, 72%	structure 99:1 CO₂Et
3[79]	SEt structure CO₂Et	Me₂Cu(SCN)Li₂, THF 25°C, 12h, 72%	structure CO₂Et 90:10

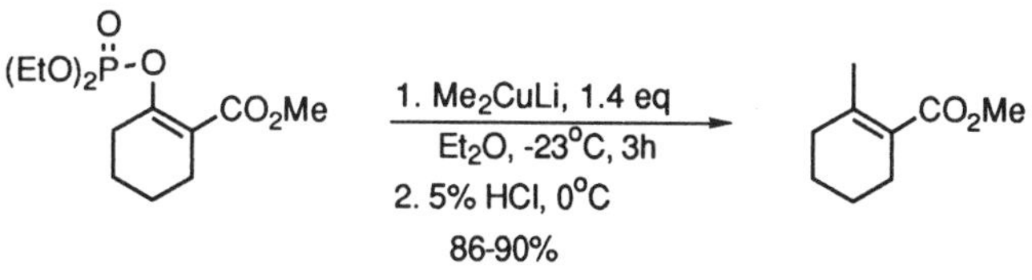

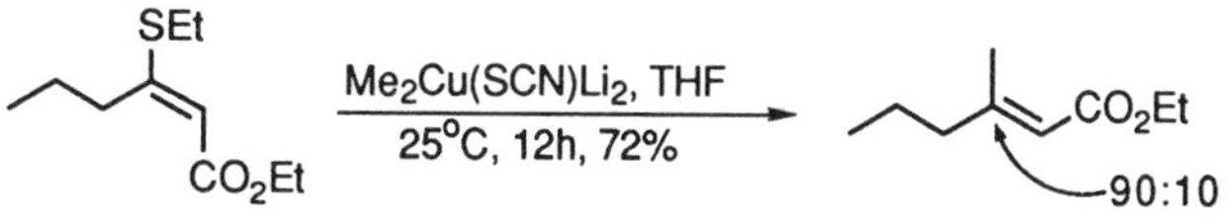

The combination of γ-alkylation of the dianion of a β-ketoester with such an addition/elimination reaction of the corresponding enol phosphate constitutes an isoprenoid chain extension process. For example, this process was used twice in the total synthesis of (*E,E*)-10-hydroxy-3,7-dimethyldeca-2,6-dienoic acid (**10**), a constituent of the male Monarch butterfly.[80,81]

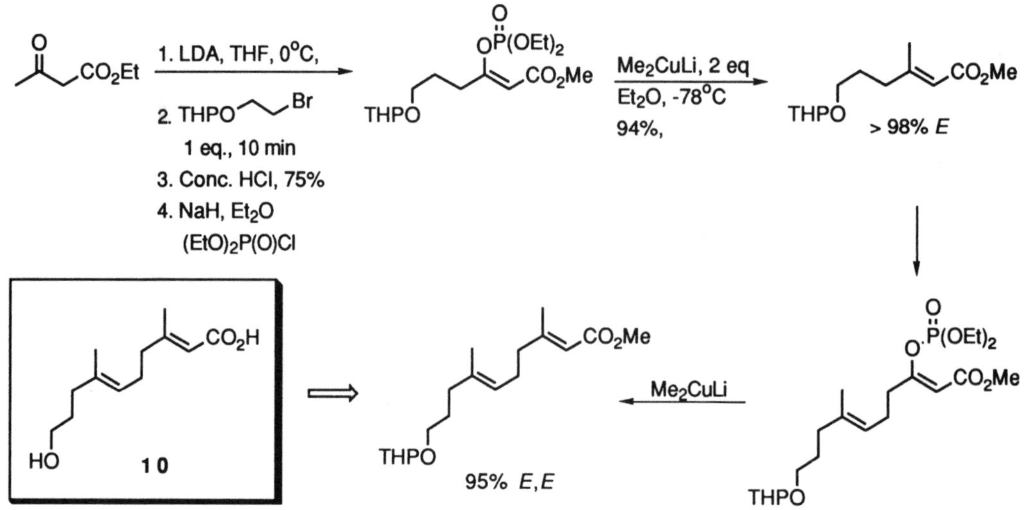

4.1.2.2   Addition of achiral organocopper reagents to chiral alkenoate derivatives

Oppolzer's group has applied its camphorsulfonamide auxiliary to a synthesis of the Southern corn rootworm pheromone. [82]

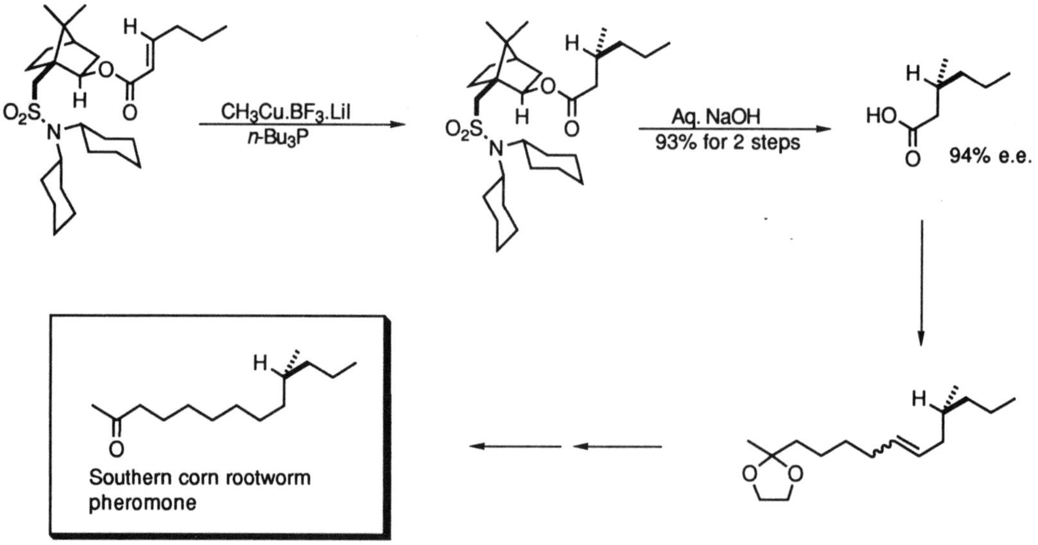

The same group has also demonstrated the high selectivity achieved in the addition of methyl, vinyl and aryl cuprates to enoyl sultams.[83] (Addition of alkyl groups, other than methyl, is best achieved using organomagnesium reagents, see section 4.1.3).

The major isomer could be obtained in almost pure form (usually ≥ 98.5% purity) after flash chromatography and recrystallization (from hexane).

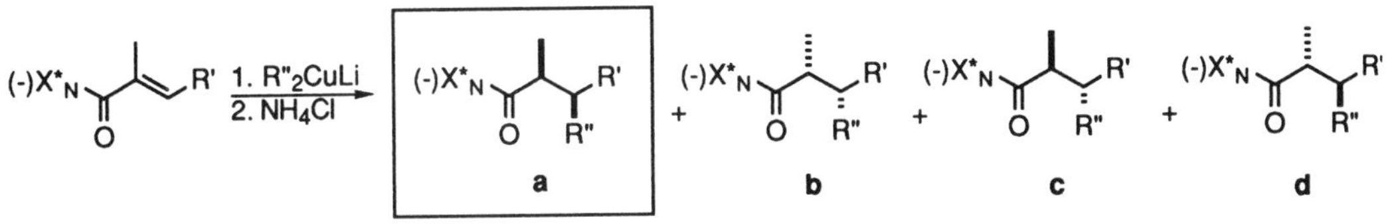

Table 4.8   Organocuprate addition followed by protonation of 2,3-disubstituted alkenoylsultams

Entry	R'	R"	Cu(I) salt	Yield(%)	Ratio (a:b:c:d)
1	Et	Me	CuI	85	85 : 15 : 0 : 0
2	Et	Me	CuI.PBu3	94	91 : 9 : 0 : 0
3	n-Bu	Me	CuI.PBu3	78	89 : 4.5 : 2.7 : 3.8
4	Ph	Me	CuI.PBu3	67	90 : 6 : (4)
5	Me	Ph	CuSCN	60	83.4 : 10.5 : 0.4 : 5.
6	Me	CH2=CH	CuI.PBu3	72	86.3 : 12.3 : 3.0 : 1.4

The principal mode of addition appears to be via a chelation enforced SO$_2$/CO *syn, s-trans* conformation.

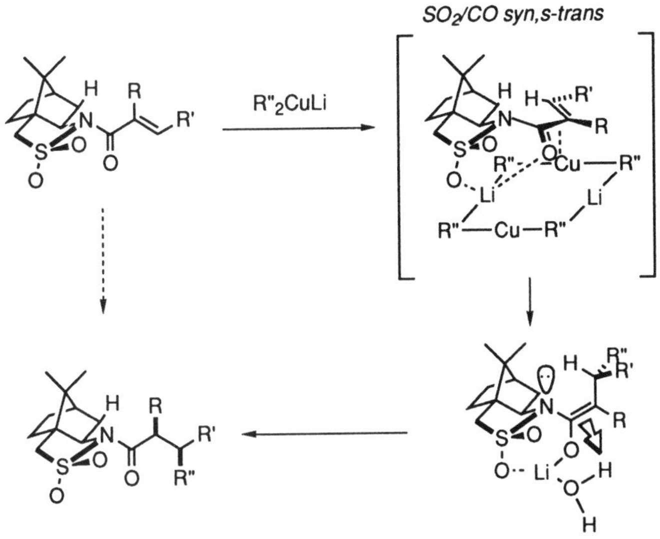

Koga's group has designed a chiral auxiliary which gives consistently good diastereoselectivity with many LO cuprate additions.[84] The predominant conformer of the

conjugate acceptor in solution, shown below, is also believed to be the reactive conformer.

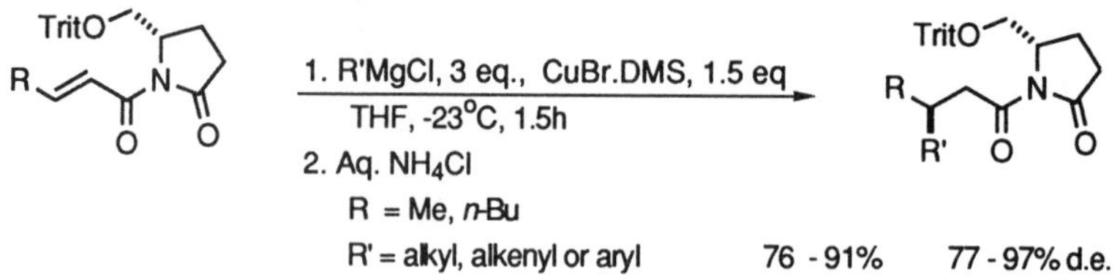

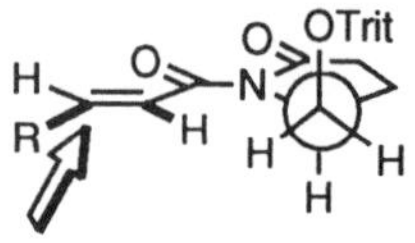

A remarkable example of double asymmetric induction involving a system containing two γ-alkoxyalkenoate moieties (**11**) has been reported.[85] The inter- followed by intra-molecular conjugate additions proceed with an overall diastereoselectivity of greater than 99% for the four different organocuprates tested.

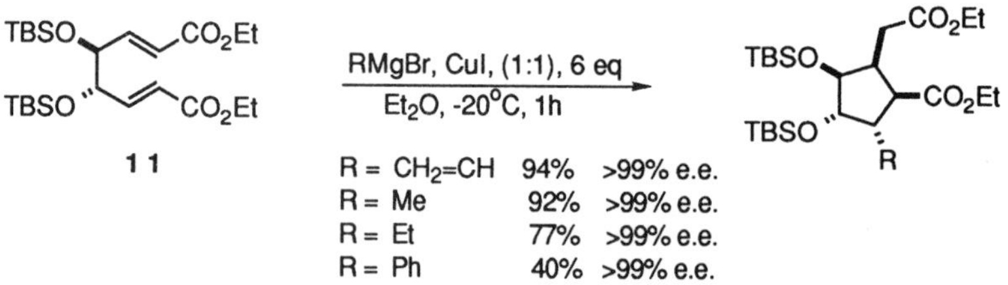

RMgBr, CuI, (1:1), 6 eq
Et$_2$O, -20°C, 1h

R = CH$_2$=CH   94%   >99% e.e.
R = Me   92%   >99% e.e.
R = Et   77%   >99% e.e.
R = Ph   40%   >99% e.e.

The authors rationalized that the first addition was occurring to the most stable ground state rotamer, which is homotopic.

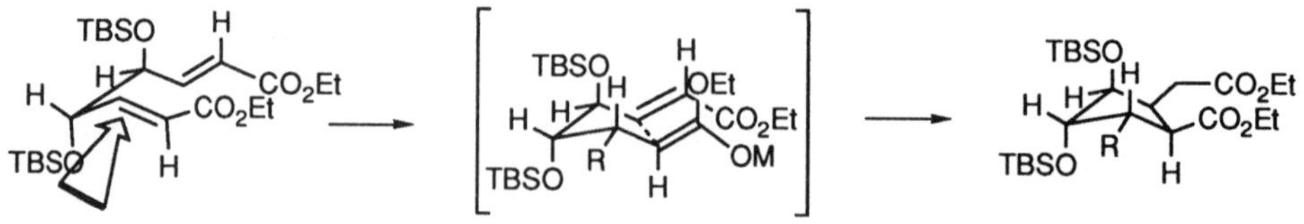

As discussed in Chapter 1, it is now well established that good stereocontrol can often be achieved in the addition of organocopper reagents to γ-substituted alkenoates.[86] For example, reaction of BuCu-BF$_3$.LiI with the *E*-isomer of ethyl 4-benzyloxy-2-pentenoate yields mainly the *anti* product (entry 1, Table 4.9). Conversely, addition to the *Z*-isomer yields mainly the *syn* product (entry 2). Addition to the corresponding diester (entry 3) is very selective, producing the *syn* product. For γ-mesyloxy systems an S_N2' reaction occurs rather than a conjugate addition.[87]

Table 4.9 Stereoselective addition of organocopper reagents to γ-alkoxyalkenoates

1	BnO─CO₂Et	RCu-BF₃.LiI, THF, -78°C then warm to r.t.	R─BnO─CO₂Et +	R─BnO─CO₂Et
	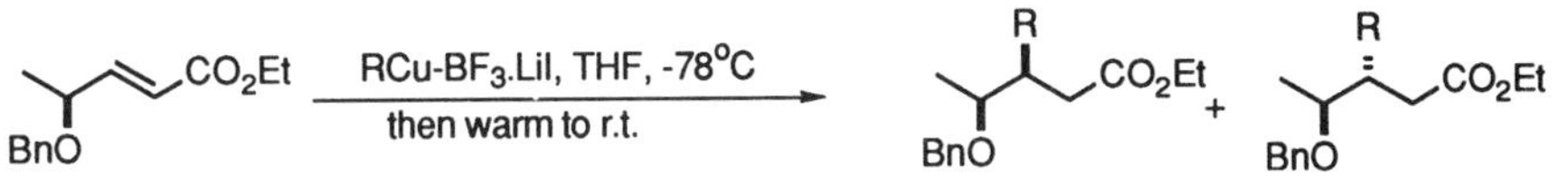	R = Me 60%	31 :	69
		R = Bu 64%	8 :	92
2	BnO CO₂Et	RCu-BF₃.LiI, THF, -78°C then warm to r.t.	R─BnO─CO₂Et +	R─BnO─CO₂Et
		R = Me 30%	78 :	22
		R = Bu 56%	78 :	22
3	BnO CO₂Et	MeCu-BF₃.LiI, THF, -78°C then warm to r.t.	R─BnO─CO₂Et CO₂Et +	R─BnO─CO₂Et CO₂Et
		54%	94 :	6

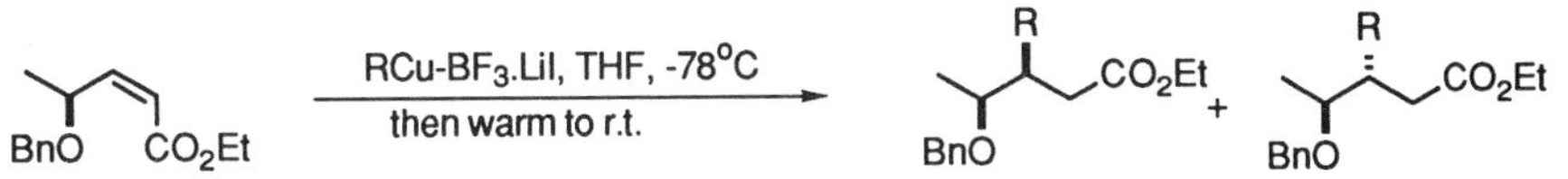

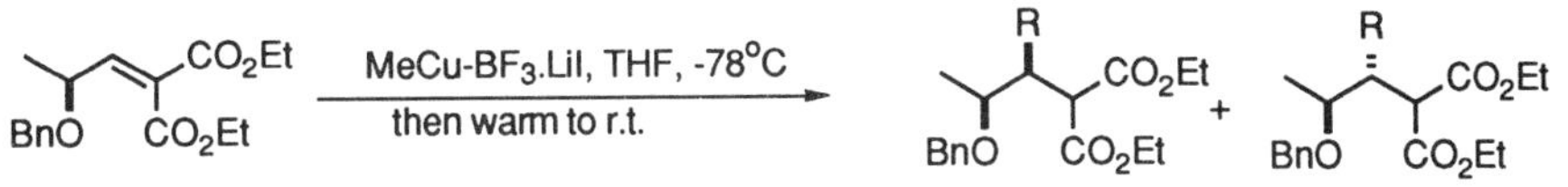

By contrast, di*alkenyl*copperlithiums give, largely, *anti* products regardless of the starting alkene stereochemistry.[86d,88]

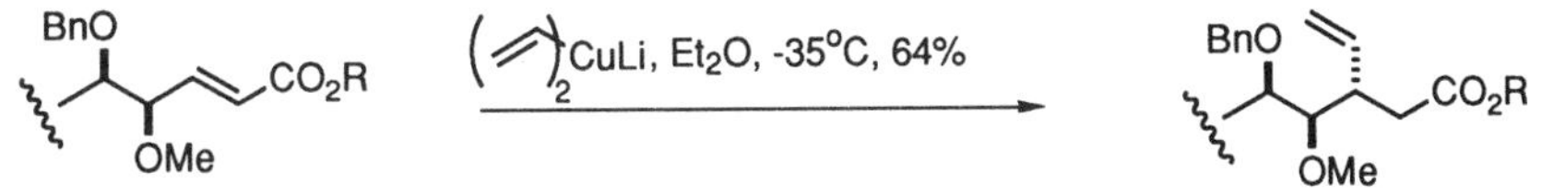

BnO ─── OMe ─── CO₂R $\left(\swarrow\right)_2$CuLi, Et₂O, -35°C, 64% ⟶ BnO ─── OMe ─── CO₂R

Additions to this class of conjugate acceptors have found application in the total synthesis of several 16-membered macrocyclic antibiotics.[89]

Carbomycin B and leucomycin A₃[90,91]

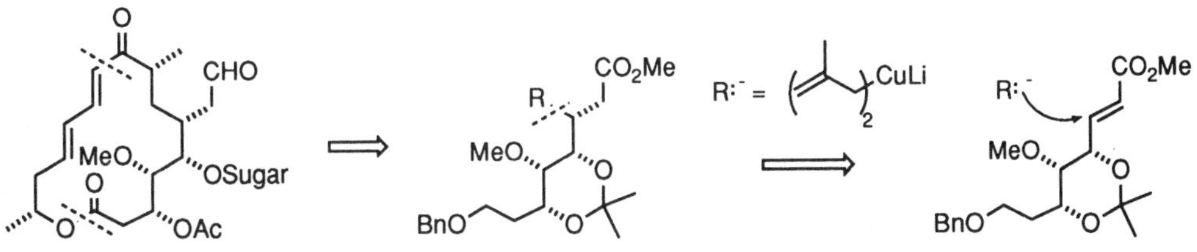

Tylonolide[92,93]

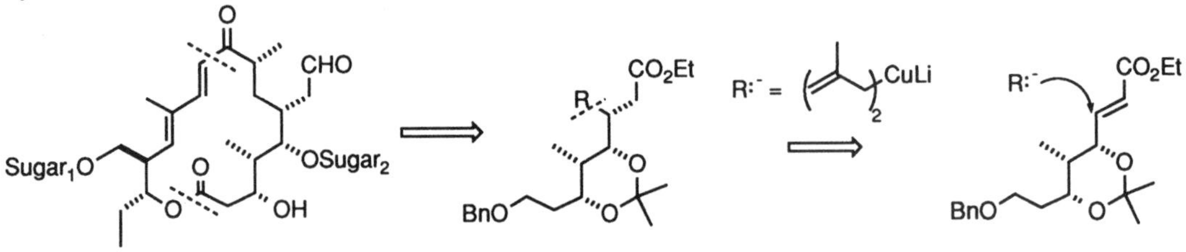

Addition of alkyl, alkenyl or aryl cuprates to protected γ-aminoalkenoates is also highly stereoselective but in favour of the *syn* diastereomer in each case.[94] Similar addition to the corresponding diesters also proceeds stereoselectively, but with the *anti* diastereomer now predominating (see Table 4.10).

Table 4.10 Stereoselective addition of organocuprates to protected γ-aminoalkenoates

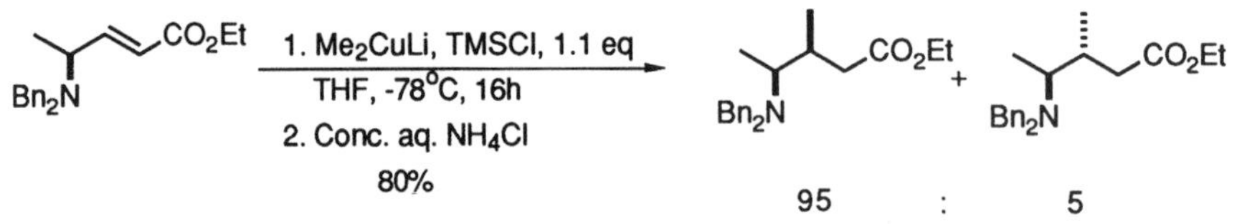

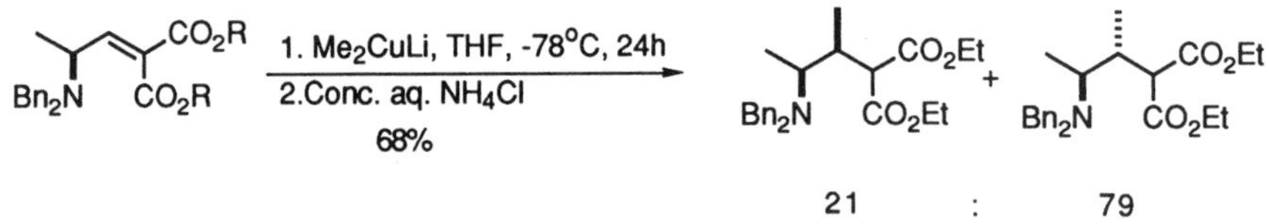

Cinnamates, bearing a chiral substituent at the ortho position, can also undergo highly diastereoselective conjugate additions with LO cuprates.[95] Chiral acetals give poor selectivity.

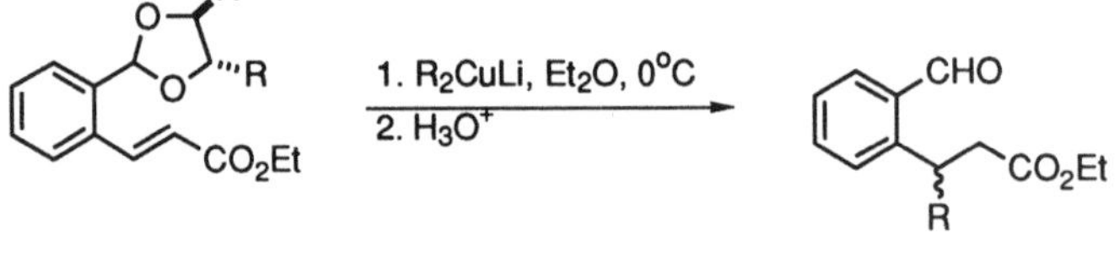

On the other hand, aminals, especially those derived from (R,R)(-)-bis N-methylaminocyclohexane (**12**), are much more effective. The authors suggested that this

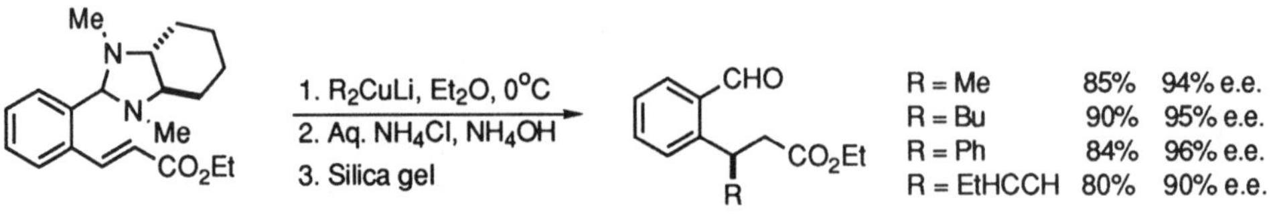

12

high selectivity is due to the reaction taking place via the following conformation.

Ullenius's group has studied LO cuprate additions to similarly substituted cinnamates.[73] They found that, in solvents of low polarity, such as toluene or dichloromethane, good diastereoselectivity could be obtained. The use of diethyl ether gave poor yields whilst tetrahydrofuran gave no reaction at all.

Conjugate addition of dialkylcopperlithiums to homochiral 2-(arylsulfinyl)alkenoates gives, after reductive removal of the sulfinyl group and ester hydrolysis, 3,3-disubstituted carboxylic acids, with moderate enantiomeric excesses.[96] A chelation-controlled transition state has been proposed to account for the stereoselectivity.

R = Me   R' = Bu   n.y.   59% e.e.
R = Bu   R' = Me   53%   65% e.e.

4.1.3   Other organometallic reagents

4.1.3.1   Addition of achiral organometallic reagents to achiral alkenoate derivatives

Reactions of organomagnesium reagents with alkenoates can give, as well as conjugate addition, reduction, dimerization and 1,2 addition products. The distribution of products is usually due to the presence of impurities in the magnesium metal used to prepare the reagent. However, these problems are not often so significant, with the availability of magnesium of improved purity, and there are now many examples of efficient conjugate additions of organomagnesium reagents to alkenoic acid derivatives.

Some organolithium reagents add to secondary amides of alkenoic acids.[97] These reactions almost certainly involve deprotonation of the amide first, which reduces the reactivity of the carbonyl towards 1,2 addition. However, this does not appear to be very general as several important reagents, such as methyllithium, phenylmagnesium bromide and 2-phenylethenyllithium don't react at all.

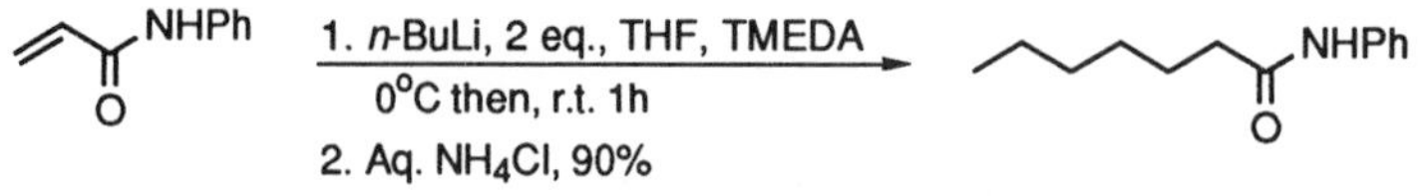

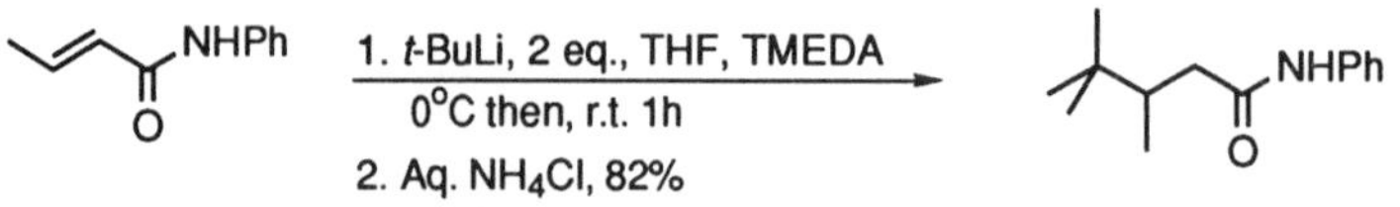

One of the most important advances in these reactions has been the discovery that thioamides of alkenoic acids are excellent conjugate acceptors for hard nucleophiles.[98] From the Table below (4.11) it is apparent that quite a variety of organometallic reagents can add to these thioamide derivatives. Under similar reaction conditions the oxoamide gives equal amounts of 1,2 and 1,4 adducts in a rather low overall yield (compare entries 1 and 2). Addition to a dialkenoic acid thioamide gave, exclusively, 1,4 rather than 1,6 addition (entry 4).

Table 4.11   Conjugate additions of organometallic reagents to alkenoic acid thioamides[98]

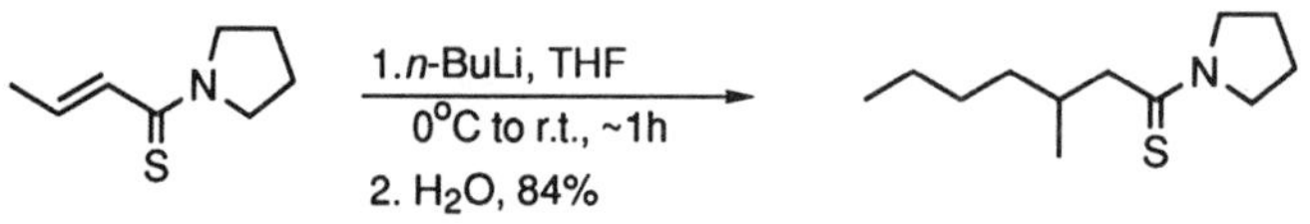

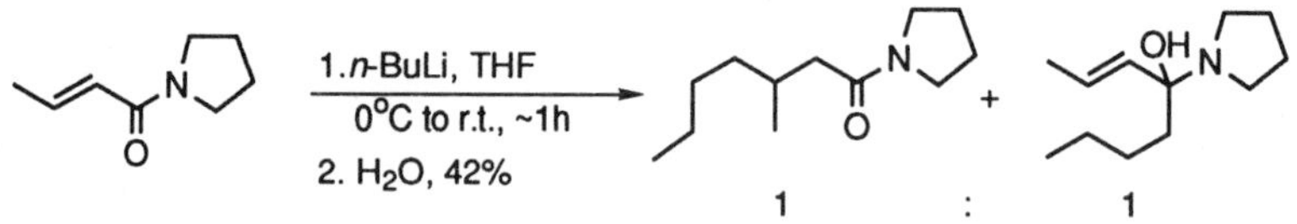

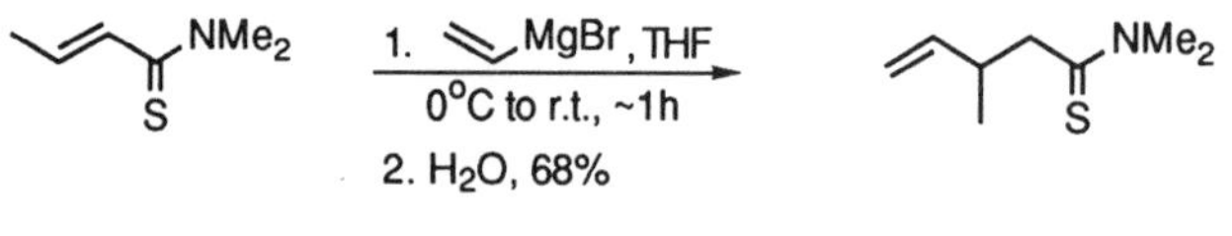

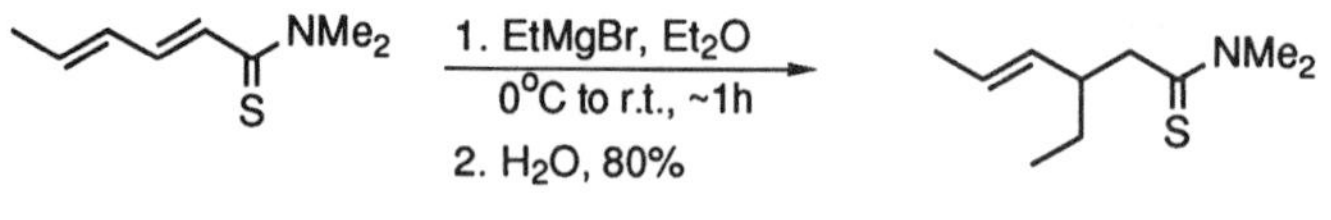

However some tertiary oxoamides of alkenoic acids react well with organolithium reagents as, for example, in reactions with dialkylamides of crotonic acid. The intermediate enolate may be quenched stereoselectively with electrophiles.[99]

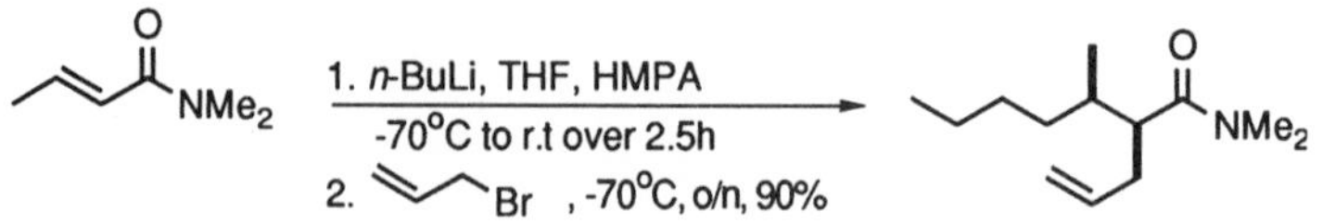

(α-Amino)alkenenitriles readily react with alkyllithiums at low temperature. The intermediate, stabilized, anion may then be quenched with an electrophile. After hydrolysis, this yields a ketone.[20]

$$\text{CN} \atop \overset{}{\diagup}\!\!\diagdown\text{NMe}_2 \quad \begin{array}{l}\text{1. } t\text{-BuLi, THF, -78°C, HMPA, 2.5h} \\ \text{2. MeI, warm to r.t.} \\ \text{3. 3N HCl, 40h, r.t., 80\%}\end{array} \longrightarrow$$

Corey and Hegedus's nickel-based masked acyl anion also adds to alkenoates.[100]

$$n\text{-BuLi + Ni(CO)}_4 \quad \xrightarrow[\text{2h}]{\text{Et}_2\text{O, pentane}} \quad \left[\text{RC-Ni(CO)}_3 \right] \quad \xrightarrow[\begin{array}{l}\text{1. Et}_2\text{O, -50°C, 4.5h} \\ \text{2. Aq. NH}_4\text{Cl} \\ \text{3. I}_2\text{, Et}_2\text{O, 76\%}\end{array}]{\text{CO}_2\text{Me}}$$

Another acyl anion equivalent is (disodium)irontetracabonyl.[101] Additions of this reagent are generally high-yielding and the reaction is completely regioselective.

$$\diagup\!\!\!\diagdown\!\!\text{CO}_2t\text{-Bu} \quad \begin{array}{l}\text{1. } n\text{-HexBr, Na}_2\text{Fe(CO)}_4 \\ \overline{\quad\text{THF, r.t., 4h}\quad} \\ \text{2. AcOH} \\ \text{3. H}_2\text{O, 83\%}\end{array} \longrightarrow \text{CO}_2t\text{-Bu}$$

β-Tosylpropenoates are readily substituted, with retention, by alkenyl- and alkynylmagnesium reagents providing a useful method for preparing dienoates and alkynenoates.[102]

$$\text{Tos}\diagup\!\!\!\diagdown\!\!\text{CO}_2\text{Et} \quad \begin{array}{l}\text{1. } \diagdown\!\!\!\diagup\text{MgBr , THF, 0°C, 3h} \\ \text{2. Aq. NH}_4\text{Cl, 79\%}\end{array} \longrightarrow \text{CO}_2\text{Et}$$

$$\text{Tos}\diagup\!\!\!\diagdown\!\!\text{CO}_2\text{Et} \quad \begin{array}{l}\text{1. } n\text{-hex}\!\!-\!\!\equiv\!\!-\!\!\text{MgBr} \\ \overline{\quad\text{THF, 0°C, 24h}\quad} \\ \text{2. Aq. NH}_4\text{Cl, 55\%}\end{array} \longrightarrow n\text{-hex}\diagdown\!\!\!\!\diagup\text{CO}_2\text{Et}$$

Allylsilanes add efficiently to alkenoate derivatives with fluoride catalysis (see Table 4.12).[103] No success was achieved using Lewis acid catalysis. Whereas allylcuprates give 1,6-addition with a 2,4-alkadienoate,[104] allylsilane addition gives exclusively 1,4 addition.

Table 4.12   Conjugate additions of allylsilanes to alkenoate derivatives[103]

1	TMS + CO₂Bn	1. TBAF, cat., DMF, HMPA, 3 eq., r.t. 10 min; 2. MeOH, HCl, 65%	CO₂Bn
2	TMS + (furyl)CN	1. TBAF, cat., DMF, HMPA, 3 eq., r.t. 10 min; 2. MeOH, HCl, 91%	CN
3	TMS + Ph CO₂Me	1. TBAF, cat., DMF, HMPA, 3 eq., r.t. 10 min; 2. MeOH, HCl, 79%	Ph CO₂Me
4	TMS + Ph CO₂Me	1. TBAF, cat., DMF, HMPA, 3 eq., r.t. 10 min; 2. MeOH, HCl, 63%	Ph CO₂Me

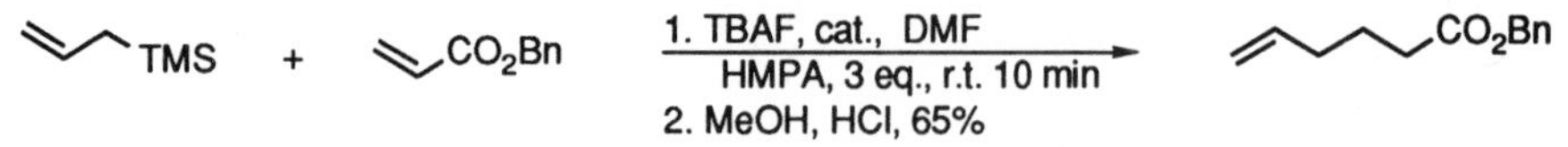

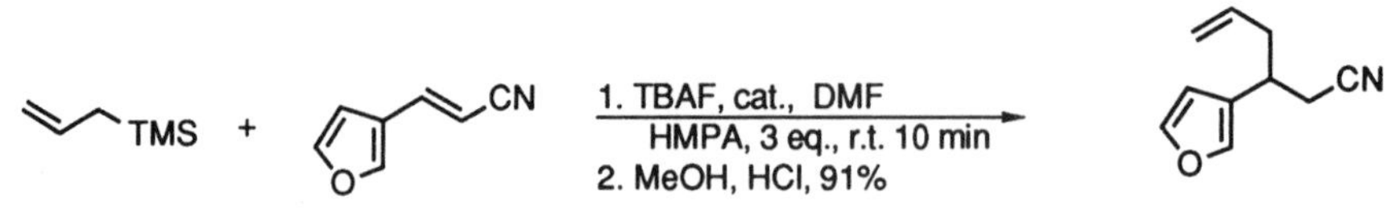

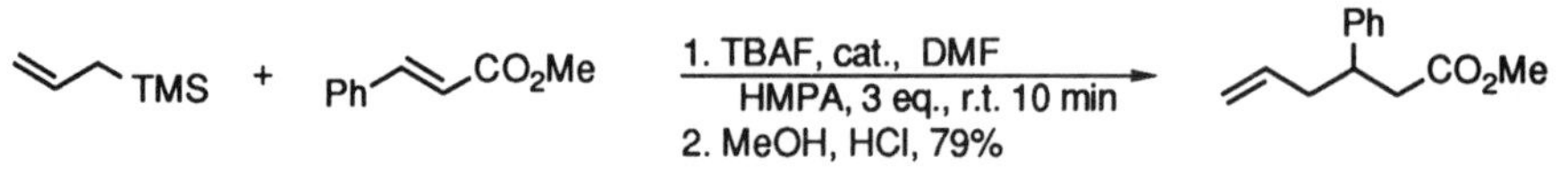

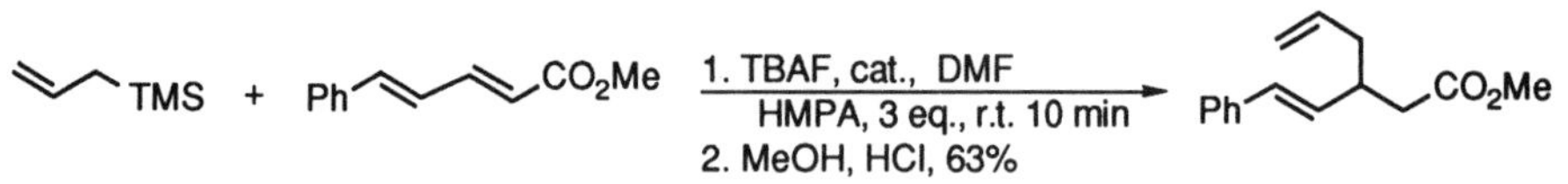

4.1.3.2   *Addition of achiral organometallic reagents to chiral alkenoate derivatives*

     Oppolzer's group has demonstrated that the addition of organomagnesium reagents to alkenoylsultams (for a definition of X^*_N, see p 227) produces the opposite stereochemical result to that for organocuprate additions (for the latter, see section 4.1.2).[105]

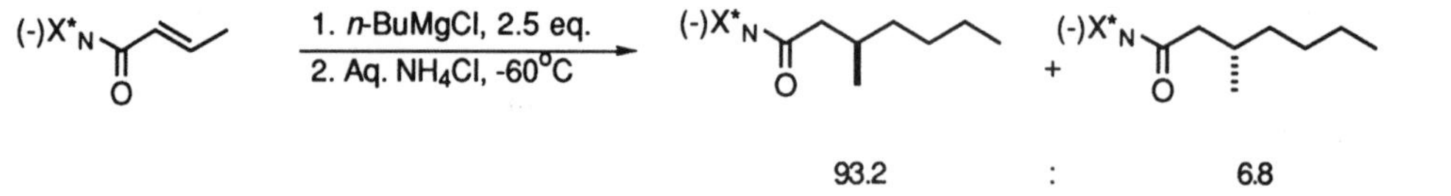

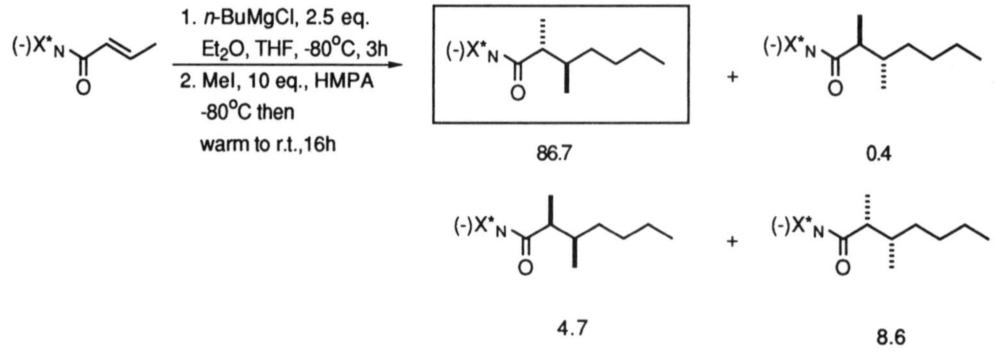

$(-)X^*_N$ —

1. *n*-BuMgCl, 2.5 eq.
2. Aq. NH₄Cl, -60°C

→ $(-)X^*_N$ + $(-)X^*_N$

         93.2   :   6.8

$(-)X^*_N$ —

1. *n*-BuMgCl, 2.5 eq. Et₂O, THF, -80°C, 3h
2. MeI, 10 eq., HMPA -80°C then warm to r.t.,16h

→ $(-)X^*_N$ + $(-)X^*_N$

         86.7   0.4

$(-)X^*_N$ + $(-)X^*_N$

         4.7   8.6

Addition to 2-substituted and 2,3-disubstituted alkenoylsultams, followed by quenching with excess methyl iodide in HMPA also yields the opposite result to that observed with organocuprate additions.

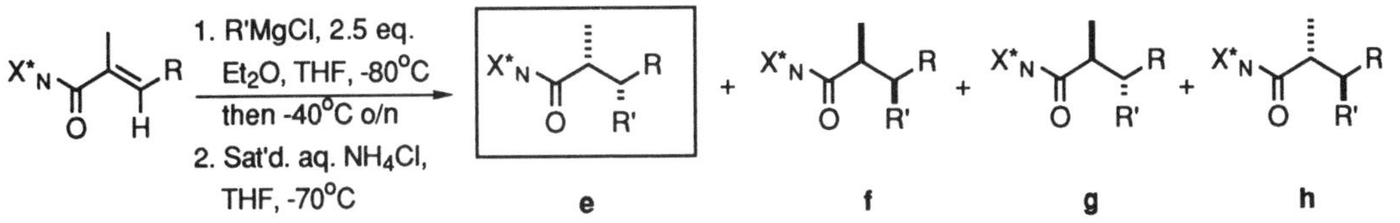

R	R'	yield	e :	f :	g :	h
Me	Et	90	99.0	0	1.0	0
Me	Bu	73	98.2	0.2	0.9	0.7
Me	Ph	n.d.	97.0	0	2.4	0.6
Et	Bu	90	97.0	0	2.6	0.4
Bu	Et	82	96.5	0	1.4	2.1

For all these organomagnesium additions, selectivity is apparently due to the reaction proceeding through a chelation-controlled conformation (>2 eq. of RMgCl required). As well as this it was assumed that a 6-membered cyclic mechanism was operating (s-cis conformation).[106] Attack then occurs from the face opposite to the nitrogen lone pair, as shown below for addition to a 2,3-disubstituted system, followed by protonation.

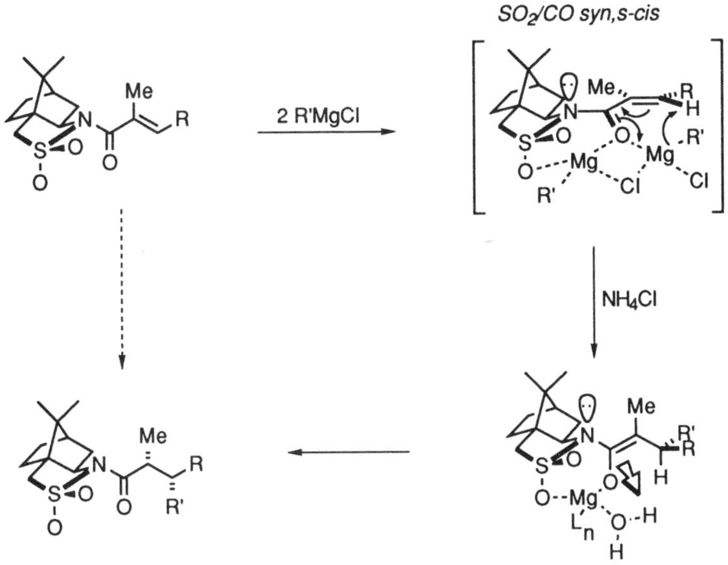

Excellent stereocontrol in the conjugate addition of organomagnesium and organolithium reagents to α-alkoxy isopropylidenemalonates, (13) and (14), has been demonstrated.[107] Either diastereomeric product may be selected by the incorporation of either a chelating or non-chelating protecting group for the α-oxygen. As shown below, best results were obtained using combinations of organomagnesium reagent/MEM protecting group for chelation control and organolithium reagent/12-crown-4/TBDMS protecting group for non-chelation control.

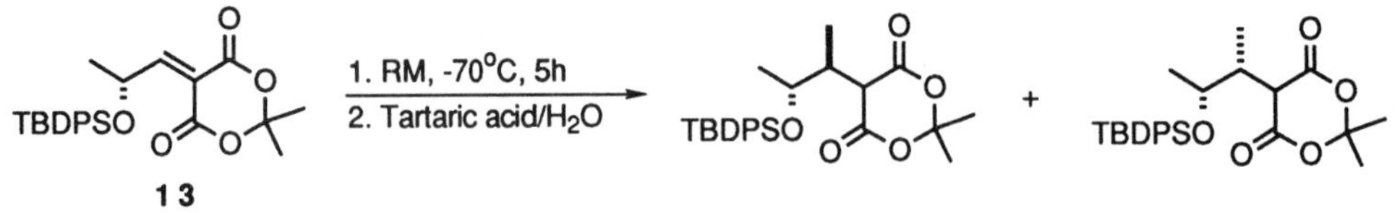

| RM = MeLi, Et$_2$O, 12-c-4, 1 eq. | 62% | 99 | : | 1 |
| RM = EtMgBr, THF | 71% | 85 | : | 15 |

| RM = MeLi, Et$_2$O | >49% | 30 | : | 70 |
| RM = MeMgBr, THF | >62% | <1 | : | >99 |

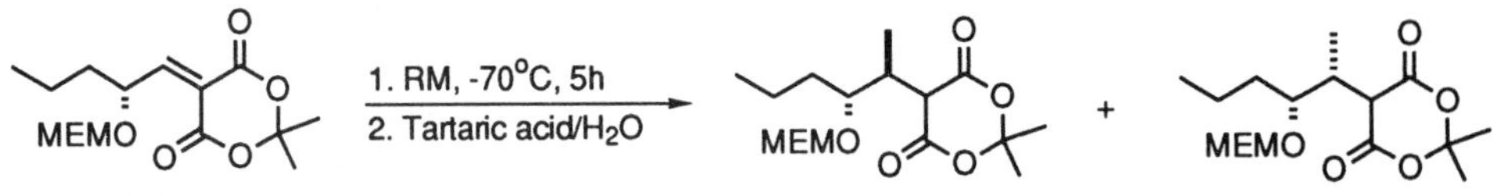

This method was applied to a total synthesis of eldanolide, the sexual pheromone of *Eldana saccharina*.[108]

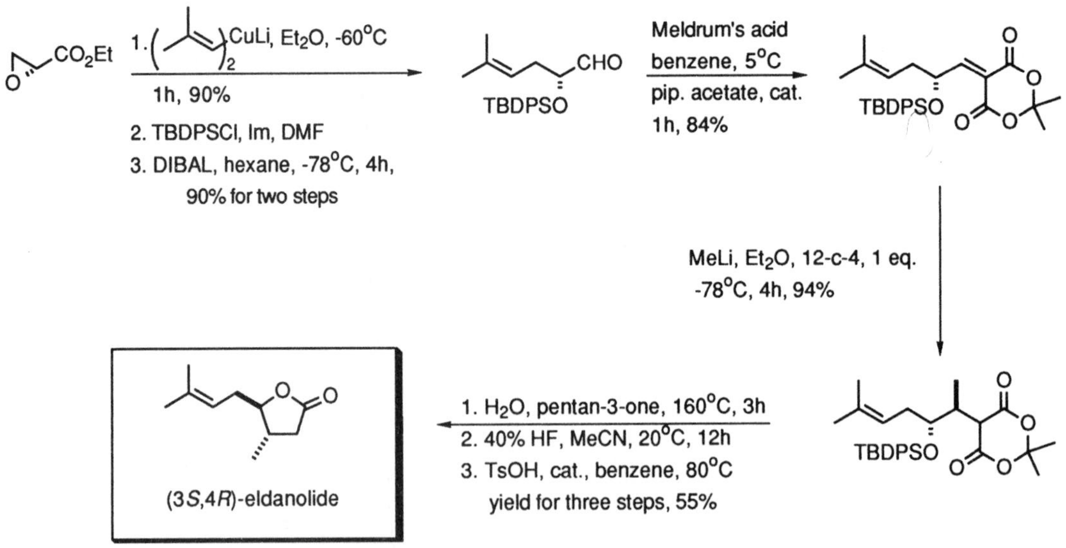

Chiral alkenyloxazolines also participate in highly stereoselective conjugate additions with organolithium reagents.[109] These oxazolines, are readily available from the corresponding 3-methyl series[110] via a Horner-Emmons alkene synthesis.[111]

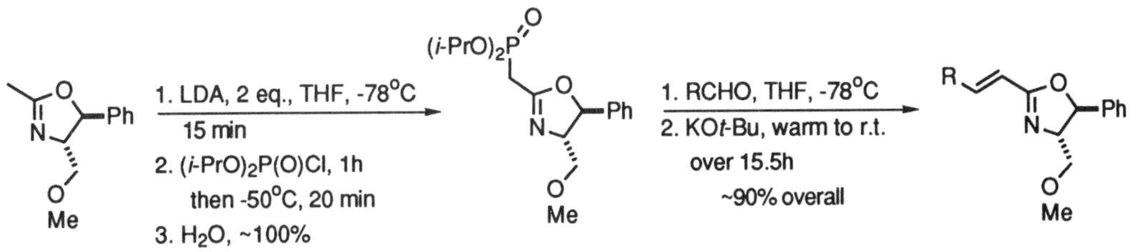

Conjugate addition of organolithium reagents, followed by hydrolysis, produces 3,3-disubstituted alkanoic acids with consistently high enantiomeric excesses. The chemical yields are not always high as some allylic deprotonation occurs. In fact, *only* allylic deprotonation occurs upon treatment of the corresponding (Z)-alkenyloxazoline with alkyllithiums.

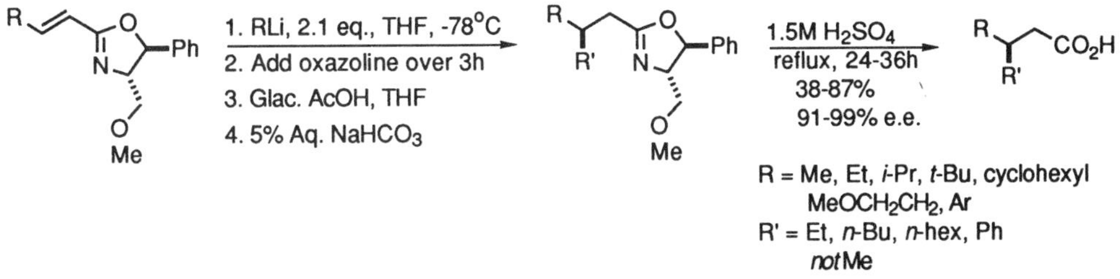

In the same paper a mechanism which accounts for the stereoselection in these additions was proposed. It is supported by the observation that identical selectivity may be obtained by carrying out similar additions to an alkenyloxazoline which bears only the 4-(S)-methoxymethyl substituent and no 5-phenyl substituent. This, therefore, excludes any possibility of the 5-phenyl group having an influence on stereoselection. In related work on additions to oxazolinylnaphthalenes (see below) the mechanism was viewed as a [1,5]-sigmatropic rearrangement. This may well apply to these additions too.

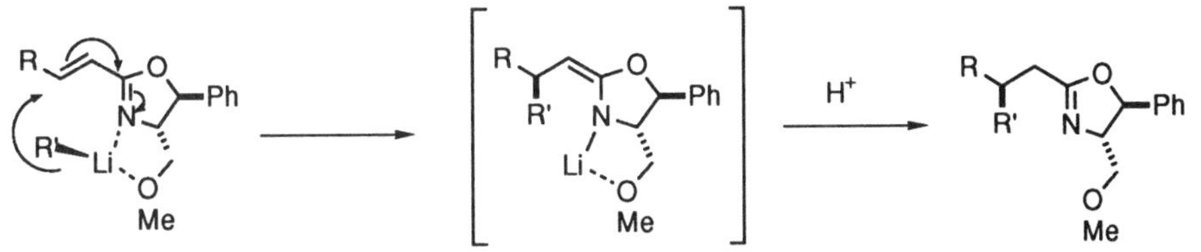

This method was applied to a total synthesis of (+)-ar-turmerone.[112]

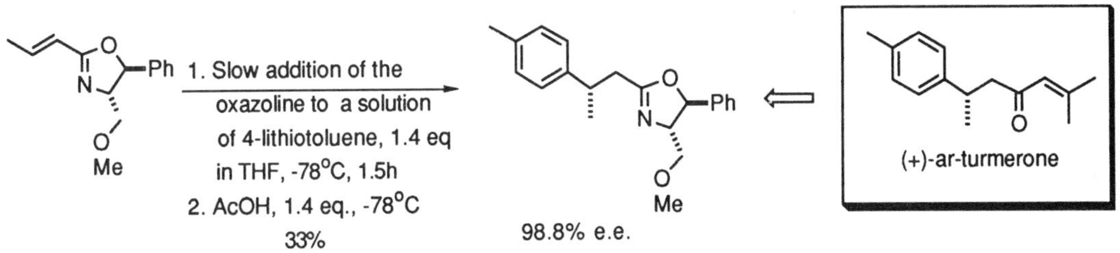

More recently, Meyers' group has demonstrated that 1,2-dihydronaphthalenes may be prepared using an extension of the above method and has used this process to synthesize the lignan, (+)-phyltetralin.[113,114]

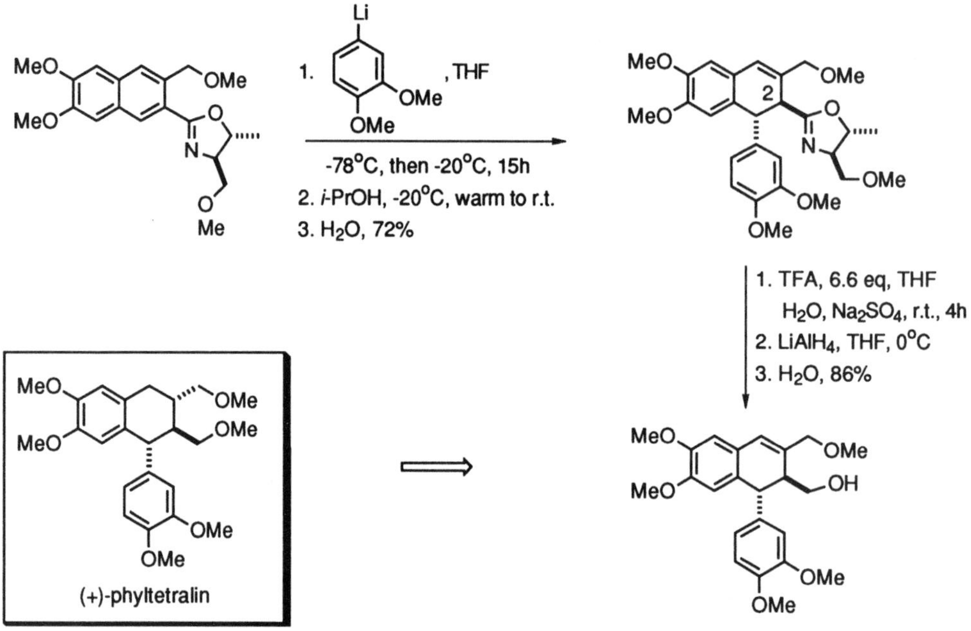

The authors offered the following [1,5]-sigmatropic shift as a mechanism which accounts for the observed selectivity. This explains the result at C1. The isolation of only one diastereomer was apparently due to complete epimerization, by *iso*-propoxide, during workup. These methods have also been applied to the synthesis of chiral 1,4-dihydropyridines[115,116] and biaryls.[117,118]

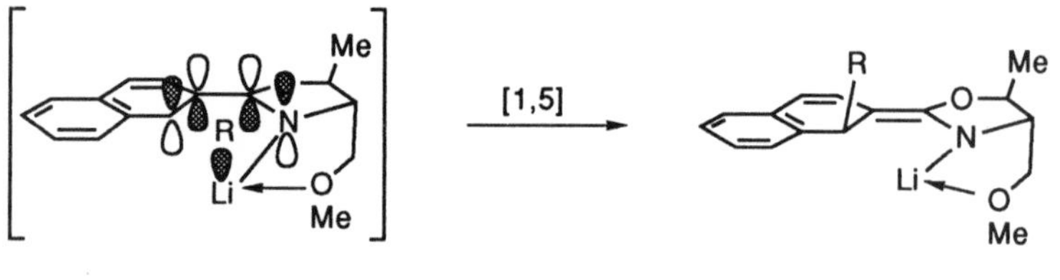

Iron acyls

The use of chiral iron acyls in synthesis has been extended to include conjugate additions to chiral α,β-unsaturated iron acyls.[119,120] Conformational analysis of both the E- and Z- isomers shows that, in each case, the s-*cis* form is preferred as it avoids steric interactions between the carbonyl ligand and a β-substituent.

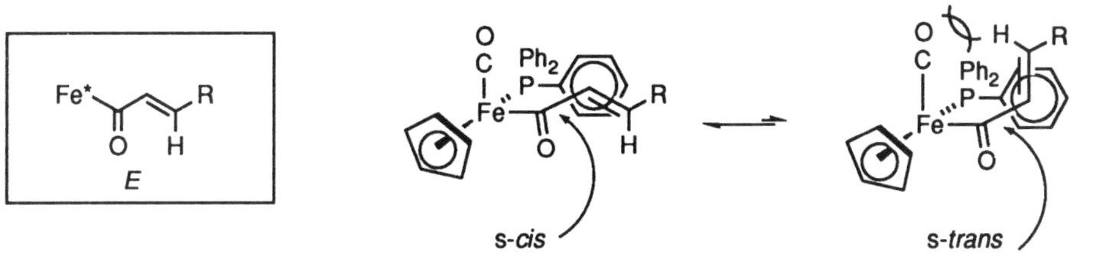

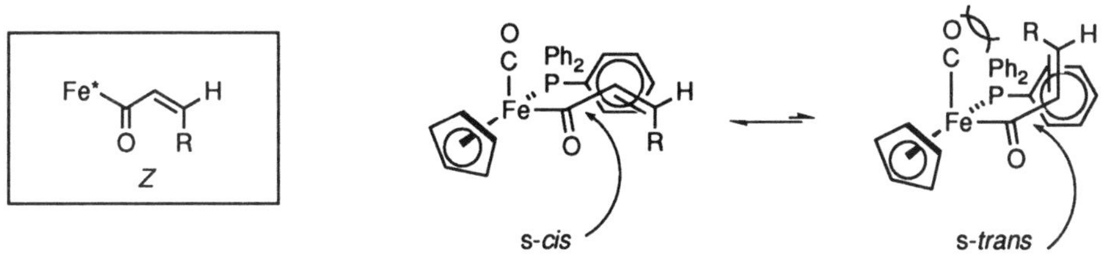

Upon exposure to an alkyllithium, the Z-isomer undergoes a chelation-controlled γ-deprotonation exclusively   and need not be discussed further here. On the other hand, the E-isomer undergoes conjugate addition, without any evidence for deprotonation. Addition is assumed to occur to the less-hindered face (of the s-*cis* conformation shown above). The conformation of the enolate intermediate is as shown, with the carbonyl ligand and the alkoxylithium *anti* to each other. Alkylation then occurs only from the less-hindered face. The diastereoselectivity of the reaction was found to be greater than 100:1:1:1.

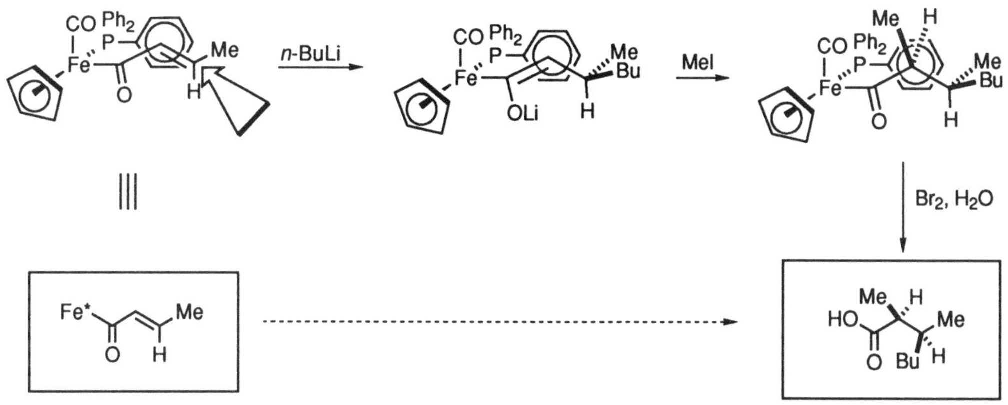

These   iron   acyls   have   also   been   used   to   prepare   homochiral cyclopropanecarboxylic acid derivatives. Either the *cis*[121] or the *trans*[122] isomer is available depending upon the geometry of the starting alkene.

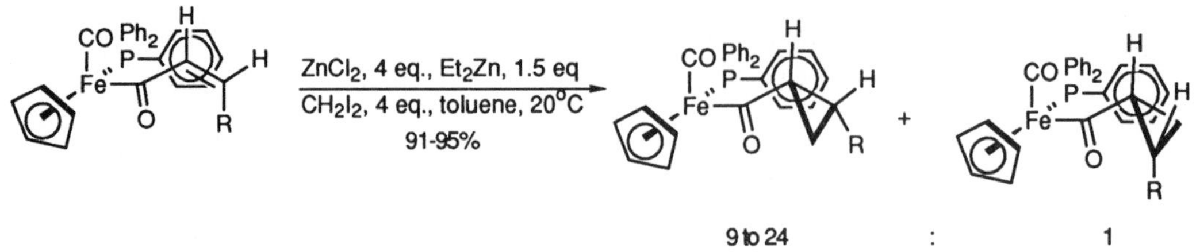

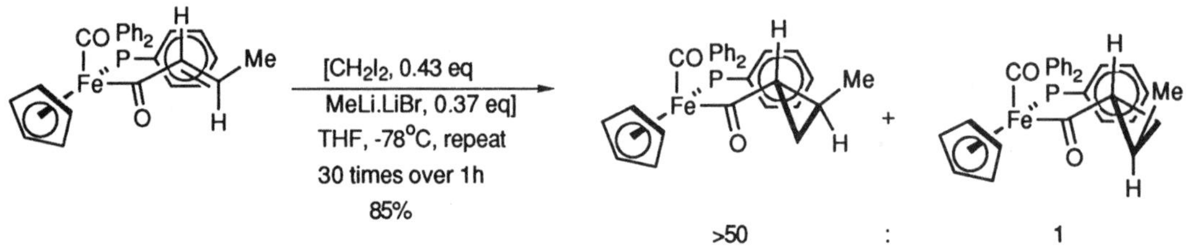

9 to 24   :   1

>50   :   1

4.1.3.3   Addition of chiral organometallic reagents to achiral alkenoate derivatives

α-Alkoxyorganolithium reagents, e.g. (**15**), may be generated from the corresponding α-alkoxyorganotributylstannane. As part of a new synthesis of γ-lactones, the addition of these reagents to a variety of alkenoyl amides and hydrazides has been studied by Chong and Mar.[123] Although conjugate addition to acrylic acid tetramethylpiperidide was efficient, attempts to convert the adduct into the desired γ-lactone failed. However, the corresponding trimethylhydrazides reacted well, although with little diastereoselectivity, and were readily converted to γ-lactones under acidic conditions.

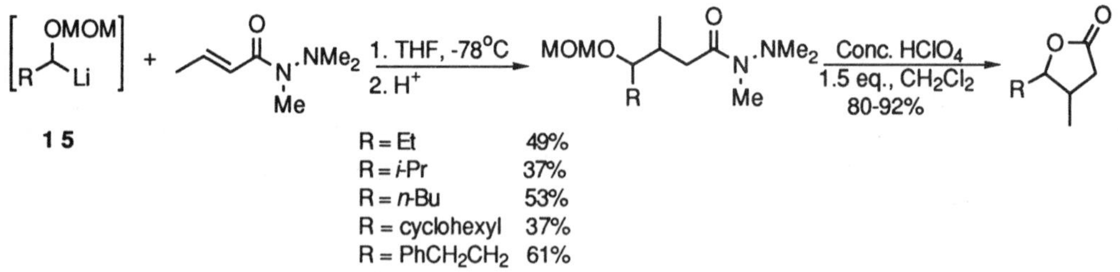

R = Et   49%
R = i-Pr   37%
R = n-Bu   53%
R = cyclohexyl   37%
R = PhCH$_2$CH$_2$ 61%

Homochiral α-alkoxyorganolithium reagents may also be generated from the corresponding α-alkoxyorganotributylstannane. These reagents are configurationally stable, at least at low temperatures, and add in a conjugate manner to alkenoyl trimethylhydrazides. This has been applied to an enantioselective synthesis of the naturally-occurring γ-lactone (S)-hexanolide.[123]

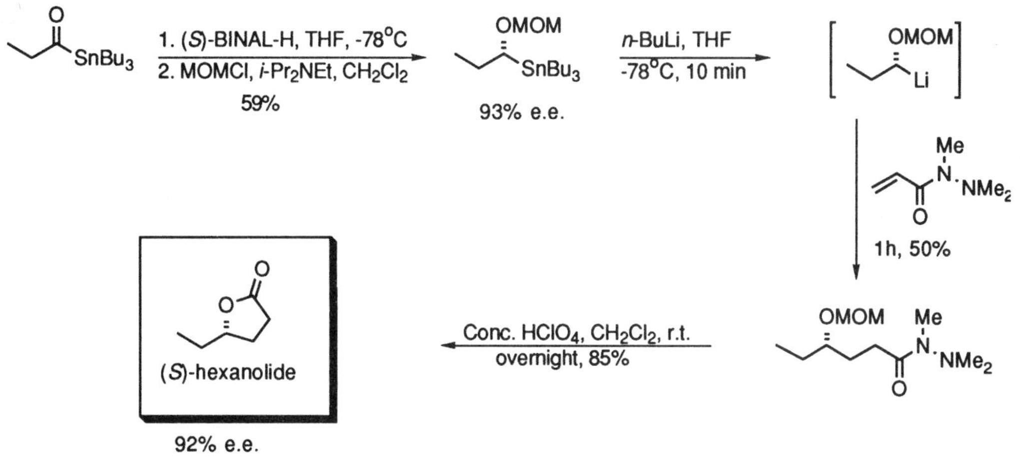

(*S*)-hexanolide

92% e.e.

4.1.4   Free radicals

Samarium iodide (prepared from samarium metal and 1,2-di-iodoethane in tetrahydrofuran)[124] is an effective catalyst for the coupling of aldehydes and ketones to alkenoates and alkenenitriles.[125] In the case of alkenoates the isolated product is a δ-lactone. The reaction rate is accelerated remarkably by the addition of hexamethylphosphoramide.[126] Only β,β-disubstituted acrylates fail to react under these conditions.

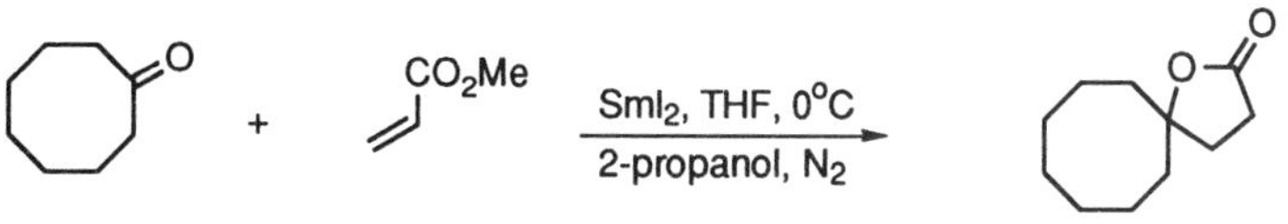

Additive	reaction time	yield
HMPA	1 min	95%
–	4 h	82%

Two mechanisms have been proposed for the reaction. The first involves electron transfer to the carbonyl carbon of the ketone or aldehyde. The resulting free radical then adds to the alkenoate (or alkenenitrile). Transfer of a hydrogen atom produces a γ-alkoxyalkenoate which spontaneously cyclizes to yield a lactone.

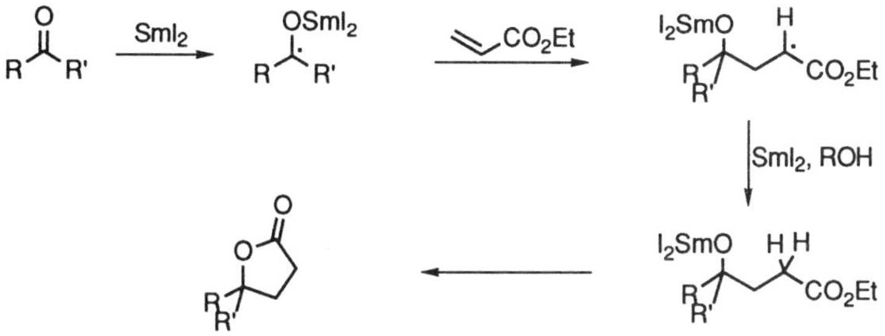

An alternative mechanism, which proceeds via a metal homoenolate was also suggested.

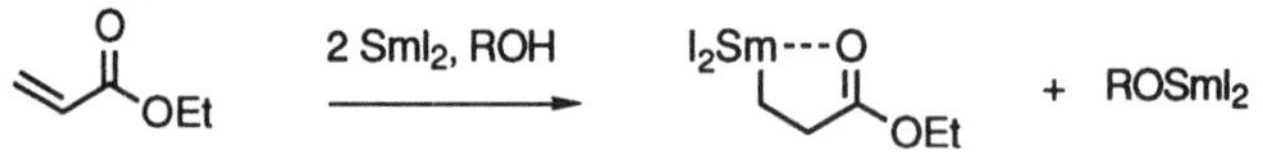

This reaction has been extended to more substituted alkenoates as part of a planned synthesis of α-alkylidene-γ-lactones.[127]

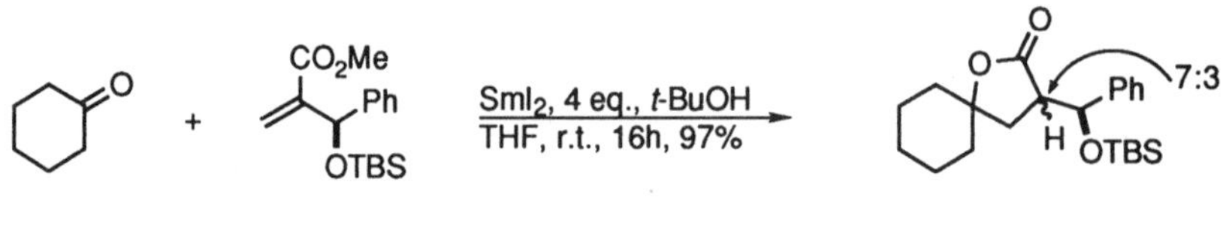

Two of the most common precursors to free radicals are haloalkanes (see entries 1 and 2) and organomercury compounds (entry 5). These are usually converted into free radicals by treatment with an initiator (light or AIBN). However, several other methods now exist and some of these are given in Table 4.13 below.[128]

Table 4.13   Conjugate additions of free radicals to alkenoic acid derivatives[128]

1[129]		
2[130]		
3[131]		
4[132]		
5[133]		

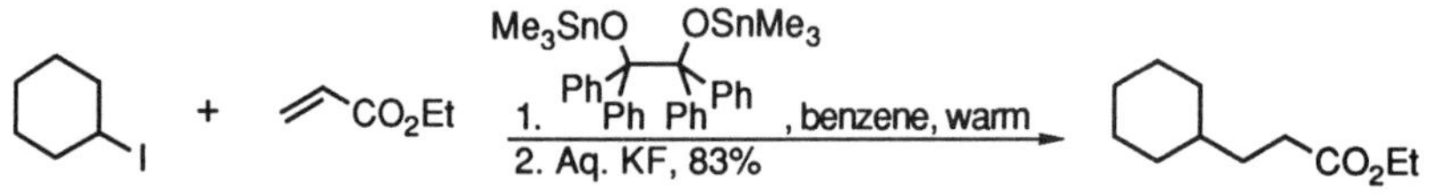

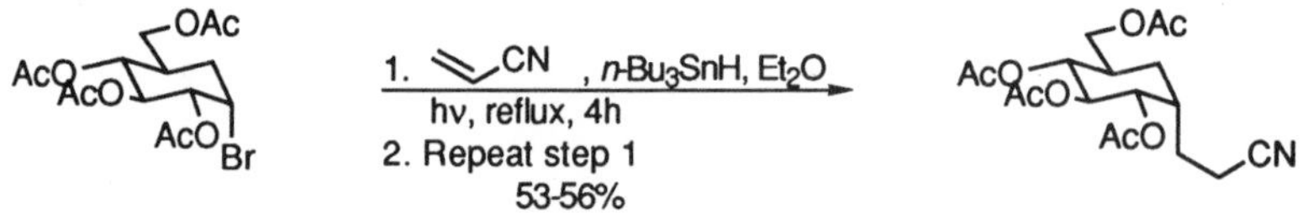

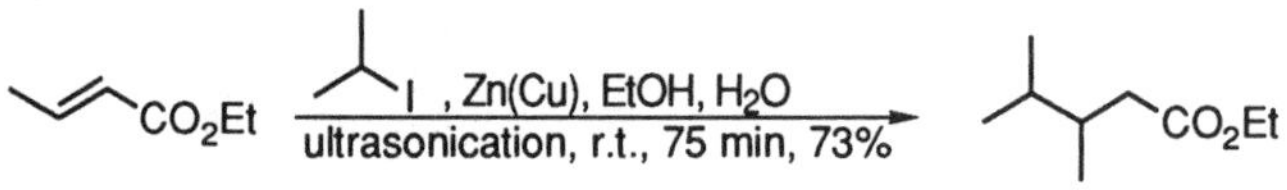

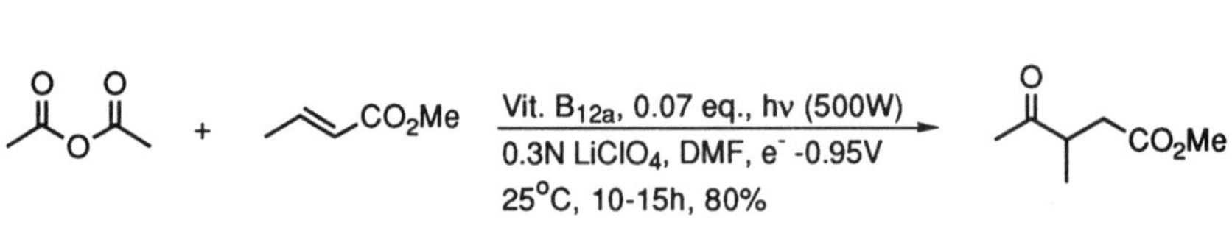

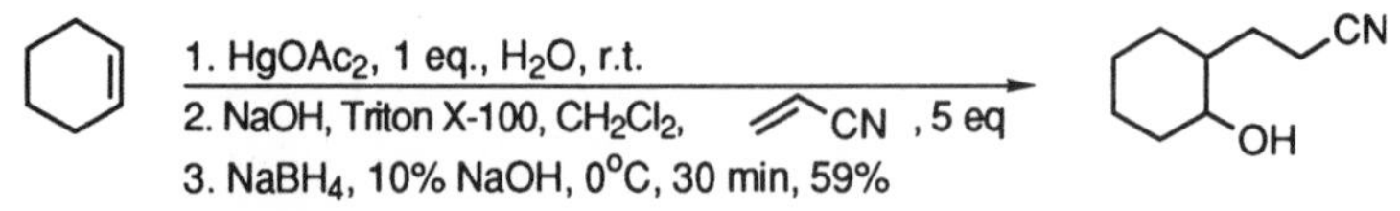

Only a few reports of stereoselective *inter*molecular conjugate additions of free radicals have so far appeared. Encouraging results have been obtained from additions to alkenoyl derivatives of (R,R)- or (S,S)-2,5-dimethylpyrrolidine. Addition of Barton's ester[273] to the chiral propenoyl amide is highly stereoselective.[274] The stereoselectivity obtained in this reaction is believed to be due to addition to the lower energy (Z) isomer of the intermediate α–radical.[275]

Free radical additions to the corresponding fumaric acid diamide derivative are also stereoselective.[134] As mentioned in section 2.1.4, high stereoselectivity may be obtained if the incoming nucleophilic radical is required to pass over the amide moiety. A consequence of this is that one of the pyrrolidine methyl substituents shields one alkene face directing the addition to occur from the opposite face. The stereoselectivity was excellent for both the n-hexyl and tert-butyl radicals.

Similar reaction with diethyl methylfumarate again showed excellent stereoselectivity.[274] In this case, asymmetric induction must be due to the newly-formed chiral centre at C3.

4.1.5   Enamines and enol ethers

4.1.5.1   Enamines

(a)   Addition of achiral enamines to achiral alkenoate derivatives

Enamines add efficiently to a variety of alkenoate derivatives. These conjugate additions are usually carried out in either polar or apolar[135] aprotic solvents sometimes at room temperature   or, occasionally[136], with acid catalysis, but, most often, simply with heating. Although N-alkylation may also occur, this is reversible with alkenoates[137,138] and the product is ultimately that of C-alkylation. Angoh and Clive have incorporated an enamine addition, to 2-(phenylseleno)-2-propenenitrile (16), as part of a new annulation procedure.[139] They found this nitrile gave clean addition as opposed to 2-chloroprop-2-

enenitrile which suffers from problems of dialkylation, cyclobutane formation and quaternization.

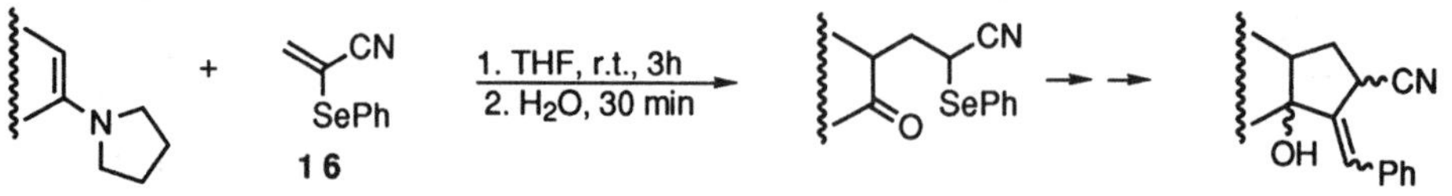

As mentioned in section 4.1.1.1, enamine chemistry is an often-used alternative method for the preparation of cycloalkanone enolate/ester adducts. Some examples are given in Table 4.14. As with other electrophiles, alkenoates react with the less-substituted position of the enamine of unsymmetrical cycloalkenones (entry 2). Clay catalysts promote enamine additions (entry 5).

Table 4.14 Enamine additions to alkenoic acid derivatives

1[140]	
2[141]	
3[141]	
4[137]	
5[142]	

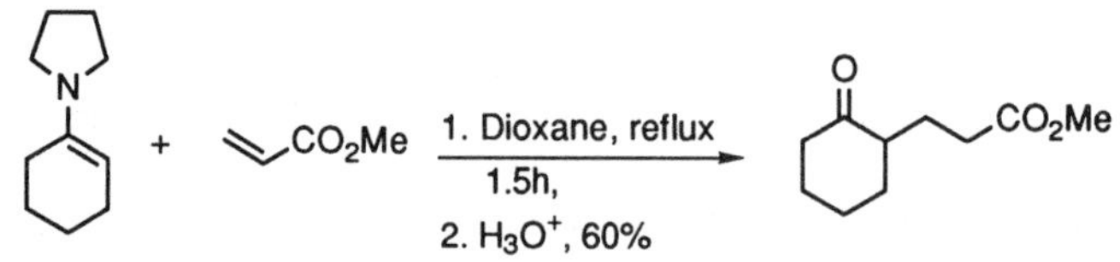

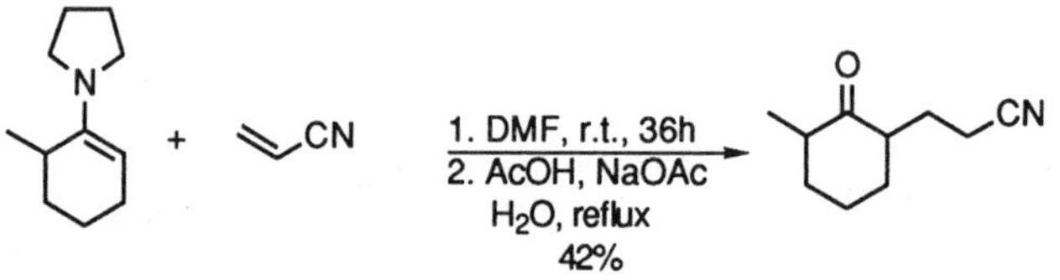

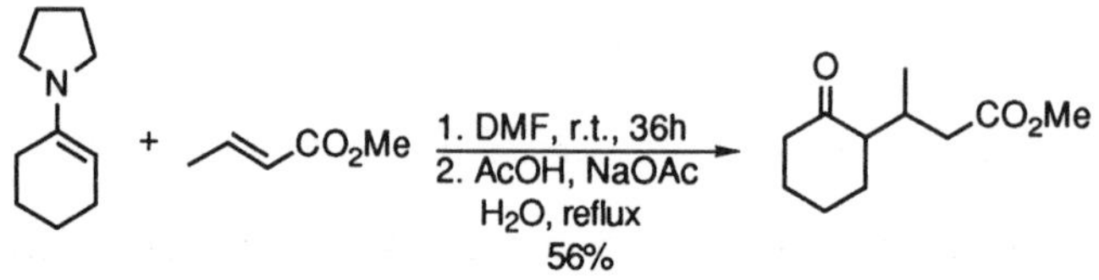

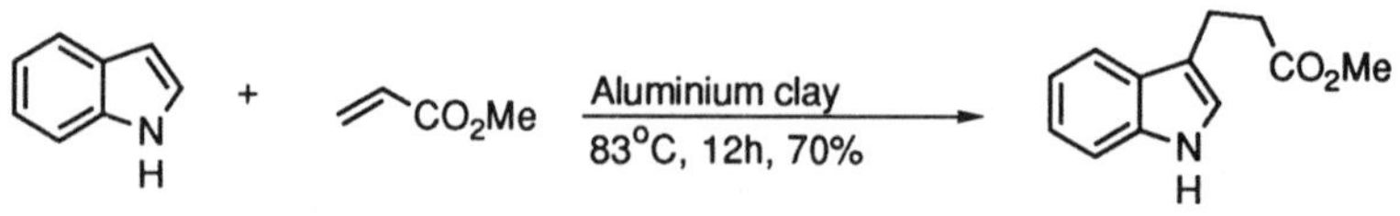

Enamines also add to cyclopropylmethylenemalonate in an exclusively 1,4 manner.[143] No evidence for attack at the cyclopropyl group was found. However, most of the product consisted of the retro-Michael compound.

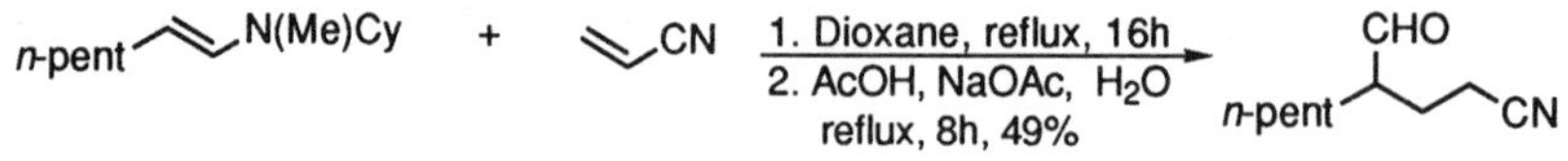

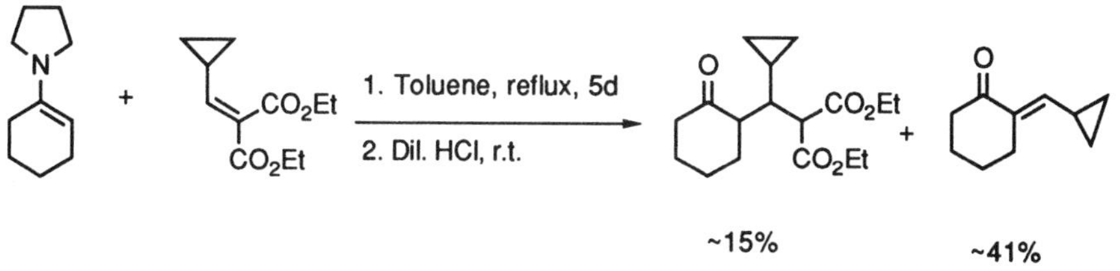

~15%   ~41%

An annulation reaction, based upon enamine addition to methyl 2,4-pentadienoate, proceeds in excellent yield and with complete regioselectivity.[144,145]

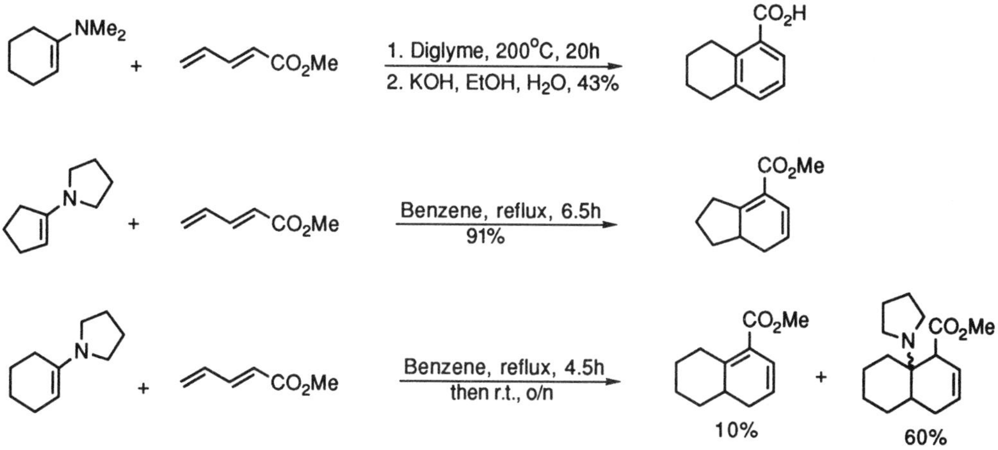

10%   60%

(b)   *Addition of chiral enamines to achiral alkenoate derivatives*

Although enamines of unsymmetrical cycloalkenones generally react at the less substituted carbon (entry 2 in Table 4.14), examples where some reaction occurs at the more-substituted carbon do exist. In fact, these reactions are quite sensitive to the solvent employed.

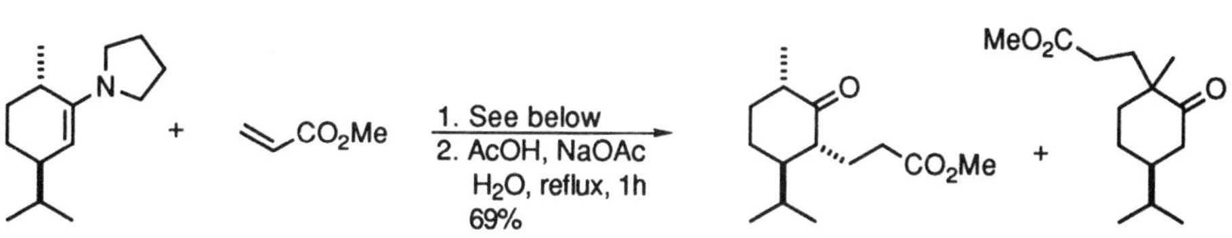

| 1 or 2 eq | benzene, 75°C, 24h | 62 | : | 38 |
| 1, 2 or 5 eq | methanol, 65°C, 3h | 100 | : | 0 |

An interesting example of enamine addition involves the following conversion, a key step in the total synthesis of (±)-vallesamidine.[4]

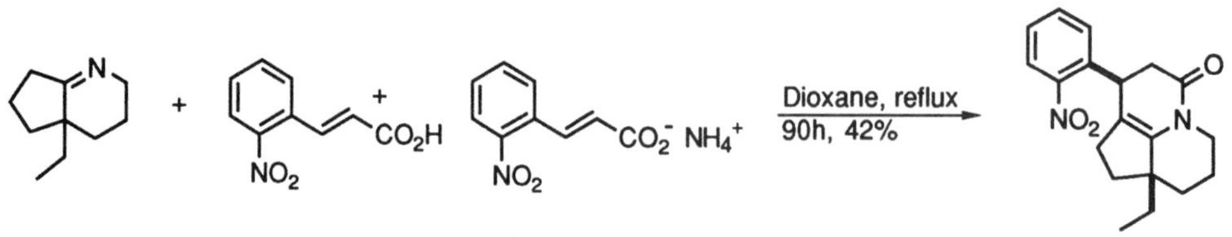

The mechanism most likely involves conjugate addition of the enamine tautomer of the bicyclic imine followed by closure to the lactam, as attempts to catalyze the intramolecular conjugate addition failed.

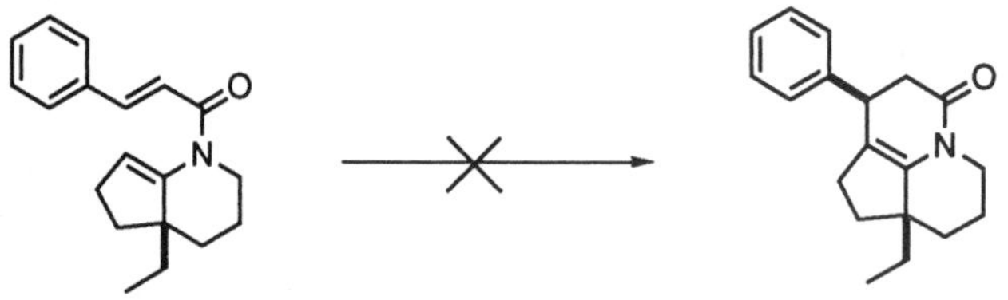

Addition of the (S)-proline derived enamine of cyclohexanone to benzylidene malonates proceeds with very high levels of both diastereo- and enantio-selectivity.[146,147] The stereochemical outcome, addition of the *Re* face of the enamine to the *Re* face of the conjugate acceptor, is again in accord with Seebach's topological rule (see section 6.1.5.1 for a description of the rule[148]).

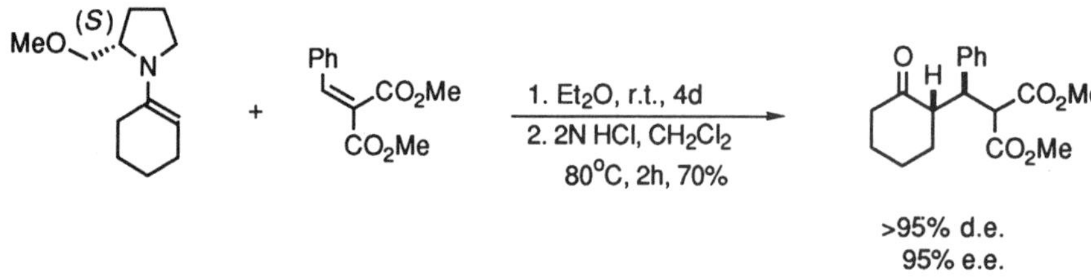

>95% d.e.
95% e.e.

This method also provides access to (S)-methyl (2-oxocyclohexyl)propanoate in excellent enantiomeric excess.

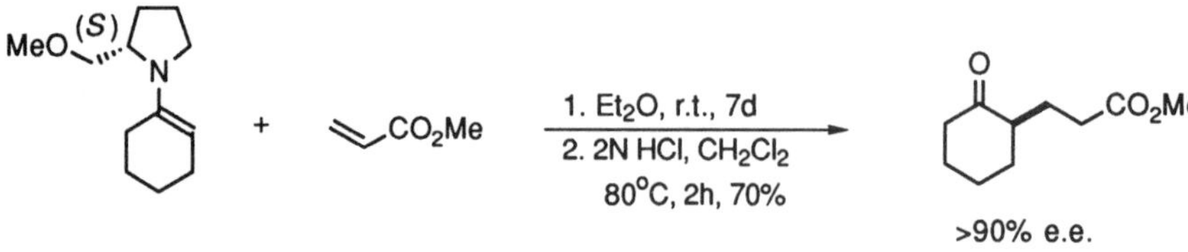

>90% e.e.

The conjugate addition of chiral organotin enamines to alkenoates offers a second efficient method for preparing these compounds in high enantiomeric excess.[149] However, similar additions to alkenenitriles are not as selective.

Diethylazodicarboxylate may be regarded as a diazaalkenedicarboxylate. Kuehne and Di Vincenzo found that it quite readily reacts with enamines.[150]

4.1.5.2   Enol ethers

The conjugate addition of trimethylsilyl enol ethers to alkenoates is the basis of Du Pont's group transfer polymerization process (GTP).[151] However, apart from this, the reaction does not appear to have been much exploited in synthesis. The yields are good and the product 1,5-dicarbonyl compounds are valuable synthetic intermediates.[152]

Mizuno et al have developed what they call a "photo-Michael reaction".[153,154] The addition proceeds via exciplex formation.[155]

* High pressure mercury arc

4.1.6   Conjugate hydrocyanation[156]

A convenient synthesis of 2,2-dimethylsuccinic acid relies upon the conjugate addition of cyanide to ethyl 2-cyano-3-methyl-2-butenoate.[157] A range of 2,2-dialkylsuccinic acids may be prepared this way.

$$\text{(CH}_3)_2\text{C=O} + \underset{\text{CN}}{\text{CH}_2\text{CO}_2\text{Et}} \xrightarrow[\text{then } 100^{\circ}\text{C, 2h, 58\%}]{\text{Pip., cat., r.t., 60h}} \underset{\text{CN}}{(\text{CH}_3)_2\text{C=C(CO}_2\text{Et})} \xrightarrow[\substack{\text{2. Conc. HCl, excess} \\ \text{reflux, 3h, 63\%}}]{\substack{\text{1. KCN, 2.2 eq., EtOH} \\ \text{H}_2\text{O, 48h, r.t.}}} \text{HO}_2\text{C}\cdots\text{CO}_2\text{H}$$

Simple alkenoates, such as methyl propenoate and propenenitrile will hydrocyanate, but only under rather forcing conditions (Table 4.15, entries 1 and 2). Interestingly, thioalkenoates tend to be more reactive than normal alkenoates. α-Substituted alkenoates react very poorly (entry 4). Succinimides are the ultimate product if a γ-alkoxycarbonyl substituent is present (entry 5). Doubly activated alkenoates react readily at room temperature under quite mild conditions (entry 7).

Table 4.15   Conjugate hydrocyanation of alkenoates and their derivatives

Entry	Substrate	Conditions	Product
1[158]	CH$_2$=CH–CO$_2$Me	HCN, KCN, neat 70-80°C, 73%	NC–CH$_2$CH$_2$–CO$_2$Me
2[158]	CH$_2$=CH–CN	HCN, pyr., neat 60-70°C, 86%	NC–CH$_2$CH$_2$–CN
3[159]	CH$_3$CH=CH–CO$_2$n-Bu	HCN, KCN, neat 150°C, 76%	NC–CH$_2$CH(CH$_3$)–CO$_2$n-Bu
4[160]	CH$_2$=C(CH$_3$)–CO$_2$Me	HCN, [Na$_4$Ni(CN)$_4$] neat, 150°C, 22%	NC–CH$_2$CH(CH$_3$)–CO$_2$Me
5[158]	MeO$_2$C–CH=CH–CO$_2$Me	HCN, KCN, neat 40°C, 72%	succinimide with CO$_2$Me side chain
6[161]	Ph–CH=C(CN)–Ph	KCN, MeOH, Et$_2$O H$_2$O, reflux, 92%	Ph–CH(CN)–C(CN)(Ph)
7[162]	(n-Hex)–CH=C(CN)–CO$_2$Et	1. HCN, aq. EtOH, r.t. 2. Aq. HCl, 85%	(n-Hex)–CH(CO$_2$H)–CH(CO$_2$H)
8[163]	O–N pyrroline with CO$_2$Me	NH$_4$Cl, NaCN, 50% aq. DMF 80°C, 12h, 81%	pyrrolidine with CO$_2$Me and CN (76 : 5)

The hydrocyanation of steroidal thioesters is highly stereoselective.[164]

1. Et$_2$AlCN, 5 eq.
   benzene, r.t.,1.8h
2. 2N HCl, 0°C
   79%

γ-Bromoalkenedinitriles react with sodium cyanide to produce cyclopropanes in moderate yield.[165]

KCN, 78% MeOH, 30°C
40%

4.1.7   Heteronucleophiles

4.1.7.1   *Nitrogen and phosphorus*

(a)   *Nitrogen*

(i)   *Addition of nitrogen to achiral alkenoates and their derivatives*[166]

Ammonia reacts at low temperature with propenoates, propenoic acid amides and propenenitriles. Reaction can be controlled to yield the mono, di, tri or tetra-alkylated product. Reactions with substituted alkenoates, including α-, β- and α,β-di-substituted systems are easier to control and the monoalkylation product is often obtained in high yield.

A variety of piperidones may be prepared by the addition of two alkenoates to a primary amine. The order of events determines the structure of the product. For example, 1,3-dimethyl-4-piperidone can be synthesized by the sequential addition of methyl 2-methylpropenoate and methyl propenoate to methylamine.[167]

MeNH$_2$, MeOH
r.t., 3d, 77%

77%

Primary and secondary amines[168] usually react readily at room temperature or with heating (see Table 4.16). Problems of further alkylation sometimes occur if the reaction is heated for extended periods. Where further reaction is required, heating a neat mixture is often useful. Aminomercuration has been used to prepare β-amino esters, although this reacton is limited to N-arylamines (e.g. entry 5).[169] Furthermore, weakly basic N-arylamines fail to add and even where the aminomercuration is successful, reduction often leads to reversion to starting materials rather than reductive demercuration (however, see entry 6).

Table 4.16   Conjugate addition of amines to alkenoic acid derivatives

| 1[170] | EtNH$_2$ + (CN alkene) | H$_2$O, r.t.,5h, then 100°C, 1h
 90% | EtHN~CN |

| 2[170] | Et$_2$NH + (CN alkene) | 50°C, 24h, r.t.,2d
 97% | Et$_2$N~CN |

| 3[171] | Et$_2$NH + (CO$_2$Et alkene) | Steam bath, 40h
 88% | Et$_2$N~CO$_2$Et |

| 4[172] | (NHMe amine) + (CO$_2$Et alkene) | 120°C, 20h
 69% | (Me, N~CO$_2$Et) |

| 5[170] | (carbazole NH) + (CN alkene) | Triton B, 0°C, then 100°C, 0.5h
 85% | (carbazole N~CN) |

| 6[169] | PhNH$_2$ + (CO$_2$Me alkene) | 1. Hg(OAc)$_2$, THF, r.t., 5h
 2. 0.5N NaOH, NaBH$_4$, 2 eq.
 r.t., 15h, 80% | PhHN~CO$_2$Me |

| 7[169] | PhNH$_2$ + Ph~CO$_2$Me | Hg(OAc)$_2$
 THF, r.t.
 5h, 89% | (PhHN, Ph~CO$_2$Me, Hg(OAc)) | 0.5N NaOH
 NaBH$_4$,2 eq.
 r.t., 15h | Ph~CO$_2$Me |

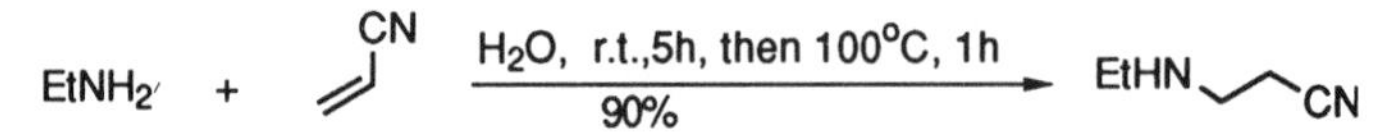

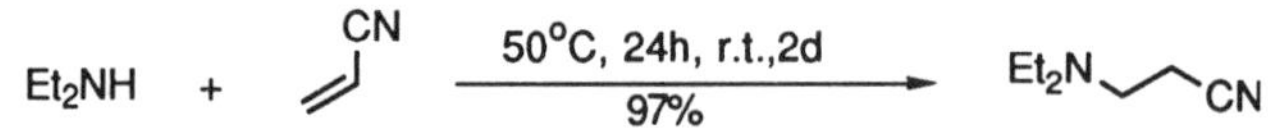

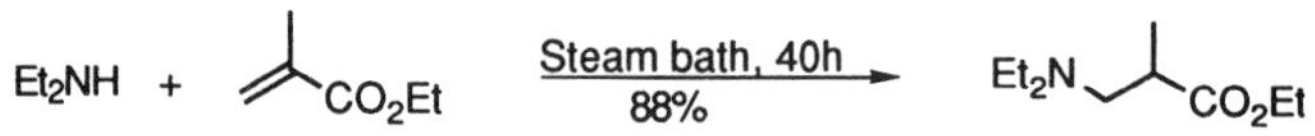

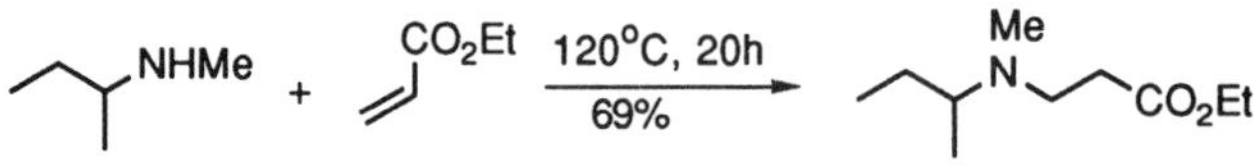

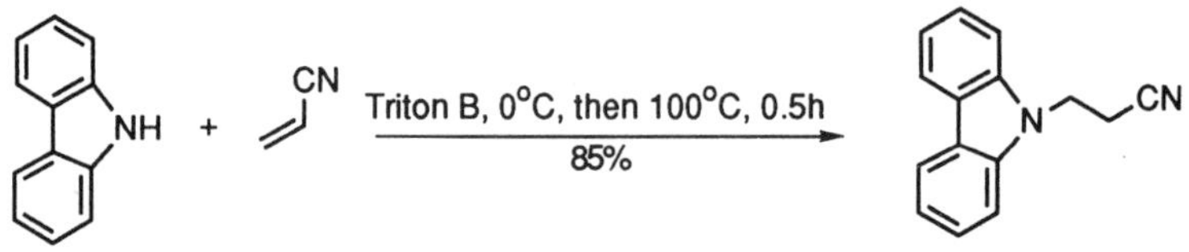

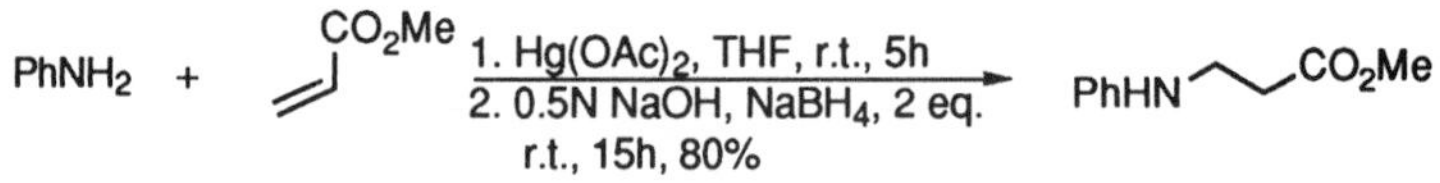

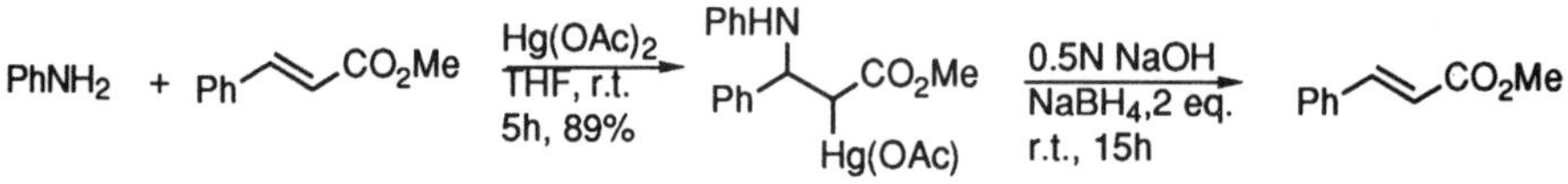

Pyrroles readily add to alkenoates as in a total synthesis of (±)-ipalbidine.

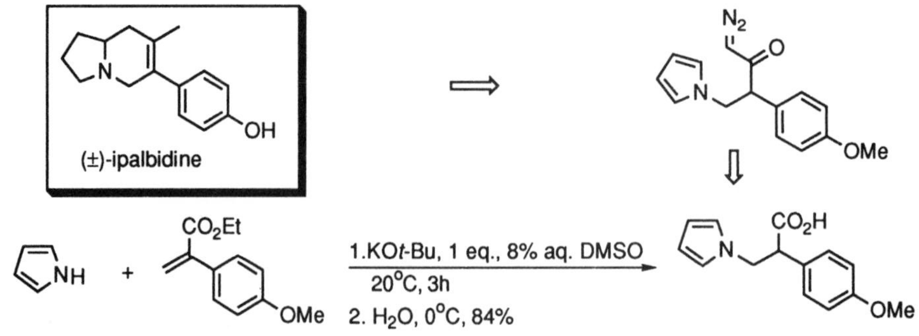

(±)-ipalbidine

$\Longrightarrow$

(pyrrole-substituted diazoketone with OMe)

$\Downarrow$

(pyrrole NH) + (CO$_2$Et, OMe arene)   1.KOt-Bu, 1 eq., 8% aq. DMSO
 20°C, 3h
 2. H$_2$O, 0°C, 84%   (pyrrole N, CO$_2$H, OMe product)

Aziridines also add easily to alkenoates. Additions to β-substituted alkenoates usually require base catalysis.[173] 2-Chloropropenenitrile is reported to react exothermically with aziridine at -20°C.[173] Pyrrolidine,[172] piperidine[174,172] and morpholine[175,172] also add efficiently to unhindered alkenoate derivatives.

Hydrolases have been used to catalyze the conjugate addition of nucleophiles to alkenoic acids with asymmetric induction.[176] Included amongst these nucleophiles is diethylamine (see also sections 4.1.7.2, 4.2.6.1 and 4.2.6.2).

Hydrolases, which have been first modified by acylation, successfully catalyze these additions in organic solvents.[177] Modest asymmetric induction was achieved (8-37% e.e. depending on the enzyme used). Unfortunately, in neither of these papers was an indication of the absolute stereochemistry given. (For examples of a trifluoromethyl group controlling regioselectivity, see page 10.)

Additions of tertiary amines are usually used to catalyze carbon-carbon bond-forming reactions. The most significant of these being the Hillman-Baylis reaction.[178] The reaction involves, usually, standing a mixture of a propenoate, an aldehyde and a catalytic amount of a tertiary amine base. When the reaction is carried out at atmospheric pressure, the base is virtually always 1,4-diazabicyclo[2.2.2]octane. The yields are usually excellent. The products have been used in the syntheses of a number of natural products.[179]

The reaction may be accelerated by the use of high pressure.[180] High pressure also makes it possible to react ketones in place of aldehydes, a reaction which is usually unsuccessful at atmospheric pressure.

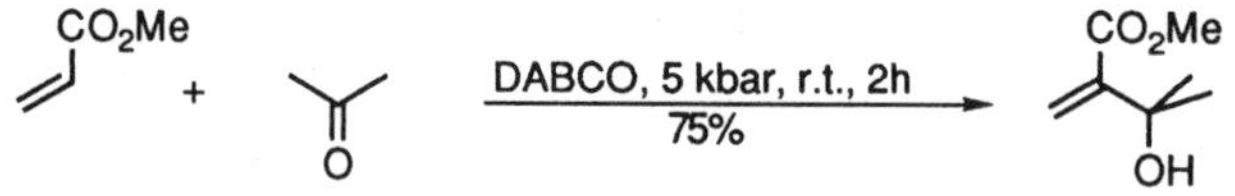

Sp^2 hybridyzed nitrogen also adds to alkenoates.[181]

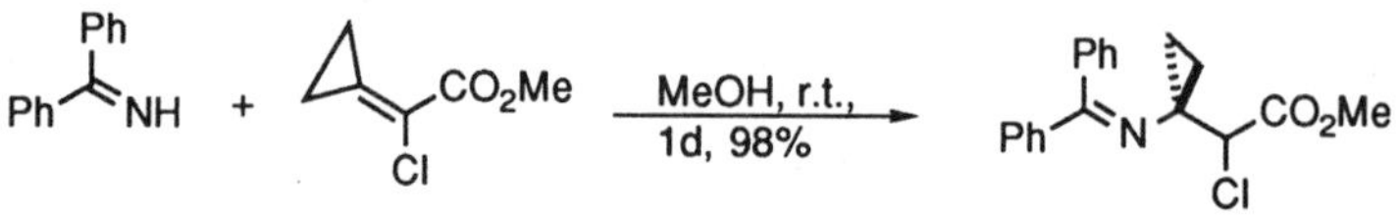

A study of various lithium salt of amines was undertaken to ascertain which of these gave the best results in conjugate additions to alkenoates.[182,183] The results for the reaction shown below are given in Table 4.17.

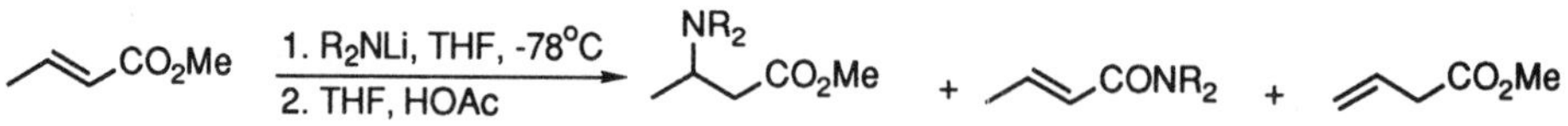

Unlike previous work,[184] this study showed that LDA not only adds in a conjugate fashion, it also causes some deconjugation of the alkenoate, producing significant amounts of methyl 3-butenoate. The most efficient conjugate addition occurred with LSA, with no significant amounts of by-products being formed. Other silylamines gave poorer results. The lithium salt of benzylamine gave mainly the product of 1,2-addition. This is to be contrasted with its behaviour with unsaturated iron acyls where only conjugate addition is observed (see below in this section).

Table 4.17   Results of the reaction of various lithium amides with methyl 2-butenoate

R$_2$NLi	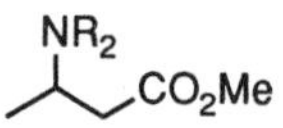 NR_2-CO$_2$Me	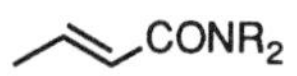 CONR$_2$	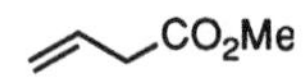 CO$_2$Me
i-Pr$_2$NLi (LDA)	44	-	14
Bn(TMS)NLi (LSA)	88	-	-
Bn(TBMS)NLi	61	7	-
Bn(DPMS)NLi	30	6	-
TMS$_2$NLi	-	-	9
BnHNLi	20	60	-
Bn$_2$NLi	18	-	-

Yamamoto's group has exploited the conjugate addition of LSA in a highly stereoselective synthesis of carbapenem-like azetidinones.[185] Using quenching experiments, with chlorotrimethylsilane, they were able to show that either the *E* or *Z* ketene acetal (and hence enolate) could be produced with ≥ 98% selectivity, depending upon the reaction conditions. They next found that each of these enolates underwent a highly stereoselective aldol reaction with aldehydes, e.g. ethanal as shown in the diagram below, with the *Z*-enolate giving mainly *syn* aldols and the *E*-enolate giving mainly *anti* aldols. The products of these aldol reactions are model precursors to the carbapenem nuclei. (The boxed product contains the correct relative stereochemistry for thienamycin.)

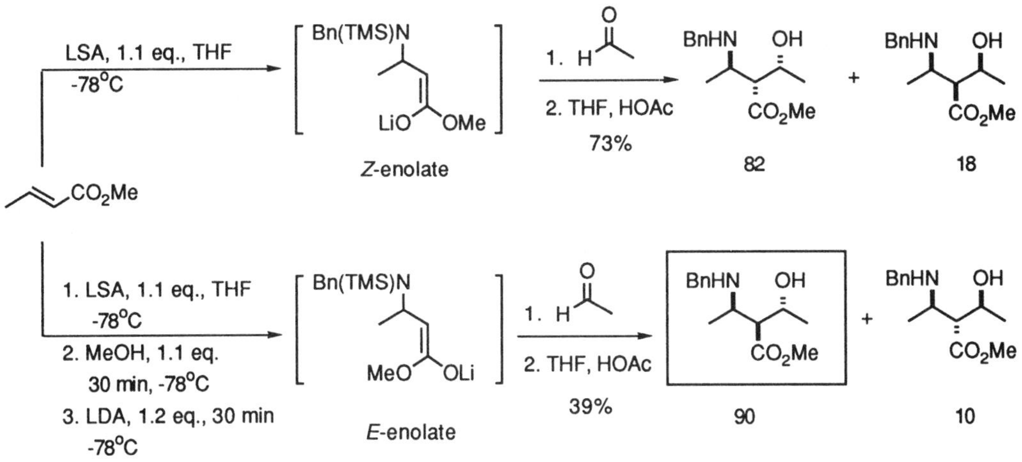

A new method, involving tandem inter-/intra-molecular conjugate additions has been developed.[186] Addition of LSA to octa-2,6-diene-1,8-dioic and nona-2,7-diene-1,9-dioic acids leads to 1,2,3-trisubstituted cyclopentane and cyclohexane products. For the larger decyl system only addition occurs, apparently closure to a cycloheptane ring is not favoured under these conditions.

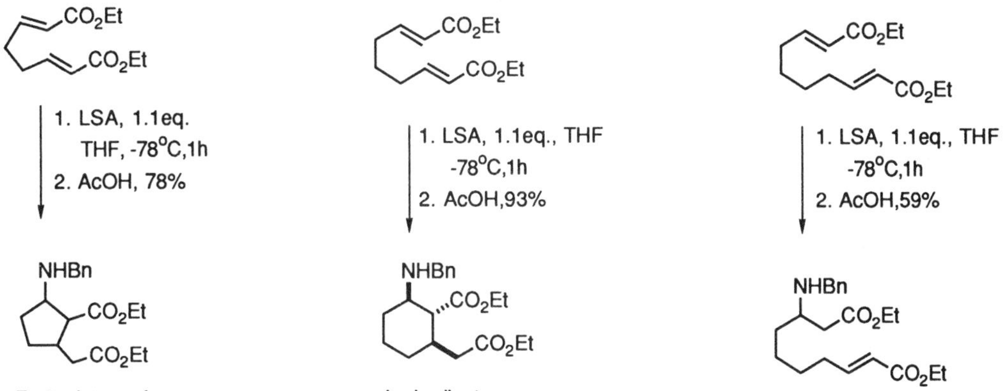

7 : 3 mixture of two diastereomers

single diastereomer

This method was applied to syntheses of (±)-dihydronepetalactone and (±)-isodihydronepetalactone. Two approaches were used. The first involved conjugate addition of LSA to the unsymmetrical system shown below. Only addition to the less-substituted enoate was observed. In the second approach, LSA was added to a symmetrical diester followed by alkylation. Each case produced a mixture of two diastereomers, although the ratios were reversed.

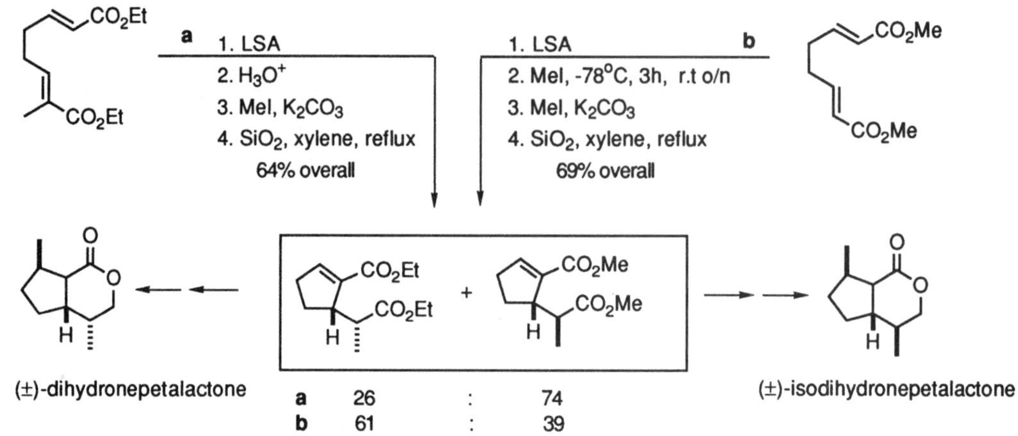

	26		74
a	26	:	74
b	61	:	39

(±)-dihydronepetalactone   (±)-isodihydronepetalactone

Snieckus's group has shown that dialkylamides of crotonic acid react well with lithium diisopropylamide. The intermediate enolate may also be quenched, stereoselectively, with an electrophile, as in the following example.[187]

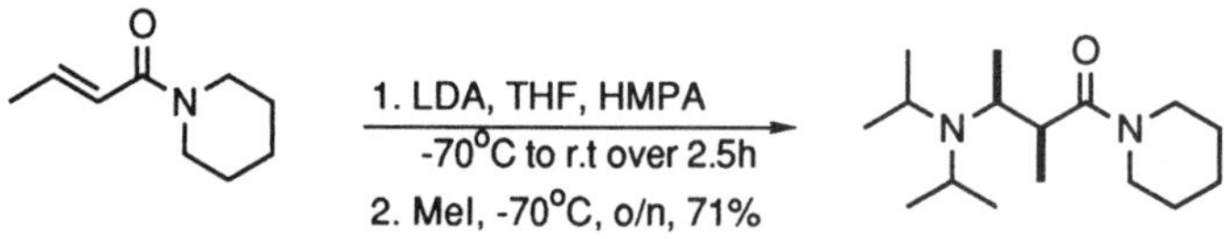

(ii)   *Addition of chiral amines to achiral alkenoates*

The use of chiral secondary amines as nucleophiles has received some attention. One remarkable example, which uses a binaphthyl-based secondary amine (**17**), provides access to either enantiomeric adduct depending on the reaction conditions employed. Thus addition of the lithium salt of the binaphthylamine yields the (R)-C3 adduct with very high selectivity (97-98% de). On the other hand, addition of the neutral amine yields the other diastereomeric adduct, with a reduced, but nevertheless significant, stereoselectivity (61% de).[188]

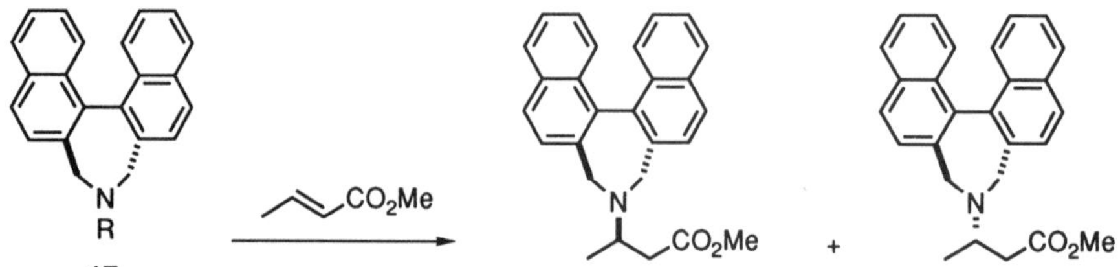

A force-field model has been developed for the likely transition states in these additions (see below for diagrams).[189] An *s-cis* conformation was chosen for the butenoate as it is slightly lower in energy.[190] For the addition of the free base ("fb", R = H) it was found that conformation **fb1** was of lower energy than **fb2**. This is possibly due to an unfavourable steric interaction between the ester group and the proximal naphthyl group. For the anionic case ("an", R = Li), a chelated six-membered transition state was assumed. The authors found that **an2** was lower in energy than **an1** and argued that this arose because the former represents a lower energy conformer of **fb2** whereas **an1** is a higher energy conformer of **fb1**. Given the simplified nature of the model used, the subtle effects which control these additions cannot be fully understood at present.

free base: conformation control   anion: chelation control

(iii)   Addition of achiral amines to chiral alkenoates

The stereoselective addition of amines to γ-alkoxyalkenoates provides access to homochiral carbapenem intermediates.[191] Addition of benzylamine to the ester derived from D-glyceraldehyde acetonide yields a β-amino ester with R stereochemistry at C3. Running the reaction at higher temperatures leads to mixtures of diastereomeric products. Interestingly, both the E- and Z- alkenes give the same stereochemical result. This method has also been applied to the synthesis of C-methyl analogues of L-acosamine.[192,193]

Double asymmetric induction also occurs in the addition of benzylamine to the *bis*-(γ-silyloxyalkenoate) (11). See section 4.1.2.2 for a discussion of organocopper additions to this system.

Addition of amines to 2-(1-silyloxyalkyl)alkenoates (18) provides an alternative pathway to carbapenems.[194] The high levels of stereoselectivity obtained in these additions occurs during protonation of the intermediate enol.

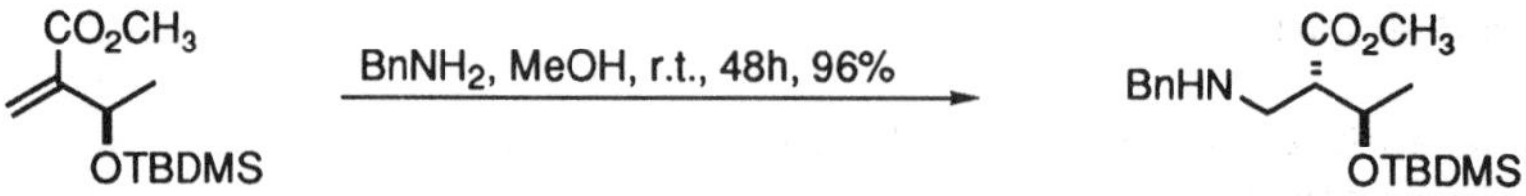

18

Addition of benzylamine to the (8-naphthyl)menthyl ester of *E*-butenoic acid under high pressure gives excellent diastereoselectivity.[195] Other 8-substituted menthyl esters were also used, however the 8-naphthyl derivative gave the highest diastereomeric excess. The authors proposed that addition occurs to the conformation shown below. Interestingly, part of the motivation for this study appears to have been the use of high pressure in light of "the sluggish behaviour" of amine additions to 2-butenoates. From some of the examples cited in this section, many amines, including especially benzylamine, add well to such alkenoates. Therefore, it seems worthwhile to repeat this study at atmospheric pressure.

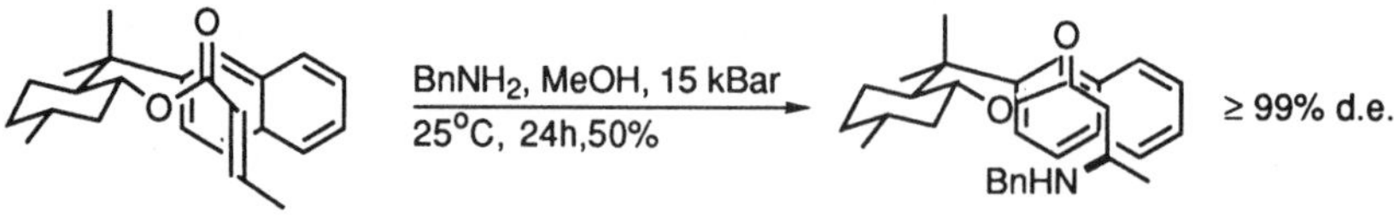

Davies's group has also applied their chiral α,β-unsaturated iron acyls to the synthesis of β-lactams.[196,197] For example, addition of the lithium salt of benzylamine to the Z-iron acyl shown below adds to the less hindered face of the s-*cis* conformation (as discussed before for alkyllithium additions to these complexes, see section 4.1.3). Quenching with an electrophile also occurs from the less-hindered face of the enolate producing the acyl complex as a single diastereomer. Oxidative decomplexation then gave the pure *cis* dimethyl β-lactam.

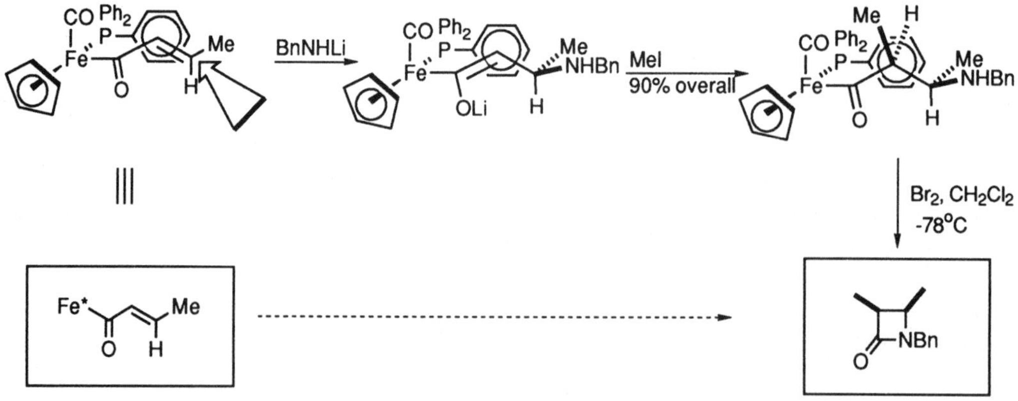

A variety of homochiral, 3- and 4- substituted as well as 3,4-disubstituted β–lactams has been prepared in this way.[197]

(b)   Phosphorus

The addition of trivalent phosphorus compounds to alkenoic acid derivatives parallels that of additions to alkenals and alkenones (section 2.1.7.1), except that, even with acid chlorides, 1,2 addition is never observed. Also because of the lower reactivity of these systems, more forcing conditions are sometimes necessary.[198] Quite a variety of phosphorus nucleophiles have been successfully used and these are summarized below.

Neutral	Anionic	
$(RO)_2PX$, X = OR[199]	X(Y)POH,	X = OH, Y = H[200]
X = OSiR'$_3$[201]		X = OR, Y = H
X = NR'$_2$[202]		X = Y = OR[203]
X = NHR		X = R, Y = OR'
X = R		X = R, Y = Ar
		X = Y = R

Reaction of tri-isopropyl phosphite with methyl 3-bromopropenoate[204] results in overall substitution at C3, whereas only addition occurs in the base-catalysed reaction of diethyl phosphite with diethyl 2-(acetamidomethylene)malonate:[205]

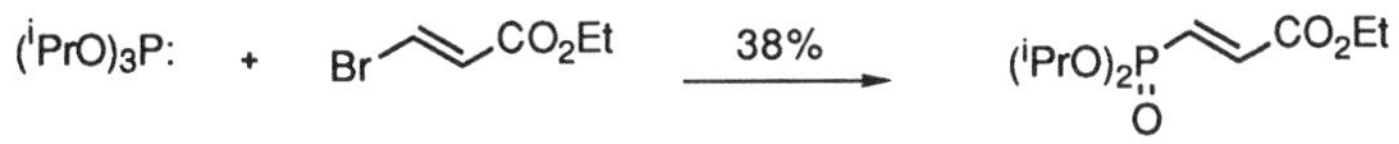

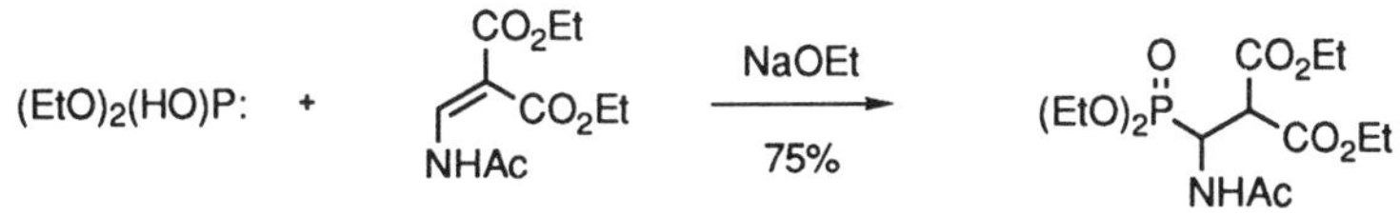

Triethylaluminium is an excellent catalyst for phosphite conjugate additions.[206]

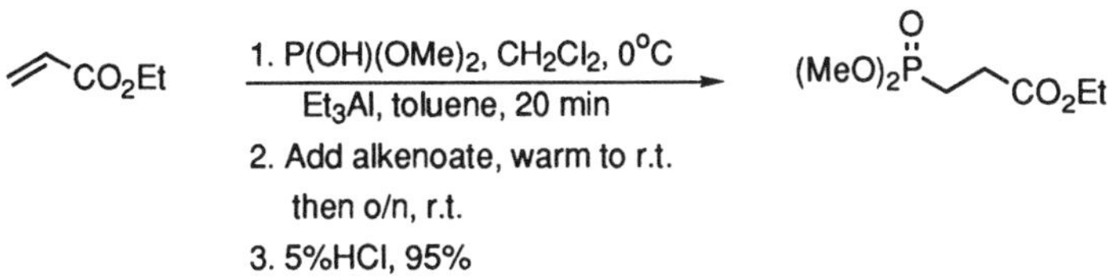

Water-soluble phosphonium salts have been prepared by the addition of phosphines to alkenoate derivatives.[207] Where the alkenoate is not water-soluble, additions may be successfully carried out in a two-phase system, e.g.:

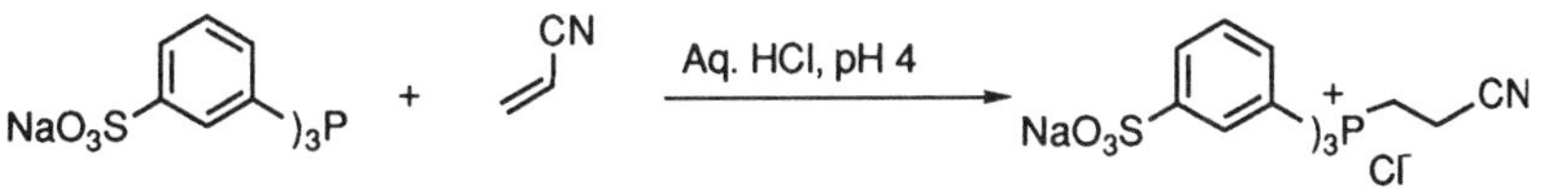

Badalski and Pietrusiewicz have developed a simple synthesis of phospholane oxides, based on the base-catalysed addition of dibenzylphosphine oxide to alkenoates:[208]

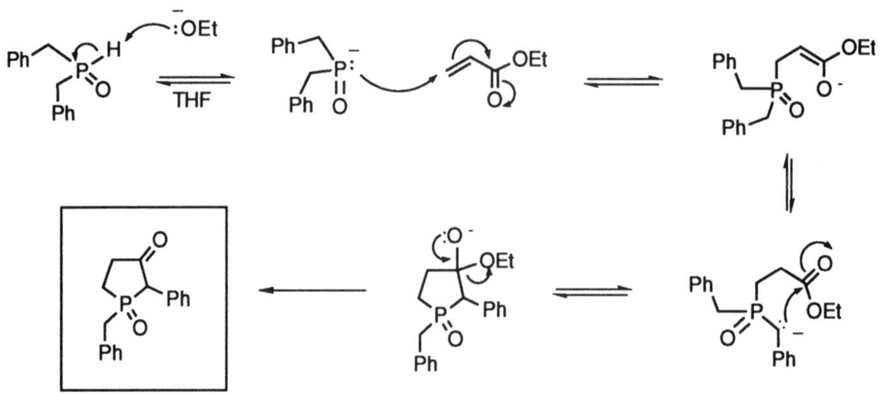

There are also a few examples in the literature of carbon-carbon bond formation catalysed by phosphorus conjugate addition. McClure has developed a triarylphosphine - catalysed dimerisation of propenenitrile which involves heating a solution of propenenitrile and a catalytic amount of tri(4-tolyl)phosphine in triethylsilanol at 160°C for 11 hours. He proposed the following reaction mechanism.[209,210]

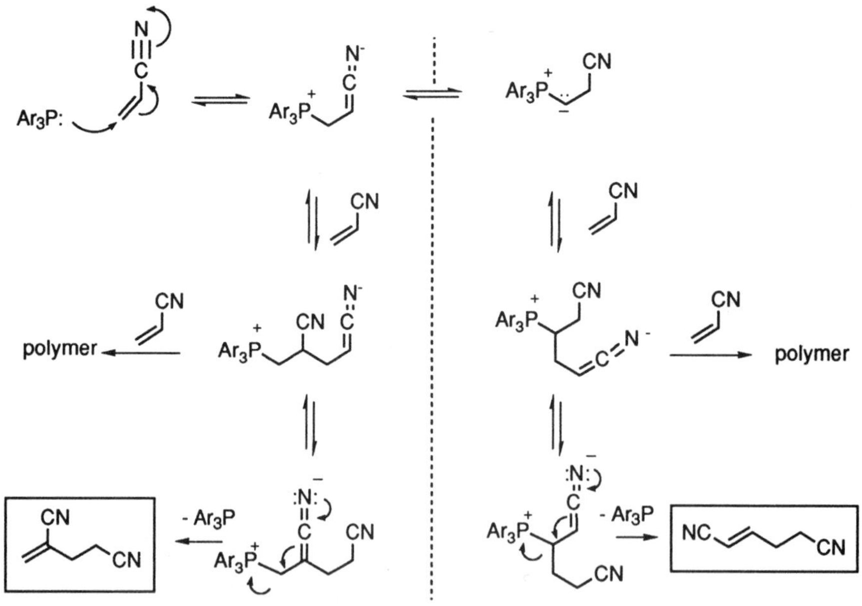

Around the same time Morita showed that the initial zwitterion formed after conjugate addition could be intercepted by aldehydes. This, in effect, was an early example of a propenenitrile α-anion equivalent.[211] They also reported that this reaction works with propenoates. More recently, it has been shown that the addition of triethylaluminium as a co-catalyst can improve the yields of this process.[212] However, the use of tertiary amines, especially DABCO, as the catalyst in this reaction appears to be far superior (see discussion in this section).

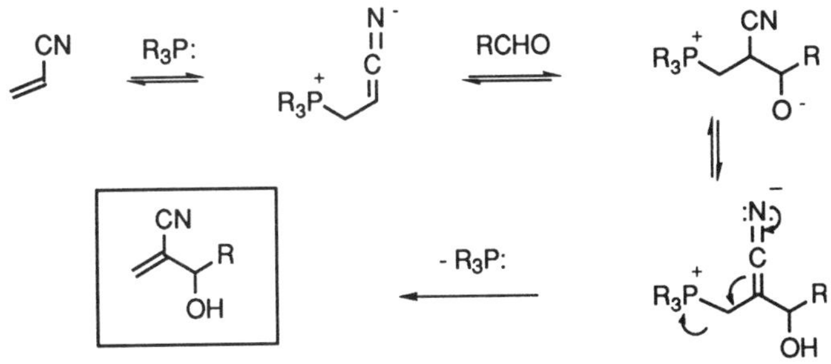

4-Phosphorinanones, e.g. (**19**), may be prepared from the commercially available phenylphosphine and propenenitrile.[213]

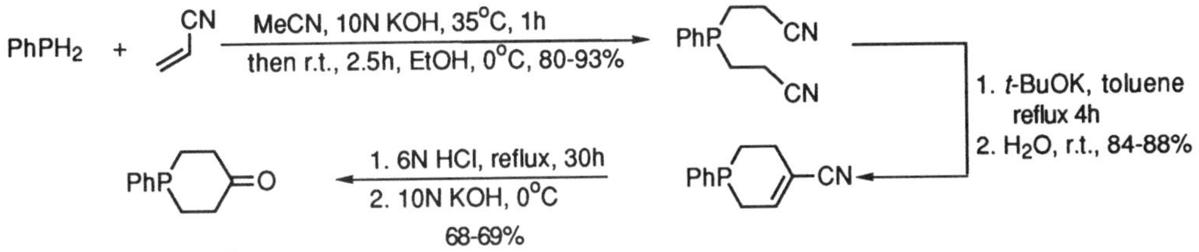

19

4.1.7.2   Oxygen and sulfur

These nucleophiles add with relative ease to alkenoate derivatives. Some examples are given in Table 4.18. Alcohols add well under base catalysis (entry 1 and below in this section). Phase transfer catalysis is especially useful in thiol additions (entries 2 to 4).

Table 4.18   Conjugate addition of oxygen and sulfur nucleophiles to alkenoic acid derivatives

	Nucleophile		Alkene	Conditions	Product
1[214]	EtOH	+	CN	Basic resin, 15-20°C, ~3h, 90%	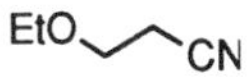 EtO~CN
2[215]	PhSH	+	CO₂Me	TBAF, 0.02 eq., THF, 25°C, 10 min, 99%	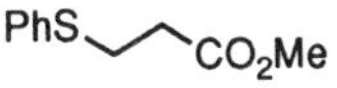 PhS~CO₂Me
3[215]	BnSH	+	~CO₂Me	BTAF, 0.02 eq., acetone, 25°C, 20 min, 100%	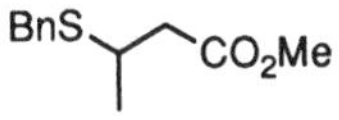 BnS~CO₂Me
4[215]	EtO₂C~SH	+	CO₂Me	TBAF, 0.03 eq., THF, 25°C, 7h, 85%	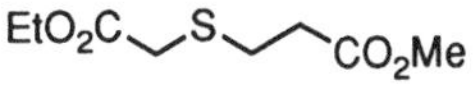 EtO₂C~S~CO₂Me
5[216,217]	PhSO₂Na	+	CN	AcOH, 1.1 eq., H₂O, 100°C, 15h, 97%	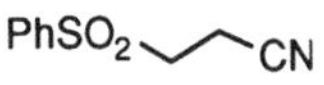 PhSO₂~CN

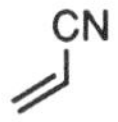

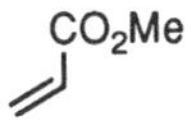

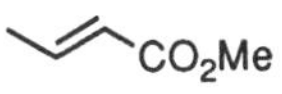

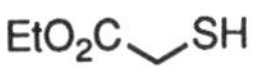

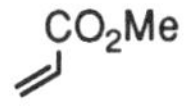

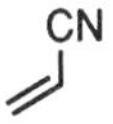

As with other nucleophiles, alkoxide adds stereoselectively to γ-alkoxyalkenoates. For example, Mulzer's group has shown that treatment of the γ-alkoxyalkenoate derived from glyceraldehyde acetonide with sodium methoxide, in methanol at low temperature, yields the *syn* diastereomer with excellent selectivity.[218] Similar treatment with other alkoxide/alkanol combinations also proceeded in high yield, however the diastereoselection was not as good. It is also important to use the same alkoxy group in both the nucleophile and the ester as rapid transesterification precedes the conjugate addition.

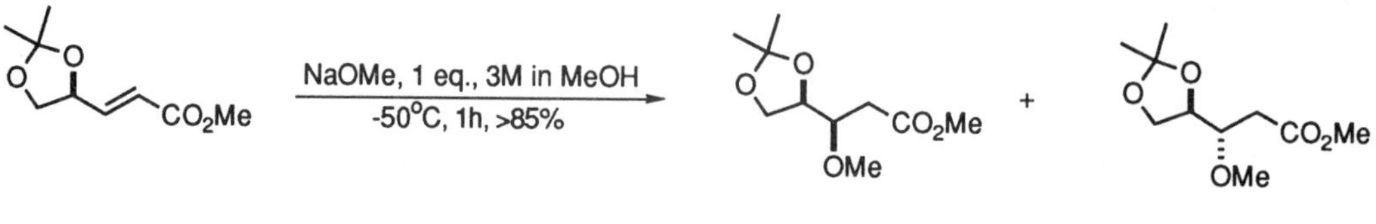

An even more adventurous study has been carried out to establish if stereoselectivity is possible in additions to δ-alkoxyalkenoates.[219] In fact, in some cases the selection is very good. However, this appears to be associated mainly with γ-alkyl-δ-alkoxyalkenoates (Table 4.19) entries 2 and 3). Therefore, the contribution made by the alkoxy substituent in stereocontrol in these cases is debatable. Where a γ-substituent is only a proton, little selectivity was found (entry 1).

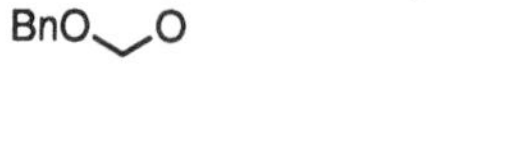

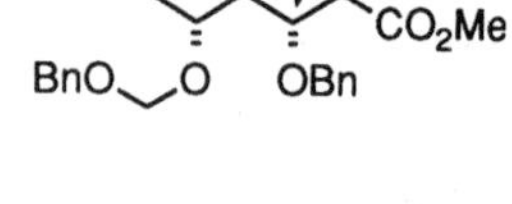

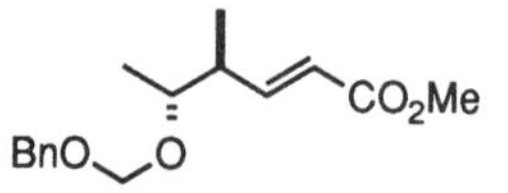

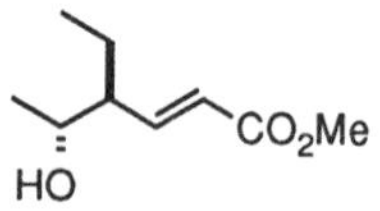

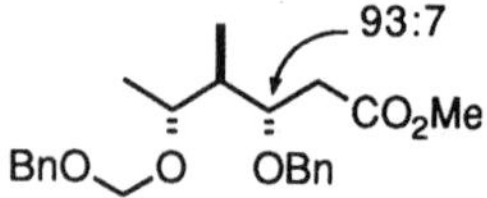

Table 4.19 Oxymercuration of γ-oxyalkenoates

1		1. BnOH, 2.5 eq, $Hg(OAc)_2$, 1.3 eq, 70% $HClO_4$, cat., CH_2Cl_2, r.t., 24h 2. pH 7 3. THF, pH 8, $NaBH_4$, 1 eq, 0°C, few min, 73%	
2		1. BnOH, 2.5 eq, $Hg(OAc)_2$, 1.3 eq, 70% $HClO_4$, cat., CH_2Cl_2, r.t., 24h 2. pH 7 3. THF, pH 8, $NaBH_4$, 1 eq, 0°C, few min, 66 %	
3		1. BnOH, 2.5 eq, $Hg(OAc)_2$, 1.3 eq, 70% $HClO_4$, cat., CH_2Cl_2, r.t., 24h 2. pH 7 3. THF, pH 8, $NaBH_4$, 1 eq, 0°C, few min, 60 %	

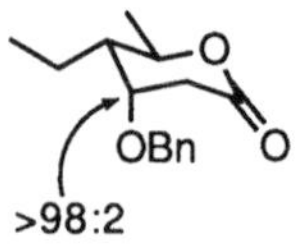

Hydrolases also catalyze the addition of water and thiophenols, but not alcohols or phenols, to 2-(trifluoromethyl)propenoate.[176]

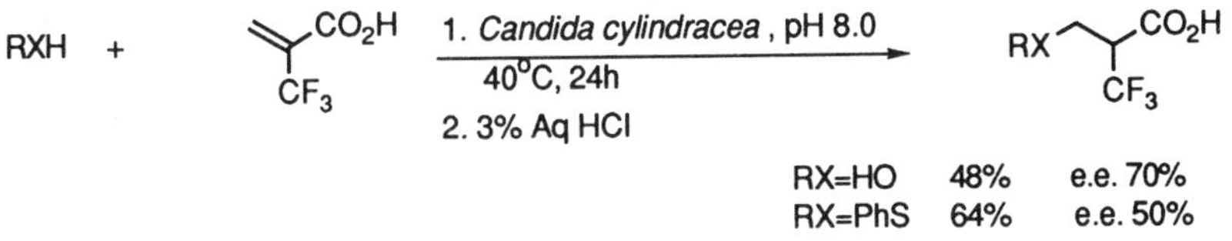

RXH +   [structure: CH2=C(CF3)CO2H]   1. *Candida cylindracea* , pH 8.0
                                            40°C, 24h
                                        2. 3% Aq HCl
                                                                    RX—CH2—CH(CF3)CO2H

                                              RX=HO   48%   e.e. 70%
                                              RX=PhS   64%   e.e. 50%

4.1.7.3   Silicon and tin

HO cuprates, derived from CuCN and two equivalents of R_3SiLi, add smoothly to a variety of alkenoic acid derivatives (Table 4.20).[220,221] The product β-silylesters are useful precursors to allylsilanes. Where the β-substituents are larger than methyl, double activation is often necessary (entry 2).

Table 4.20   Conjugate addition of silicon nucleophiles to alkenoic acid derivatives

1[222]	[structure: R"–C(R')=CH–CO2R] → 1. 2PhMe₂SiLi, CuCN, THF, -23°C / 2. CH₃I, -23°C to r.t., 3h / 88%	[structure: R"·–C(R')(PhMe₂Si)–CH(CH₃)–CO₂R]
2[222]	[cyclohexyl structure with CH2CO2R and SiPhMe₂, ( )ₙ] ←R'=H— X — [cyclohexylidene structure =C(R')CO₂R, ( )ₙ] →R'=CO₂R, 2PhMe₂SiLi, CuCN→ [cyclohexyl structure with C(CO₂R)(CO₂R) and SiPhMe₂, ( )ₙ]		
3[223]	Ph–CH=CH–CO₂Me 1. Me₂PhSiZnMe₂Li, 1.2 eq / THF, -78°C, 1h / 2. 1N HCl, 48%	Ph–CH(Me₂PhSi)–CH₂–CO₂Me

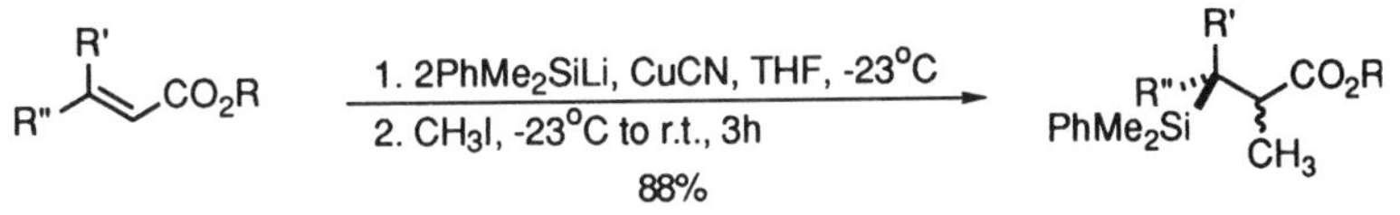

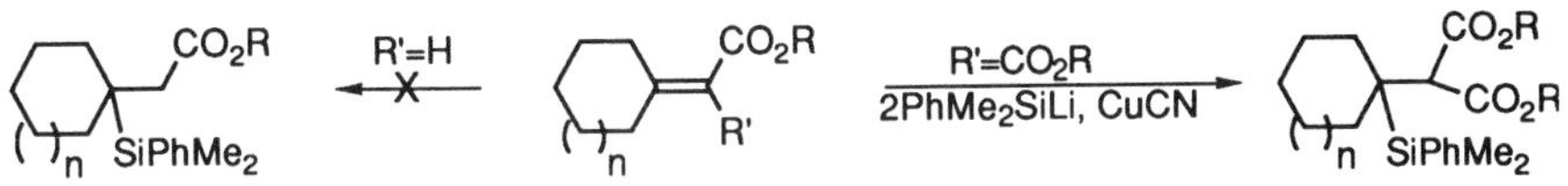

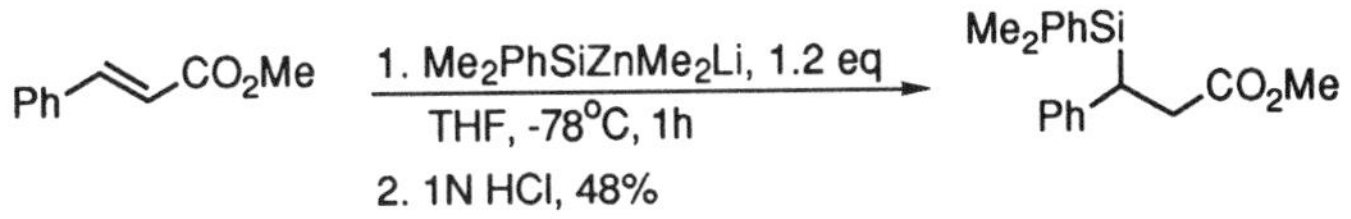

Conjugate addition of HO silylcuprates to alkenoates, followed by quenching with an electrophile, is highly diastereoselective.[224]

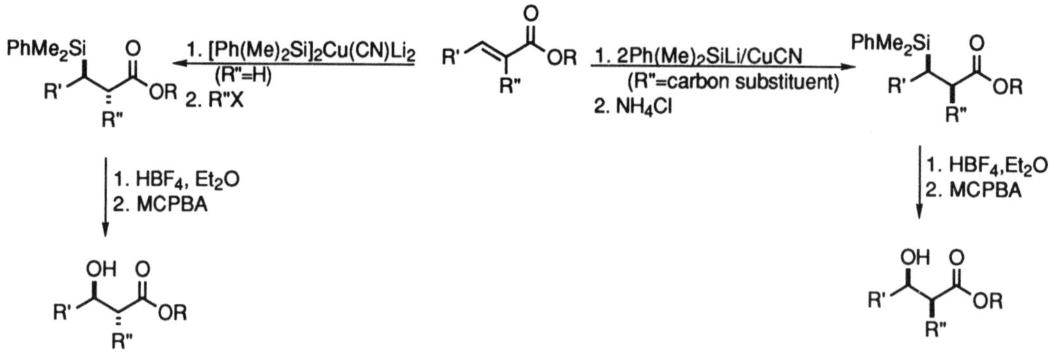

The stereoselectivity was attributed to electrophilic attack on the following conformation. In this conformation, the allylic hydrogen is in-plane and the electrophile

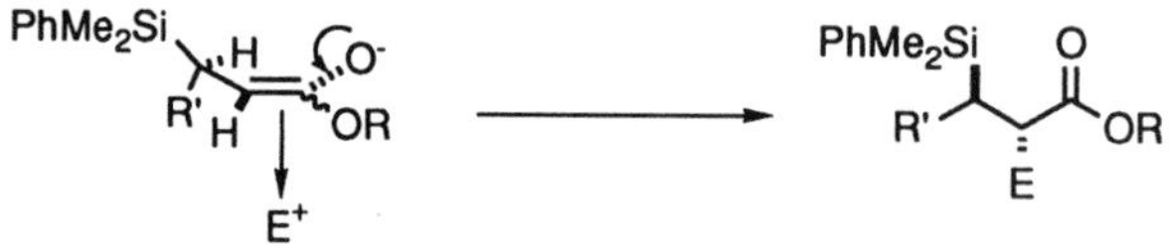

attacks the π-system from the face away from the silicon group.[225,226] There is evidence that, in addition to any steric screening, an electronic component is also operating in the electrophilic attack. Variation in the bulk of R', up to the *tert*-butyl group, does not change the π-facial selectivity, i.e. the electrophile, MeI or H+ in this study, always attacks the face away from the silicon group, even though *tert*-butyl is most likely larger than phenyldimethylsilyl.

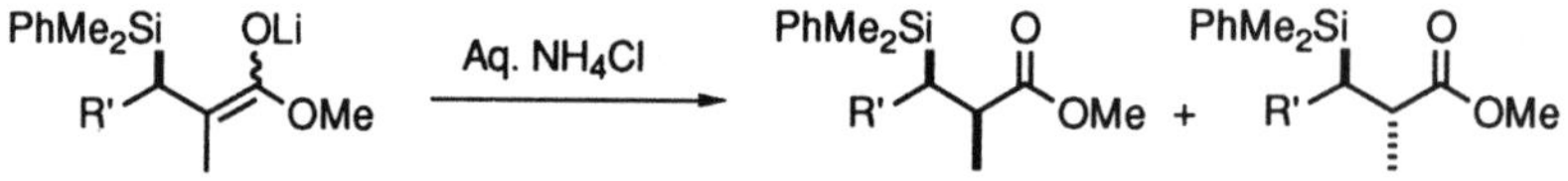

R' = Me	82%	87 :	13
R' = *i*-Pr	56%	96 :	4
R' = *t*-Bu	38%	96 :	4
R' = Ph	84%	85 :	15

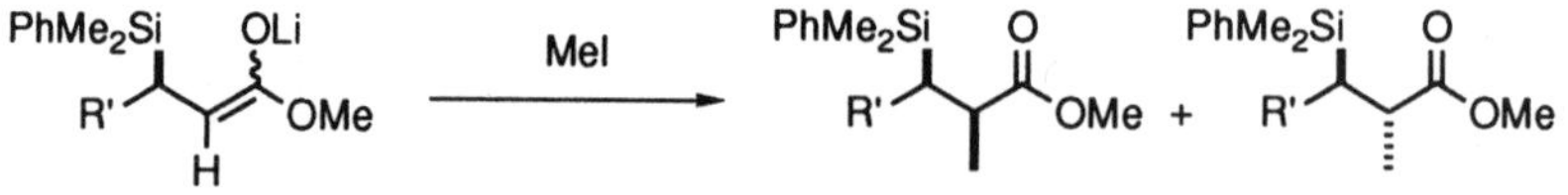

R' = Me	78%	9 :	91
R' = *i*-Pr	95%	15 :	85
R' = *t*-Bu	83%	34 :	66
R' = Ph	88%	3 :	97

Consistent with this analysis is the finding that electrophilic attack of the corresponding nitrile intermediate is less selective. As allylic control is no longer necessarily operating (there is no substituent , H or Me as above, *cis* to the chiral centre), the corresponding conformation is not necessarily the lowest energy conformation.

The same group has also examined the stereoselectivity associated with tandem silylcuprate conjugate addition/aldol reactions. π-Facial selectivity is maintained and the relative stereochemistry at the carbinol carbon can be correlated with the enolate geometry.[227]

This methodology has been applied to total syntheses of thienamycin and the Prelog-Djerassi lactone.[228]

Quite an extensive study of the asymmetric induction obtained from addition of HO silylcuprates to chiral alkenoates has produced some interesting and useful conclusions.[229] Several different chiral auxiliaries were tested and most produced disappointing results, in terms of diastereomeric excesses. In fact, addition of either a silyl cuprate to a chiral, 3-phenyl propenoate or a phenyl cuprate to the corresponding chiral, 3-silylpropenoate, produced the *same* diastereomer. This, of course, implies that each reagent is selecting a different face of the chiral alkenoate during the addition!

This highlights the preference of silylcuprates to react with the s-*cis* conformer of alkenoates (and alkenones[225,230]), whereas the camphor-based alkenoate reacts with other cuprates in the s-*trans* conformer. Of the six different chiral auxiliaries tested, the most successful was that designed by Koga's group[68] (see section 4.1.2.2).

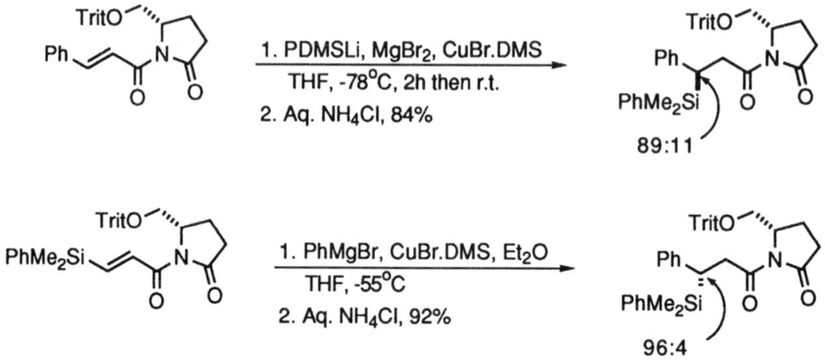

Tributylstannanes also add well to alkenoates and alkenenitriles under either thermal conditions (Table 4.21, entries 1 and 2) or as the lithium salt (entry 3).

Table 4.21   Conjugate addition of tin nucleophiles to alkenoic acid derivatives

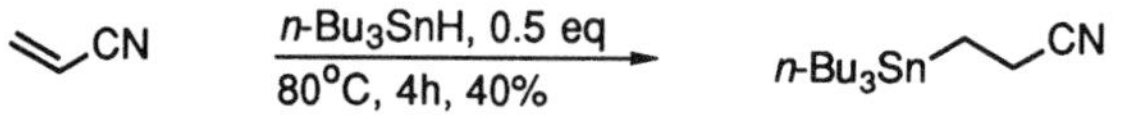

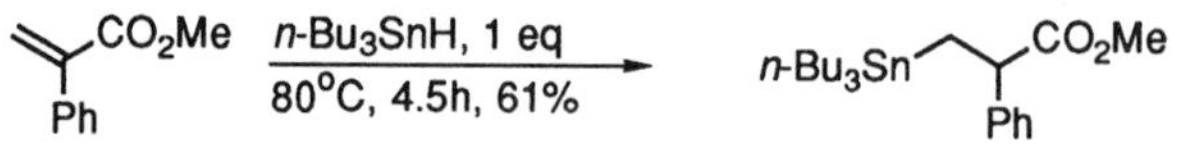

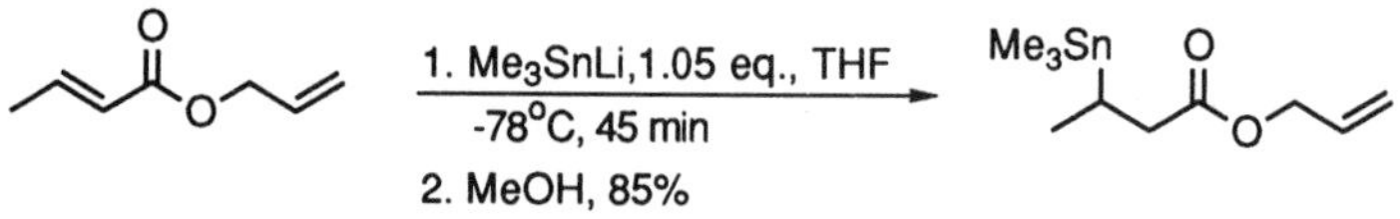

4.1.7.4   *Conjugate reduction*

A number of hydride reagents reduce alkenoic acid derivatives in a conjugate fashion (see Table 4.22).

Table 4.22   Conjugate reduction of alkenoic acid derivatives

1[233]	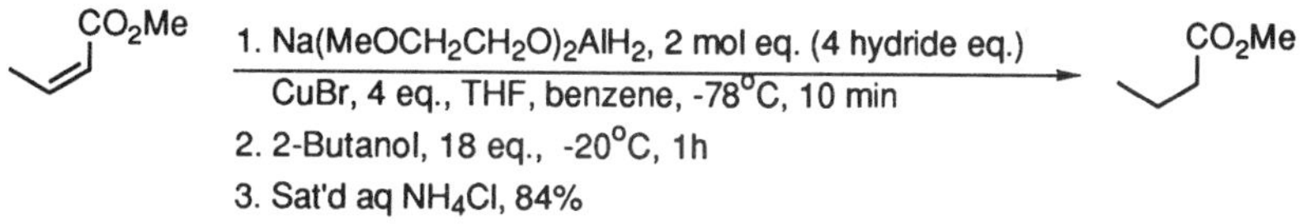 CO₂Me	1. Na(MeOCH₂CH₂O)₂AlH₂, 2 mol eq. (4 hydride eq.) → CO₂Me
		CuBr, 4 eq., THF, benzene, -78°C, 10 min
		2. 2-Butanol, 18 eq., -20°C, 1h
		3. Sat'd aq NH₄Cl, 84%
2[234]	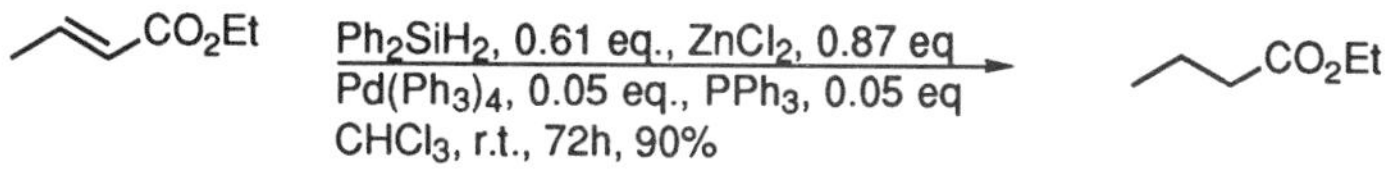 CO₂Et	Ph₂SiH₂, 0.61 eq., ZnCl₂, 0.87 eq → CO₂Et
		Pd(Ph₃)₄, 0.05 eq., PPh₃, 0.05 eq
		CHCl₃, r.t., 72h, 90%
3[235]	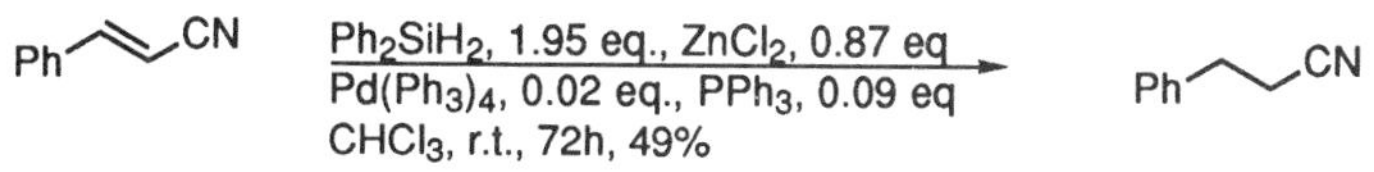 Ph~CN	Ph₂SiH₂, 1.95 eq., ZnCl₂, 0.87 eq → Ph~CN
		Pd(Ph₃)₄, 0.02 eq., PPh₃, 0.09 eq
		CHCl₃, r.t., 72h, 49%
4[235,236]	CO₂Bn	[(Ph₃P)CuH]₆, 0.24 eq → CO₂Bn
		H₂O, benzene, r.t., 30 min
		95%

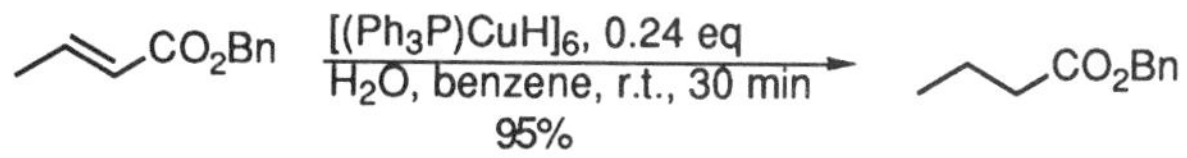

Conjugate hydride additions are also amenable to steric or chelation control. For example inclusion of a remote oxygen functional group into cyclopentylidene malonates allows for some variation in the stereoselection during reduction.[237]

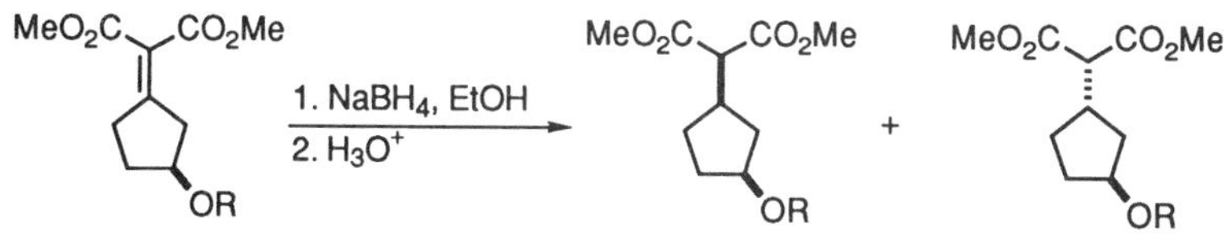

R = MEM	5	:	1
R = H (EtOH)	0.7	:	1
R = H (THF)	0.4	:	1

In the same paper, the authors reported an application of this method to the reduction of a key intermediate in their synthesis of the spatane nucleus. In this case, the reduction is highly stereoselective, probably as a result of a combination of chelation control and steric screening of one face of the π-bond by some of the other substituents in the molecule, as shown (**20**).

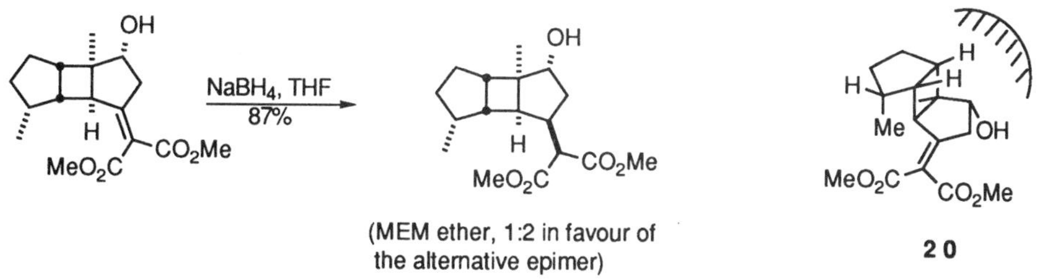

(MEM ether, 1:2 in favour of
the alternative epimer)

20

4.2   Intramolecular reactions

4.2.1   Stabilized carbanions

4.2.1.1   Carbanions stabilized by π-conjugation with one heteroatom

An excellent example of an application of Stetter's methodology (see section 2.1.1.1) is found in Trost's total synthesis of (±)-hirsutic acid.[238] (For the synthesis of the aldehydo-ester (21), see section 7.3.1.)

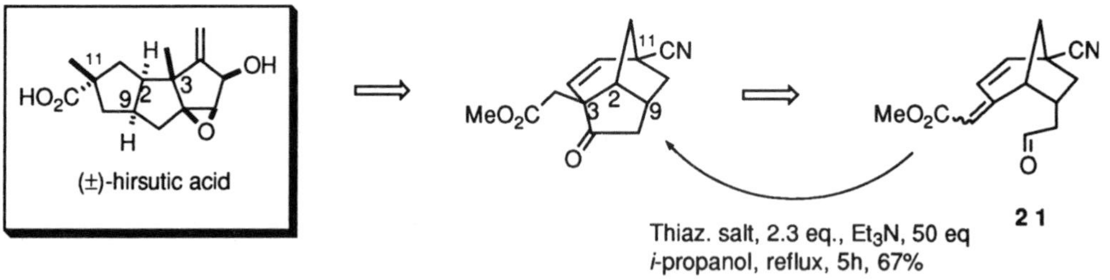

(±)-hirsutic acid

Thiaz. salt, 2.3 eq., Et₃N, 50 eq
i-propanol, reflux, 5h, 67%

2 1

The 2-azabicyclo[2.2.2]octane ring system, found in many naturally-occurring alkaloids, may be prepared by an intramolecular conjugate addition. For example, ibogamine and epi-ibogamine have been synthesized by closure of a dihydropyridone bearing an alkenoyl side chain at C5.[239] The authors found that potassium carbonate in ethanol as the base gave better results than sodium hydride in benzene.

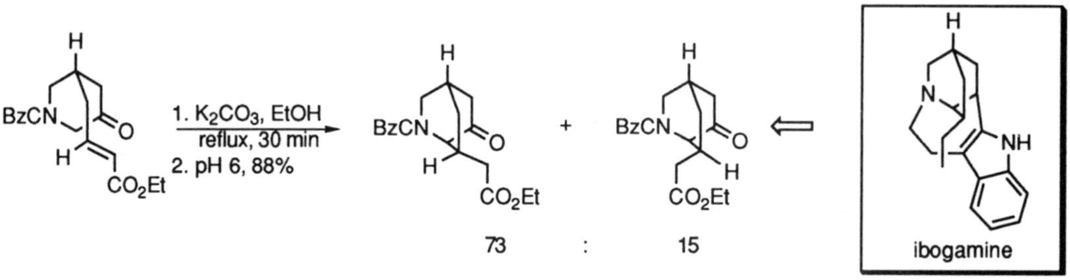

1. K₂CO₃, EtOH
   reflux, 30 min
2. pH 6, 88%

73   :   15

ibogamine

Intramolecular double conjugate additions have been used to prepare some highly complex multi-ring systems. An intermediate in a projected total synthesis of some *Aconitum* alkaloids was prepared from the alkenone/alkenoate shown below.[240] Selective deprotonation at the α-carbon was followed by addition to the alkenoate. The newly formed carbanion then closed onto the alkenone.

The authors chose to represent the transition structure as that shown on the next page. The conformation implies the possibility of an anion accelerated [4π + 2π] cycloaddition as the overall process yields a single diastereomer. The yield of this reaction was extremely sensitive to the reaction conditions. Several lithium amide bases were tried and lithium hexamethyldisilazide, in conjunction with slow addition of a solution of the substrate to a solution of the base, gave the best results.

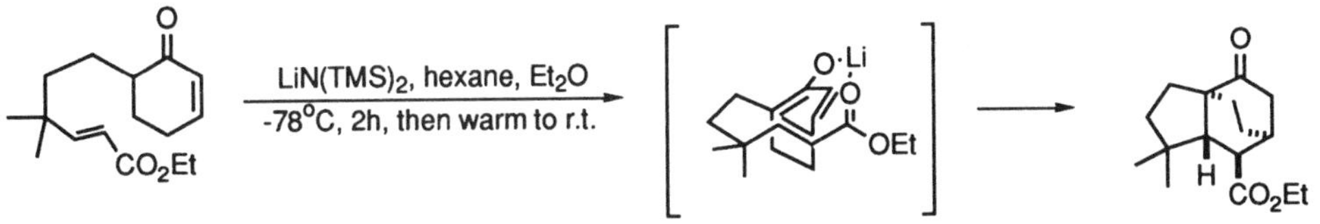

The same method can be used to prepare [6.2.2.0[1.6]] systems such as (+)-atiserine, a diterpene from *Erythroxylon monogynum*.[241,242]

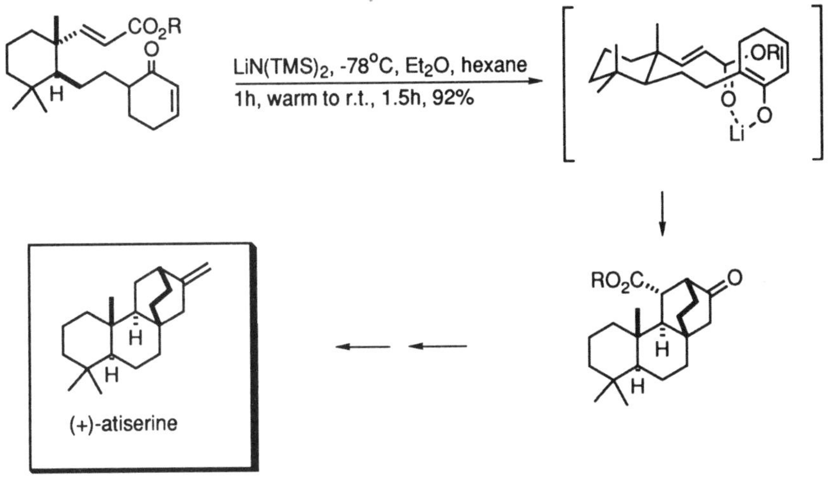

In their total synthesis of the gibberellins, Mander's group employed an intramolecular conjugate addition as a key step in generating two stereocentres, those at C5 and C6, with complete selectivity.[243] However, the product was a 2:1 mixture of epimers at C4.

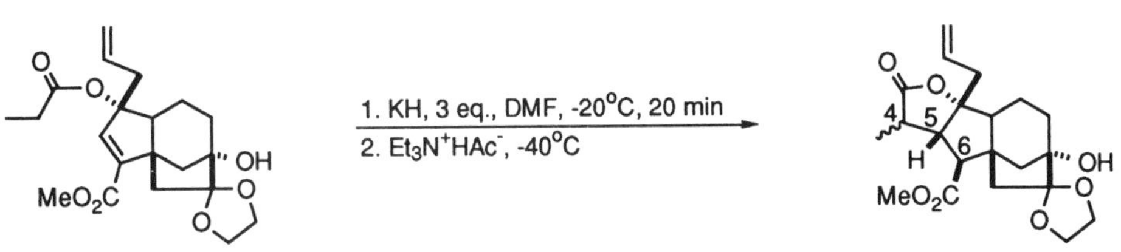

4.2.1.2   *Carbanions stabilized by π-conjugation with more than one heteroatom*

In a study related to their earlier work on intramolecular conjugate additions (see section 2.2.1) Stork's group has developed a highly stereoselective synthesis of 1,2-disubstituted cyclopentanes and cyclohexanes.[244] They found that ring closures of β-ketoester anions onto alkenoates was not very selective. However, switching to benzene as solvent improved the stereoselectivity dramatically as shown below.

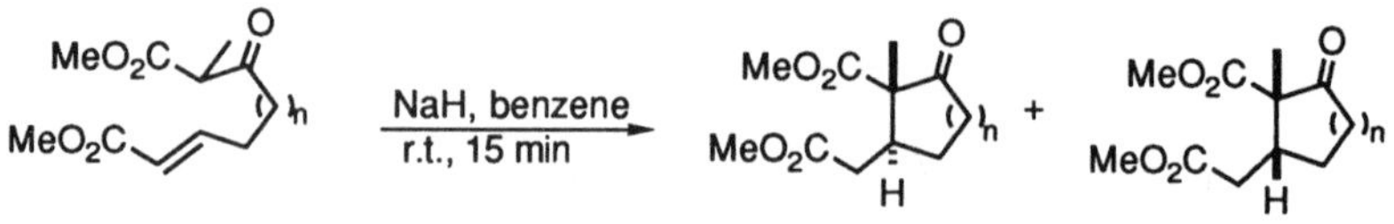

n = 1	85%	22	:	1
n = 2	88%	30	:	1

Apparently the system prefers to adopt a reactive conformation which places the alkenoate ester group away from, rather than over, the chelate.

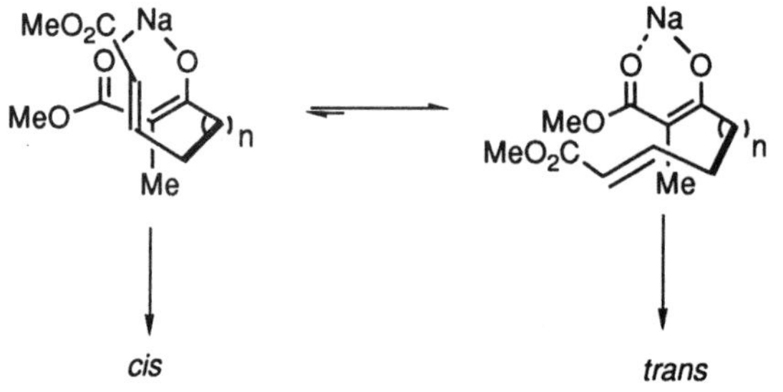

cis   trans

This approach was then successfully applied to an enantioselective synthesis of 2,2,3-trisubstituted cyclopentanones using a chiral auxiliary in the β-keto ester moiety.[245]

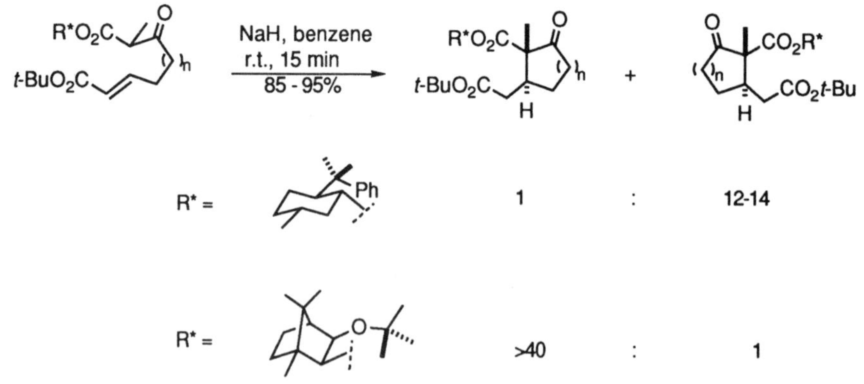

R* =		1	:	12-14
R* =		>40	:	1

The major product from such a cyclization is a useful homochiral intermediate for the construction of adrenosterone and other 11-ketosteroids.[246,247]

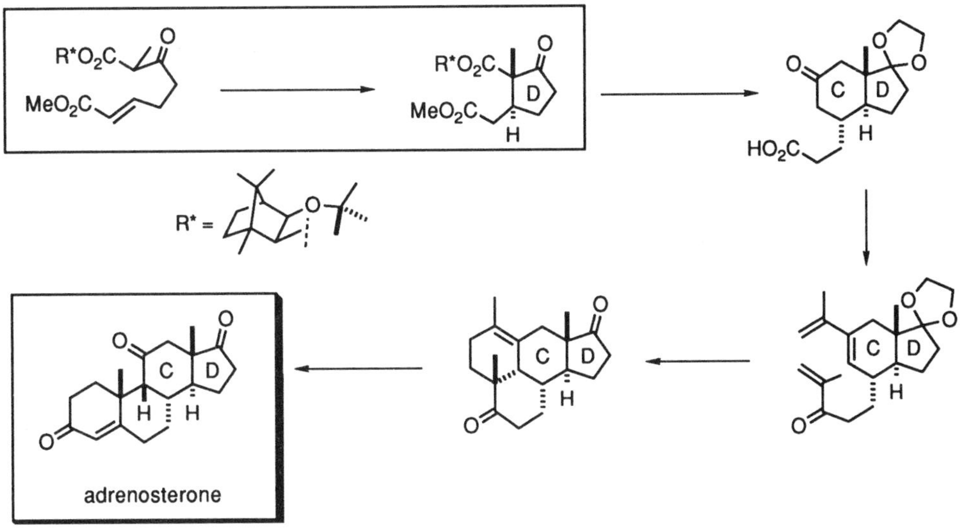

adrenosterone

Similar cyclizations have been used in the total synthesis of (±)-emetine.[248,249] Apparently, in each case, a single diastereomer is produced.

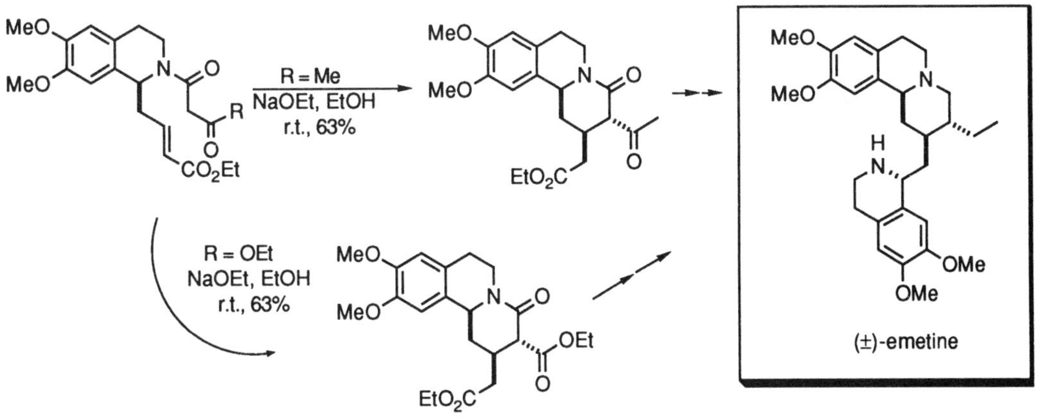

(±)-emetine

4.2.2   Free radicals

Newcomb's group has studied the effect on the rate of free radical cyclization caused by the attachment of a cyano group to the terminus of an alkene. They found that both 4-[250] and 5-[251] membered ring closures were faster then for the corresponding unactivated alkenes, e.g.:

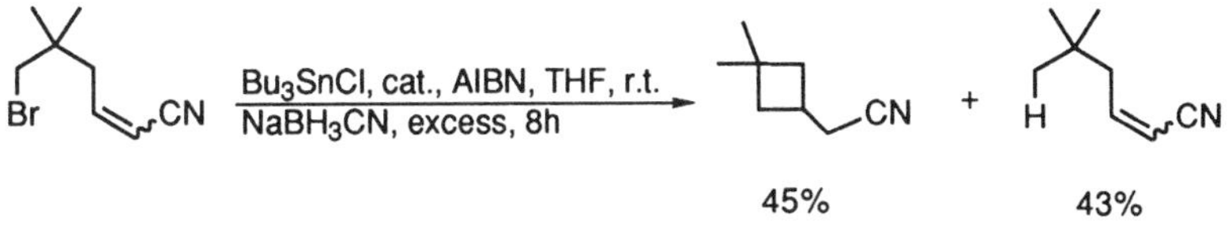

45%   43%

The use of free radicals in ring closure reactions represents one of the most successful applications of free radical chemistry in synthesis. Many structurally complex precursors, often carbohydrate based, can be prepared quite rapidly.

In one approach, an acyclic precursor is constructed along the following lines. An alkenoate, bearing a remote functional group is prepared. An extra unit is then attached to the remote functional group. This unit typically contains a halogen or other radical precursor. This is then treated with a reducing reagent, such as a stannane, generating , *in situ*, a free radical which then closes onto the alkenoate. The final step is then quenching of the newly formed radical with a hydrogen atom to propagate a radical chain reaction.

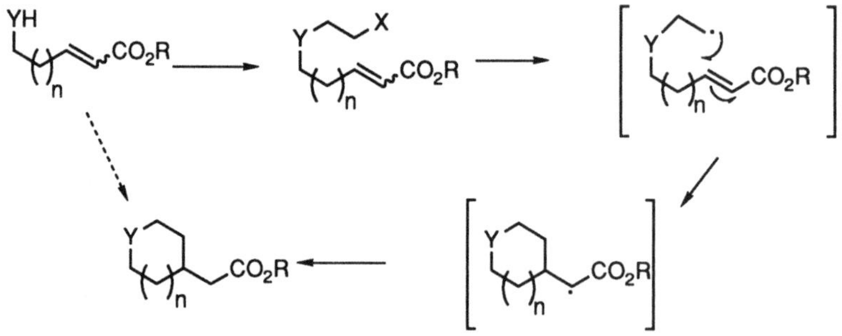

One trend which is emerging is that often a *cis*-alkene leads to much higher stereoselectivity than the corresponding *trans*-isomer. For example, in a synthesis of (-)-protoemetinol, each isomer was subjected to a photoinitiated free radical cyclization.[252] The *E*-isomer gave a mixture of all four possible stereoisomers, whereas the *Z*-isomer gave only one diastereomer in 88% yield.

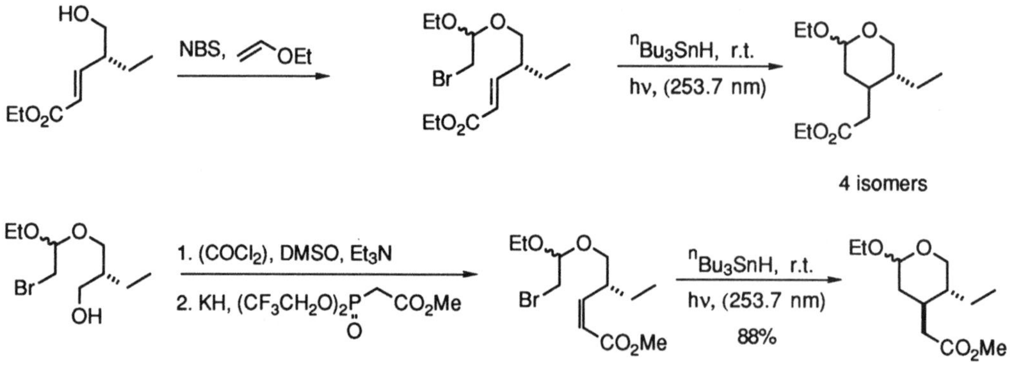

This selectivity, in the *Z*-case, is probably due to allylic control. The following conformation minimizes any steric interactions between allylic substituents and the ethoxycarbonyl group.

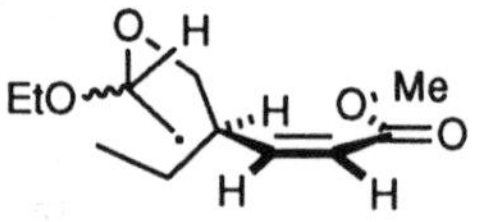

In a related study, a key step in a synthesis of methyl elenolate involved a stereoselective ring closure onto an *E*-alkenoate.[253]

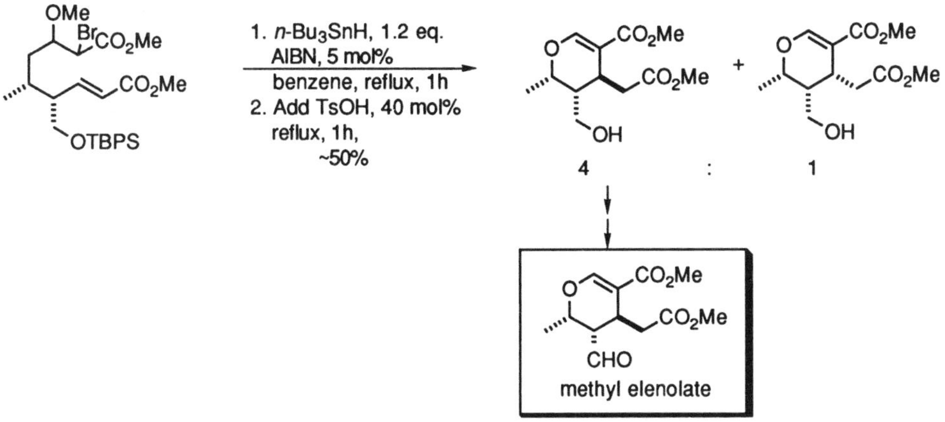

Mariano's group has also begun studying photosensitized ring closures with alkenoates.[254]

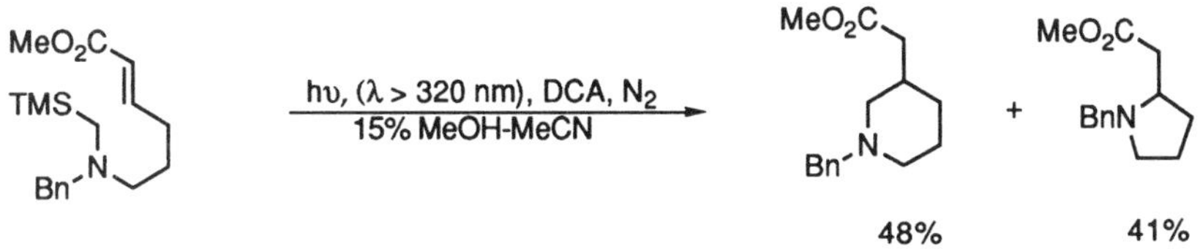

Retrosynthetic analysis of phyllantocin yielded a free radical ring closure onto an alkenoate constructed from D-glucose.[255] The *Z*-alkene gave a 9:1 selectivity in favour of the desired C2 epimer, whereas the *E*-alkene only gave a 2:1 ratio.

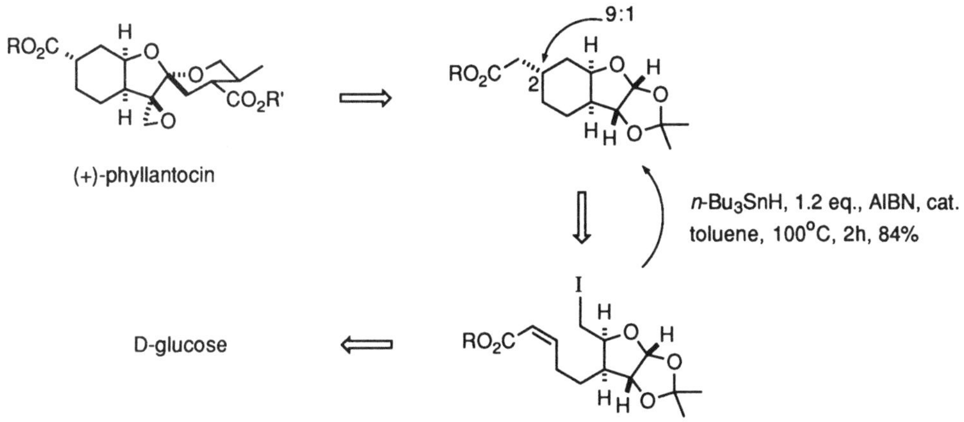

In a variation on this theme, several samarium iodide promoted ring closures have been reported.[256] Once again, the *Z*-alkene displays far greater stereoselectivity.

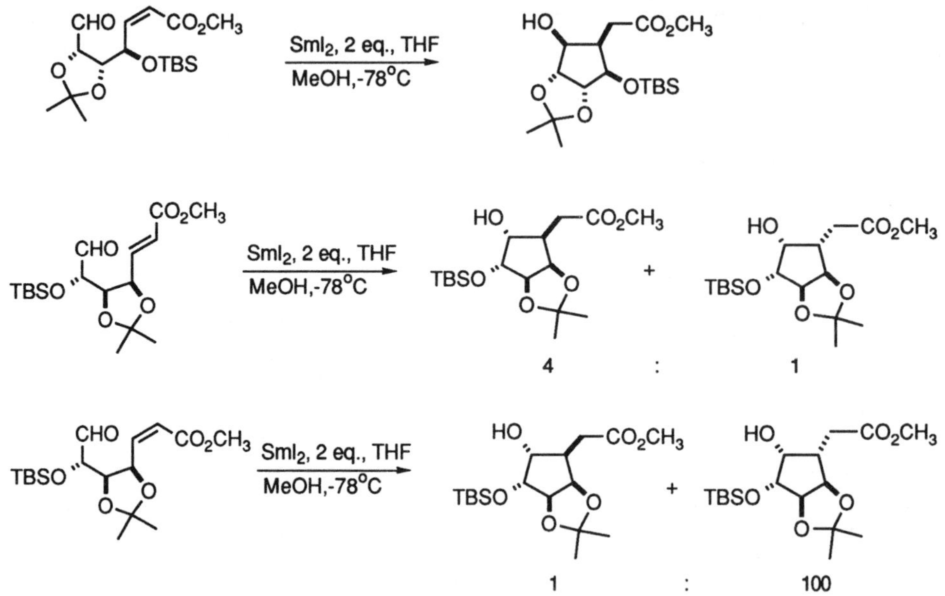

In a continuation of their studies on the total synthesis of emetine and other benzo[a]quinolizine derivatives, Hirai's group has studied various radical cyclization reactions.[249] The use of tributylstannane led to a mixture of ring-closed products as well as substantial amounts of reduction. A better method was to use copper catalysis. Only cyclized products were formed and in very high yield. Reductive dehalogenation then gave a mixture of diastereomeric benzo[a]quinolizine derivatives, (22) and (23), in high yield and modest stereoselectivity.

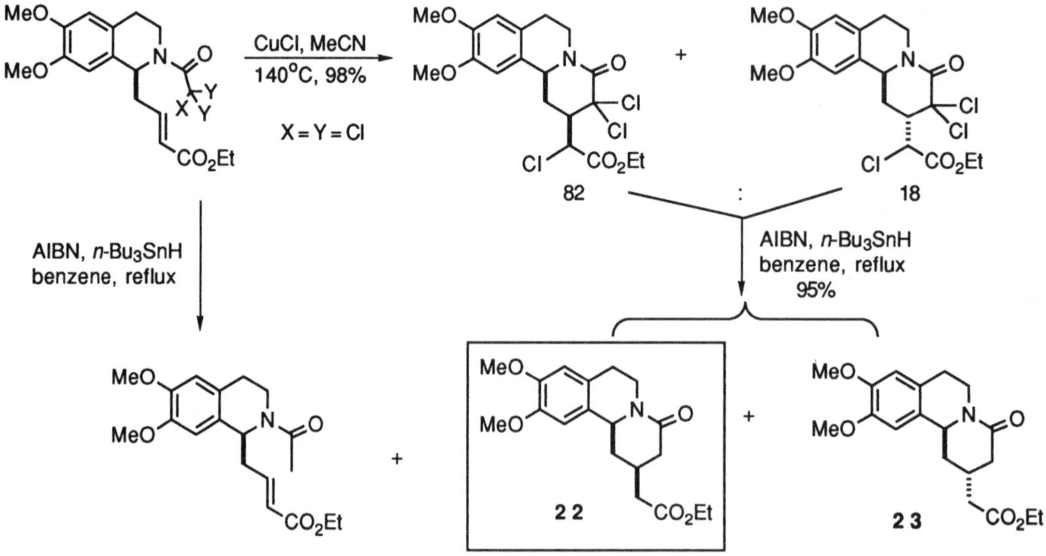

This general approach has also been applied to free radical precursors attached to a ring, i.e.:

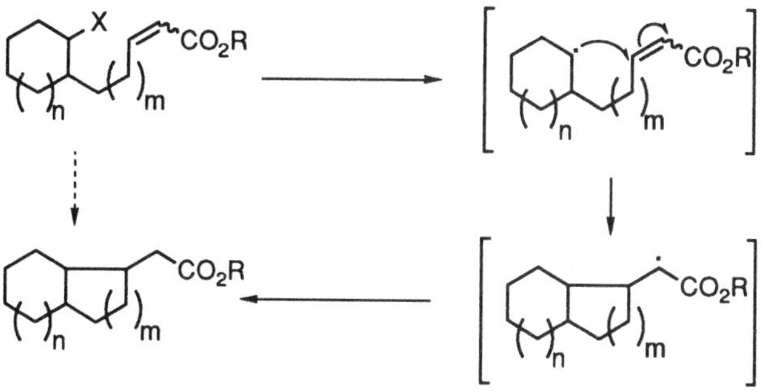

In their synthesis of *dl*-pleurotin, Hart's group developed a highly efficient and stereoselective synthesis of the perhydroindane nucleus via the following cyclization.[257] Considering that three new asymmetric centres are being generated in this cyclization, the selectivity was remarkable. Firstly, only two pairs of diastereomers were formed in a ratio of ~6:1. The major diastereomeric pair was formed in a ratio of 40:1. The authors proposed that the major reaction pathway is via the conformations shown below.

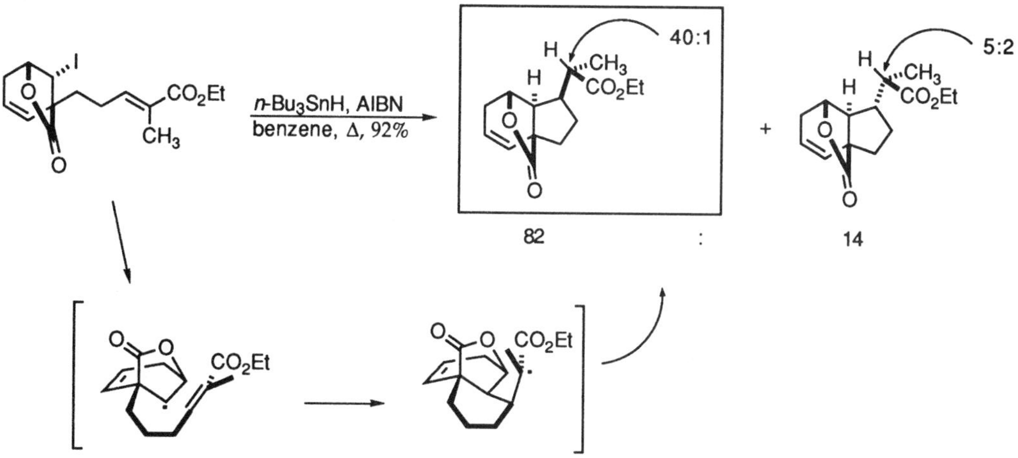

Free radical *endo* macrolactonizations work for rings containing more than ten members.[258] The reaction is especially good for rings greater than fifteen-membered. The

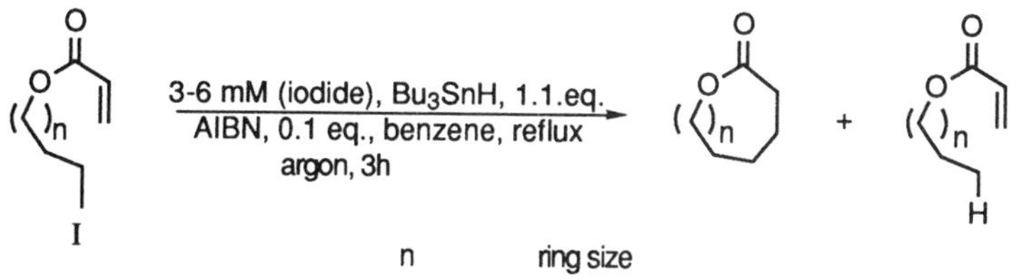

n	ring size		
5	11	15-25%	23-28%
14	20	68-76%	17-19%

authors also found that a tertiary iodide cyclizes even more efficiently and at significantly lower concentrations (<1 mM) than a primary iodide (≥3 mM). The value for k_{cyc} for the following reaction was calculated to be 3×10^8 s^{-1}. Rao found, in their total synthesis of (+)-seychellene that, whereas an allylic free radical, formed from (24) failed to close, a related alkenyl radical, from(25) closed quite efficiently.[259]

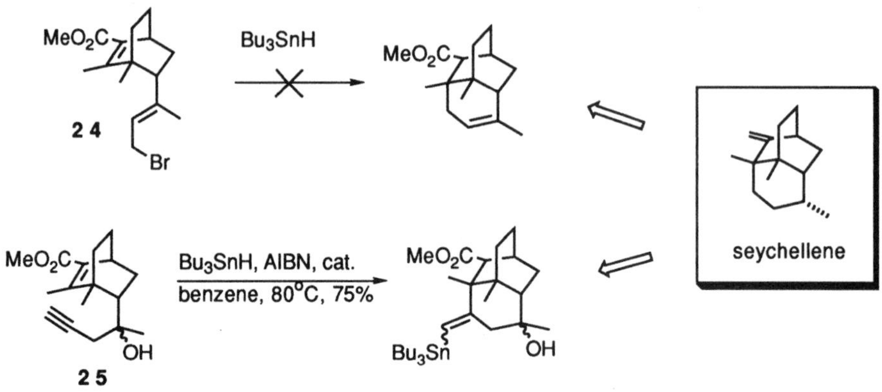

4.2.3   Enamines

Dunham and Lawton have developed a spiro-annulation process which involves sequential intermolecular alkylation and intramolecular conjugate addition.[260]

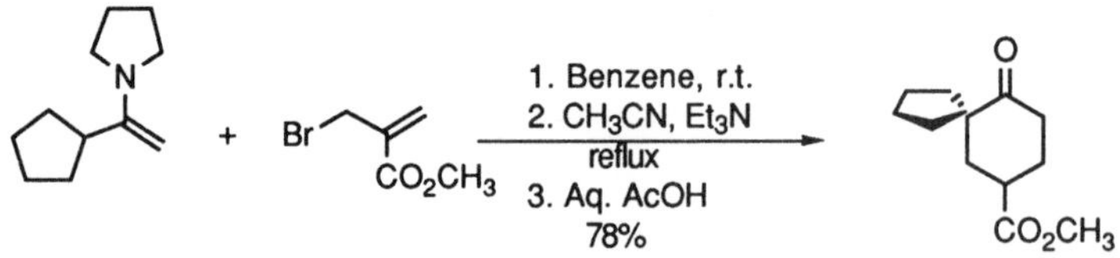

As mentioned in Section 4.1.5.1, enamine additions to indolylacrylates have been used to prepare several classes of alkaloids. Many of these are intramolecular additions. However they probably follow a concerted rather than a conjugate addition mechanism, e.g.:[261]

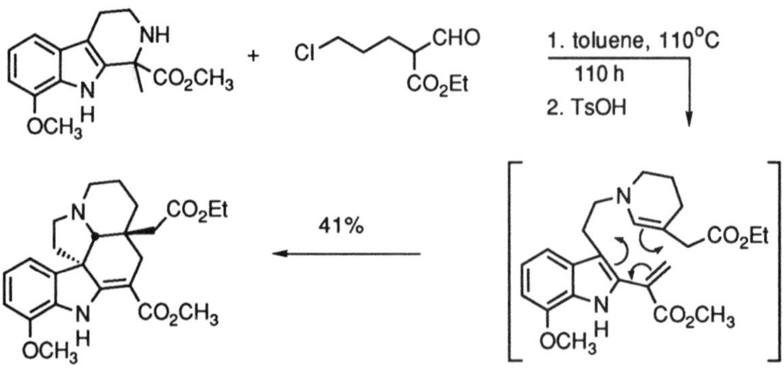

4.2.4   Heteronucleophiles

4.2.4.1   *Nitrogen*

Oxazolidinones (5-membered cyclic carbamates) may be prepared using a base-catalysed 5-*exo*-trig ring closure of allylic carbamates.[262,263] As shown below these reactions are highly selective for the 4,5 *trans* product. KO-*t*-Bu proved to be the most satisfactory base. Sodium hydride also works well although with a diminished reaction rate. The best stereoselection was observed when R was large and the alkene geometry was Z.

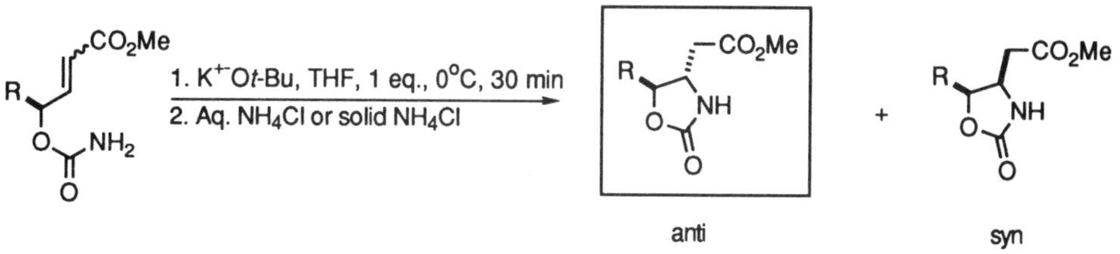

Reagents: 1. K^+Ot-Bu, THF, 1 eq., 0°C, 30 min; 2. Aq. NH_4Cl or solid NH_4Cl

anti   syn

For the *E* alkene:

	anti		syn
R = Me, 66%	5	:	1
R = Ph, 85%	12	:	1

For the *Z* alkene:

R = Ph, 75%	>100	:	1

These ring closures also work well with homoallylic carbamates, producing oxazinones. As with the examples cited above, during the ring closure an *anti* relationship is set up between the newly created chiral centre and the adjacent existing one.[264]

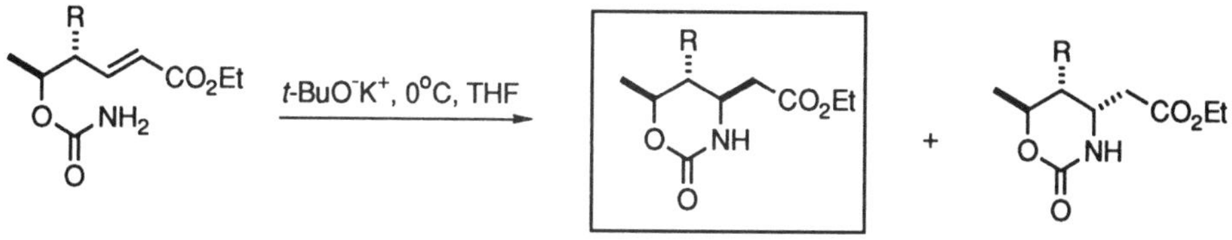

Reagents: *t*-BuO⁻K⁺, 0°C, THF

R=H[a]	10	:	1
R = *t*-BuO	7	:	1
R = MeO	18	:	1
R = AcO	19	:	1
R = TBDMSO	36	:	1
R = TESO	>50	:	1

[a] Methyl instead of ethyl ester

Several of the 3-amino-2,3,6-trideoxyhexoses, including L-*N*-acetylacosamine and L-*N*-benzoylristosamine[265] and L-*N*-benzoyldaunosamine[266] have been synthesized in this way. The synthesis of L-*N*-acetylacosamine is especially interesting as the much faster rate of 5- versus 6-exo-trig ring closures was exploited in the key reaction.

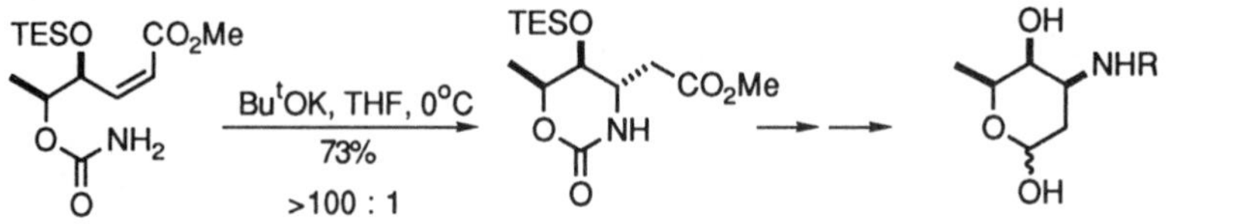

L-*N*-benzoyldaunosamine

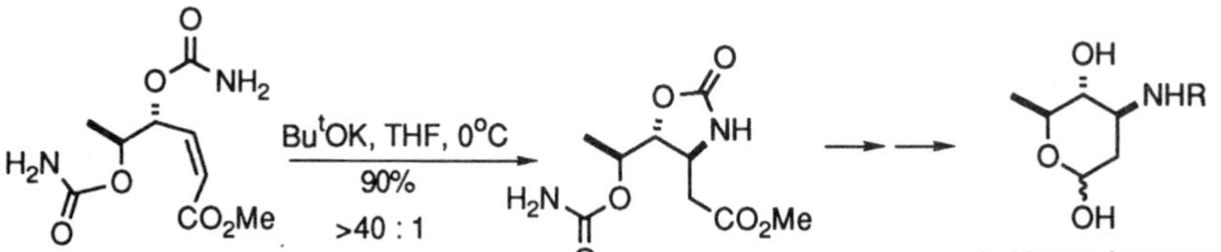

L-*N*-acetylacosamine

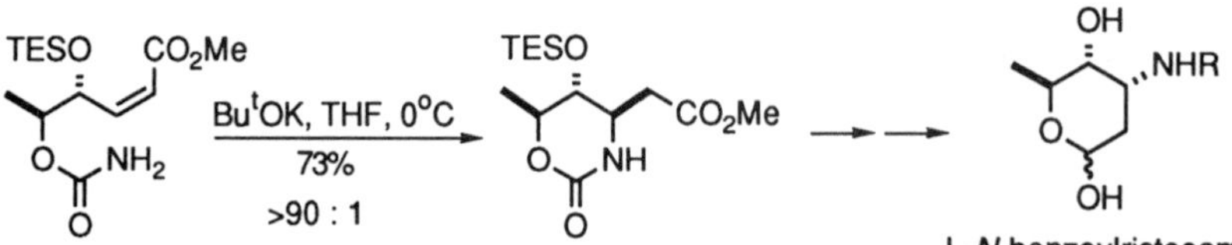

L-*N*-benzoylristosamine

Kametani's group has reported what they believe is an intramolecular double conjugate addition in their total synthesis of the phenanthrolindolizidine alkaloid, (-)-tylophorine.[267] In this case a chiral ester auxiliary was used. Although no evidence was given regarding the mechanism, the result is nevertheless remarkable as only one diastereomer was produced, in good yield. Somewhat ironically, three of the four new asymmetric centres were subsequently destroyed in order to complete the synthesis.

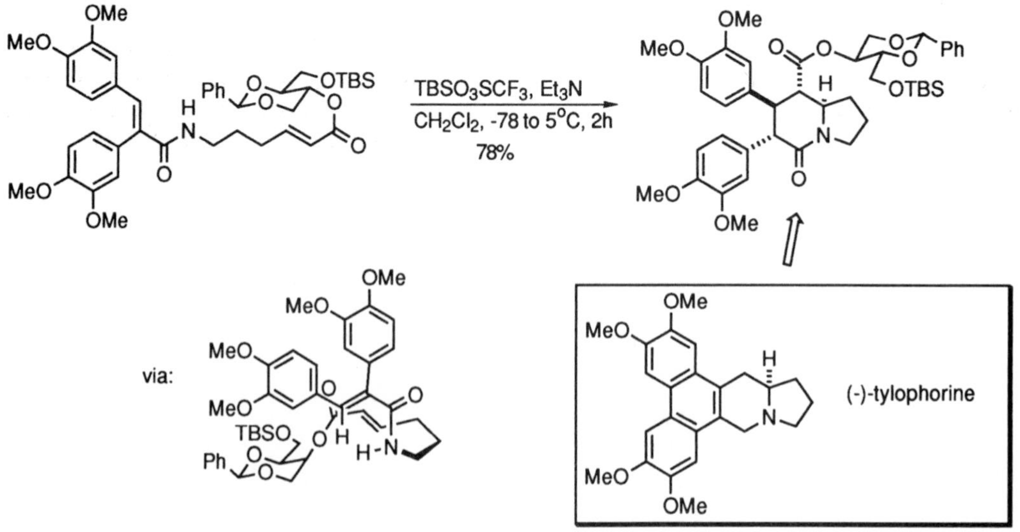

(-)-tylophorine

Hydrolases catalyze what is apparently an intramolecular conjugate addition of primary amines to 2-(trifluoromethyl)propenamides.[176]

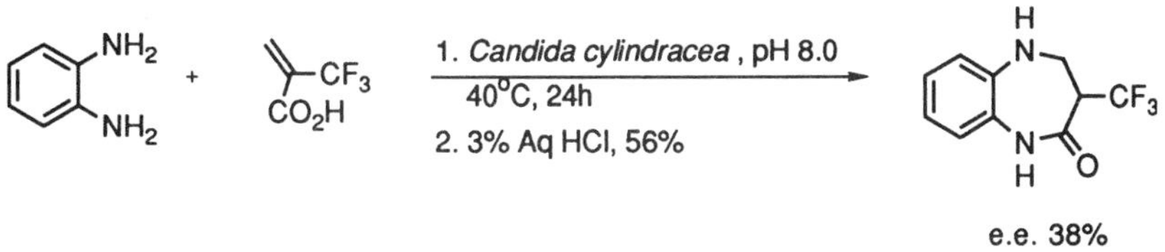
NH2
NH2
+
CF3
CO2H
1. Candida cylindracea, pH 8.0
40°C, 24h
2. 3% Aq HCl, 56%
H
N
CF3
N
H
O
e.e. 38%

4.2.4.2 *Oxygen and sulfur*

Two intramolecular conjugate additions of this class were used in the synthesis of the C101 to C115 segment of the marine toxin, palytoxin.[268]

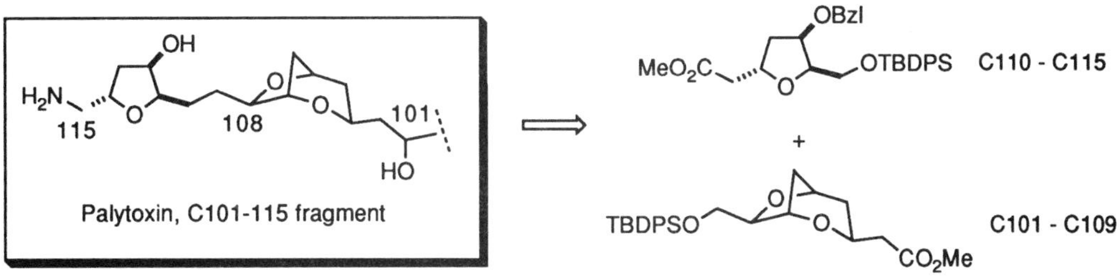
OH
H2N
115
108
101
HO
Palytoxin, C101-115 fragment
OBzl
MeO2C
OTBDPS
C110 - C115
+
TBDPSO
CO2Me
C101 - C109

The C110 to C115 segment was prepared by a straightforward ring closure of a D-xylose derived 5-hydroxyalkenoate. A single diastereomer was produced.

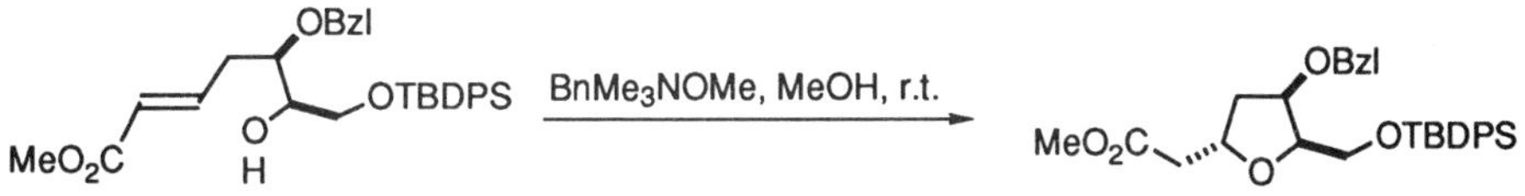
OBzl
OTBDPS
MeO2C
O
H
BnMe3NOMe, MeOH, r.t.
OBzl
MeO2C
OTBDPS

The C101 to C109 portion was prepared from a 2-deoxy-D-ribose derived 7,8-dihydroxydialkenoate by consecutive ring closures in the sequence shown. The product was an equilibrium mixture of the desired bicyclic system as well as a diastereomeric mixture of monocyclic compounds. The latter could be isolated and recycled, if so desired.

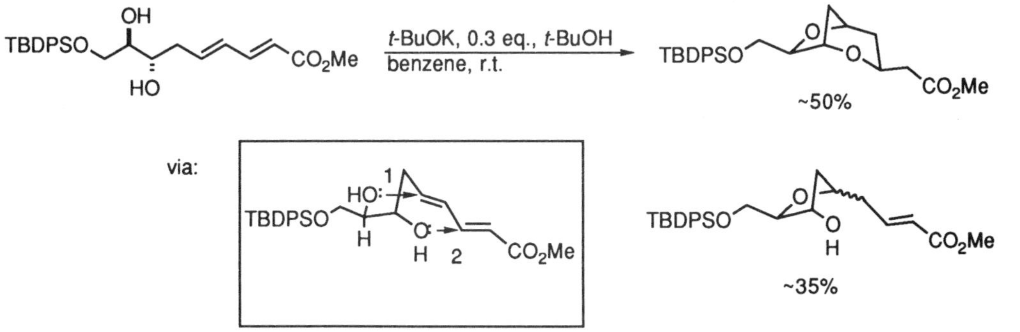
OH
TBDPSO
CO2Me
HO
t-BuOK, 0.3 eq., t-BuOH
benzene, r.t.
TBDPSO
CO2Me
~50%
via:
HO:
1
TBDPSO
H
H 2
CO2Me
TBDPSO
O
H
CO2Me
~35%

The total synthesis of several leukotrienes has been achieved using the "chiron" approach.[269] Central to this approach is the use of a retro-conjugate addition from a suitably substituted 2-deoxy-D-ribose derivative.[270] Some examples are given below.

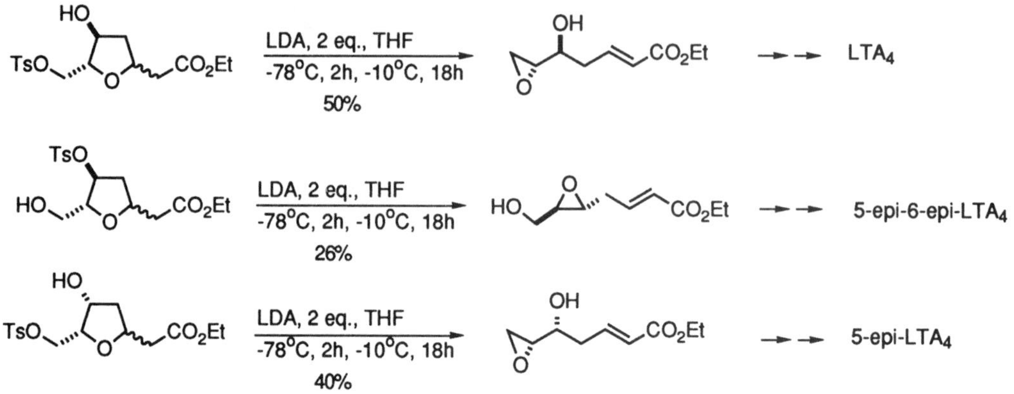

This same approach was used to prepare an intermediate in the total synthesis of some leukotriene B$_4$ analogues.[271]

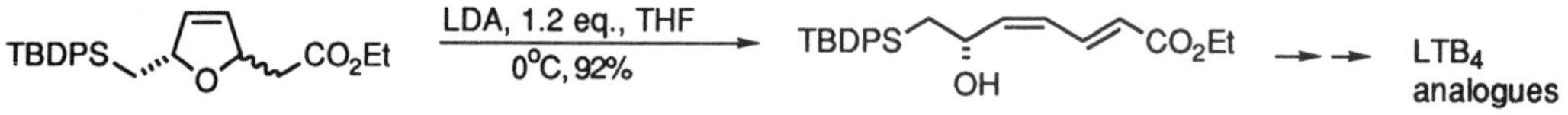

The synthesis of (±)-Δ-(9,12)-capnellene, a member of the family of non-isoprenoid sesquiterpenes, also uses a retro-conjugate addition reaction.[272]

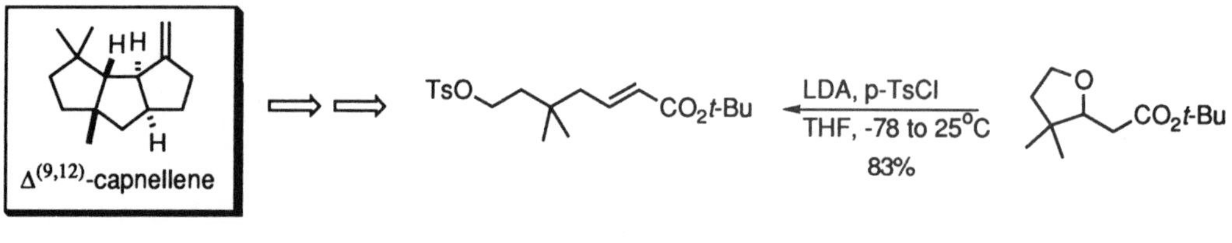

Hydrolases, such as *Candida cylindracea*, catalyze what is thought to be an intramolecular conjugate addition in the reaction between 2-aminophenols (or thiophenols) and 2-(trifluoromethyl)propenoate.[176]

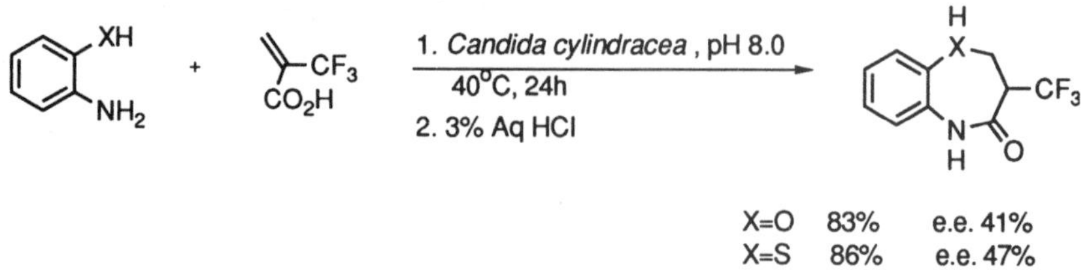

X=O   83%   e.e. 41%
X=S   86%   e.e. 47%

References

1. H. O. House and M. Schellenbaum, *J. Org. Chem.*, 1963, **28**, 34
2. For an excellent discussion of these reactions, see H. O. House, "Modern Synthetic Reactions", Benjamin, Menlo Park, 1972, 595-623
3. E. S. Binkley and C. H. Heathcock, *J. Org. Chem.*, 1975, **40**, 2156
4. D. A. Dickman and C. H. Heathcock, *J. Amer. Chem. Soc.*, 1989, **111**, 1528
5. S. Kim and P. H. Lee, *Tetrahedron Lett.*, 1988, **29**, 5413
6. G. B. Mpango, K. K. Mahalanabis, Z. Mahdavi-Damghani and V. Snieckus, *Tetrahedron Lett.*, 1980, **21**, 4823
7. C. Najera and M. Yus, *Tetrahedron Lett.*, 1989, **30**, 173
8. See ref. 6
9. G. H. Posner, K. A. Canella and E. F. Silversmith, *Proc. Indian. Acad. Sci. (Chem. Sci.)*, 1988, **100**, 81
10. G. H. Posner, *Chem. Rev.*, 1986, **86**, 831
11. R. A. Lee, *Tetrahedron Lett.*, 1973, 3333
12. D. Spitzner, *Tetrahedron Lett.*, 1978, 3349 and D. Spitzner, P. Wagner, A. Simon and K. Peters, *Tetrahedron Lett.*, 1989, **30**, 547
13. D. Spitzner, A. Engler, T. Liese, G. Splettstosser and A. de Meijere, *Angew. Chem. Int. Ed. Engl.*, 1982, **21**, 791
14. J. E. Baldwin, R. M. Adlington, J. C. Bottaro, A. U. Jain, J. N. Kohle, M. D. Perry and I. M. Newington, *J. Chem. Soc., Chem. Commun.*, 1984, 1095
15. B. B. Snider, R. S. E. Conn and S. Sealfon, *J. Org. Chem.*, 1979, **44**, 218
16. M. Yamaguchi, M. Tsukamoto and I. Hirao, *Chem. Lett.*, 1984, 375. See also, M. Yamaguchi, *Yuki Gosei Kagaku*, 1986, **44**, 405
17. M. Yamaguchi, M. Tsukamoto, S. Tanaka and I. Hirao, *Tetrahedron Lett.*, 1984, **25**, 5661
18. G. H. Posner and E. M. Shulman-Roskes, *J. Org. Chem.*, 1989, **54**, 3514
19. H. Ahlbrecht and H-M. Kompter, *Synthesis*, 1983, 645
20. H. Ahlbrecht and K. Pfaff, *Synthesis*, 1978, 897 (contains the method for preparing these conjugate acceptors)
21. H. Ahlbrecht and M. Ibe, *Synthesis*, 1985, 421
22. H. Ahlbrecht and M. Dietz, *Synthesis*, 1985, 417
23. H. Stetter, *Angew. Chem. Int. Ed. Engl.*, 1976, **15**, 639
24. H. Stetter and H. Kuhlmann, *Chem. Ber.*, 1976, **109**, 2890
25. H. Stetter and B. Rajh, *Chem. Ber.*, 1976, **109**, 534
26. J. Nogami, T. Sonoda and S. Wakabayashi, *Synthesis*, 1983, 763
27. W. T. Monte, M. M. Baizer and R. D. Little, *J. Org. Chem.*, 1983, **48**, 803
28. K. Matsumoto, *Angew. Chem. Int. Ed. Engl.*, 1981, **20**, 770
29. M. Mikolajczyk and P. Balczewski, *Synthesis*, 1984, 691
30. N. Ono, A. Kamimura, H. Miyake, I. Hamamoto and A. Kaji, *J. Org. Chem.*, 1985, **50**, 3692
31. G. R. Newkome, C. N. Moorefield and K. J. Theriot, *J. Org. Chem.*, 1988, **53**, 5552
32. J. E. Richman, J. L. Herrmann and R. H. Schlessinger, *Tetrahedron Lett.*, 1973, 3271
33. For a review of camphor based chiral auxiliaries in synthesis, see W. Oppolzer, *Tetrahedron*, 1987, **43**, 1969; for a (complete) Erratum, see *ibid.*, 1987, **43**, 4057
34. W. Oppolzer, C. Chapuis and G. Bernardinelli, *Tetrahedron Lett.*, 1984, **25**, 5885

35. From Aldrich Chemical Co., Inc.

36. W. Oppolzer, C. Chapuis and G. Bernardinelli, *Helv. Chim. Acta*, 1984, **67**, 1397

37. M. Vandewalle, J. Van der Eycken, W. Oppolzer and C. Vullioud, *Tetrahedron*, 1986, **42**, 4035

38. For certain other clases of addition to these substrates, such as nitrile oxide additions, approach of the incoming reactant is *syn* to the nitrogen lone pair. See D. P. Curran, B. H. Kim, J. Daugherty and T. A. Heffer, *Tetrahedron Lett*, 1988, **29**, 3555

39. D. Enders and K. Papadopoulos, *Tetrahedron Lett.*, 1983, **24**, 4967

40. D. Enders, K. Papadopoulos and B. E. M. Rendenbach, *Tetrahedron Lett.*, 1986, **27**, 3491

41. D. Enders, P. Gerdes and H. Kipphardt, *Angew. Chem. Int. Ed. Engl.*, 1990, **29**, 179

42. D. Desmaële and J. d'Angelo, *Tetrahedron Lett.*, 1989, **30**, 345

43. See also J. d'Angelo and A. Guingant,*Tetrahedron Lett.*, 1988, **29**, 2667

44. E. J. Corey and H. E. Ensley, *J. Amer. Chem. Soc.*, 1975, **97**, 1234

45. E. J. Corey and R. T. Peterson, *Tetrahedron Lett.*, 1985, **26**, 5025

46. M. Takasu, H. Wakabayashi, K. Furuta and H. Yamamoto, *Tetrahedron Lett.*, 1988, **29**, 6943

47. See M. Alonso-Lopéz, M. Bernabé, A. Fernandez-Mayoralas, J. Jiménez-Barbero, M. Martin-Lomas and S. Penadés, *Carbohydr. Res.*, 1986, **150**, 103

48. M. Alonso-Lopéz, Jiménez-Barbero, M. Martin-Lomas and S. Penadés, *Tetrahedron*, 1988, **44**, 1535

49. M. Yamaguchi, K. Hasebe, S. Tanaka and T. Minami, *Tetrahedron Lett.*, 1986, **27**, 959

50. U. Schollkopf, U. Groth, K-O. Westphalen and C. Deng, *Synthesis*, 1981, 969 and U. Schollkopf, U. Groth and C. Deng, *Angew. Chem. Int. Ed. Engl.*, 1981, **20**, 798

51. For reviews of the application of bislactim ethers to amino acid synthesis, see U. Schollkopf, *Chem. Scripta,* 1985, **25**, 105; *Ibid*, *Pure Appl. Chem.*, 1983, **55**, 1799 and *ibid*, in J, Streith, H. Prinzbach and G. Schill (eds), "Organic Synthesis: An Interdisciplinary Challenge", Blackwell Scientific Publications, 1985, 101

52. U. Schollkopf and J. Schroder, *Liebig's Ann. Chem.*, 1988, 87

53. J. Mulzer, A. Chucholowski, O. Lammer, I. Jibril and G. Huttner, *J. Chem. Soc., Chem. Commun.*, 1983, 869. The structures of the major products shown in this paper have the incorrect relative configuration, see footnote 123 in D. A. Oare and C. H. Heathcock, in E. L. Eliel and S. H. Wilen (eds), "Topics in Stereochemistry", Wiley, New York, 1989, 227

54. M. Casey, A. C. Manage and L. Nezhat, *Tetrahedron Lett.*, 1988, **29**, 5821

55. B. S. Furniss, A. J. Hannaford, P. W. G. Smith and A. R. Tatchell, "Vogel's Textbook of Practical Organic Chemistry", Longman, Burnt Mill, 1989, 718

56. D. Cartier, D. Patigny and J. Levy, *Tetrahedron Lett.*, 1982, **23**, 1897

57. B. Bernet, P. M. Bishop, M. Caron, T. Kawamata, B. L. Roy, L. Ruest, G. Sauve P. Soucy and P. Deslongchamps, *Can. J. Chem.*, 1985, **63**, 2810

58. J. L. Zeistra and J. B. F. N. Egberts, *J. Org. Chem.*, 1974, **39**, 3215

59. B. S. Furniss, A. J. Hannaford, P. W. G. Smith and A. R. Tatchell, "Vogel's Textbook of Practical Organic Chemistry", Longman, Burnt Mill, 1989, 687

60. M. Nakada, S. Kobayashi and M. Ohno, *Tetrahedron Lett.*, 1988, **29**, 3951

61. S. Iwazaki, M. Namikoshi, H. Kobayashi, J. Furukawa, S. Okuda, A. Itai, A. Kasuya, Y. Iitaka and Z. Sato, *J. Antibiotics*, 1986, **39**, 424

62. R. Lawrence and P. Perlmutter, *J. Org. Chem.*, 1990, submitted for publication

63. D. H. Hua, S. N. Bharathi, F. Takusagawa, A. Tsujimoto, J. A. K. Panangadan, M-H. Hung, A. A. Bravo and A. M. Erpelding, *J. Org. Chem.*, 1989, **54**, 5659

64. K. Katsuura and V. Snieckus, *Can. J. Chem.*, 1987, **65**, 124

65. J. Luchetti and A. Krief, *J. Organomet. Chem.*, 1980, **194**, C49

66. J. Mulzer and M. Kappert, *Angew. Chem. Int. Ed. Engl.*, 1983, **22**, 63

67. G. P. Mpango and V. Snieckus, *Tetrahedron Lett.*, 1980, **21**, 4827

68. K. Tomioka, M. Sudani, Y. Shinini and K. Koga, *Chem. Lett.*, 1985, 329

69. J. P. Marino and L. C. Katterman, *J. Chem. Soc., Chem. Commun.*, 1979, 946

70. E-L. Lindstedt and M. Nilsson, *Acta Chem. Scand. B*, 1986, **40**, 466

71. E-L. Lindstedt, M. Nilsson and T. Olsson, *J. Organomet. Chem.*, 1987, **334**, 255

72. H. Malmberg and M. Nilsson, *Tetrahedron*, 1982, **38**, 1509

73. B. Christenson, G. Hallnemo and C. Ullenius, *Tetrahedron*, 1990, **46**, submitted for publication

74. For an extensive compilation and review, see M. J. Chapdelaine and M. Hulce, *Org. React.*, 1990, **38**, 225

75. T. Fujisawa, A. Noda, T. Kawara and T. Sato, *Chem. Lett.*, 1981, 1159

76. F. Naf and R. Decorzant, *Helv. Chim. Acta*, 1971, **54**, 1939

77. W. A. Nugent and F. W. Hobbs, *J. Org. Chem.*, 1986, **51**, 3376

78. M. Alderdice, F. W. Sum and L. Weiler, *Org. Synth.*, 1984, **62**, 14

79. R. K. Dieter and L. A. Silks, *J. Org. Chem.*, 1986, **51**, 4687

80. F. W. Sum and L. Weiler, *J. Chem. Soc., Chem. Commun.*, 1978, 985

81. See also F. W. Sum and L. Weiler, *Tetrahedron Lett.*, 1979, 707 and F. W. Sum and L. Weiler, *J. Amer. Chem. Soc.*, 1979, **101**, 4401

82. W. Oppolzer, P. Dudfield, T. Stevenson and T. Godel, *Helv. Chim. Acta*, 1985, **68**, 212. See also W. Oppolzer, R. Moretti and G. Bernardinelli, *Tetrahedron Lett.*, 1986, **27**, 4713

83. W. Oppolzer, A. J. Kingma and G. Poli, *Tetrahedron*, 1989, **45**, 479

84. K. Tomioka, T. Suenaga and K. Koga, *Tetrahedron Lett.*, 1986, **27**, 369

85. S. Saito, Y. Hirohara, O. Narahara and T. Moriwake, *J. Amer. Chem. Soc.*, 1989, **111**, 4533

86. X = OR, see (a) M. Larchevegue, G. Tamagnan and Y. Petit, M. Hirama, *J. Chem. Soc., Chem. Commun.*, 1989, 31; (b) T. Iwakuma and S. Ito, *J. Chem. Soc., Chem. Commun.*, 1987, 1523; (c) M. Isobe, Y. Ichawa, Y. Funabashi, S. Mio and T. Goto, *Tetrahedron*, 1986, **42**, 2863; (d) W. R. Roush and B. M. Lesur, *Tetrahedron Lett.*, 1983, **24**, 2231
     X = NR$_2$, M. T. Reetz and D. Röhrig, *Angew. Chem. Int. Ed. Engl.*, 1989, **28**, 1706

87. T. Ibuka, M. Tanaka, H. Nemoto and Y. Yamamoto, *Tetrahedron*, 1989, **45**, 435

88. See also, W. R. Roush, M. R. Michaelides, D. F. Tai, B. M. Lesur, W. K. M. Chong and D. J. Harris, *J. Amer. Chem. Soc.*, 1989, **111**, 2984

89. See also I. W. Lawston and T. D. Inch, *J. Chem. Soc., Perkin Trans. I*, 1983, 2629

90. K. C. Nicolau, M. R. Pavia and S. P. Seitz, *J. Amer. Chem. Soc.*, 1981, **103**, 1224

91. K. Tatsuta, Y. Amemiya, S. Maniwa and M. Kinoshita, *Tetrahedron Lett.*, 1980, **21**, 2837

92.   K. Tatsuta, Y. Amemiya, Y. Kanemura and M. Kinoshita, *Tetrahedron Lett.*, 1981, **22**, 3997

93.   K. C. Nicolau, M. R. Pavia and S. P. Seitz, *J. Amer. Chem. Soc.*, 1982, **104**, 2027

94.   M. T. Reetz and D. Rohring, *Angew. Chem. Int. Edn. Engl.*, 1989, **28**, 1706

95.   A. Alexakis, R. Sedrani, P. Mangeney and J. F. Normant, *Tetrahedron Lett.*, 1988, **29**, 4411

96.   G. H. Posner, J. P. Mallamo and K. Miura, *J. Amer. Chem. Soc.*, 1981, **103**, 2886

97.   J. E. Baldwin and W. A. Dupont, *Tetrahedron Lett.*, 1980, **21**, 1881. This paper also contains some useful references to early examples of conjugate additions of organomagnesium reagents to alkenoic derivatives

98.   Y. Tamaru, T. Harada, H. Iwamoto and Z-i. Yoshida, *J. Amer. Chem. Soc.*, 1978, **100**, 5221

99.   G. B. Mpango, K. K. Mahalanabis, Z. Mahdavi-Damghani and V. Snieckus, *Tetrahedron Lett.*, 1980, **21**, 4823

100. E. J. Corey and L. S. Hegedus, *J. Amer. Chem. Soc.*, 1969, **91**, 4926

101. M. P. Cooke, Jr. and R. M. Parlman, *J. Amer. Chem. Soc.*, 1977, **99**, 5222

102. C. Najera and M. Yus, *Tetrahedron Lett.*, 1989, **30**, 173

103. G. Majetich, A. M. Casares, D. Chapman and M. Behnke, *Tetrahedron Lett.*, 1983, **24**, 1909

104. G. Daviaud and P. Miginiac, *Tetrahedron Lett.*, 1973, 3345

105. W. Oppolzer, G. Poli, A.J. Kingma, C. Starkemann and G. Bernardinelli, *Helv. Chim.Acta*, 1987, **70**, 2201

106. T. Kindt-Larsen, V. Bitsch, I. G. K. Andersen, A. Jart and J. Munch-Petersen, *Acta Chem. Scand.*, 1963, **17**, 1426

107. M. Larcheveque, G. Tamagnan and Y. Petit, *J. Chem Soc., Chem. Commun.*, 1989, 31

108. J. P. Vigneron, R. Meric, M. Larcheveque, A. Debal, J. Y. Lallemond, G. Kunesch, P. Zagatti and M. Gallois, *Tetrahedron*, 1984, **40**, 3521

109. For a review of the applications of stereogenic oxazolines in synthesis, see K. A. Lutomski and A. I. Meyers in J. D. Morrison (ed.), *Asymmetric Synthesis*, Academic Press, New York, 1984, Vol. 3, Part B, Chapter 3, 259 to 274

110. A. I. Meyers, G. Knaus, K. Kamata and M. E. Ford, *J. Amer. Chem. Soc.*, 1976, **98**, 567

111. A. I. Meyers, R. K. Smith and C. E. Whitten, *J. Org. Chem.*, 1979, **44**, 2250

112. A. I. Meyers and R. K. Smith, *Tetrahedron Lett.*, 1979, **25**, 2749

113. A. I. Meyers, G. P. Roth, D. Hoyer, B. A. Barner and D. Laucher, *J. Amer. Chem. Soc.*, 1988, **110**, 4611

114. See also A. I. Meyers and K. Higashiyama, *J. Org. Chem.*, 1987, **52**, 4592

115. A. I. Meyers, N. R. Natale, D. G. Wettlaufer, S. Rafii and J. Clardy, *Tetrahedron Lett.*, 1981, **22**, 5123

116. A. I. Meyers and N. R. Natale, *Heterocycles*, 1982, **18**, 13

117. A. I. Meyers and K. A. Lutomski, *J. Amer. Chem. Soc.*, 1982, **104**, 879

118. J. M. Wilson and D. J. Cram, *J. Amer. Chem. Soc.*, 1982, **104**, 881

119. The unsaturated derivatives are prepared from the iron acyl, [(η-C5H5)Fe(CO)(PPh3)COMe]. Both enantiomers of this complex are commercially available

120.   For reviews, see S. G. Davies, I. M. Dordor-Hedgecock, R. J. C. Easton, K. H. Sutton and J. C. Walker, *Bull. Soc. Chem. France*, 1987, 608 and S. G. Davies, R. J. C. Easton, J. C. Walker and P. Warner, *Tetrahedron*, 1986, **42**, 175

121.   P. W. Ambler and S. G. Davies, *Tetrahedron Lett.*, 1988, **29**, 6979

122.   P. W. Ambler and S. G. Davies, *Tetrahedron Lett.*, 1988, **29**, 6983

123.   J. M. Chong and E. K. Mar, *Tetrahedron Lett.*, 1990, **31**, 1981

124.   P. Girard, J. L. Namy and H. B. Kagan, *J. Amer. Chem. Soc.*, 1980, **102**, 2693

125.   S-i. Fukuzawa, A. Nakanishi, T. Fujinami and S. Sakai, *J. Chem. Soc., Chem. Commun.* 1986, 624

126.   K. Otsubo, J. Inanaga and M. Yamaguchi, *Tetrahedron Lett.*, 1986, **27**, 5763

127.   F. W. Eastwood, R. Mifsud and P. Perlmutter, unpublished results

128.   For an extensive survey and thorough review of the generation and reactions of carbon radicals, see A. Ghosez, B. Giese and H. Zipse in M. Regitz and B. Giese (eds), "Houben-Weyl. Methoden der organischen Chemie", Georg Thieme, Stuttgart, 1989, Band E19a/Teil 2. See also B. Giese, "Radicals in Organic Synthesis: Formation of Carbon-Carbon Bonds", Pergamon Press, Oxford, 1986

129.   D. J. Hart and F. L. Seely, *J. Amer. Chem. Soc.*, 1988, **11**, 1631

130.   B. Giese, J. Dupuis and M. Nix, *Org. Synth.*, 1987, **65**, 236

131.   J. L. Luche and C. Allavena, *Tetrahedron Lett.*, 1988, **29**, 5369

132.   R. Scheffold and L. R. Orlinski, *J. Amer. Chem. Soc.*, 1983, **105**, 7200

133.   J. Barluenga, J. Lopez-Prado, P. J. Campos and G. Asensio, *Tetrahedron*, 1988, **39** 2863

134.   N. A. Porter, D. M. Scott, B. Lacher, B. Giese, H. G. Zeitz and H. J. Lindner, *J. Amer. Chem. Soc.*, 1989, **111**, 8311

135.   S. Danishefsky and G. Rovnyak, *J. Org Chem.*, 1971, **93**, 2074.

136.   O. Cervinka, *Coll. Czech. Chem. Commun.*, 1960, **25**, 2675 (use of perchloric acid as catalyst)

137.   G. Stork, A. Brizzolara, H. Landesman, J. Szmuszkovicz and R. Terrell, *J. Amer. Chem. Soc.*, 1963, **85**, 207

138.   See reference 2, page 616

139.   A. G. Angoh and D. L. J. Clive, *J. Chem. Soc., Chem. Commun.*, 1985, 941.

140.   H. Nemoto, E. Shitara, K. Fukumoto and T. Kametani, *Heterocycles*, 1985, **23**, 1911

141.   G. Stork and H. Landesman, *J. Amer. Chem. Soc.*, 1956, **78**, 5128

142.   Z. Iqbal, A. H. Jackson and K. R. N. Rao, *Tetrahedron Lett.*, 1988, **29**, 2577

143.   S. Danishefsky, G. Rovnyak and R. Cavanaugh, *J. Chem. Soc., Chem. Commun.*, 1969, 636

144.   S. Danishefsky and R. Cunningham, *J. Org. Chem.*, 1965, **30**, 3676

145.   G. A. Berchtold, J. Ciabattoni and A. A. Tunick, *J. Org. Chem.*, 1965, **30**, 3679

146.   S. J. Blarer and D. Seebach, *Chem. Ber.*, 1983, **116**, 2250.

147.   See also, K. Hiroi, K. Achiwa and S. I. Yamada, *Chem. Pharm. Bull. Jpn.*, 1972, **20**, 246

148.   D. Seebach and J. Golinski, *Helv. Chim. Acta*, 1981, **64**, 1413

149.   C. Stetin, B. De Jeso and J-C. Pommier, *J. Org. Chem.*, 1985, **50**, 3863

150.   M. E. Kuehne and G. Di Vincenzo, *J. Org. Chem.*, 1972, **37**, 1023

151.   O. W. Webster and G. D. Y. Sogah in M. Fantaville and A. Guyot, (eds), "Recent Advances in Mechanistic and Synthetic Aspects of Polymerization", Reidel, Dordrecht, 1987, 3

152.   K. Narasaka, K. Soai, Y. Aikawa and T. Mukaiyama, *Bull. Soc. Chem. Jap.*, 1976, **49**, 779

153.   C. Pak, H. Mizuno, H. Okamoto and H. Sakurai, *Synthesis*, 1978, 589

154.   H. Mizuno, *Chem. Lett.*, 1975, 237

155.   T. Majima, C. Pak and H. Sakurai, *Bull. Soc. Chem. Jap.*, 1978, **51**, 1811

156.   For a review, see W. Nagata and M. Yoshioka, *Org. Reactions*, 1977, **25**, 255

157.   B. S. Furniss, A. J. Hannaford, P. W. G. Smith and A. R. Tatchell, "Vogel's Textbook of Practical Organic Chemistry", Longman, Burnt Mill, 1989, 686

158.   P. Kurtz, *Ann. Chem.*, 1951, **572**, 23

159.   W. Franke and H. Weber, Ger. Pat. 808,835 (1951), *Chem. Abstr.*, 1953, **47**, 5427d

160.   G. R. Coraor and W. Z. Heldt, U. S. Pat. 2,904,581 (1959), *Chem. Abstr.*, 1960, **54**, 4393f

161.   A. Lapworth and J. A. McRae, *J. Chem. Soc.*, 1922, **121**, 1699

162.   A. Lapworth and J. A. McRae, *J. Chem. Soc.*, 1922, **121**, 2741

163.   H. O. Hankovsky, K. Hideg, P. C. Sár, M. J. Lovas and G. Jerkovich, *Synthesis*, 1990, 59

164.   W. Nagata, T. Okumura and M. Yoshioka, *J. Chem. Soc., C*, 1970, 2347

165.   H. J. Storesund and P. Kolsaker, *Tetrahedron Lett.*, 1972, 2255

166.   For a review, see S. I. Suminov and A. N. Kost, *Russ. Chem. Rev.*, 1969, **38**, 884

167.   D. R. Howton, *J. Org. Chem.*, 1945, **10**, 277

168.   For some very early examples, see W. F. Holcomb and C. S. Hamilton, *J. Amer. Chem. Soc.*, 1942, **64**, 1309

169.   J. Barluenga, J. Villamana and M. Yus, *Synthesis*, 1981, 375

170.   F. C. Whitmore, H. S. Mosher, R. R. Adams, R. B. Taylor, E. C. Chapin, C. Weisel and W. Yanko, *J. Amer. Chem. Soc.*, 1944, **66**, 725

171.   C. A. Weisel, R. B. Taylor, H. S. Mosher and F. C. Whitmore, *J. Amer. Chem. Soc.*, 1945, **67**, 1071

172.   O. Hromatka, *Chem. Ber.*, 1942, 75B, 131

173.   D. Rosenthal, G. Brandrup, K. Davis and M. Wall, *J. Org. Chem.*, 1965, **30**, 3689

174.   C. A. Weisel, R. B. Taylor, H. S. Mosher and F. C. Whitmore, *J. Amer. Chem. Soc.*, 1945, **67**, 1071

175.   F. Critchfield, G. Funk and W. Johnson, *Analytic Chem.*, 1956, **28**, 76

176.   T. Kitazume, T. Ikeya and K. Murata, *J. Chem. Soc., Chem. Commun.*, 1986, 1331

177.   T. Kitazume and K. Murata, *J. Fluorine Chem.*, 1987, **36**, 339

178.   A. B. Baylis and M. E. D. Hillman, German Patent 2,155,113, 1972. *Chem. Abstr.*, 1972, **77**, 34174q

179.   For a review of this reaction, see S. E. Drewes and G. H. P. Roos, *Tetrahedron*, 1988, **44**, 4653

180.   J. S. Hill and N. S. Isaacs, *Tetrahedron Lett.*, 1986, **27**, 5007

181.   L. Wessjohann, G. McGaffin and A. de Meijere, *Synthesis*, 1989, 359

182.   T. Uyehara, N. Asao and Y. Yamamoto, *J. Chem. Soc., Chem. Commun.*, 1987, 1410

183.   T. Uyehara, N. Asao and Y. Yamamoto, *Tetrahedron*, 1988, **44**, 4173

184.   For previous reports of successful additions of LDA, see M. W. Rathke and D. Sullivan, *Tetrahedron Lett.*, 1972, 4249; J. L. Herrmann, G. R. Kieczykowski and

R. H. Schlessinger, *Tetrahedron Lett.*, 1973, 2433; R. D. Little and J. R. Dawson, *Tetrahedron Lett.*, 1980, 2609 and T. A. Hase and P. Kukkola, *Synth. Commun.*, 1980, **10**, 451

185.   T. Uyehara, N. Asao and Y. Yamamoto, *J. Chem. Soc., Chem. Commun.*, 1989, 753

186.   T. Uyehara, N. Shida and Y. Yamamoto, *J. Chem. Soc., Chem. Commun.*, 1989, 113

187.   See reference 99

188.   J. M. Hawkins and G. C. Fu, *J. Org. Chem.*, 1986, **51**, 2820

189.   K. Rudolf, J. M. Hawkins, R. J. Loncharich and K. N. Houk, *J. Org. Chem.*, 1988, **53**, 3879.

190.   R. J. Loncharich, T. R. Swartz and K. N. Houk, *J. Amer. Chem. Soc.*, 1987, **109**, 14

191.   H. Matsunaga, T. Sakamaki, H. Nagaoka and Y. Yamada, *Tetrahedron Lett.*, 1983, **24**, 3009

192.   G. Fronza, C. Fuganti, P. Grasselli, L. Majori, G. Pedrocchi-Fantoni and F. Spreafico, *J. Org. Chem.*, 1982, **47**, 3289

193.   See also, J. T. Welch and B-M. Svahn, *J. Carbohydr. Chem.*, 1985, **4**, 421

194.   P. Perlmutter and M. Tabone, *Tetrahedron Lett.*, 1988, **29**, 949

195.   J. d'Angelo and J. Maddaluno, *J. Amer. Chem. Soc.*, 1986, **108**, 8112

196.   S. G. Davies, I. M. Dordor-Hedgecock, R. J. C. Easton, K. H. Sutton and J. C. Walker, *Bull. Soc. Chem. France*, 1987, 608

197.   S. G. Davies, R. J. C. Easton, J. C. Walker and P. Warner, *Tetrahedron*, 1986, **42**, 175

198.   For a useful review and compilation of conjugate additions where a new C-P bond is formed in the product, see R. Engel, "Synthesis of Carbon-Phosphorus Bonds", CRC Press, Boca Raton, 1988

199.   H. B. F. Dixon, and M. J. Sparkes, *Biochem. J.*, 1974, **141**, 715; J. R. Chambers and A. F. Isbell, *J. Org. Chem.*, 1964, **29**, 832; R. G. Harvey, *Tetrahedron*, 1966, **22**, 2561

200.   L. A. Cates and V-s. Li, *Phosphorus and Sulfur*, 1984, **21**, 187

201.   Z. S. Novikova and I. F. Lutsenko, *J. Gen. Chem. USSR*, 1970, **40**, 2110

202.   B. E. Ivanov, L. A. Kudreyavtseva, S. V. Samurina, A. B. Ageeva and T. I. Karpova, *J. Gen. Chem. USSR*, 1979, **49**, 1552

203.   A. F. Isbell, J. P. Berry and L. W. Tansey, *J. Org. Chem.*, 1972, **37**, 4399; A. N. Pudovik and B. A. Arbuzov, *Dokl. Akad. Navk. S.S.S.R*, 1950, **73**, 327; *Chem. Abstr.*, 1951, **45**, 2853b; V. A. Shokol, V. F. Gamaleya and L. I. Molyavko, *J. Gen. Chem. USSR*, 1974, **44**, 87; J. R. Chambers and A. F. Isbell, *J. Org. Chem.*, 1964, **29**, 832

204.   E. Ohler, E. Haslinger and E. Zbiral, *Chem. Ber.*, 1982, **115**, 1028

205.   M. Soroka and P. Mastalerz, *Rocz. Chem.*, 1976, **50**, 661

206.   K. Green, *Tetrahedron Lett.*, 1989, **30**, 4807

207.   C. Larpent and H. Patin, *Tetrahedron*, 1988, **44**, 6107

208.   R. Bodalski and K. Pietrusiewicz, *Tetrahedron Lett.*, 1972, 4209

209.   J. D. McClure, *J. Org. Chem.*, 1970, **35**, 3045

210.   See also M. M. Baizer and J. D. Anderson, *J. Org. Chem.*, 1965, **30**, 1357 and C. C. Price, *J. Amer. Chem. Soc.*, 1962, **84**, 489

211.   K. Morita, Z. Suzuki and H. Hirose, *Bull. Chem. Soc. Jpn.*, 1968, **41**, 2815

212.   T. Imagawa, K. Uemura, Z. Nagai and M. Kawasani, *Synthetic Commun.*, 1984, **14**, 1267

213.   T. E. Snider, D. L. Morris, K. C. Srivastava and K. D. Berlin, *Org. Synth.*, 1988, **Coll. Vol. 6**, 932; this paper contains a useful series of references on the synthetic applications of phosphorinanones

214.   B. S. Furniss, A. J. Hannaford, P. W. G. Smith and A. R. Tatchell, "Vogel's Textbook of Practical Organic Chemistry", Longman, Burnt Mill, 1989, 719

215.   I. Kuwajima, T. Murobushi and E. Nakamura, *Synthesis*, 1978, 602

216.   V. N. Mikhailova, N. Borisova and D. Stankevitch, *Zh. Organ. Khim.*, 1966, **2**, 1437

217.   S. De Lombaert, I. Nemery, B. Roekens, J. C. Carretero, T. Kimmel and L. Ghosez, *Tetrahedron Lett.*, 1986, **42**, 5099

218.   J. Mulzer, M. Kappert, G. Huttner and I. Jibril, *Angew. Chem. Int. Ed. Engl.*, 1984, **23**, 704

219   S. Thaisrivongs and D. Seebach, *J. Amer. Chem. Soc.*, 1983, **105**, 7407

220.   These were originally formulated as LO cuprates but are most probably HO cuprates. See following reference

221.   B. H. Lipshutz, *Synthesis*, 1987, 325

222.   I. Fleming and D. Waterson, *J. Chem. Soc., Perkin Trans. 1*, 1984, 1809

223.   W. Tuckmantel, K. Oshima and H. Nozaki, *Chem. Ber.*, 1986, **119**, 1581

224.   W. Bernhard, I. Fleming and D. Waterson, *J. Chem. Soc., Chem. Commun.*, 1984, 28

225.   I. Fleming, J. H. M. Hill, D. Parker and D. Waterson, *J. Chem. Soc., Chem. Commun.*, 1985, 318

226.   M. N. Paddon-Row, N. G. Rondan and K. N. Houk, *J. Amer. Chem. Soc.*, 1982, **104**, 7162

227.   I. Fleming and J. D. Kilburn, *J. Chem. Soc.*, 1986, 305

228.   H-F. Chow and I. Fleming, *Tetrahedron Lett.*, 1985, **26**, 397

229.   I. Fleming and N. D. Kindon, *J. Chem. Soc., Chem. Commun.*, 1987, 1177

230.   See also, M. Bergdahl, M. Nilsson and T. Olsson, *J. Organomet. Chem.*, 1990, **391**, C19

231.   I. Fleming and C. J. Urch, *J. Organomet. Chem.*, 1985, **285** (1-3), 173. See also G. J. M. Van der Kerk, J. G. Noltes and J. G. A. Luitjen, *J. Appl. Chem.*, 1957, **7**, 356

232.   I. Fleming and M. Rowley, *J. Chem. Soc., Perkin Trans. I*, 1987, 2259

233.   M. F. Semmelhack, R. D. Stauffer and A. Yamashita, *J. Org. Chem.*, 1977, **42**, 3180

234.   E. Keinan and N. Greenspoon, *J. Amer. Chem. Soc.*, 1986, **108**, 7314

235.   W. S. Mahoney, D. M. Brestensky and J. M. Stryker, *J. Amer. Chem. Soc.*, 1988, **110**, 291

236.   For an improved preparation of this reagent, see D. M. Brestensky, D. E. Huseland, C. McGettigan and J. M. Stryker, *Tetrahedron Lett.*, 1988, **29**, 3749

237.   R. G. Salomon, N. D. Sachinvala, S. R. Raychaudhuri and D. B. Miller, *J. Amer. Chem. Soc.*, 1984, **106**, 2211

238.   B. M. Trost, C. D. Shuey and F. DiNinno, Jr., *J. Amer. Chem. Soc.*, 1979, **101**, 1284

239.   T. Imanishi, N. Yagi and H. Miyoji, *Chem. Pharm. Bull.*, 1985, **33**, 4202

240.   M. Ihara, Y. Ishida, M. Abe, M. Toyota, K. Fukumoto and T. Kametani, *J. Chem. Soc., Perkin Trans. 1*, 1988, 1155

241.   M. Ihara, M. Toyota, K. Fukumoto and T. Kametani, *J. Chem. Soc., Perkin Trans. 1*, 1986, 2151

242.   This method has also been applied to an asymmetric total synthesis of atisine, see M. Ihara, M. Suzuki, K. Fukumoto and C. Kabuto, *J. Amer. Chem. Soc.*, 1990, **112**, 1164

243.   L. Lombardo, L. N. Mander and J. V. Turner, *J. Amer. Chem. Soc.*, 1980, **102**, 6626

244.   G. Stork, J. D. Winkler and N. A. Saccomano, *Tetrahedron Lett.*, 1983, **24**, 465

245.   G. Stork and N. A. Saccomano, *Nouv. J. Chim.*, 1986, **10**, 677

246.   G. Stork and N. A. Saccomano, *Tetrahedron Lett.*, 1987, **28**, 2087

247.   The few percent of the minor diastereomeric product were removed by conversion of the products to their dioxolan derivatives followed by chromatographic purification

248.   Y. Hirai, A. Hagiwara and T. Yamazaki, *Heterocycles*, 1986, **24**, 571

249.   Y. Hirai, A. Hagiwara, T. Terada and T. Yamazaki, *Chem. Lett.*, 1987, 2417

250.   S-U. Park, T. R. Varick and M. Newcomb, *Tetrahedron Lett.*, 1990, **31**, 2975

251.   S-U. Park, S-K. Chung and M. Newcomb, *J. Amer. Chem. Soc.*, 1986, **108**, 240

252.   M. Ihara, K. Yasui, N. Taniguchi and K. Fukumoto, *Tetrahedron Lett.*, 1988, **29**, 4963

253.   S. Hatakeyama, N. Ochi, H. Numata and S. Takano, *J. Chem. Soc., Chem. Commun.* 1988, 1202

254.   W. Xu, Y. T. Jeon, E. Hasegawa, U. C. Yoon and P. S. Mariano, *J. Amer. Chem. Soc.*, 1989, **111**, 406

255.   B-W. A. Yeung, J. L. M. Contelles and B. Fraser-Reid, *J. Chem. Soc., Chem. Commun.* 1989, 1160

256.   E. J. Enholm and A. Trivellas, *J. Amer. Chem. Soc.*, 1989, **111**, 6463

257.   D. J. Hart, H-H. Huang, R. Krishnamurthy and T. Schwartz, *J. Amer. Chem. Soc.*, 1989, **111**, 7507

258.   N, A. Porter and V. H-T. Chang,, *J. Amer. Chem. Soc.*, 1987, **109**, 4976

259.   K. V. Bhaskar and G. S. R. S. Rao, *Tetrahedron Lett.*, 1989, **30**, 225

260.   D. J. Dunham and R. G. Lawton, *J. Amer. Chem. Soc.*, 1971, **93**, 2074

261.   J. P. Brennan and J. E. Saxton, *Tetrahedron*, 1986, **42**, 6719

262.   M. Hirama, T. Shigemoto, Y. Yamazaki and S. Ito, *J. Amer. Chem. Soc.*, 1985, **107**, 1797

263.   See also K. Shishido, Y. Sukegawa, K. Fukumoto and T. Kametani, *J. Chem. Soc., Perkin Trans. I*, 1987, 993 and *idem*, *Heterocycles*, 1986, **24**, 641

264.   See also M. Hirama, T. Iwakuma and S. Ito, *J. Chem. Soc., Chem. Commun.*, 1987, 1523

265.   M. Hirama, T. Shigemoto and S. Ito, *J. Org. Chem.*, 1987, **52**, 3342

266.   M. Hirama, I. Nishizaki, T. Shigemoto and S. Ito, *J. Chem. Soc., Chem. Commun.*, 1986, 393

267.   M. Ihara, Y. Takino, K. Fukumoto and T. Kametani, *Tetrahedron Lett.*, 1988, **29**, 4135

268.   S. S. Koo, L. L. Klein, K-P. Pfaff and Y. Kishi, *Tetrahedron Lett.*, 1982, **23**, 4415

269.   S. Hanessian, "Total Synthesis of Natural Products: The Chiron Approach", Pergamon Press, Oxford, 1983

270.   J. Rokach, C-K. Lau, R. Zamboni and Y. Guindon, *Tetrahedron Lett.*, 1981, **22**, 2763

271.   Y. Guindon and D. Delorme, *Can. J. Chem.*, 1987, **65**, 1438

272.   J. R. Stille and R. H. Grubbs, *J. Amer. Chem. Soc.*, 1986, **108**, 855

273.   D. H. R. Barton and S. Z. Zard, *Pure Appl. Chem.*, 1986, **58**, 675

274.   B. Giese, M. Zehnder, M. Roth and H-G. Zeitz, *J. Amer. Chem. Soc.*, 1990, **112**, 6741

275.   N. A. Porter, E. Swann, J. Nally and A. T. McPhail, *J. Amer. Chem. Soc.*, 1990, **112**, 6740. See also D. P. Curran, W. Shen, J. Zhang and T. A. Heffner, *ibid.*, 1990, **112**, 6738

Chapter Five α,β–Unsaturated lactones and lactams

5 Introduction

Compared with some of the other classes of conjugate acceptors, α,β-unsaturated lactones and lactams have received little attention. However, there have been several important studies on stereoselective conjugate additions carried out on both γ-lactones and lactams as well as larger-sized lactones. Naturally-occurring γ-lactones have been the targets of several total syntheses employing conjugate additions, some with *in situ* enolate-trapping. Studies on conjugate additions to 5-substituted γ-lactones have been popular and quite a variety of nucleophiles add with excellent *anti*-stereoselectivity.

5.1 Stabilized carbanions

5.1.1 *Carbanions stabilized by π-conjugation with one heteroatom*

Not many studies have been reported on these additions. However, Yamada's group has demonstrated that additions of enolates to 5-substituted γ-lactones, e.g. (**1**) (which are available in homochiral form[1,2,3]), can be a high yielding, highly stereoselective process as in the following example.[4]

1

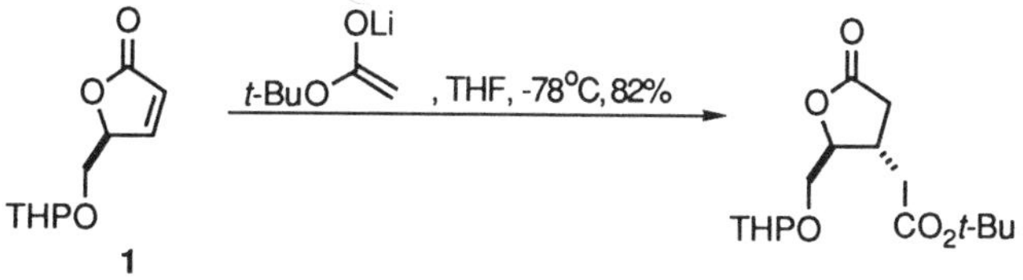

Holton's group has explored the use of conjugate additions to γ-lactones in their synthesis of aphidicolin, a diterpene with anti-viral activity isolated from *Cephalosporium aphidicolia* Petch.[5] In their analysis they settled upon the following tricyclic lactone (**2**) as a precursor to two of aphidicolin's rings.

2

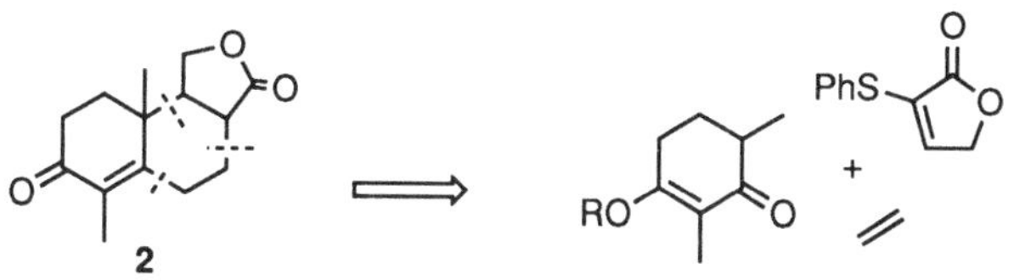

Accordingly, they found that addition of the enolate of the partially-blocked cyclohexenone, 2,6-dimethyl-3-methoxy-2-cyclohexenone (3) to 3-(phenylthio)-2(5H)-furanone, gives only one diastereomer, with the required relative stereochemistry.

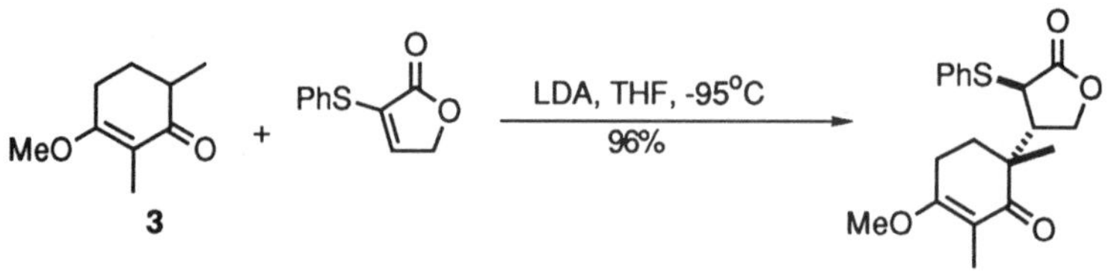

This was exploited in a total synthesis of aphidicolin, whose retrosynthesis is outlined below.[6] By employing a chiral sulfoxide auxiliary in place of the phenylthio substituent the synthesis proved to be enantioselective as well as diastereoselective.

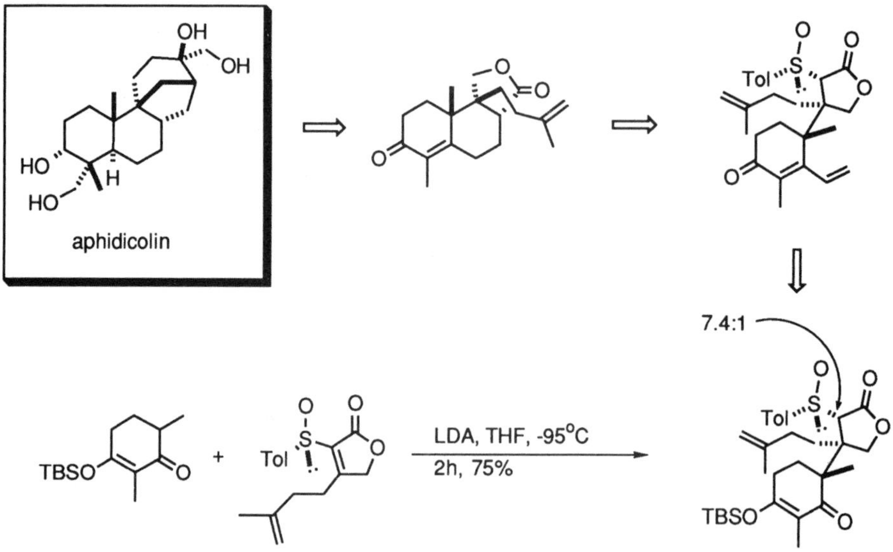

aphidicolin

7.4:1

As with similar additions to other chiral sulfoxide-containing acceptors, a chelated transition state adequately accounts for the stereoselectivities associated with this reaction.

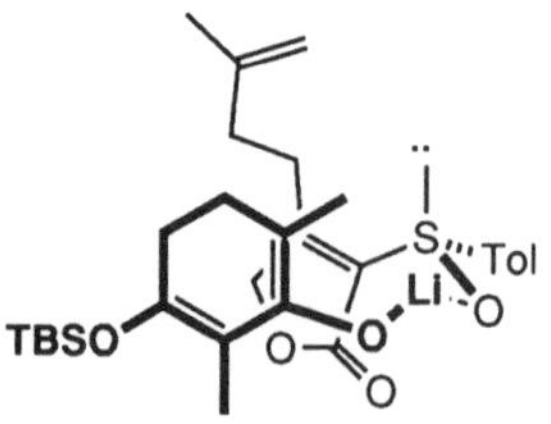

Masked glyoxylate anions add well to many conjugate acceptors, including 5-methoxy-2(5H)-furanone.[7]

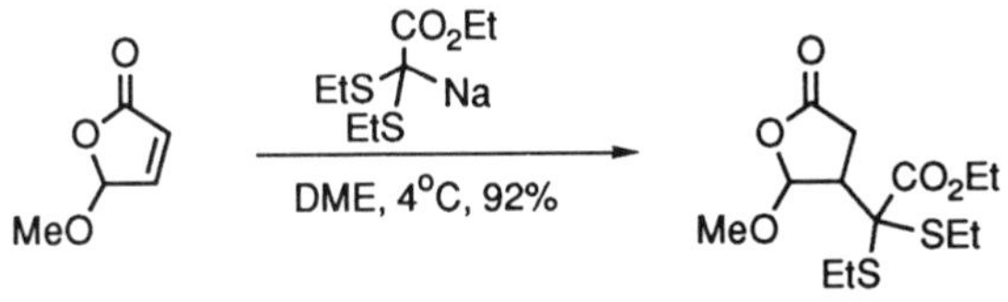

Additions to unsaturated δ-lactams also work well as in the case below.[8] Sometimes, where ester enolates fail to react successfully with unsaturated lactams, reaction with the corresponding amide enolate gives an excellent yield of adduct.[9]

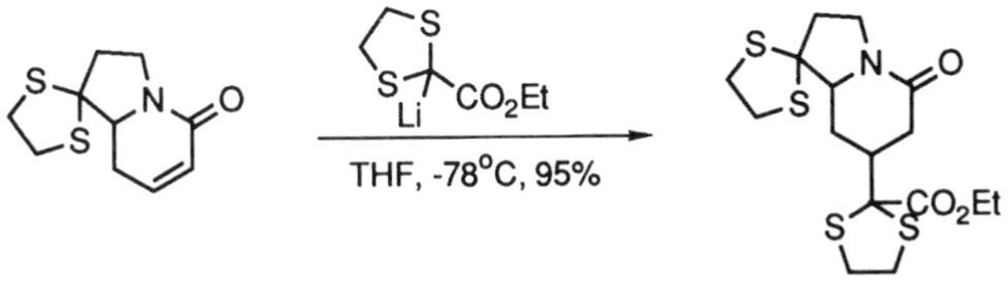

The addition of a benzylic anion (in conjugation with an ester carbonyl, e.g. (4)) to a 5,6-disubstituted δ-lactone followed by an intramolecular Claisen condensation provided an intermediate in a total synthesis of olivin trimethyl ether.[10]

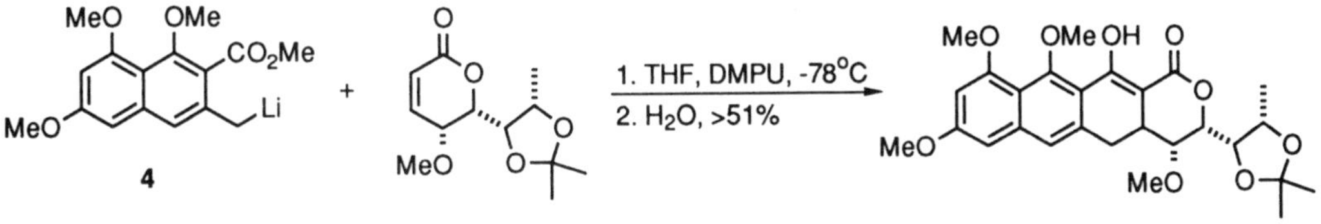

5.1.2   Carbanions stabilized by one or more α-heteroatoms

The conjugate addition of the carboxyl anion equivalent, tris(methylthio)methyllithium, to 5-*tert*-butyl(dimethyl)silyloxymethyl-2(5H)-furanone provides the basis for a synthesis of several chiral (polyhydroxy)-fragments of natural products.[11] The addition is completely stereoselective and the intermediate may be trapped, *in situ*, by alkylating agents, such as iodomethane. In this way, the configuration at C5 may be used to control the relative stereochemistry at C3 and C4.

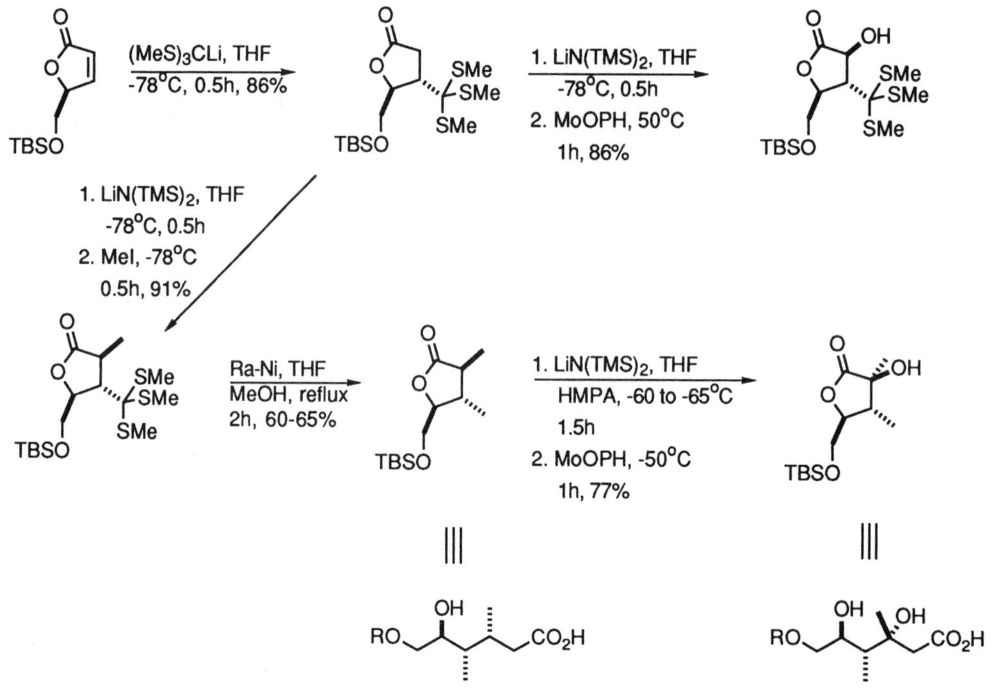

Zeigler and Schwartz used a conjugate addition of a 2-aryl-2-lithio-1,3-dithiane (**5**) to an unsaturated lactone in their total synthesis of (±)-podorhizol (see also section 5.3).[12]

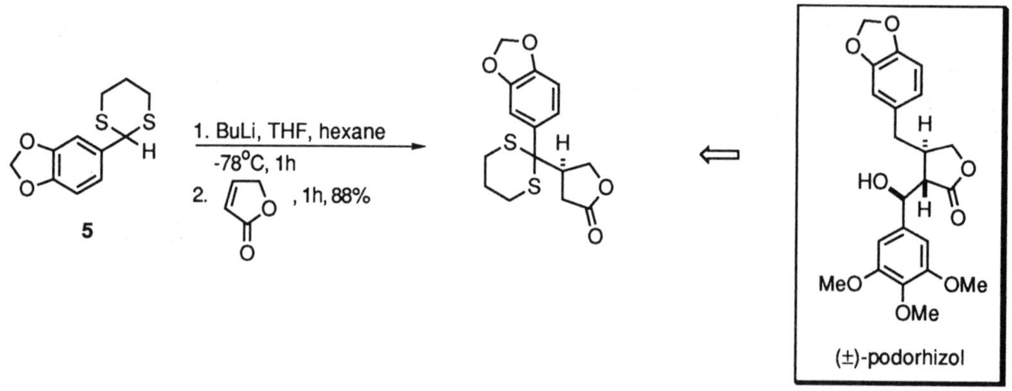

(±)-podorhizol

A similar approach was used in a synthesis of the antileukemic lignan, isostegane. The presence of a substituent at C5 directs the incoming nucleophile to the opposite face. Trapping of the enolate with piperonyl bromide gave the isostegane precursor (**6**) in excellent yield.[13]

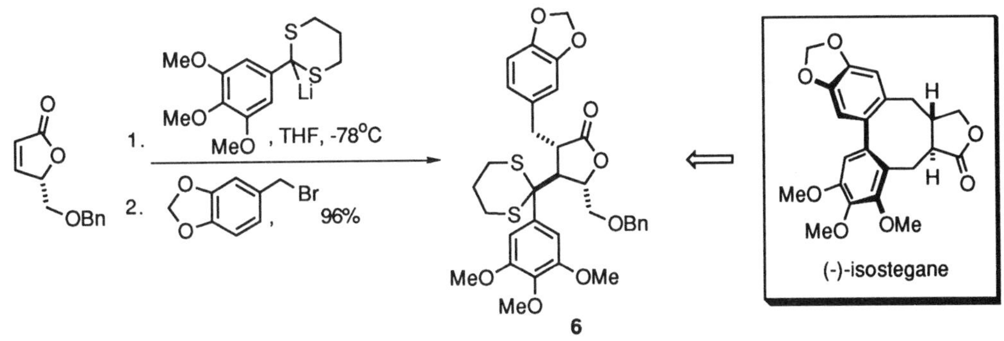

The stereoselective introduction of a methyl group at C4 of 5-menthyloxy-2(5H)-furanone (7) has also been reported. The first step involves addition of the lithium salt of tris-(methylthio)methane. Raney-nickel reduction then completes the process.[14] In this report the authors did not mention the obvious alternative of adding a cuprate. However, in light of the problems of dimer and trimer formation in these reactions with LO cuprates (see next section), this two-step procedure appears to be quite valuable.[15]

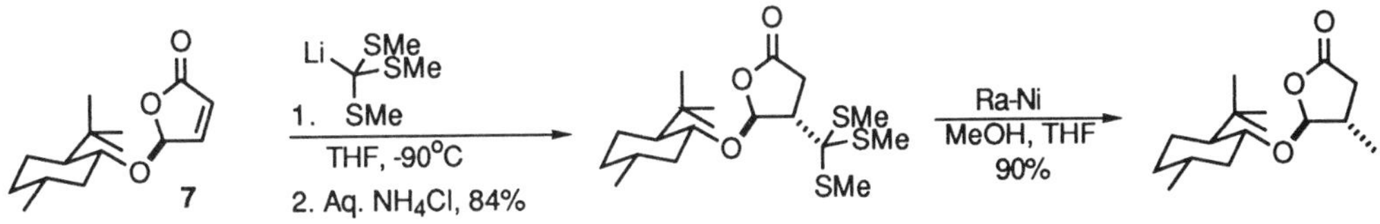

5.2   Organocopper reagents

LO organocuprates sometimes react rather poorly with unsaturated lactones as in the following example. The stereoselectivity was good, however the yield, whilst not optimized, is apparently poor. Certainly, under similar conditions, additions of HO silyl cuprates are much better yielding (see section 5.5.3).[16]

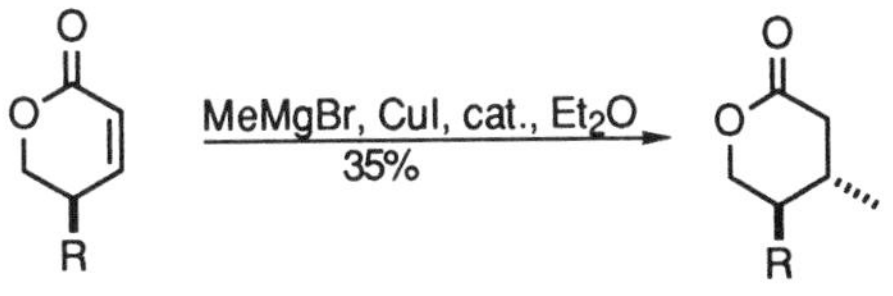

This variable behaviour of cuprate additions to unsaturated lactones has led to the use of a two-step alternative (see section 5.1.3).This behaviour is even more puzzling when the addition of the far-less conventional (and therefore less-studied) "acylcopper" reagents from Seyferth's group add so efficiently, as in the following example.[17]

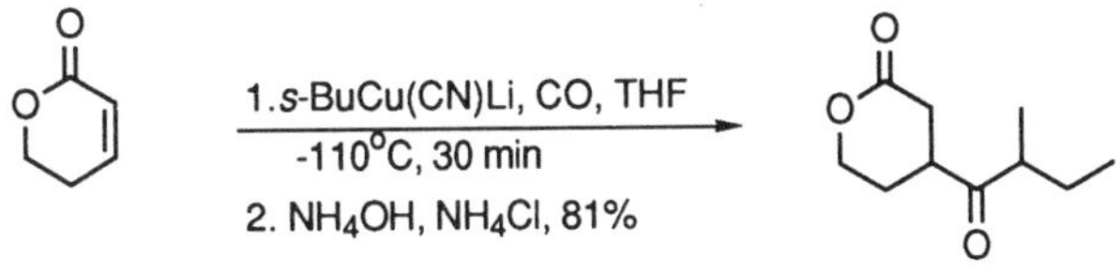

Still and Galynker have studied the conjugate addition of dimethylcopperlithium to a variety of monosubstituted 9- to 12-membered unsaturated lactones.[18] As in the case of medium size cycloalkenones (see section 3.1.2), these additions proceed with high levels of stereoselectivity.

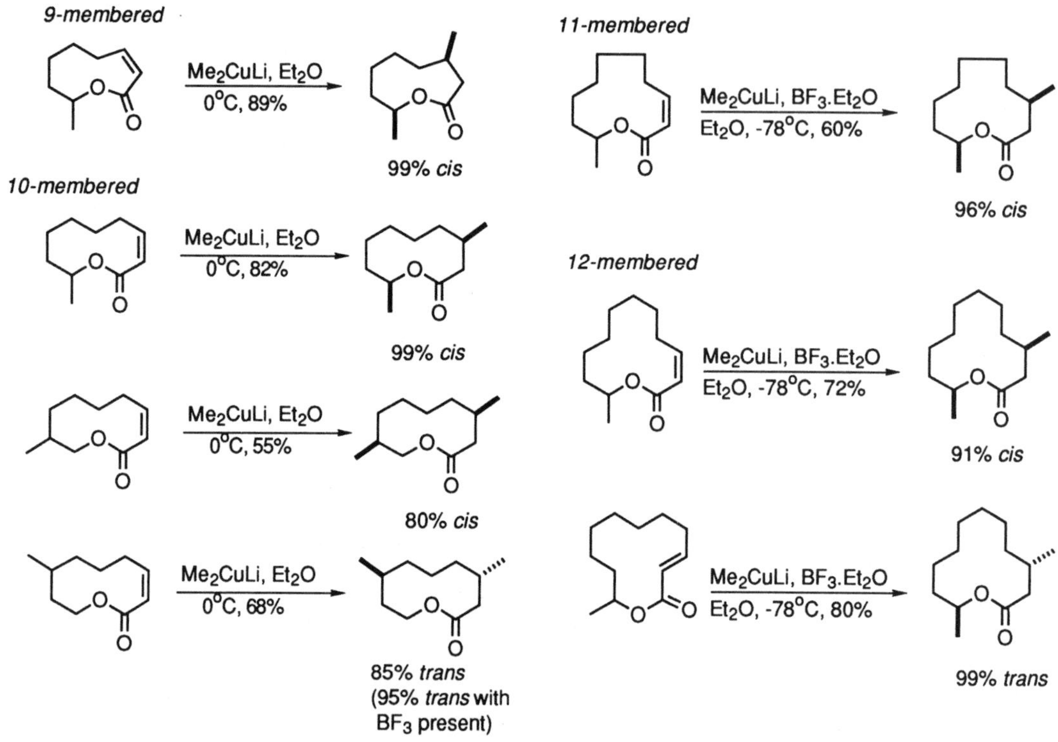

9-membered

Me₂CuLi, Et₂O
0°C, 89%

99% cis

11-membered

Me₂CuLi, BF₃.Et₂O
Et₂O, -78°C, 60%

96% cis

10-membered

Me₂CuLi, Et₂O
0°C, 82%

99% cis

12-membered

Me₂CuLi, Et₂O
0°C, 55%

80% cis

Me₂CuLi, BF₃.Et₂O
Et₂O, -78°C, 72%

91% cis

Me₂CuLi, Et₂O
0°C, 68%

85% trans
(95% trans with
BF₃ present)

Me₂CuLi, BF₃.Et₂O
Et₂O, -78°C, 80%

99% trans

Addition of dimethylcopperlithium to the bicyclic lactones shown below was used in a total synthesis of (±)-dihydronepetalactone and (±)-isodihydronepetalactone.[19]

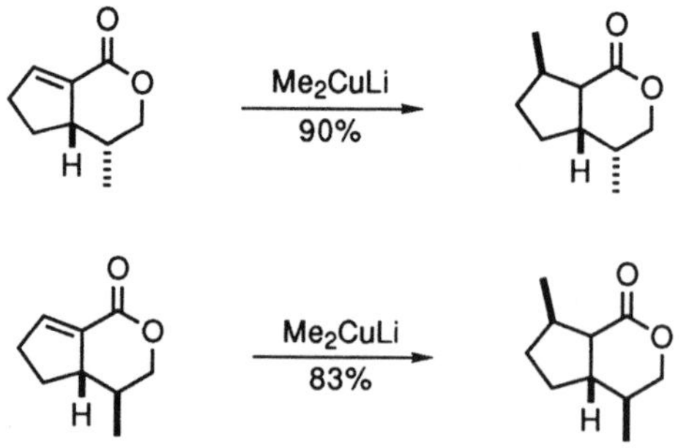

Me₂CuLi
90%

Me₂CuLi
83%

As mentioned at the beginning of this section, organocuprate additions to unsaturated lactones sometimes present problems. For example, rather than simple addition, they can initiate the cyclotrimerization of γ-methylenebutenolides, by a series of 1,6-additions.[20]

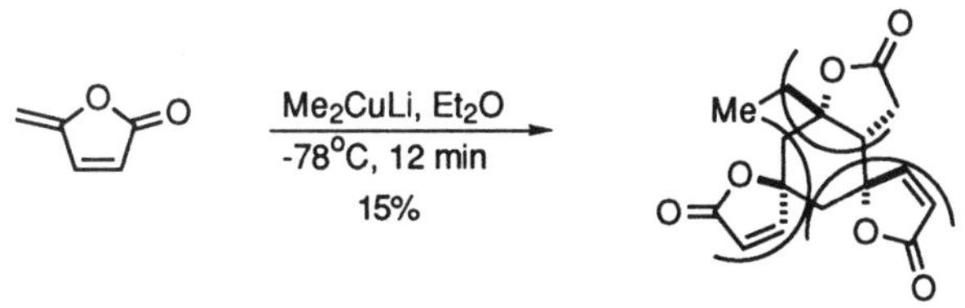

Another problem was found in the addition of LO cuprates to the D-ribonolactone-derived 4-(benzyloxymethyl)butenolide where the adduct is always accompanied by some dimeric product.[21] The product is an intermediate in an economical synthesis of some naturally-occurring lactones, including (+)-*trans*-cognac lactone. The use of lower temperatures than ~15°C or long reaction times led to the production of only dimer and no normal addition product.

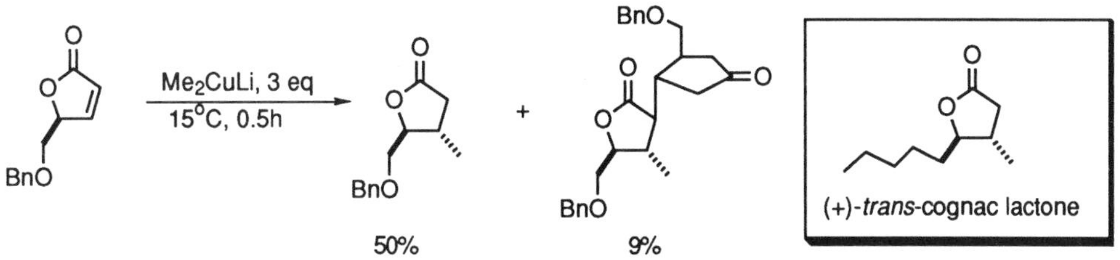

50%   9%   (+)-*trans*-cognac lactone

Conjugate addition/elimination of 3-benzoylamino-4-bromo-5(2H)-furanone (8) has been studied.[22] It was found important to use three equivalents of the organolithium to prepare the cuprate. Presumably the third equivalent was able to deprotonate the benzamide. The result of this may have been some chelation between the LO cuprate and the amide. This seems probable as the excellent yield in this case contrasted sharply with the poorer yields obtained from additions of dimethylcopperlithium with no excess methyllithium present.

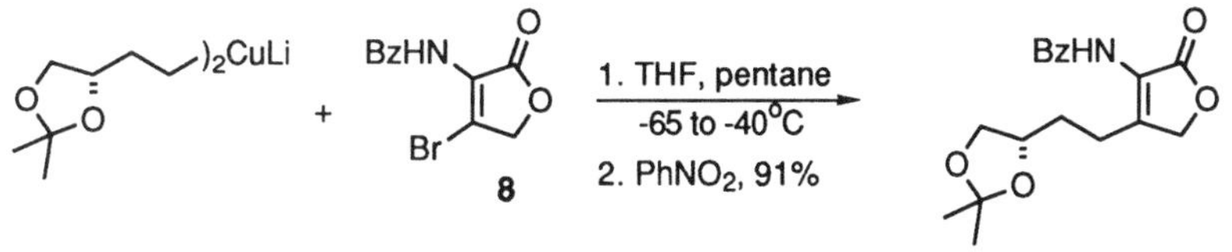

2,6-Disubstituted dioxinones undergo highly stereoselective additions with cuprates (both stoichiometric and catalytic in copper).[23,24] As the dioxinones are fairly flat, the observed stereoselection was attributed not to steric screening but rather to a kinetic

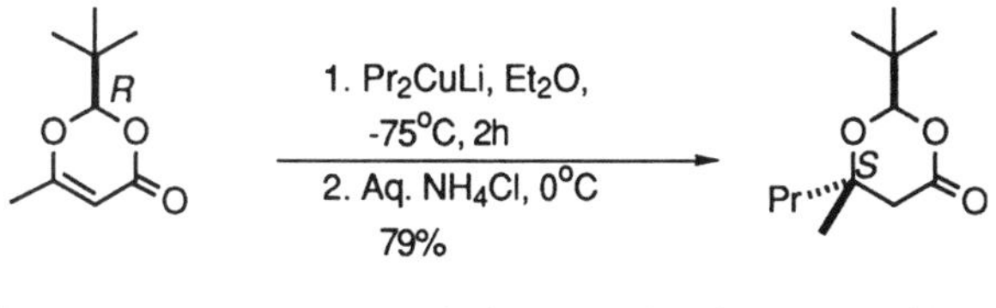

stereoelectronic effect. From a survey of the Cambridge crystallographic data base, ~40 related molecules, each possessing a sofa-type conformation, were examined and it was found that the trigonal centre of each molecule was pyramidalized in the same direction. In all cases pyramidalization occurs in the direction from which addition occurs in the dioxinone series.

Comins's group has developed a stereoselective route to piperidine and quinolizidine alkaloids based on LO cuprate additions to 2,3-dihydro-4-pyridones. For example, (±)-lasubine II has been prepared in the following manner.[25] The authors argued that the stereoselectivity derives from addition to the preferred ring conformation which has the aryl group oriented axially .

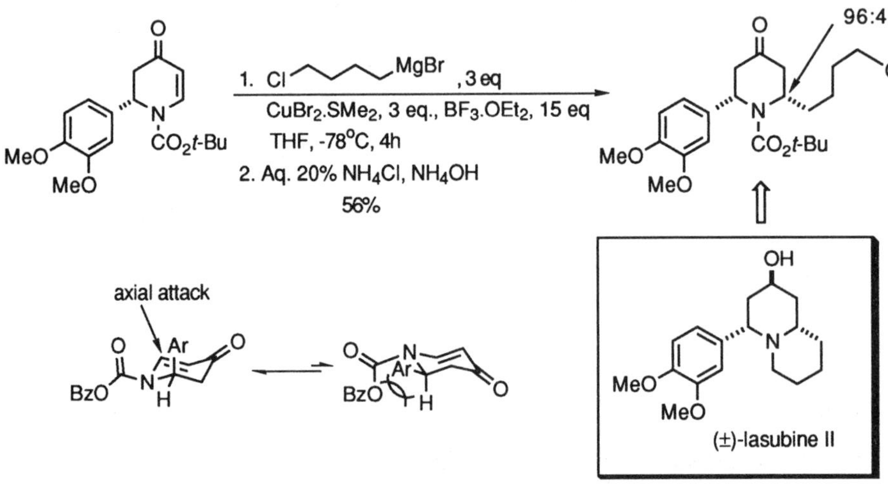

5.3   Other organometallic reagents

Organomagnesium reagents add to homochiral 2-(tolyl)sulfinyl-γ-lactones, e.g. (9) with very high stereoselectivity. This was used in a short total synthesis of (-)-podorhizon, another member of the podophyllotoxin anti-cancer family (see also section 5.1.3).[26]

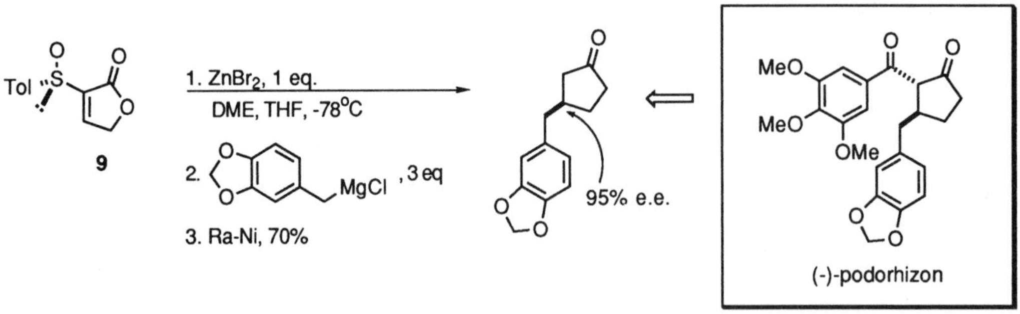

The addition of an organo*titanium* reagent, methyl titaniumtriisopropoxide, to 2-(S)-(+)-(4-methoxyphenylsulfinyl)-5,6-dihydro-2H-pyran-2-one was used to introduce a centre of asymmetry, critical to the total synthesis of fragrant sesquiterpene ovetiver oil, (-)-

β-vetivone.[27] The 4-methoxyphenyl derivative produces the adduct in dramatically improved enantioselectivity compared to the 4-methylphenyl derivative. If one assumes that some of the excess of the titanium reagent is being used to form a chelate, then it follows that the 4-methoxy derivative should form a more tightly bound chelate than the 4-methyl derivative. This in turn would then lead to a more effective screening of the si-face at C3 (see section 3.1.3).

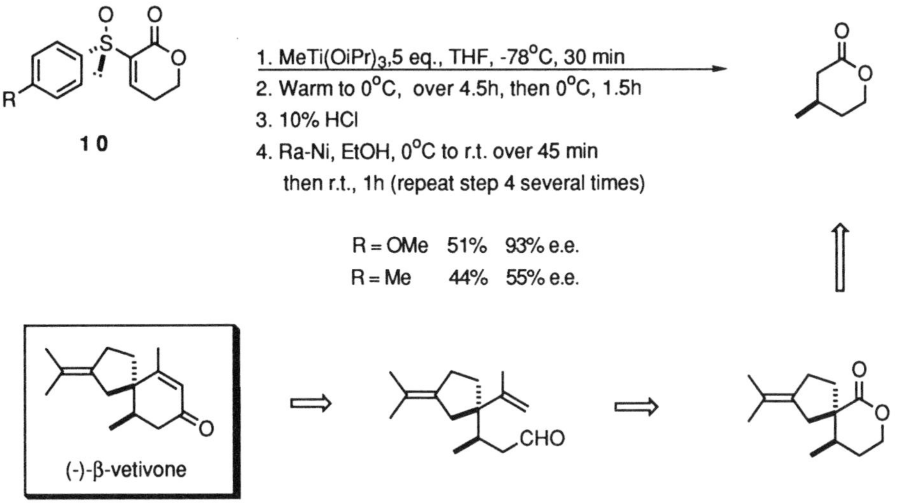

10

1. MeTi(OiPr)$_3$,5 eq., THF, -78°C, 30 min
2. Warm to 0°C, over 4.5h, then 0°C, 1.5h
3. 10% HCl
4. Ra-Ni, EtOH, 0°C to r.t. over 45 min
   then r.t., 1h (repeat step 4 several times)

R = OMe 51% 93% e.e.
R = Me   44% 55% e.e.

(-)-β-vetivone

CHO

Allylsilanes add stereoselectively to 5-substituted δ-lactones.[16]

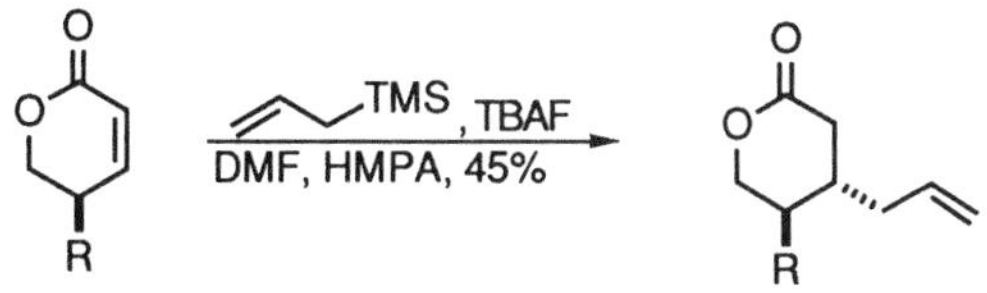

TMS , TBAF
DMF, HMPA, 45%

Thiolactams, like thioamides (see section 4.1.3), react well with hard nucleophiles, such as an alkyllithium.[28]

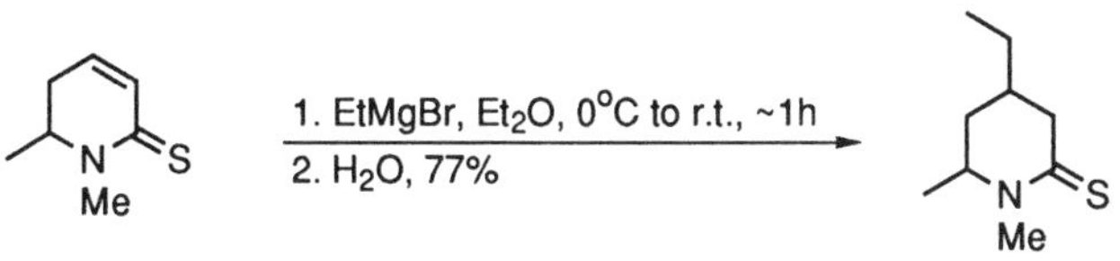

1. EtMgBr, Et$_2$O, 0°C to r.t., ~1h
2. H$_2$O, 77%

Me

Me

5.4   Free radicals

Although there are, so far, not many examples of intramolecular free radical cyclizations with unsaturated lactones, they constitute a valuable synthetic method. The following reaction, which was used to synthesize the aflatoxin B$_1$ skeleton, was more successful using palladium catalysis than the more conventional combination of AIBN and irradiation.[29] It may be viewed as a free radical 5-*exo*-trig process.

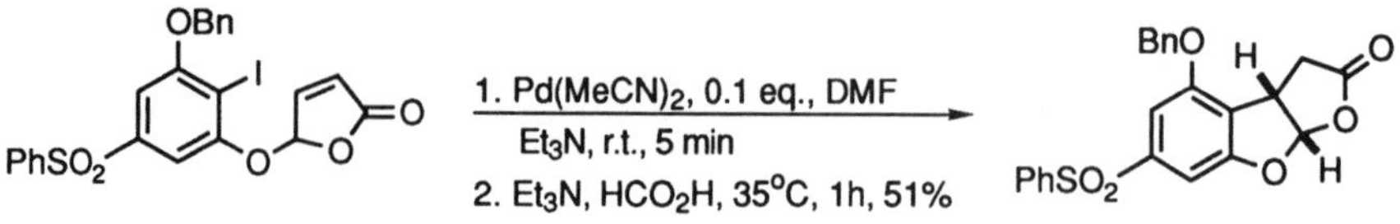

OBn
I
PhSO2
O
O
O
1. Pd(MeCN)2, 0.1 eq., DMF
Et3N, r.t., 5 min
2. Et3N, HCO2H, 35°C, 1h, 51%
BnO
H
O
O
PhSO2
O
H

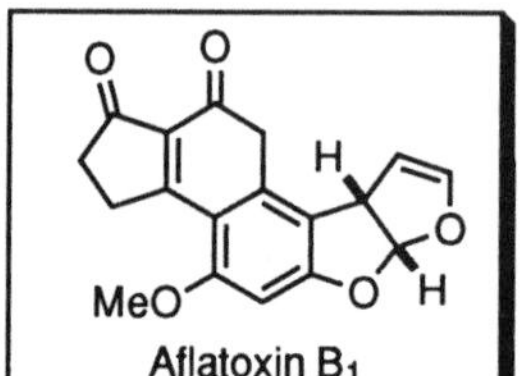

O
O
H
O
MeO
O
H

Aflatoxin B_1

5.5 Heteronucleophiles

5.5.1 *Nitrogen*

Feringa's group has developed syntheses of both racemic and chiral 2-amino-1,4-diols using the basic strategy outlined below.

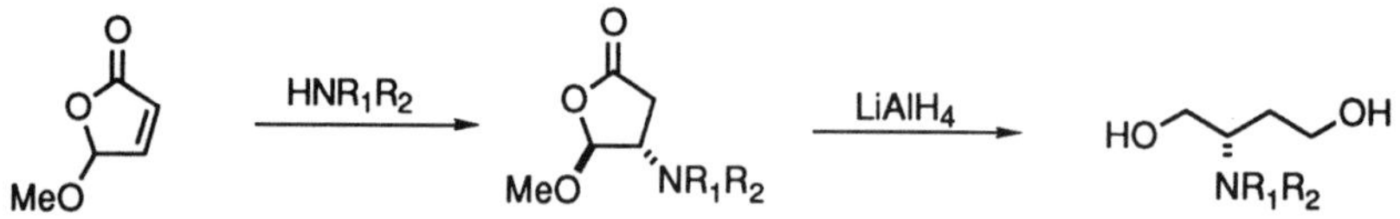

O
O
MeO
HNR1R2
O
O
MeO
NR1R2
LiAlH4
HO
OH
NR1R2

They found that addition of amines to 5-alkoxy-2(5*H*)-furanones may be carried out simply by mixing the reactants together at room temperature in either dichloromethane or dimethyl formamide.[30]

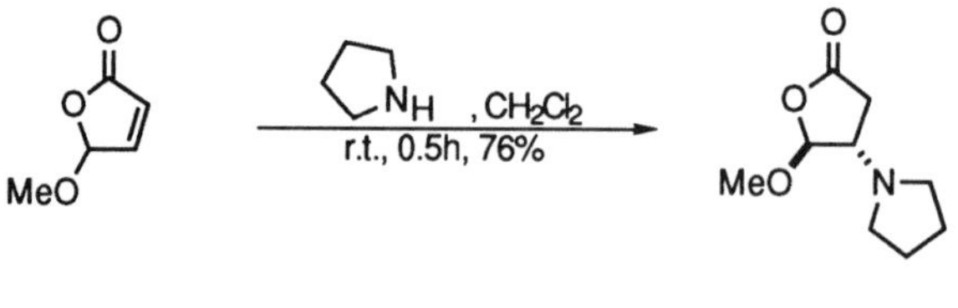

O
O
MeO
NH , CH2Cl2
r.t., 0.5h, 76%
O
O
MeO
N

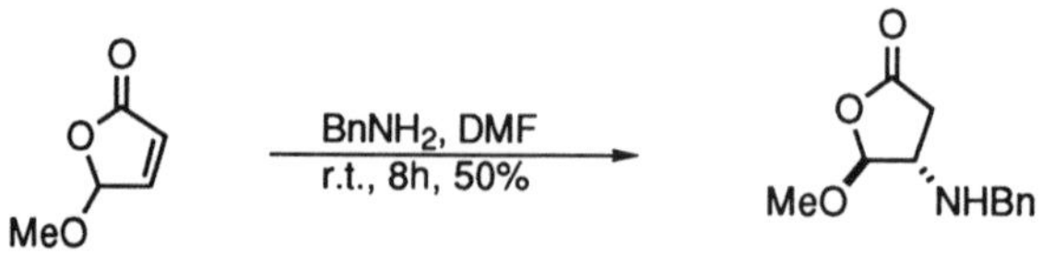

O
O
MeO
BnNH2, DMF
r.t., 8h, 50%
O
O
MeO
NHBn

By using the 5-menthyloxy derivative, they were able to prepare these aminodiols in homochiral form.

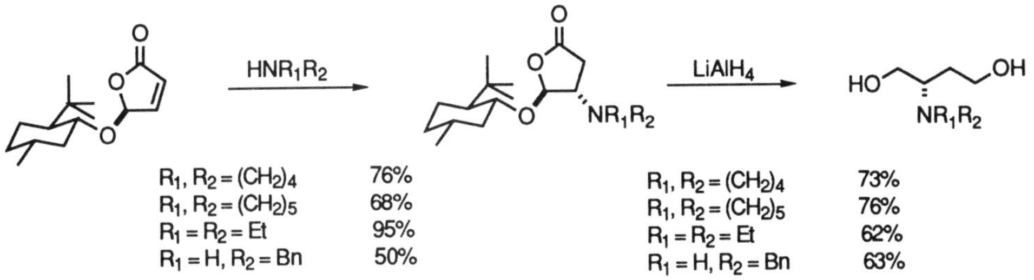
HNR1R2
LiAlH4
NR1R2
HO
OH
NR1R2
R1, R2 = (CH2)4 76%
R1, R2 = (CH2)5 68%
R1 = R2 = Et 95%
R1 = H, R2 = Bn 50%
R1, R2 = (CH2)4 73%
R1, R2 = (CH2)5 76%
R1 = R2 = Et 62%
R1 = H, R2 = Bn 63%

Conjugate additions to 6-substituted α,β-unsaturated δ-lactones have been little studied. As opposed to the γ-lactone counterparts, addition of benzylamine or N-benzylhydroxylamine to these δ-lactones yields products from attack at the carbonyl. However, switching to O-benzylhydroxylamine produces the conjugate adduct.[31]

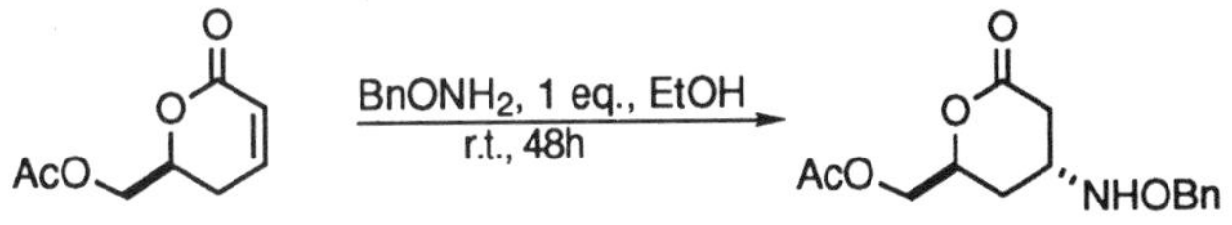
AcO
BnONH2, 1 eq., EtOH
r.t., 48h
AcO
NHOBn

5.5.2 *Oxygen and sulfur*

In their synthesis of the lactone portions of both compactin and mevinolin, Roth and Roark investigated the addition of alkoxide to 6-substituted δ-lactones.[32] Their retrosynthesis is shown below.

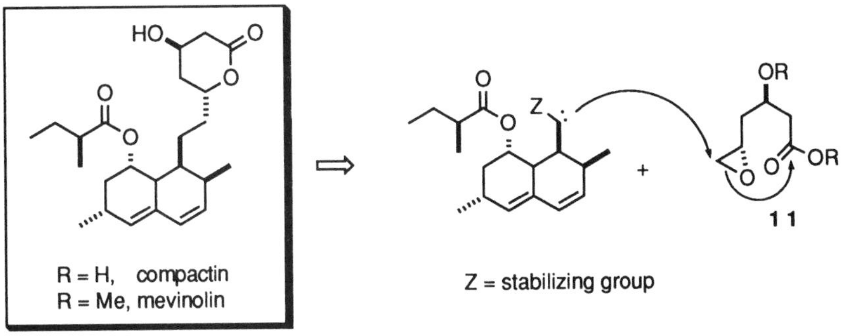
HO
R = H, compactin
R = Me, mevinolin
Z = stabilizing group
OR
OR
11

Thus, addition of each of several sodium alkoxides in their corresponding alcohol to 6-tosyloxymethyl-5,6-dihydro-2H-pyran-2-one produced the epoxy-ester (**11**) directly, in good yields and with excellent stereoselection. In light of the high degree of stereoselectivity, it seems likely that conjugate addition precedes epoxide formation.

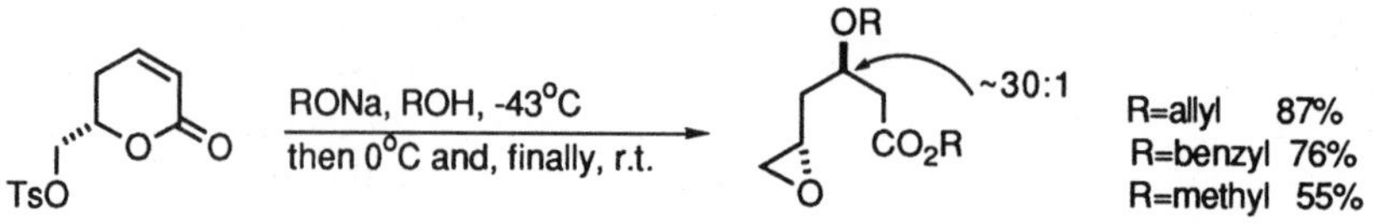

RONa, ROH, -43°C
then 0°C and, finally, r.t.

~30:1

R=allyl   87%
R=benzyl 76%
R=methyl 55%

PTC additions of thiols are extremely efficient as exemplified by entry 1 in the following Table. Two groups have shown that the use of more powerful sulfur nucleophiles leads to enolate intermediates which may be trapped by aldehydes, entries 2 and 3.

Table 5.1 Conjugate addition of sulfur nucleophiles to α,β-unsaturated lactones

1[33]

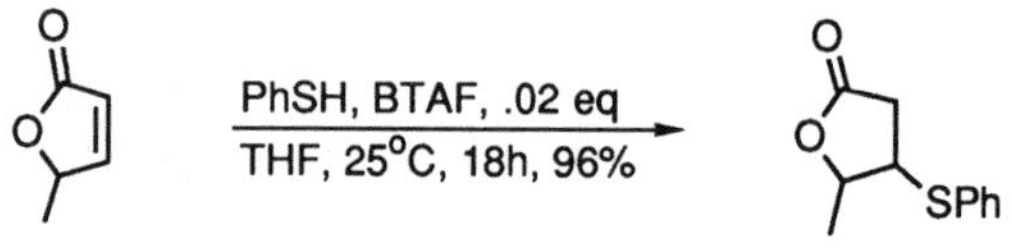

PhSH, BTAF, .02 eq
THF, 25°C, 18h, 96%

2[34]

1. Me₂AlSPh, CH₂Cl₂, hexane
   -78°C, 20 min

2. PhCHO, THF, -78°C, 20 min

3. H₂O, 77%

3[35]

1.PhSLi, THF, -50°C
2. PhCHO
3. H₂O, 92%

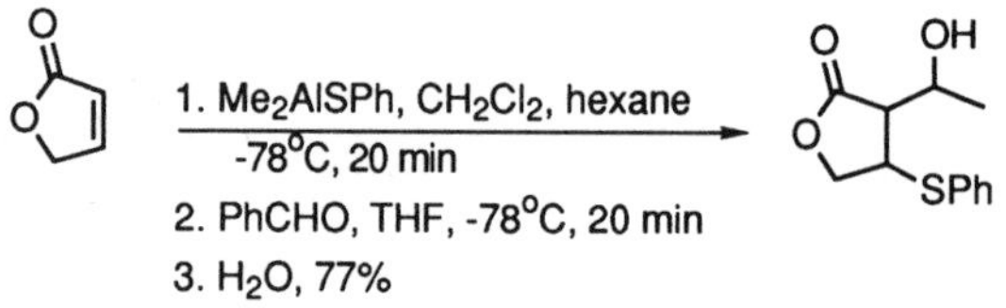

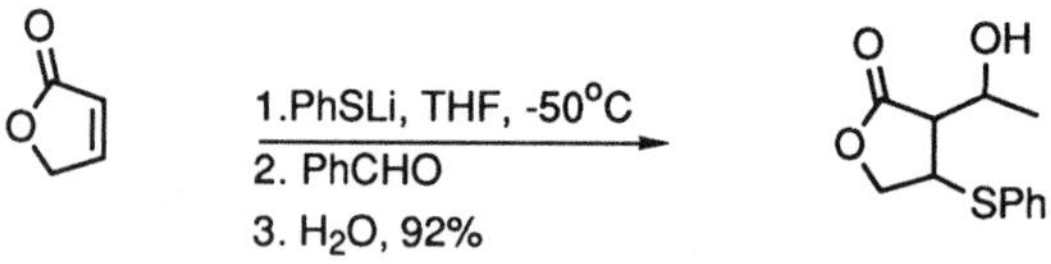

5.5.3   Silicon and tin

HO silylcuprates add in high yield and with excellent stereoselectivity to 4-substituted δ-lactones and lactams.[16] In the case of the lactone adducts, replacement of the silyl substituent with oxygen (which occurs with retention) leads to the lactone moiety of compactin.

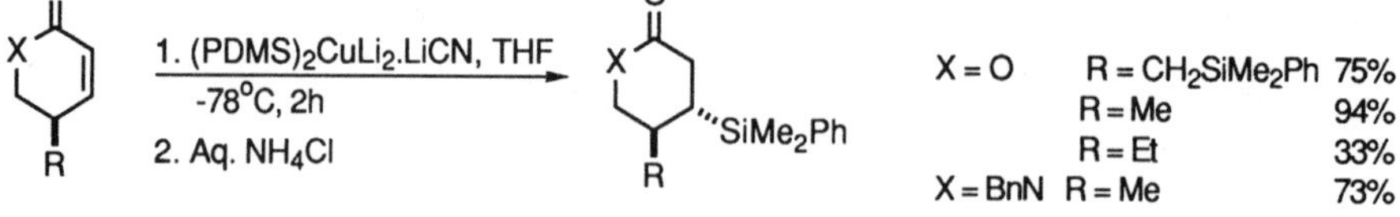

1. (PDMS)₂CuLi₂.LiCN, THF
   -78°C, 2h

2. Aq. NH₄Cl

X = O   R = CH₂SiMe₂Ph 75%
          R = Me   94%
          R = Et   33%
X = BnN   R = Me   73%

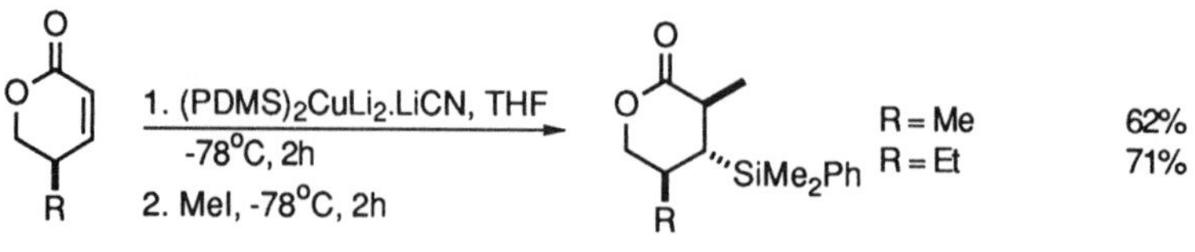

1. (PDMS)₂CuLi₂.LiCN, THF
   -78°C, 2h

2. MeI, -78°C, 2h

R = Me   62%
R = Et   71%

A relatively simple method for the generation and addition of silicon nucleophiles to exocyclic alkenyllactones has been discovered by Seebach's group.[36,37] Addition of the HO cuprate, $Bu_2Cu(CN)Li_2$, to a solution containing the conjugate acceptor and a chlorosilane resulted in clean conjugate addition of the *silyl* rather than the butyl group. These reactions are highly stereoselective,

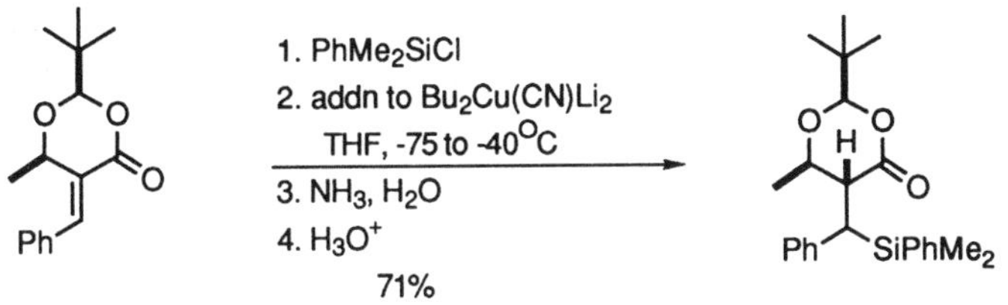

Ph
1. PhMe2SiCl
2. addn to Bu2Cu(CN)Li2
THF, -75 to -40°C
3. NH3, H2O
4. H3O+
71%
H
Ph
SiPhMe2

Macrolides have been prepared using a multicomponent annulation procedure beginning with a 1,4 addition of tributylstannyllithium to 5,6-dihydropyranone[38]

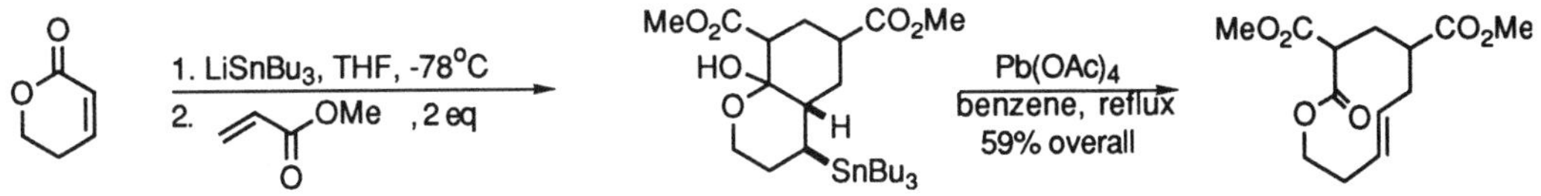

1. LiSnBu3, THF, -78°C
2. OMe ,2 eq
MeO2C
CO2Me
HO
H
SnBu3
Pb(OAc)4
benzene, reflux
59% overall
MeO2C
CO2Me

5.5.4 *Conjugate reduction*

As mentioned in section 5.2, organocuprates deliver their transfer ligand axially to 2,3-dihydro-4-pyridones. The same group which had developed this chemistry has extended this idea to the stereoselective reduction of such pyridones and applied this to a total synthesis of (±)-myrtine.[39] Several reducing systems were examined. A combination of a Lewis acid with a lithium hydride reagent and a copper salt at low temperature gave good yields of the 1,4 reduction product with excellent stereoselectivity.

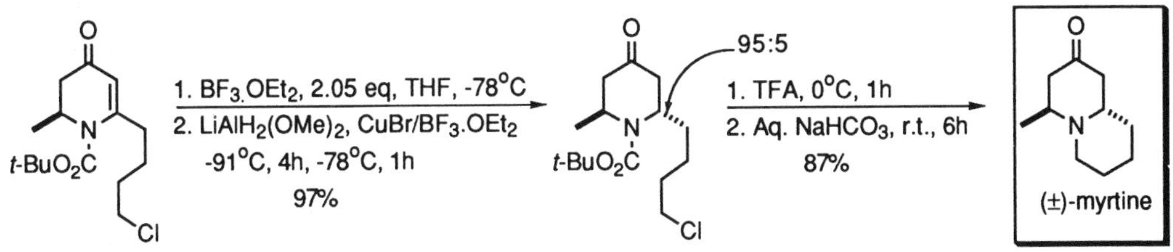

t-BuO2C
Cl
1. BF3.OEt2, 2.05 eq, THF, -78°C
2. LiAlH2(OMe)2, CuBr/BF3.OEt2
-91°C, 4h, -78°C, 1h
97%
t-BuO2C
Cl
95:5
1. TFA, 0°C, 1h
2. Aq. NaHCO3, r.t., 6h
87%
(±)-myrtine

References

1. S. Kano, S. Shibaya and T. Ebata, *Heterocycles*, 1980, **14**, 661
2. S. Takano, E. Goto, M. Hirama and K. Ogasawara, *Heterocycles*, 1981, **15**, 951
3. P. Camps, J. Cardellach, J. Font, R. M. Ortuno and O. Ponsati, *Tetrahedron*, 1982, **38**, 2395
4. H. Nagaoka, M. Iwashima, H. Abe and Y. Yamada, *Tetrahedron Lett.*, 1989, **30**, 5911
5. M. E. Krafft, R. M. Kennedy and R. A. Holton, *Tetrahedron Lett.*, 1986, **27**, 2087
6. R. A. Holton, R. M. Kennedy, H-B. Kim and M. E. Krafft, *J. Amer. Chem. Soc.*, 1987, **109**, 1597
7. J. L. Herrmann, J. E. Richman and R. H. Schlessinger, *Tetrahedron Lett.*, 1973, 2599
8. See also J. E. Richman, J. L. Herrmann and R. H. Schlessinger, *Tetrahedron Lett.*, 1973, 3271
9   T. Naito, O. Miyata and I. Ninomiya, *Heterocycles*, 1987, **26**, 1739
10. R. W. Franck, V. Bhat and C. S. Subramanian, *J. Amer. Chem. Soc.*, 1986, **108**, 2455
11. S. Hanessian and P. J. Murray, *J. Org. Chem.*, 1987, **52**, 1170
12. F. E. Zeigler and J. A. Schwartz, *Tetrahedron Lett.*, 1975, 4643
13. K. Tomioka, T. Ishigura and K. Koga, *J. Chem. Soc., Chem. Commun.*, 1979, 652
14. J. F. G. A. Jansen and B. L. Feringa, *Tetrahedron Lett.*, 1989, **30**, 5481
15. See also, B. L. Feringa, B. de Lange and J. C. de Jong, *J. Org. Chem.*, 1989, **54**, 2471
16. I. Fleming, N. L. Reddy, K. Takaki and A. C. Ware, *J. Chem. Soc., Chem. Commun.*, 1987, 1472
17. D. Seyferth and R. C. Hui, *Tetrahedron Lett.*, 1986, **27**, 1473
18. W. C. Still and I. Galynker, *Tetrahedron*, 1981, **37**, 3981
19. T. Uyehara, N. Shida and Y. Yamamoto, *J. Chem. Soc., Chem. Commun.*, 1989, 113
20. J. Bigorra, J. Font, R. M. Ortuna, F. Sanchez-Ferrando, F. Florencio, S. Martinez-Carrera and S. Garcia-Blanco, *Tetrahedron*, 1985, **41**, 5577 and 5589
21. R. M. Ortuño, R. Merce and J. Font, *Tetrahedron*, 1987, **43**, 4497
22. R. K. Olsen, W. J. Hennen and R. B. Wardle, *J. Org. Chem.*, 1982, **47**, 4605
23. D. Seebach, J. Zimmermann, U. Gysel, R. Ziegler and T-K Ha, *J. Amer. Chem. Soc.*, 1988, **110**, 4763.
24. This is to be contrasted with reactions of cuprates with dioxanones which result in clean S_N2 substitution at the acetal carbon: S. L. Schreiber and J. Reagan, *Tetrahedron Lett.*, 1986, **27**, 2945; D. Seebach, R. Imwinkelried and G. Stucky, *Angew. Chem. Int. Ed. Engl.*, 1986, **25**, 178 and *Helv. Chim. Acta*, 1987, **70**, 448.
25. D. L. Comins and D. H. LaMunyon, *J. Amer. Chem. Soc.*, 1988, **110**, 7445
26. G. H. Posner, T. P. Kogan, S. R. Haines and L. L. Frye, *Tetrahedron Lett.*, 1984, **25**, 2627
27. G. H. Posner and T. G. Hamill, *J. Org. Chem.*, 1988, **53**, 6031
28. Y. Tamaru, T. Harada, H. Iwamoto and Z-i. Yoshida, *J. Amer. Chem. Soc.*, 1978, **100**, 5221
29. S. Wolff and H. M. R. Hoffmann, *Synthesis*, 1988, 760
30. B. de Lange, F. van Bolhuis and B. L. Feringa, *Tetrahedron*, 1989, **45**, 6799
31. I. Panfil, C. Belzecki, M. Chmielewski and K. Suwinska, *Tetrahedron*, 1989, **45**, 233

32.   B. D. Roth and W. H. Roark, *Tetrahedron Lett.*, 1988, **29**, 1255

33.   I. Kuwajima, T. Murobushi and E. Nakamura, *Synthesis*, 1978, 602

34.   A. Itoh, S. Ozawa, K. Oshima and H. Nozaki, *Tetrahedron Lett.*, 1980, **21**, 361

35.   M. Watanabe, K. Shirai and T. Kumamoto, *Bull. Chem. Soc. Jap.*, 1979, **52**, 3318

36.   W. Amberg and D. Seebach, *Angew. Chem. Int. Ed. Engl.*, 1988, **27**, 1718.

37.   For the preparation of enantiomerically pure dioxanones like that given in the
        example, see D. Seebach and J. Zimmermann, *Helv. Chim. Acta*, 1986, **69**, 1147

38.   G. H. Posner, *Chem. Rev.*, 1986, **86**, 831, footnote 64

39.   D. L. Comins and D. H. LaMunyon, *Tetrahedron Lett.*, 1989, **30**, 5053

Chapter Six   Other Systems

6   Introduction

Nitroalkenes are highly reactive conjugate acceptors. As a result, many studies have been carried out on their reactions with various nucleophiles. The synthetic value of the adducts lies mainly in the areas of heterocycle synthesis, the synthesis of 1,4-dicarbonyls and some new annulation reactions.

6.1   Nitroalkenes

6.1.1   Stabilized carbanions[1]

A variety of stabilized carbanions have been added to nitroalkenes (see Table 6.1). The additions are usually quite efficient and after Nef hydrolysis offer simple access to a wide range of 4-oxoalkanoates. The sodium salts of nitroalkanes may be added to nitroalkenes by simply mixing the two together in aqueous methanol (entry 4).

Table 6.1 Conjugate addition of some stabilized carbanions to nitroalkenes

1[2]	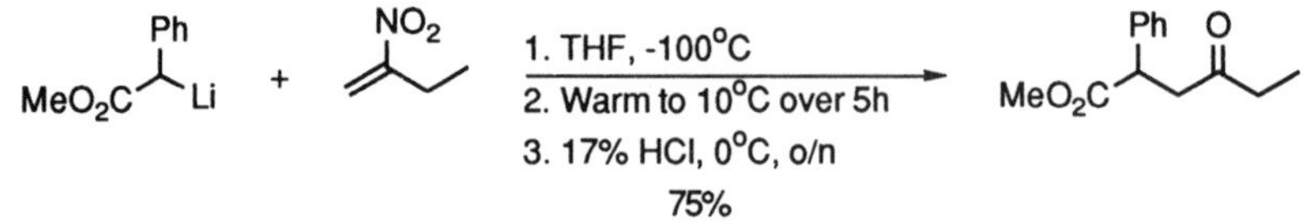 Ph / MeO₂C–C–Li + NO₂ (nitroalkene)	1. THF, -100°C 2. Warm to 10°C over 5h 3. 17% HCl, 0°C, o/n 75%	Ph O / MeO₂C~~
2[2]	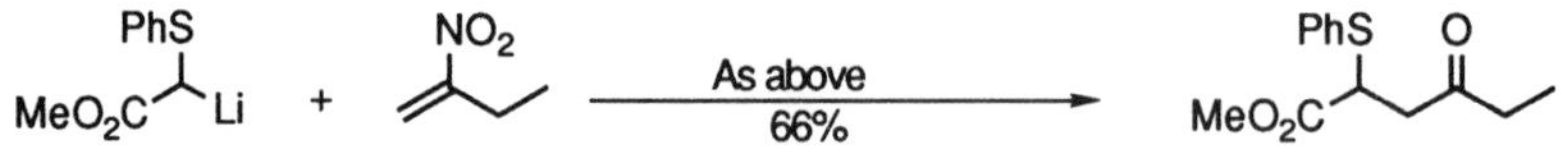 PhS / MeO₂C–C–Li + NO₂	As above 66%	PhS O / MeO₂C~~
3[2,3]	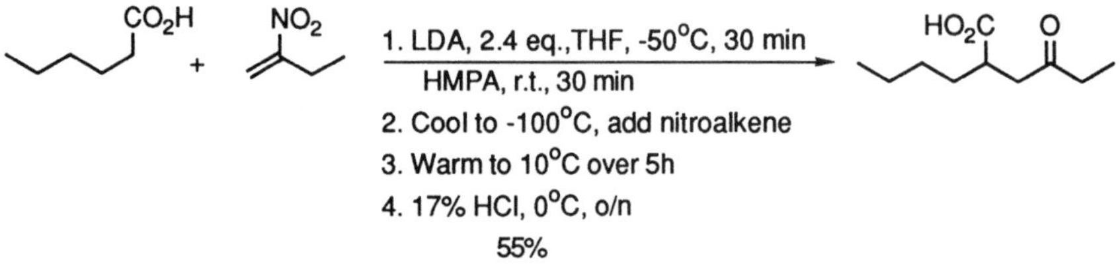 CO₂H + NO₂	1. LDA, 2.4 eq.,THF, -50°C, 30 min HMPA, r.t., 30 min 2. Cool to -100°C, add nitroalkene 3. Warm to 10°C over 5h 4. 17% HCl, 0°C, o/n 55%	HO₂C O
4[4]	cyclohexenyl-NO₂ + acetoacetate CO₂Et	1. BnEt₃N⁺OMe, cat., dioxane 25°C, 30 min 2. NaOH, 7.2 eq., H₂O, EtOH 3. 3N HCl, 20 eq., 70°C, 20 min 70%	bicyclic ketone
5[5]	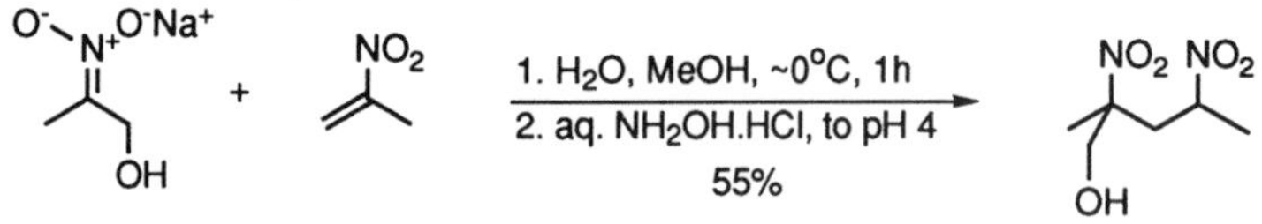 O⁻ N⁺ O⁻Na⁺ / OH + NO₂	1. H₂O, MeOH, ~0°C, 1h 2. aq. NH₂OH.HCl, to pH 4 55%	NO₂ NO₂ / OH
6[6]	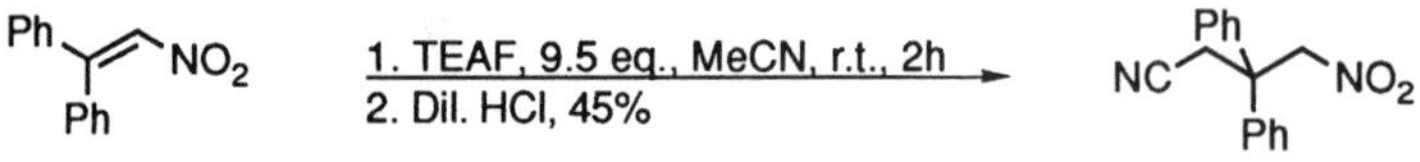 Ph / Ph NO₂	1. TEAF, 9.5 eq., MeCN, r.t., 2h 2. Dil. HCl, 45%	Ph / NC NO₂ / Ph
7[7]	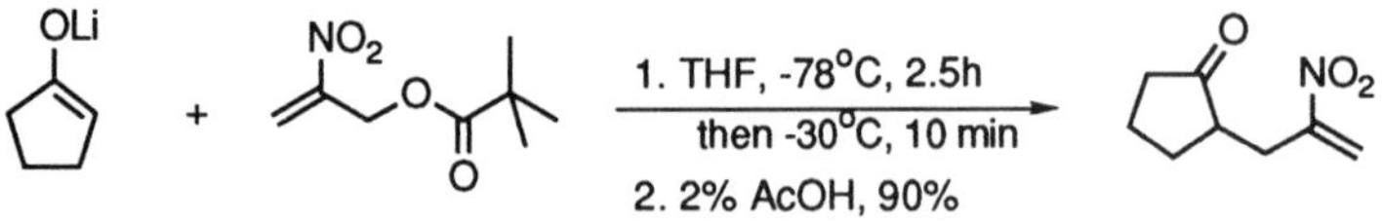 OLi (cyclopentenyl) + NO₂ ester	1. THF, -78°C, 2.5h then -30°C, 10 min 2. 2% AcOH, 90%	O NO₂
8[8]	MeO₂C MeO₂C + NO₂ / SPh	1. DMF, KOt-Bu, -20°C, 30 min 2. O₃, MeOH, -78°C 3. pH 7, 60%	MeO₂C O / MeO₂C SPh

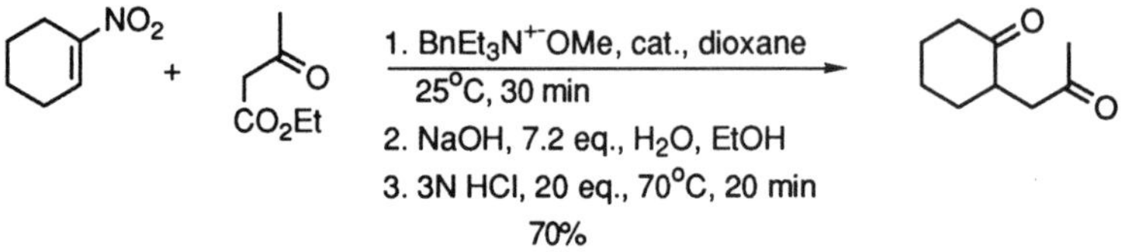

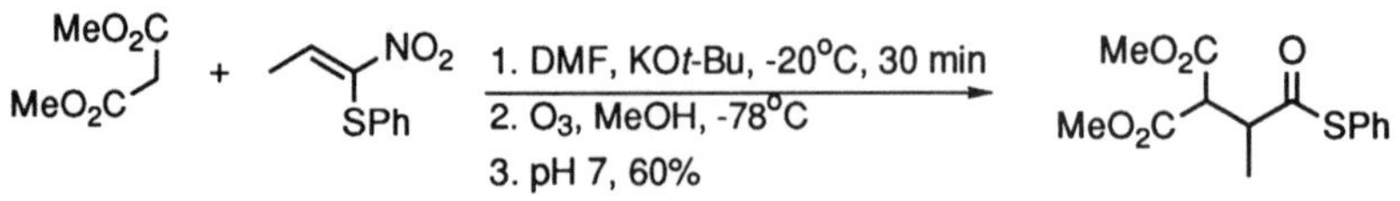

Cory's group has developed several methods for synthesizing tricyclic ketones from the addition of dienolates to nitroalkenes,[9] (and alkenylsulfones[10] and alkenylphosphonium salts[11] as well). Addition, followed by quenching with acetic acid gives good yields of the adduct, although with little stereoselectivity.

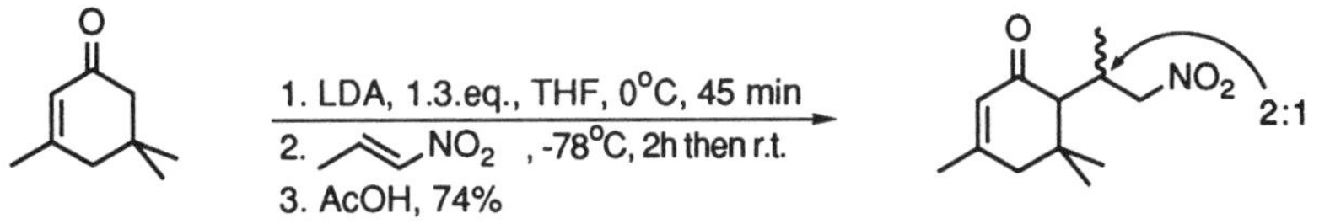

However, if the intermediate is not quenched, but rather heated in hexamethylphosphoramide, a tricyclic ketone is formed with good stereoselectivity. Unfortunately, the reaction failed in all attempts with enolates of acyclic alkenones.

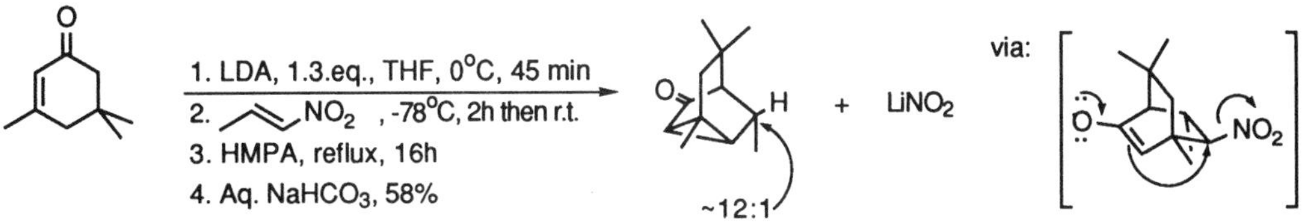

Several pyrrole syntheses are based on the conjugate addition of a stabilized carbanion to a nitroalkene. For example, the addition of a β-ketoester to 2-nitro-1-phenylpropene gave the furan product (1) which could be rearranged in almost quantitative yield to the 2,3,4,5-tetrasubstituted pyrrole.[12,13]

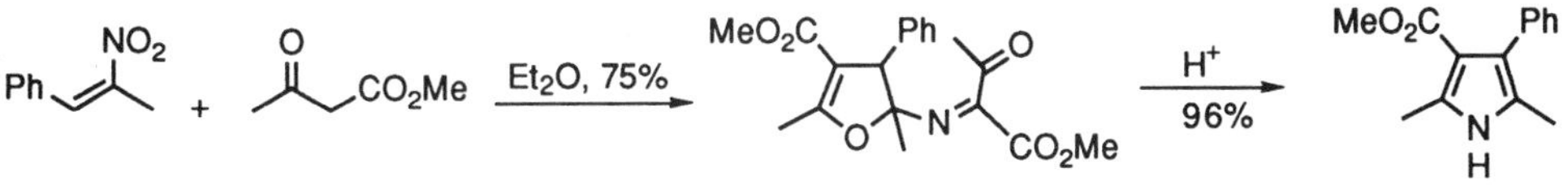

The conjugate addition of carbanions, generated from *iso*cyanoacetates, to nitroalkenes leads to a pyrrolic product under very mild conditions.[14] The suggested mechanism involves the following steps, shown for the synthesis of the trail marker pheromone of the Texas leaf-cutting ant, *Atta texana* (Buckley).

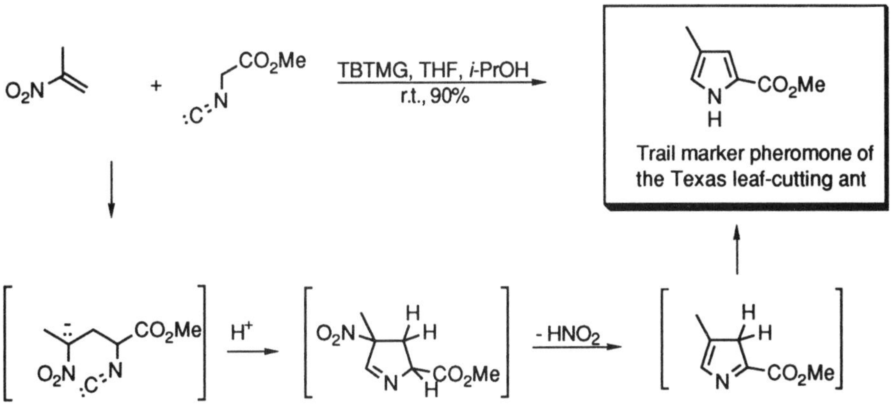

Trail marker pheromone of the Texas leaf-cutting ant

With β-nitrostyrene derivatives, reaction sometimes only occurs if the temperature is kept considerably lower, typically -70°C, than in other cases. Under these conditions, the initial anion (**3**), produced after addition, can be intercepted by another conjugate acceptor such as methyl propenoate.

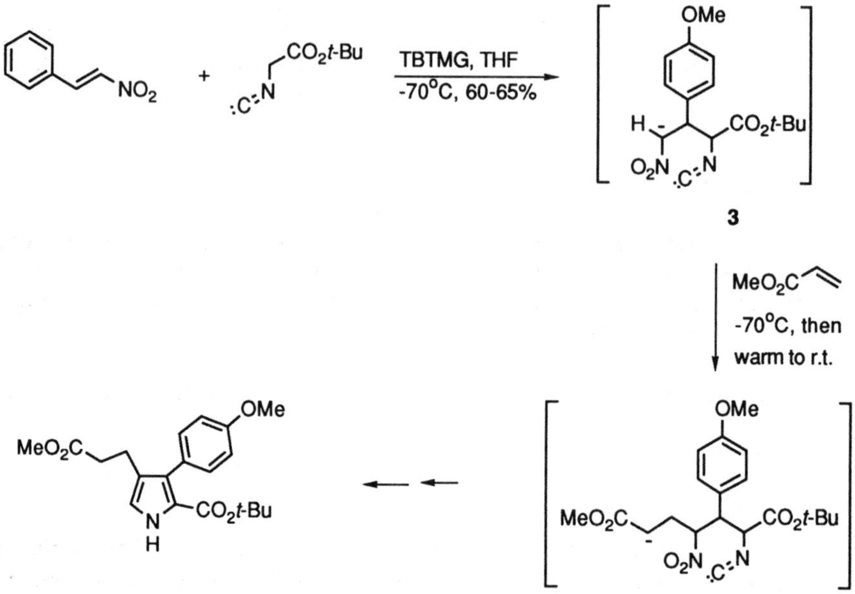

The intermediate nitronate, obtained from addition of a ketone enolate to a nitroalkene, may be isolated in excellent yield as a mixed anhydride. This can then be reduced to the pyrrole with a zinc-copper couple.[15] This method provides a direct synthesis of polyalkylpyrroles and obviates the need for proceeding via a pyrrolecarboxylate.

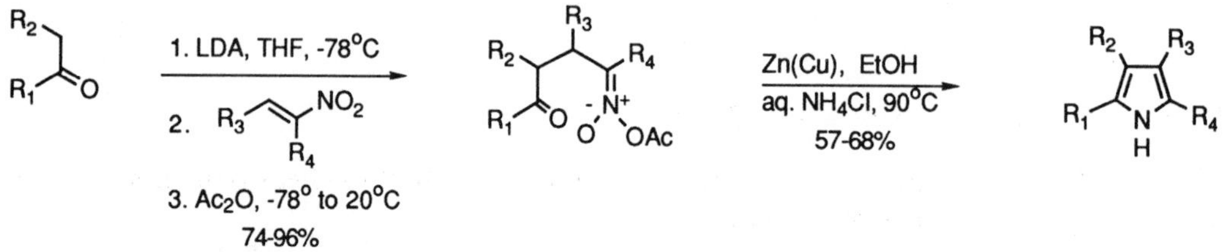

The following synthesis of 3,4,5-trisubstituted pyrazoles, e.g.(**4**) uses a nitro group both in the nucleophile and the conjugate acceptor.[16]

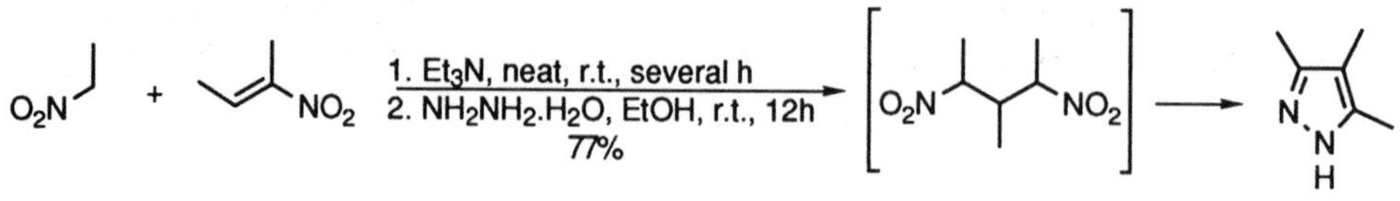

Seebach's nitroallylating reagent reacts efficiently with 2-lithio-1,3-dithianes.[17]

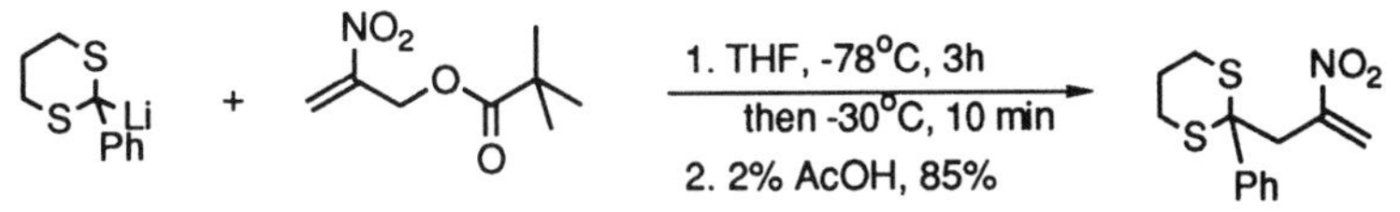

6.1.2   Organocopper reagents

The addition of organocopper reagents to nitroalkenes has not proven to be popular, so far. However, some examples do exist. Early studies showed that additions of LO cuprates to 1-nitro-2-phenylethene only gave low yields of the conjugate adduct, accompanied by considerable polymer formation.[18]

Mono-organocopper reagents add reasonably well as in the example below.[19]

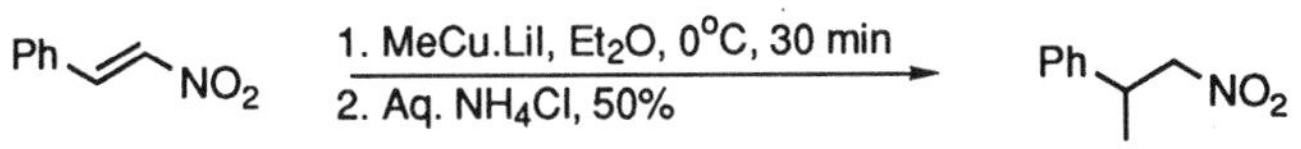

If a nitroaromatic group is present in the acceptor, then no conjugate addition occurs.[19] Rather, an azoxy compound is formed. In fact, nitrobenzene is sometimes used to quench such reactions.

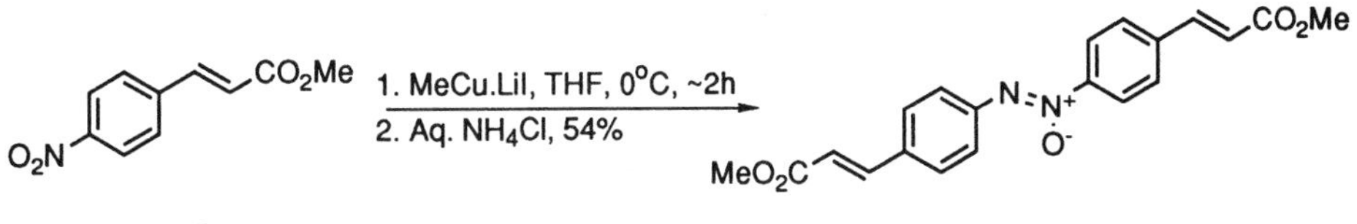

6.1.3   Other organometallic reagents

Barrett's synthesis of thioesters from 2-(phenylthio)-1-nitroalkenes has been successfully applied to an aryllithium.[20]

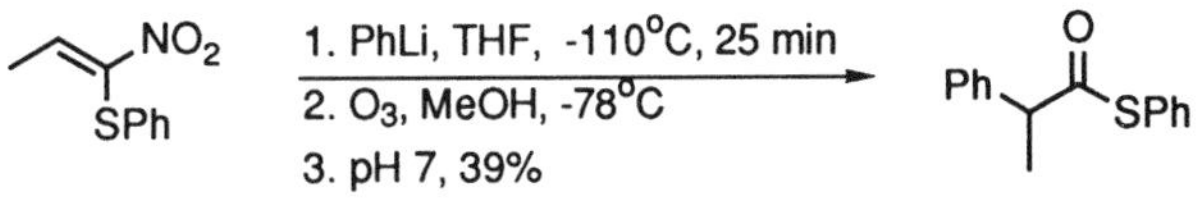

Alkenylmagnesium reagents add cleanly to 1-nitrocyclohexene. The intermediate nitronate (5) may then be quenched stereoselectively with a variety of electrophiles.[24]

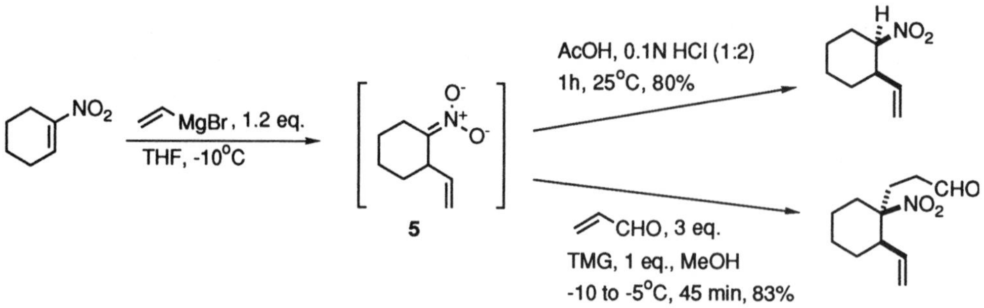

Seebach's nitroallylation reaction also works with alkynyllithiums.[21]

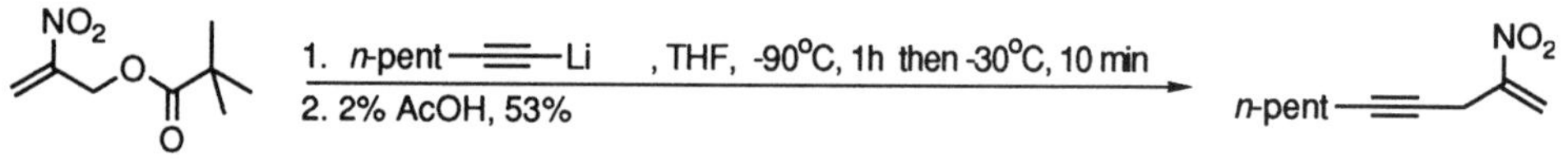

6.1.4   Free radicals

There have been relatively few reports of free radical additions to nitroalkenes. However, Barton's group has shown that thiopyridone-based esters react well, adding across the double bond of nitroalkenes.[22]

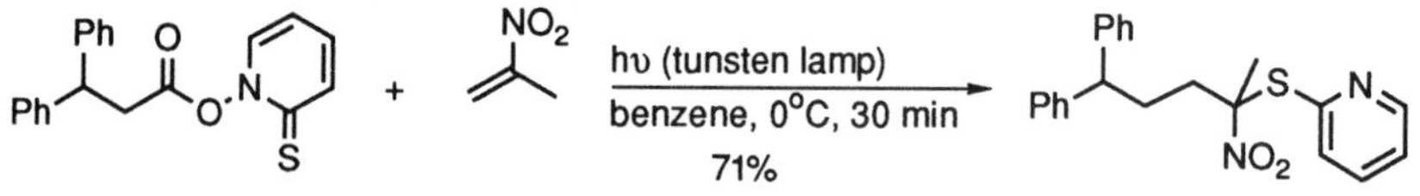

6.1.5   Enamines and enol ethers

6.1.5.1   *Enamines*

Enamines react readily with many nitroalkenes. In non-polar solvents, cyclobutanes are formed, whereas in polar solvents only the open chain adduct could be isolated.[23] Subsequent treatment with dilute acid provided the δ-nitroketone in good yield.

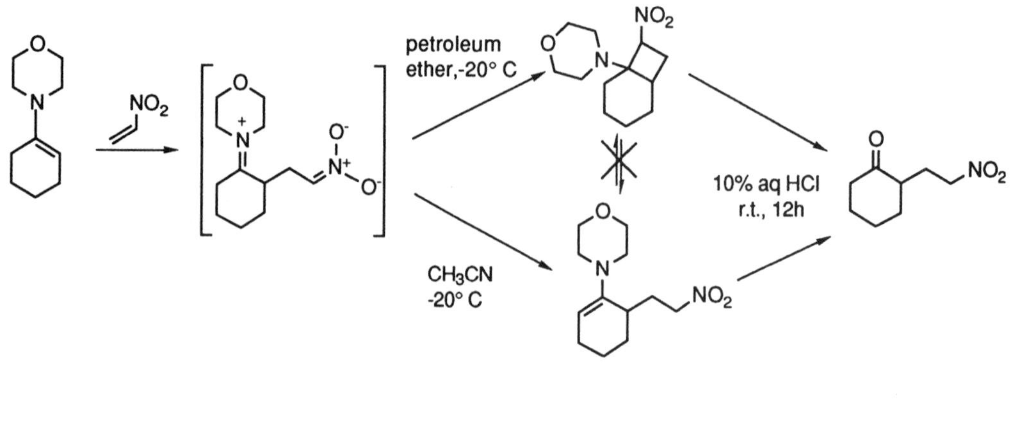

They also showed that, for nitroethene where base-catalyzed polymerization often competed with conjugate addition, the use of 2-acetoxynitroethane was an effective alternative.

Stereoselective conjugate additions of enamines to nitroalkenes provided the experimental foundations for Seebach's topological rule developed in the early 1980's.[24] This rule requires that the donor and acceptor moieties of the two reactants are close to each other and that the larger substituents, if there are any, will be pointing away from each other. For example, this accounts for why addition of the morpholine enamine of cyclohexanone adds with such high *anti*-selectivity to *E*-1-nitro-2-phenylethene.

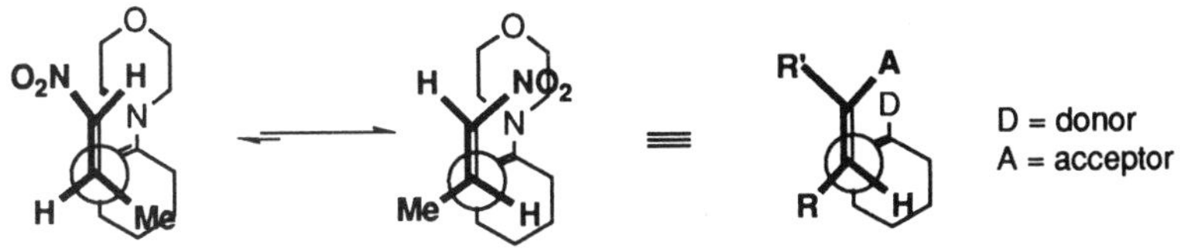

D = donor
A = acceptor

In all these cases (assuming aprotic media and kinetic control) the mode of approach is *Re* with respect to both the donor and the acceptor. The rule breaks down when R and R' are both large and other topologies need to be considered. For those cases where the enamine is derived from an unsymmetrical cyclic ketone, the usual preference for "anti-parallel" attack is over-ridden by steric requirements leading to products from "parallel attack" (see Chapter 1). Both Seebach's and Risalti's groups[25] discovered that these additions often proceed with very high levels of diastereoselection. Some examples are given in Table 6.2.

Table 6.2 Stereoselective conjugate additions of enamines to nitroalkenes

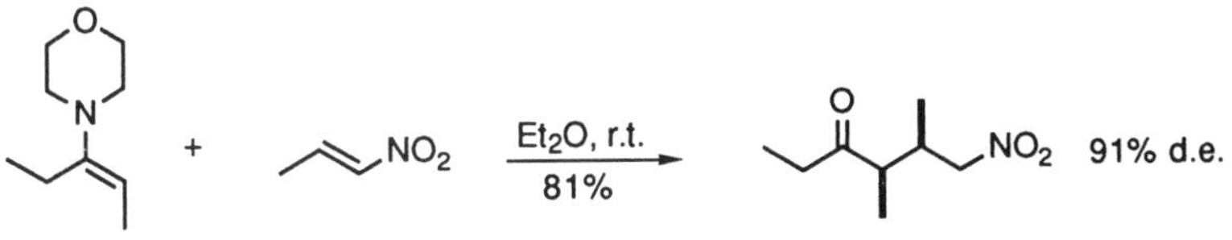

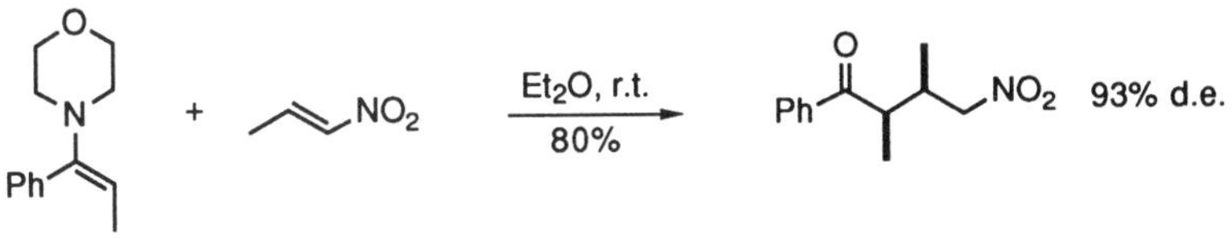

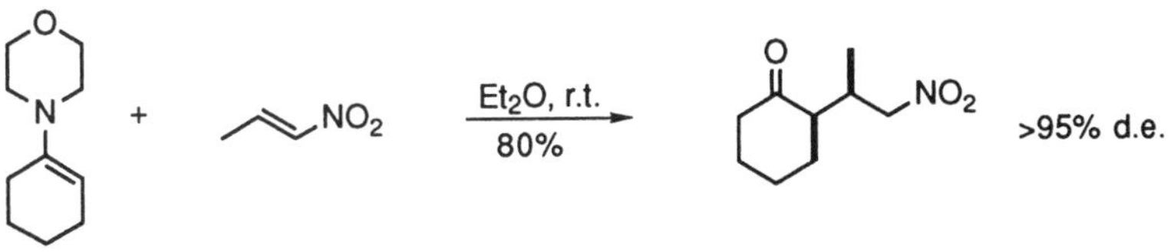

One application of such additions is in an early synthesis of α-methylene-γ-lactones such as (6).[26]

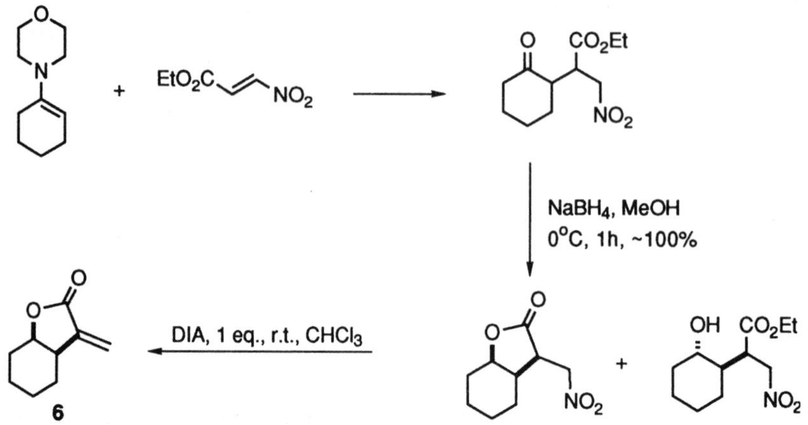

Whereas enamine addition to β-nitrostyrene gives the usual product (Table 6.2, entry 2), addition to α-nitrostyrene produces a mixture of cyclic nitronates with modest stereoselectivity.[27]

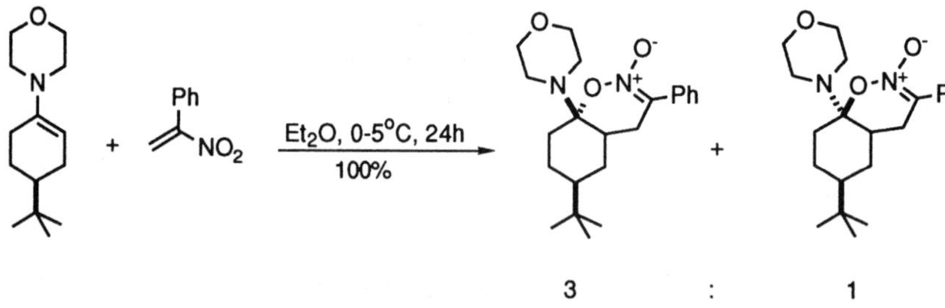

Extension of this reaction to the addition of enediamines to nitroalkenes provides an extremely high yielding synthesis of branched polycyclic ketones.[28] The authors demonstrated that, even though the reactions are highly stereoselective, a stepwise, rather than concerted, mechanism is operating.

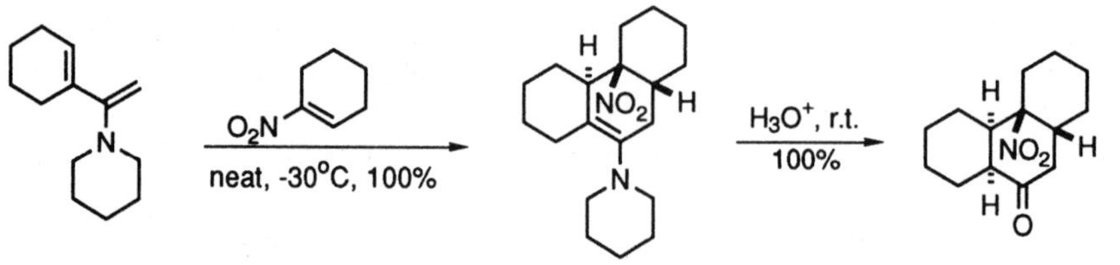

6.1.5.2 *Enol ethers*

Yoshikoshi's group has shown that silyl enol ethers add well to nitroalkenes using several different Lewis acids.[29] The acid present in the reaction mixture is sufficient, after the addition of water, to enable direct conversion of the adduct to the corresponding keto product. This, in turn is readily transformed into the cyclic alkenone using base catalysis.

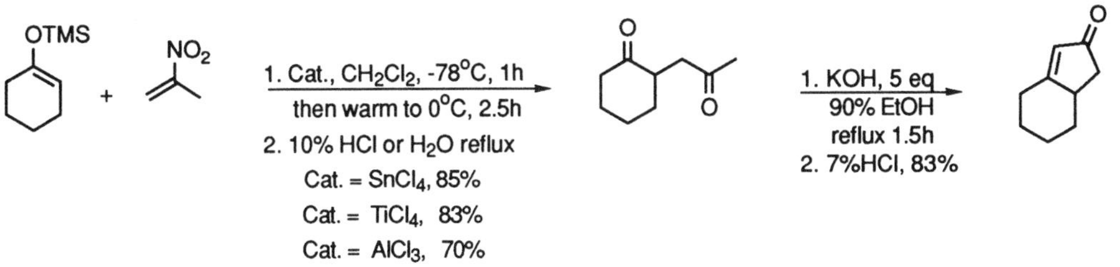

OTMS
NO2
+
1. Cat., CH2Cl2, -78°C, 1h
then warm to 0°C, 2.5h
2. 10% HCl or H2O reflux
Cat. = SnCl4, 85%
Cat. = TiCl4, 83%
Cat. = AlCl3, 70%
O
O
1. KOH, 5 eq
90% EtOH
reflux 1.5h
2. 7%HCl, 83%
O

In the same paper it was shown that ketene acetals also add well under Lewis acid catalysis.

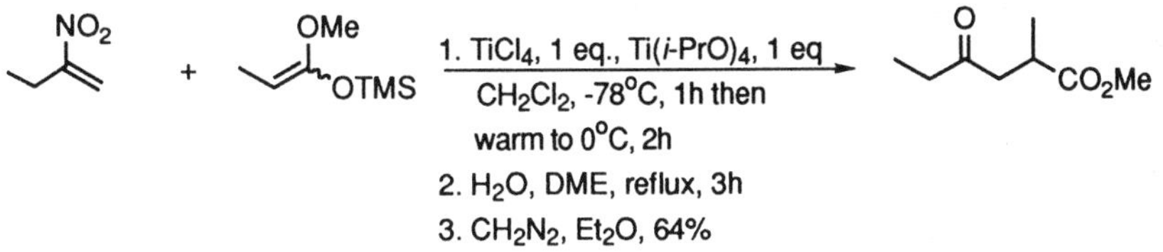

NO2
+
OMe
OTMS
1. TiCl4, 1 eq., Ti(i-PrO)4, 1 eq
CH2Cl2, -78°C, 1h then
warm to 0°C, 2h
2. H2O, DME, reflux, 3h
3. CH2N2, Et2O, 64%
O
CO2Me

Fleming has also used this reaction as part of a two-step synthesis to prepare 4-ethyl-4-methyl-2-cyclopentenone (7).[30]

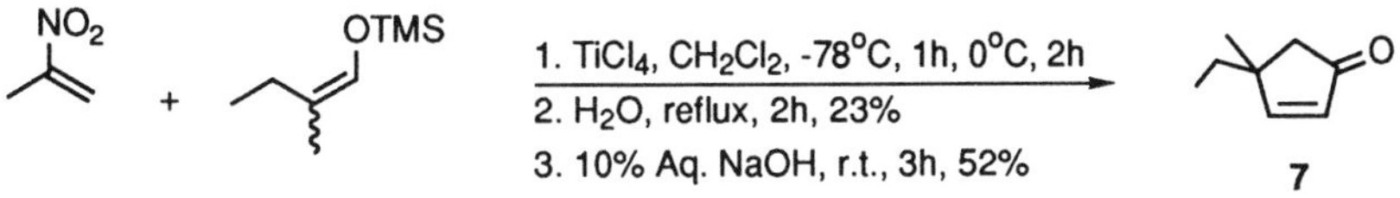

NO2
+
OTMS
1. TiCl4, CH2Cl2, -78°C, 1h, 0°C, 2h
2. H2O, reflux, 2h, 23%
3. 10% Aq. NaOH, r.t., 3h, 52%
O
7

6.1.6 Heteronucleophiles

Addition of ammonia to 1-nitrocyclohexene is achieved simply by using concentrated ammonia.[31] Shorter reaction times produced some of the *cis*-isomer. In the same paper it was stated that the *cis*-isomer may be obtained selectively using hydrazoic acid/sodium azide instead of ammonia, however no reaction details were provided.

NO2
Conc. NH4OH, 4 eq., THF
45°C, 24h, 95%
NO2
NH2

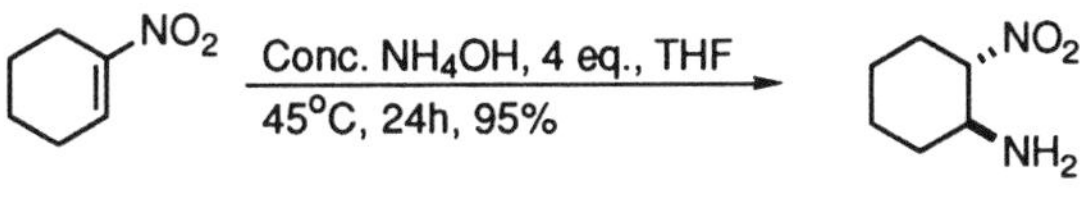

Both intra- and inter-molecular fluoride-catalyzed additions of silylamines have proved to be valuable. For example, intramolecular closure of the nitroalkene (8) provides a useful route to the azetidinone antibiotics.[32,33]

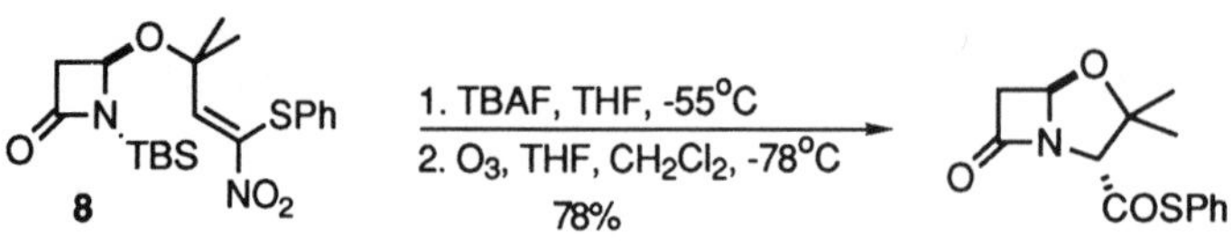

Table 6.3 Conjugate addition of nitrogen and phosphorus nucleophiles

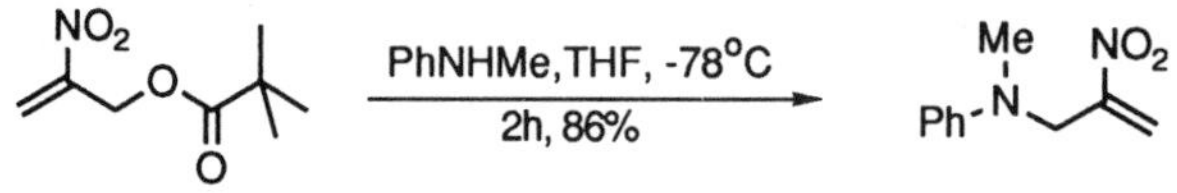

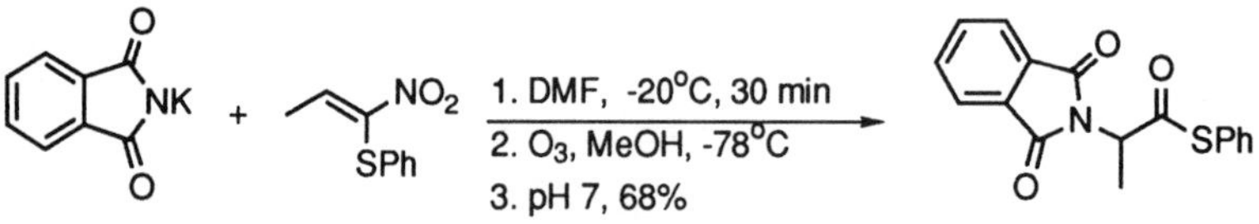

Addition/elimination reactions of pyrrolidine with several β-substituted nitroalkenes have been carried out. Retention of stereochemistry was observed in each case.[37]

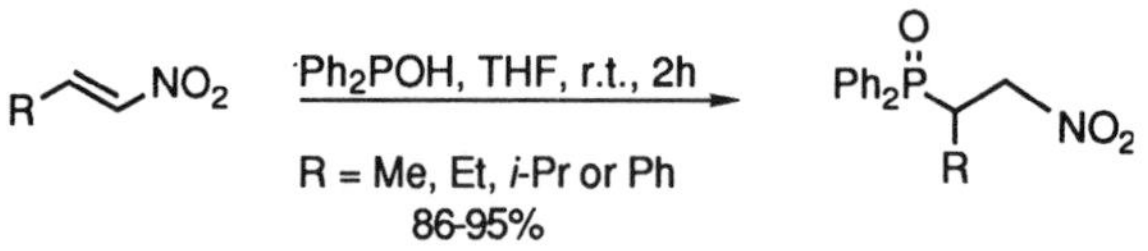

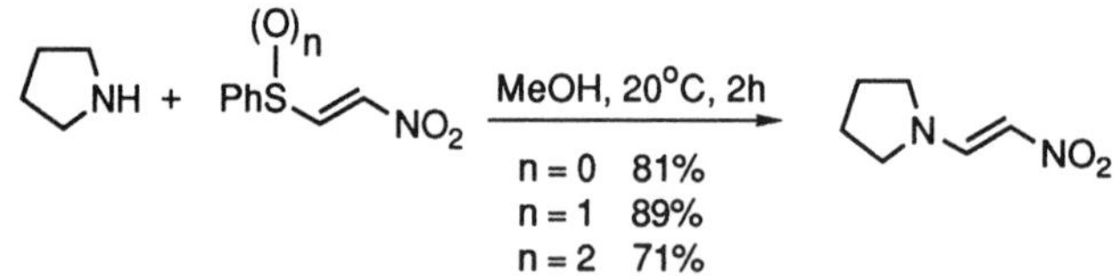

Kamimura and Ono have shown that the addition of oxygen nucleophiles to nitroalkenes (Table 6.5, entry 1) proceeds with good stereoselectivity. This was accounted for by assuming a kinetically-controlled protonation of the intermediate nitronate (9). Addition of a thiolate ion followed by *in situ* trapping with an aldehyde is equally stereoselective.[38] This is also a useful method for preparing *E*-allylic alcohols.

R⌒NO₂

1. PhSLi, 1.5 eq., THF, r.t., 1h
2. 37% Aq. HCHO, 2 eq
3. AcOH, 2 eq., -78°C, 1h

R = alkyl or aryl

→

NO₂ / R—CH(SPh)—CH—CH₂OH   ~9:1

1. (PhCO)₂O
2. Bu₃SnH, 3.5 eq., AIBN, 1 eq., toluene
   110°C, 30 min

R⌒OBz

E : Z (99:1)

via:

$$\left[\text{HO—} \underset{R \quad H}{\overset{SPh}{\bigcirc}} =N^+\!\!\!\overset{O^-}{\underset{O^-}{}} \right]$$

↓ H⁺   **9**

Other heteronucleophiles also add easily to nitroalkenes (see Table 6.4). Seebach's nitroallylation methodology works well with sulfur nucleophiles (entry 2). The use of a chiral base catalyst leads to significant enantioselectivity in the addition of thioglycolic acid under very mild conditions (entry 4).

Table 6.4   Conjugate addition of oxygen and sulfur nucleophiles to nitroalkenes

1[39]	Ph⌒⌒NO₂ → 1. NaOMe, THF, r.t., 24h / 2. AcOH, -78°C, 2h, 42%	Ph⌒⌒(MeO)—⌒NO₂ 9:1
2[40]	(NO₂)CH₂=C—CH₂—O—C(=O)—C(CH₃)₃ → 1. PhSLi, THF, hexane -78°C / 30 min, then -30°C, 10 min / 2. 2% AcOH, 62%	PhS⌒—C(=CH₂)NO₂
3[41]	⌒—C(SPh)=CH—NO₂ → 1. NaOMe, MeOH -20°C, 30 min / 2. O₃, MeOH, -78°C / 3. pH 7, 79%	MeO—CH(—)—C(=O)—SPh
4[42]	HO₂C⌒SH + Ph⌒NO₂ → Quinine, 1.1.eq., toluene / r.t., 15 min, 86%	HO₂C⌒S—CH(Ph)—CH₂—NO₂ 58% e.e.

Sulfur and oxygen nucleophiles add to 1-nitrocyclohexene. Oxidation of the adducts provides an effective method for preparing 2-alkoxy or 2-alkylthio-cyclohexanone.[43]

(1-nitrocyclohexene) NO₂ → 1. KOMe, 1.5 eq., THF, r.t., 2h / 2. KMnO₄, MgSO₄, H₂O, 0°C / 85% → (cyclohexanone)—OMe

(1-nitrocyclohexene) NO₂ → 1. i-PrSLi, 1.5 eq., THF, r.t., 2h / 2. KMnO₄, MgSO₄, H₂O, 0°C / 81% → (cyclohexanone)—S-i-Pr

Alternatively, the intermediate anion can be quenched with ethyl propenoate.

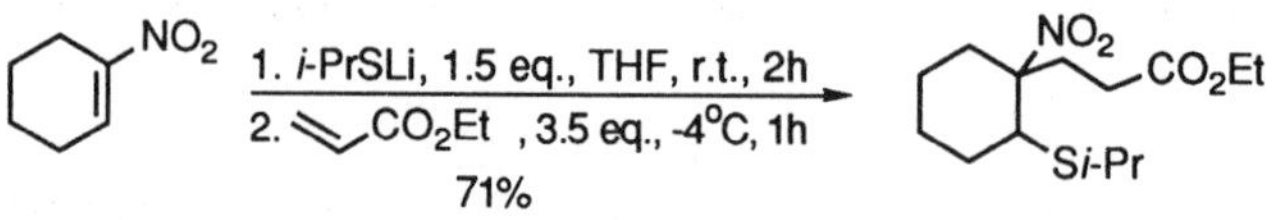

NO2
1. i-PrSLi, 1.5 eq., THF, r.t., 2h
2. CO2Et , 3.5 eq., -4°C, 1h
71%
NO2
CO2Et
Si-Pr

An example of the addition of oxygen nucleophiles to *nitroso*alkenes has been reported.[44]

NOTMS
Br
TBAF
H
N
O
OH
OH
N
O
H
NaOCl
67% overall
O
O
N

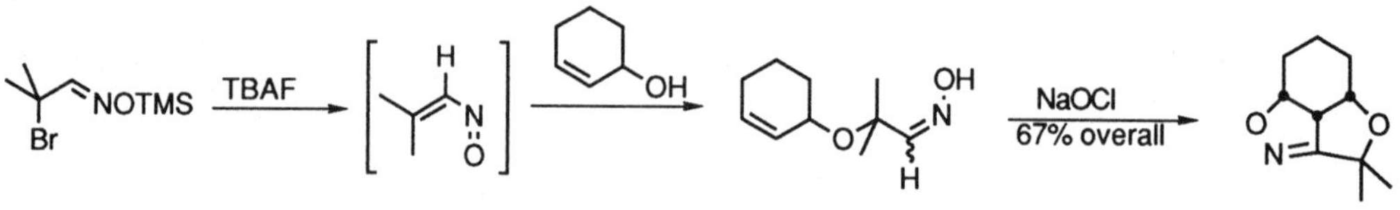

As part of their extensive work on the reduction of nitro-containing compounds, Kabalka's group has developed a polymer-bound hydride reagent which efficiently reduces the double bond in nitroalkenes.[45,46]

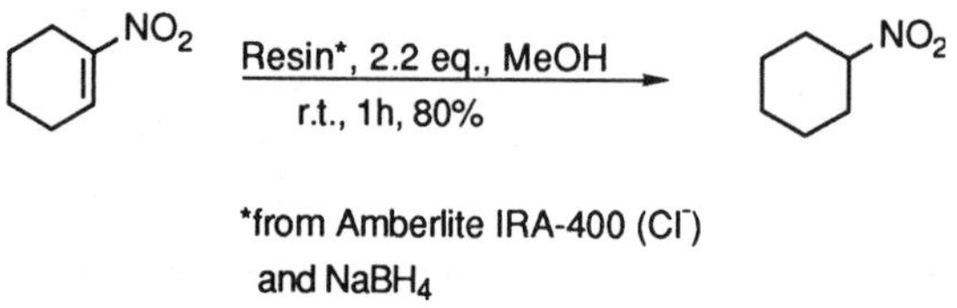

NO2
Resin*, 2.2 eq., MeOH
r.t., 1h, 80%
NO2
*from Amberlite IRA-400 (Cl⁻)
and NaBH4

Baker's yeast also reduces a variety of 3,3-disubstituted (and 3-monosubstituted) nitroalkenes with good stereoselectivity.[47]

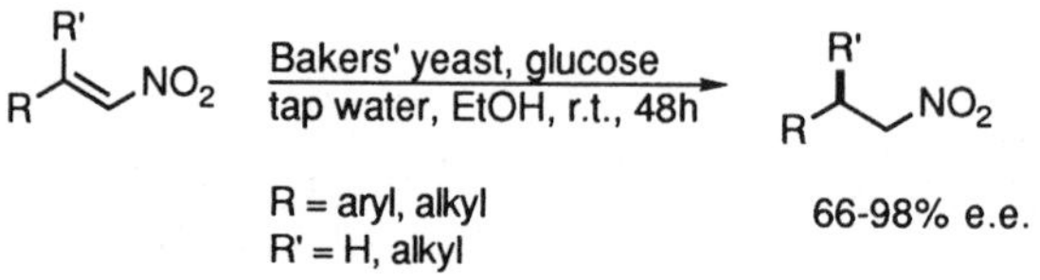

R'
R
NO2
Bakers' yeast, glucose
tap water, EtOH, r.t., 48h
R'
R
NO2
R = aryl, alkyl
R' = H, alkyl
66-98% e.e.

6.2 Alkenylsulfoxides, phosphonates and related compounds

Conjugate additions to alkenylsulfoxides have become quite popular because of, firstly, their applications as chiral auxiliaries, especially when attached to cycloalkenones (see chapter 3) and second, because of the high stereoselectivity often obtained in these additions. They have also found use in annulation reactions. Additions to alkenyl phosphonates have only been reported occasionally.

6.2.1 Stabilized carbanions

Many stabilized carbanions add to alkenylsulfoxides and their congeners. Some examples are collected in Table 6.5.

Table 6.5 Conjugate additions of stabilized carbanions to alkenylsulfoxides and their congeners

148		THF, 20°C, 12h, 63%	
249		n-BuLi, THF, -60°C then 0°C, 1h, 100%	
349		n-BuLi, THF	35% 65%
450		1. Benzene, r.t., 24h, 2. Ice, ~100%	

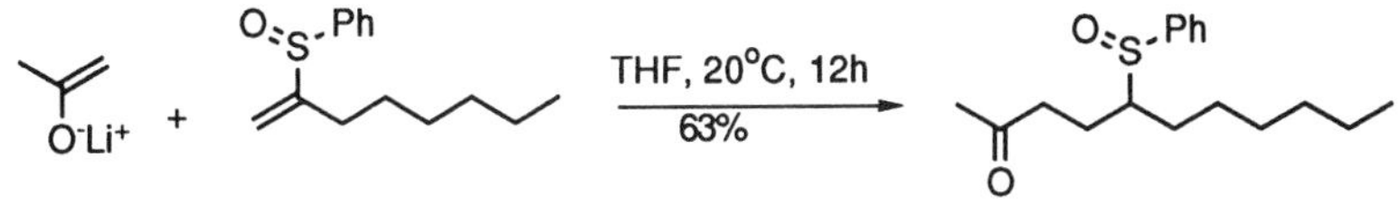

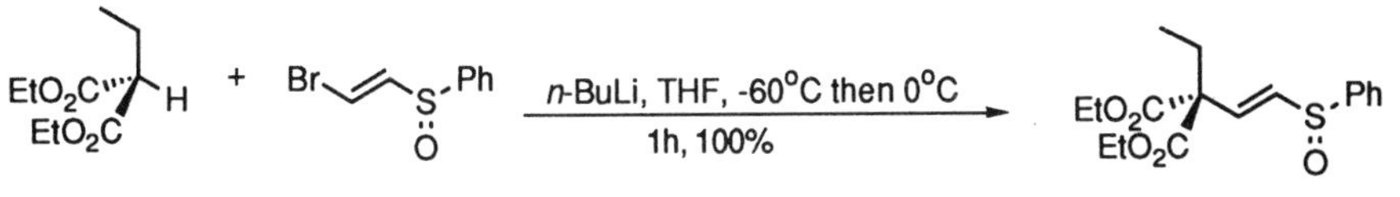

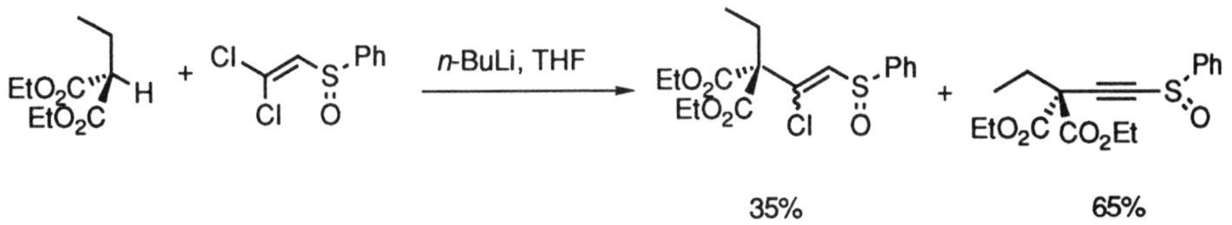

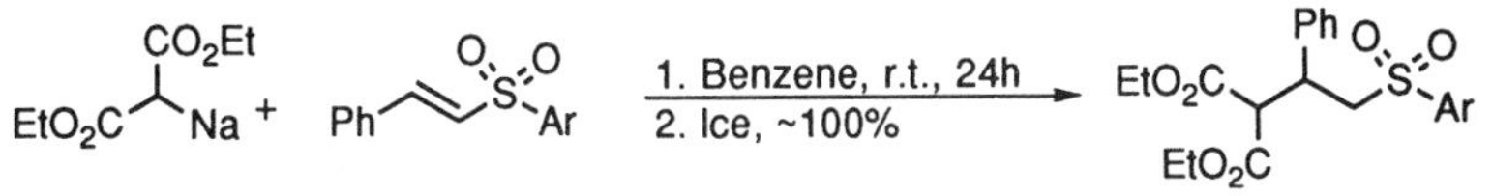

As was mentioned in section 2.1.1.1, the addition of nitro-stabilized carbanions is a particularly useful method as reductive removal is often very efficient, leading to overall conjugate addition of an alkyl group.[51,52] This is a powerful alternative to the use of organocopper and other organometallic reagents.[53]

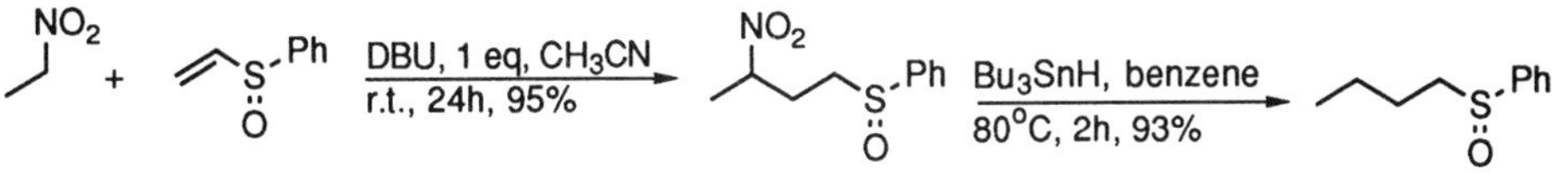

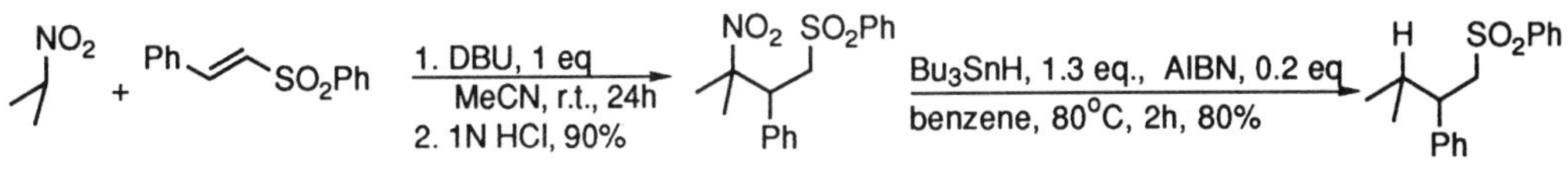

(scheme: NO_2 + alkenylsulfoxide, DBU, 1 eq, CH_3CN, r.t., 24h, 95% → then Bu_3SnH, benzene, 80°C, 2h, 93%)

(scheme: NO_2 + Ph–SO_2Ph, 1. DBU, 1 eq, MeCN, r.t., 24h; 2. 1N HCl, 90% → then Bu_3SnH, 1.3 eq., AIBN, 0.2 eq, benzene, 80°C, 2h, 80%)

Additions of nitro-stabilized carbanions to 2-(alkylthio)alkenylsulfoxides, such as (10), followed by oxidation and elimination provides a route to alkenals.[53]

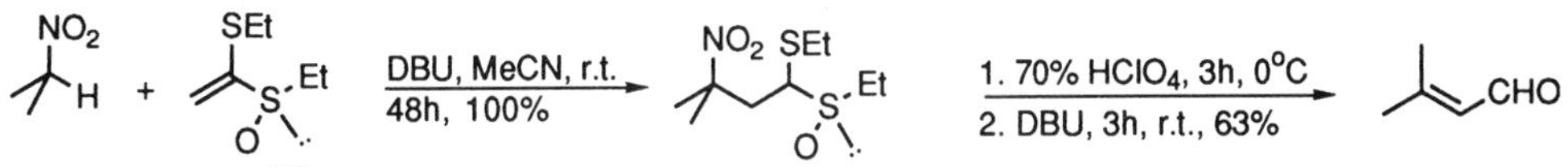

(scheme: NO_2 + **10**, DBU, MeCN, r.t., 48h, 100% → then 1. 70% $HClO_4$, 3h, 0°C; 2. DBU, 3h, r.t., 63% → CHO)

Double conjugate additions to alkenylsulfones have been used to construct ring F of the aspidospermine nucleus, as in Ohnuma's total synthesis of the alkaloid, (+)-4-hydroxyaspidofractinine.[54]

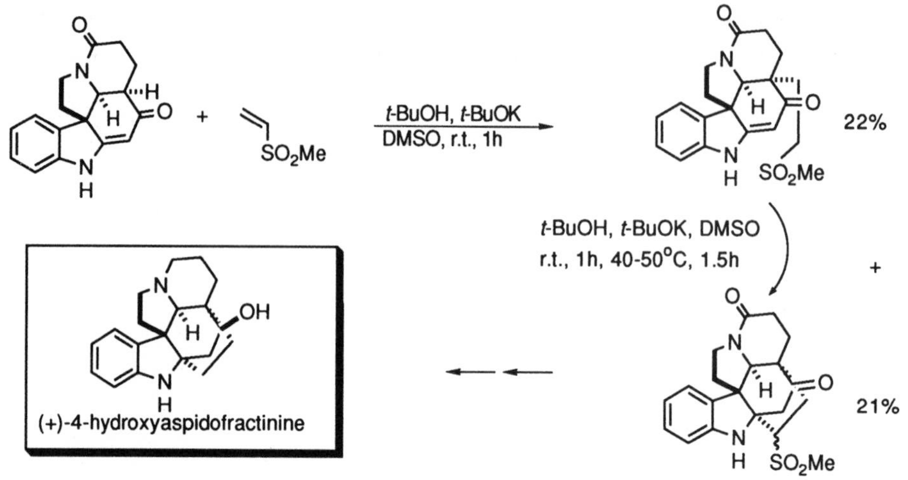

Additions to (3-silyl)alkenylsulfones, e.g. (**11**) yield intermediates which carry a β-silylsulfone moiety as a latent double bond. Once introduced, this group can be retained during subsequent manipulations and then the double bond may be "released" by treatment with tetrabutylammonium fluoride in tetrahydrofuran.

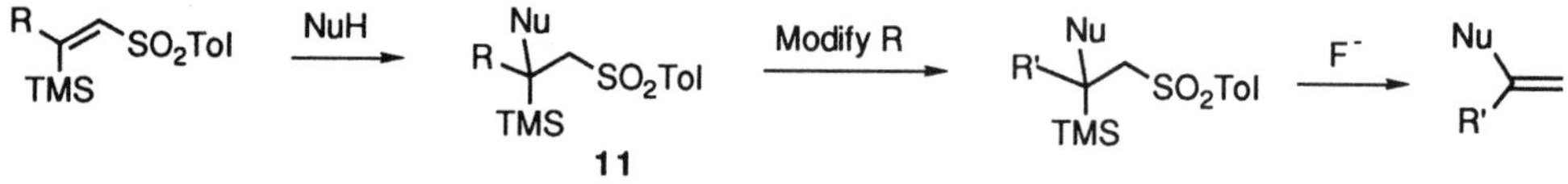

This idea was applied to a formal total synthesis of bicyclomycin, an antibiotic which is used in animal foodstuffs.[55]

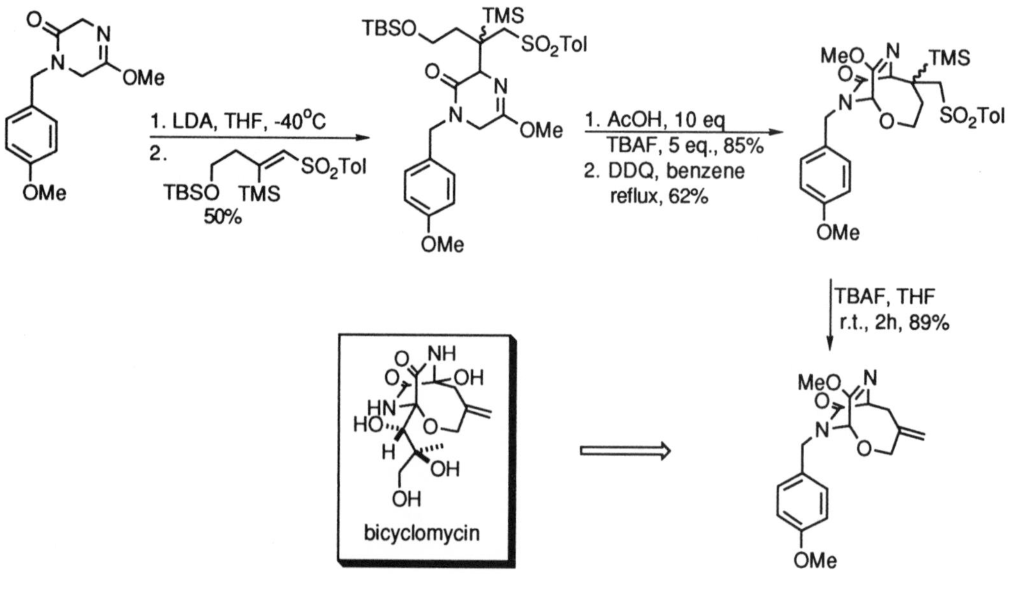

An alkene synthesis, based on sequential conjugate addition/Ramberg-Backlund rearrangement, has been demonstrated by de Waard (Nu: = sodium phenylsulfinate).[56]

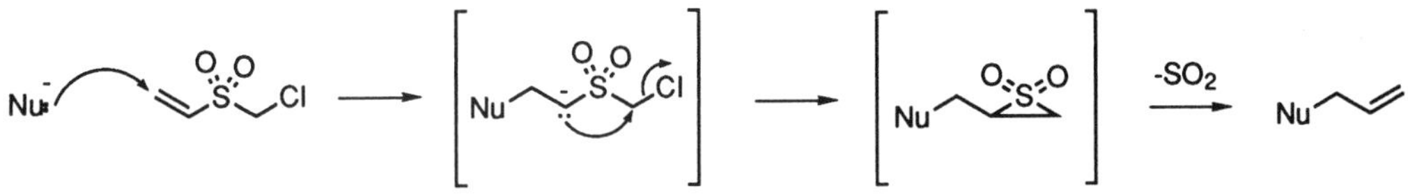

6.2.2   Organocopper reagents

A relatively early report of cuprate additions to alkenylsulfoxides demonstrated that the intermediate could be quenched with reactive electrophiles (or simply water).[57] However, these additions are not always very efficient.[58]

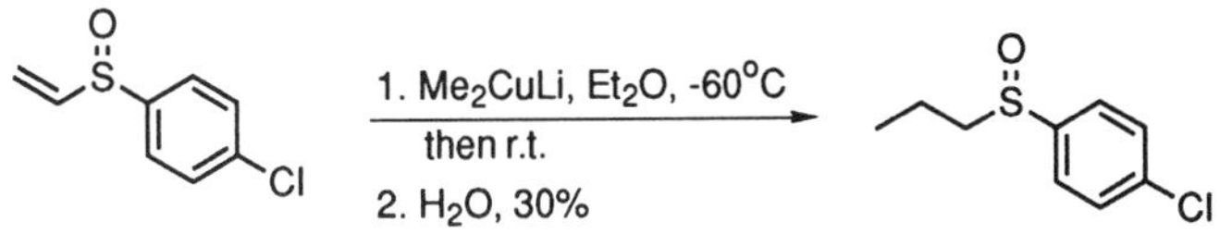

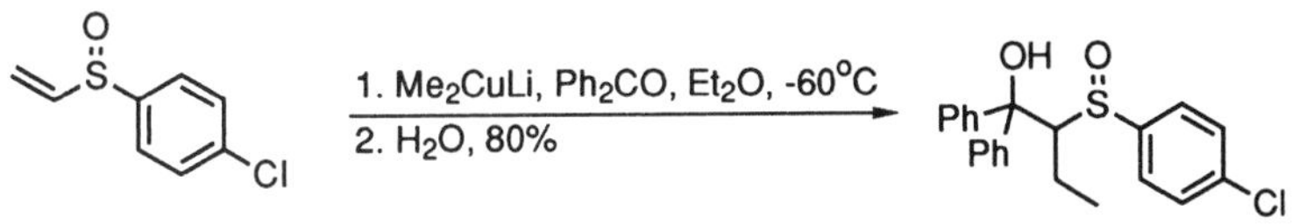

1. Me$_2$CuLi, Et$_2$O, -60°C
then r.t.
2. H$_2$O, 30%

1. Me$_2$CuLi, Ph$_2$CO, Et$_2$O, -60°C
2. H$_2$O, 80%

Fuchs' group has developed a synthesis of cis-3,5-disubstituted cyclopentenylsulfones, based upon conjugate additions of LO cuprates to the ammonium salt (12) shown below.

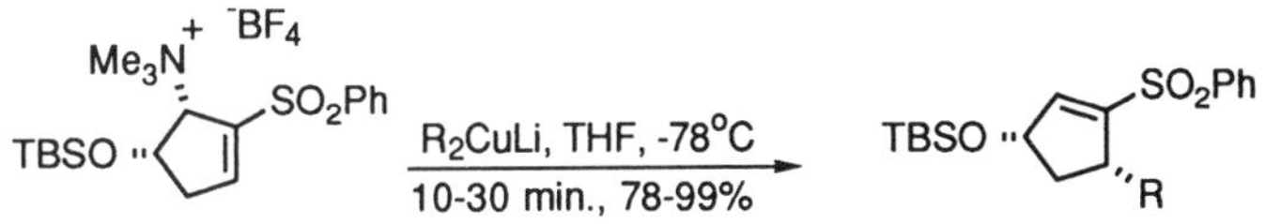

12

97.4 to >99% cis

R$_2$CuLi, THF, -78°C
10-30 min., 78-99%

R = Me, Ph, allyl
R = t-Bu gave excellent selectivity (96% cis)
however the yield was only 25%

Pyne has studied the addition of a number of different organocopper reagents to chiral (+)-norephedrine-derived alkenylsulfoximines.[59] He found that addition of either an LO cuprate or a mono-organocopper led to a reversal in stereoselectivity.

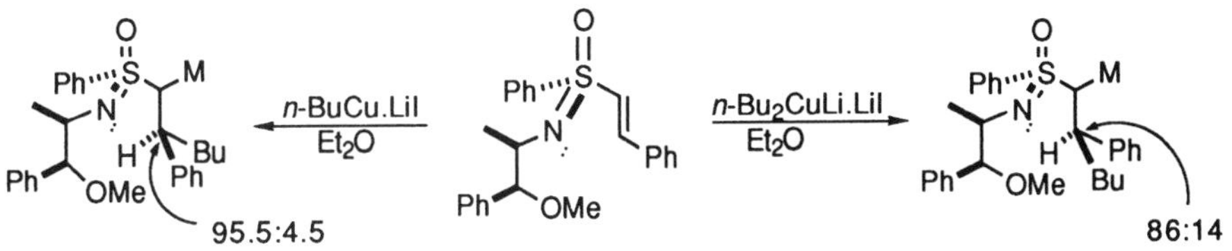

n-BuCu.LiI
Et$_2$O

95.5:4.5

n-Bu$_2$CuLi.LiI
Et$_2$O

86:14

Assuming that the LO cuprate chelates to the aminoether, then addition to the conformation shown below leads to the correct diastereomer.

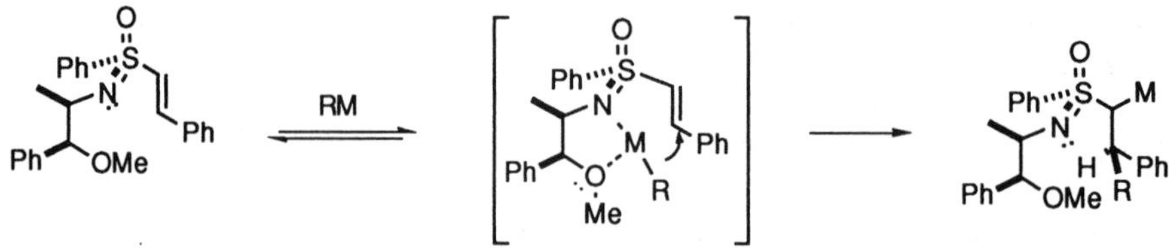

On the other hand, it appears that the lithium iodide succesfully competes with the mono-organocopper in chelating to the aminoether. As a result, addition to the new lithium chelate occurs from the less-hindered face of the alkenylsulfoximine, leading to excellent selection for the opposite diastereomer.

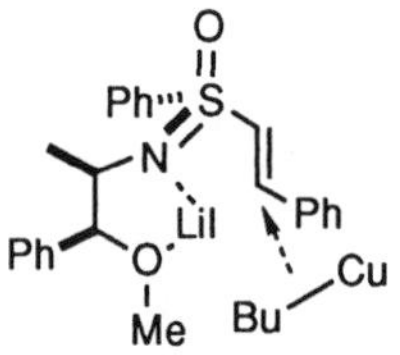

Alkenylcoppers add well to alkenylphosphonates as long as both hexamethylphosphoramide and triethylphosphite are present.[60]

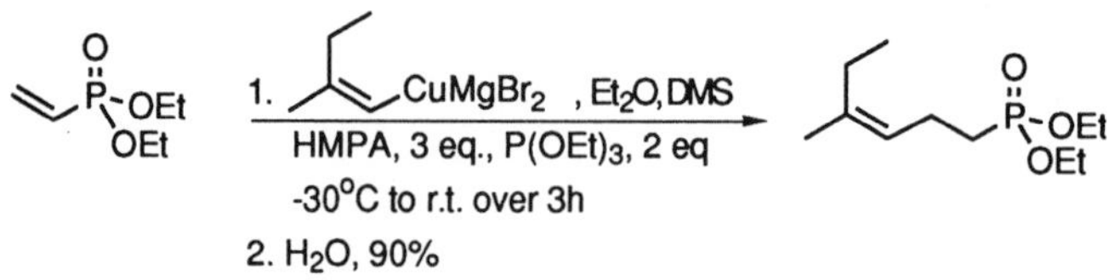

As part of its extensive work on phosphorus compounds, Pietrusiewicz's group has shown that LO cuprates add well to alkenylphosphine oxides.[61] They also

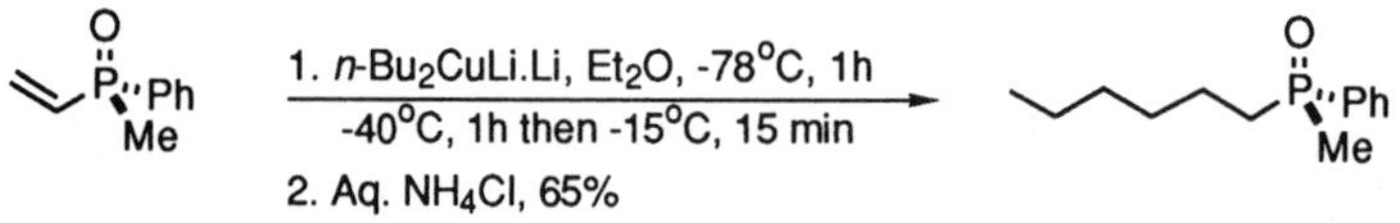

showed that cuprate addition to a *cyclic* alkenylposphinate, followed by *in situ* alkylation, is completely stereoselective.[62]

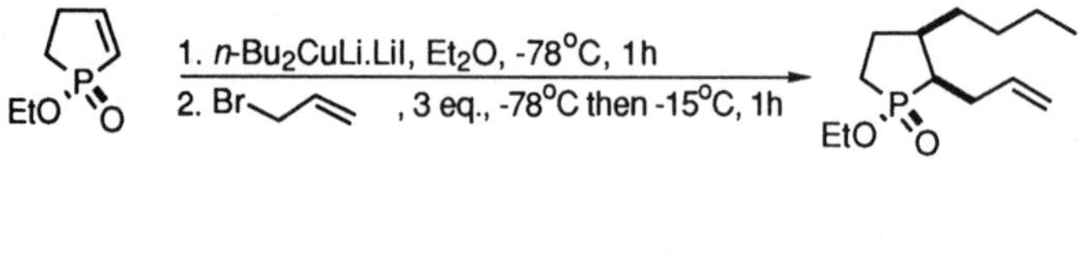

6.2.3   Other organometallic reagents

Organomagnesium reagents were first demonstrated to add in a conjugate fashion to alkenylsulfones over fifty years ago.[50] However, in this early work, only mixtures were obtained when alkenylsulfoxides were similarly treated. (For studies of conjugate additions to alkenones bearing an α-sulfoxido moiety, refer to the relevant sections in Chapters 2 and 3).

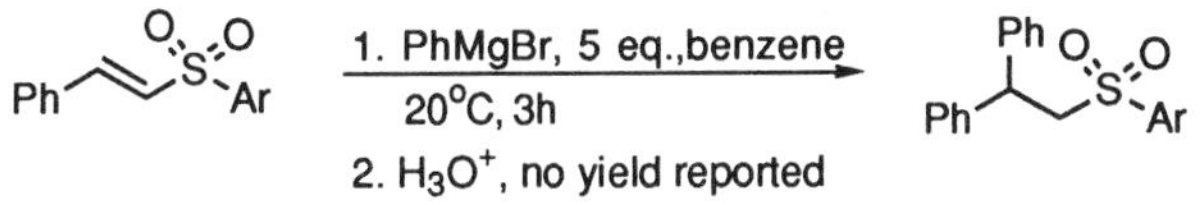

Organolithium reagents have proved to be more popular in conjugate additions to alkenylsulfones and their congeners. For example, γ-oxido-alkenylsulfones react with organolithium reagents giving an intermediate γ-lithiooxy-α,β-alkenylsulfone (13).[63,64] This intermediate then reacts with a second equivalent of the organolithium yielding the doubly alkylated products after quenching with an alkylating agent. Chelation directs the incoming nucleophile to add *syn* to the alkoxide. This method has been used to synthesize a variety of 2,3-disubstituted cycloalkenones.

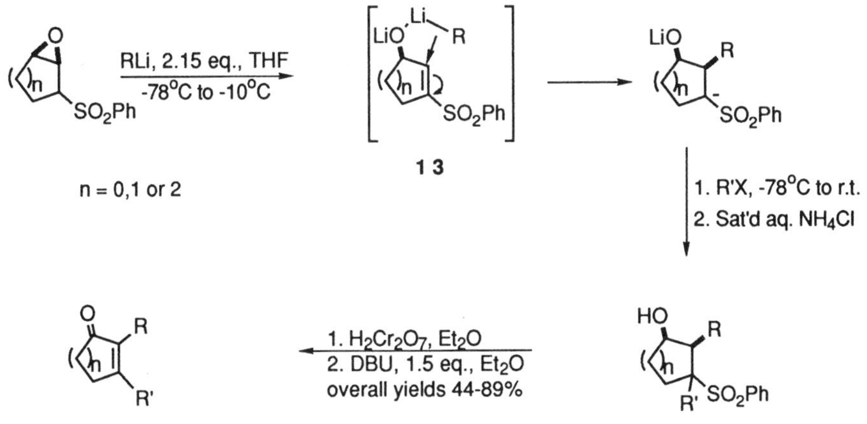

Isobe's group has examined the use of chelation control in acyclic systems, using α–trimethylsilylalkenylsulfones, e.g. (14), as conjugate acceptors.[65] These are readily prepared in two steps (Peterson olefination followed by oxidation) from bis(trimethylsilyl)-phenylthiomethane and the appropriate aldehyde.

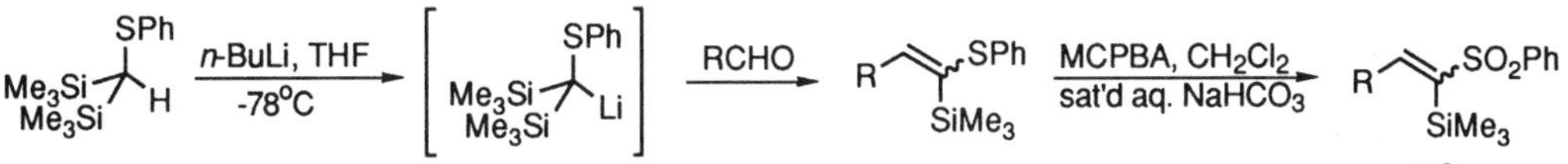

Most remarkable is the excellent stereoselection obtained, especially when the γ-oxy substituent was left unprotected. Only methlymagnesium bromide failed to react well, giving a low yield and poor stereoselectivity.

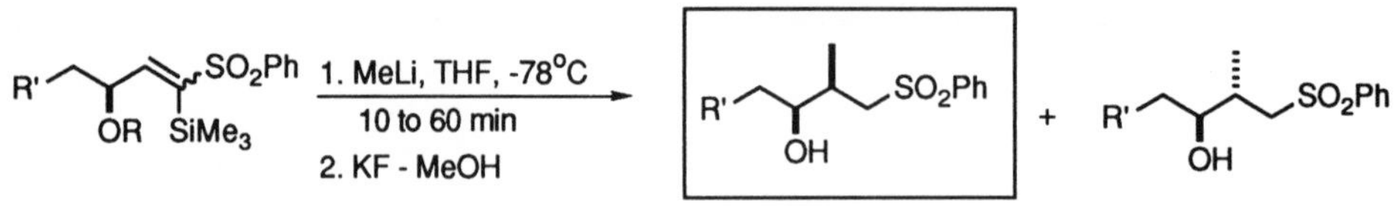

R = H*	R' = CH2Ph	85%	>99	:	<1
R = H*	R' = CH2Ph**	23%	80	:	20
R = TBS	R' = CH2Ph§	52%	95	:	5
R = MEM	R' = CCCH2OCH2OMe	78%	90	:	10
R = H*	R' = CCCH2OTHP	90%	97	:	3
R = H*	R' = CCCH2OH	65%	>99	:	<1
R = H*	R' = (CH2)3OCH2OMe	95%	>99	:	<1

*	2 eq, of MeLi used
**	MeMgBr, 18°C, 72h used instead
§	Reaction carried out at -40°C

The stereoselection may be accounted for by assuming a chelation-controlled transition state, as shown below. The opposite product stereochemistry often predominates

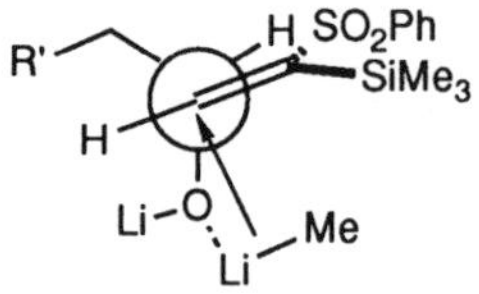

when additions are carried out with δ-oxyalkenylsulfones.[66] By varying the organolithium reagents, as well as the subsequent manipulations, the following pure, diastereomeric, 2-methoxy-6-alkyltetrahydropyrans were prepared.[67,68]

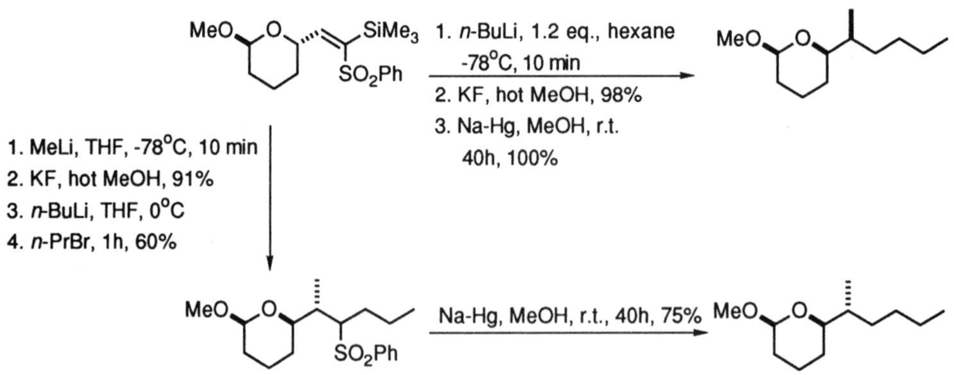

As with organocopper reagents, additions of alkyllithiums to Pyne's nor-ephedrine-derived alkenylsulfoximine (15) are highly stereoselective.[59,69]

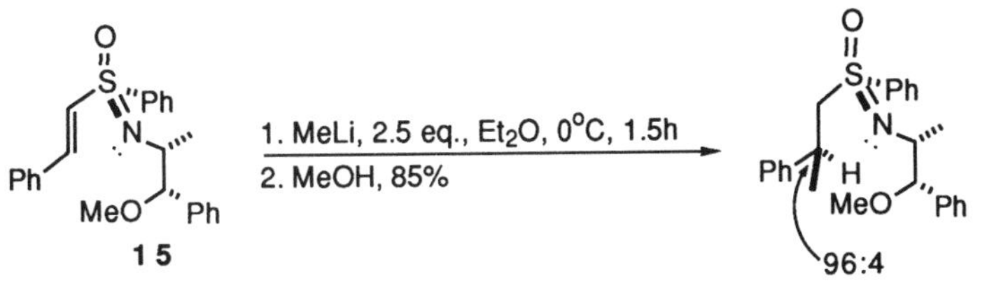

Another use for α-silylalkenylsulfones is in the synthesis of sulfines. Conjugate addition of an organolithium, followed by quenching with sulfur dioxide at low temperature, provides a simple route to sulfines.[70]

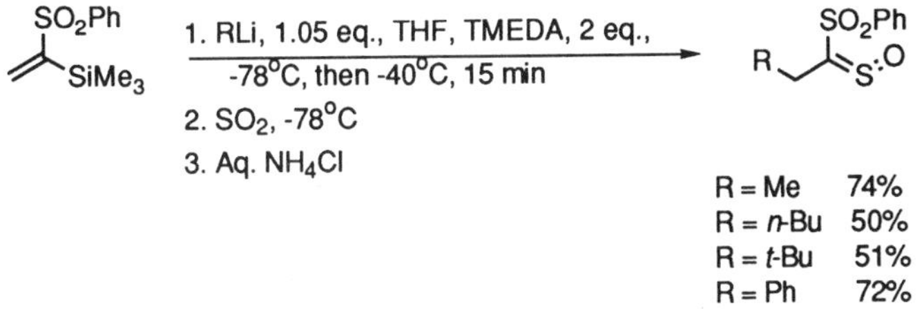

R = Me	74%
R = n-Bu	50%
R = t-Bu	51%
R = Ph	72%

Fuchs' group has studied the competition between conjugate addition and α-deprotonation in a series of γ-oxycycloalkenylsulfones.[71] They found that only rarely is deprotonation significant. The problem only develops with hindered aryllithium reagents. The authors reasoned that, as addition is perpendicular to the alkenyl plane whereas deprotonation is parallel to it, increasing the size of the sulfonyl substituent should inhibit deprotonation. This was shown to be the case. In equivalent intramolecular examples deprotonation can become much more significant.

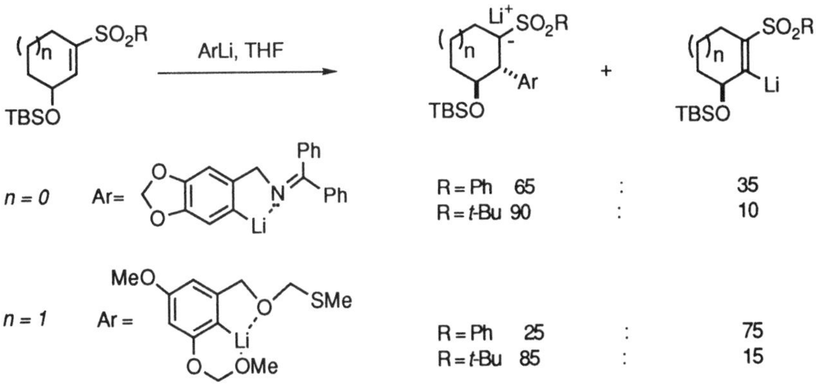

n = 0		
R = Ph 65	:	35
R = t-Bu 90	:	10

n = 1		
R = Ph 25	:	75
R = t-Bu 85	:	15

Fuchs' group has also developed several synthetic methods which are based upon conjugate additions of organolithium reagents to (mainly cyclo-)alkenylsulfones.[72] For example, chiral epoxycyclopentenylsulfones may be reacted with dimethylcopperlithium to give either the *cis* or *trans* product, depending upon the reaction conditions.[73]

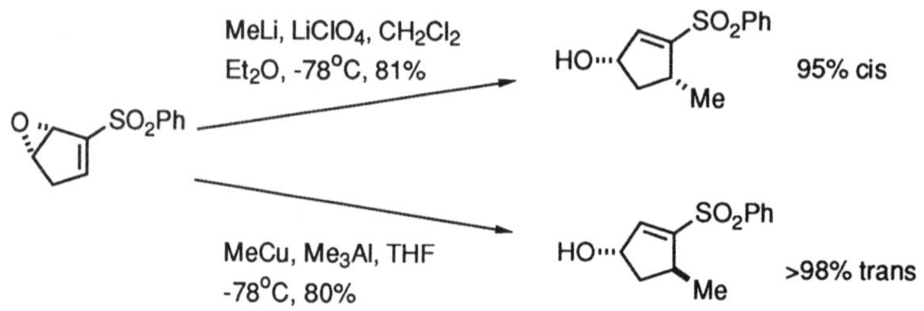
MeLi, LiClO4, CH2Cl2
Et2O, -78°C, 81%
95% cis
MeCu, Me3Al, THF
-78°C, 80%
>98% trans
SO2Ph
HO
Me

Alternatively, the chiral 4-silyloxy-5-dimethylaminocyclopentenylsulfone[74,75] may be reacted to give a stereochemically pure adduct.[76] N-Methylation followed by elimination then affords the *trans*-disubstituted cyclopentenylsulfone.

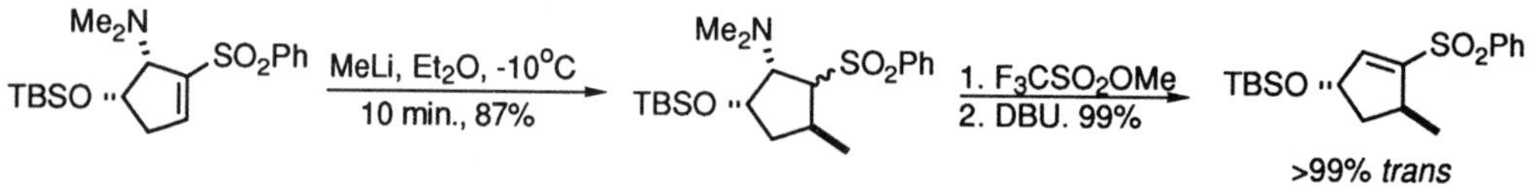
Me2N
SO2Ph
TBSO
MeLi, Et2O, -10°C
10 min., 87%
1. F3CSO2OMe
2. DBU. 99%
>99% trans

A similar strategy was used in a triply convergent synthesis of *l*-(-)-PGE$_2$.[77] Firstly, the 4-oxy-5-aminocyclopentene system had to be converted into the 3-oxy-5-amino system (**16**), an operational equivalent of 4-alkoxycyclopentenones.

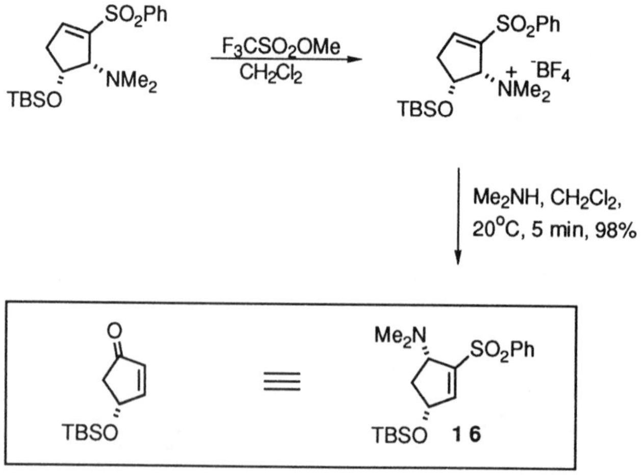
SO2Ph
NMe2
TBSO
F3CSO2OMe
CH2Cl2
BF4
Me2NH, CH2Cl2,
20°C, 5 min, 98%
Me2N
16

This was achieved by N-methylation followed by treatment with dimethylamine at low temperature. Conjugate addition of an alkenyllithium reagent proceeded smoothly producing the adduct in good yield and with excellent stereocontrol, after quenching with the allylating reagents shown below.

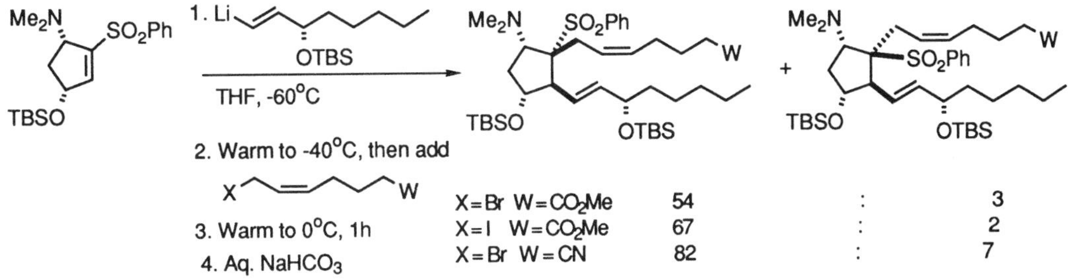

X = Br W = CO$_2$Me	54	:	3
X = I W = CO$_2$Me	67	:	2
X = Br W = CN	82	:	7

The ammonium tetrafluoroborate also reacts with organolithium reagents to give the opposite stereochemical result to that for the free amine discussed above, although in rather poor yields and stereoselectivity. This is to be contrasted with the excellent reactivity shown by this ammonium salt toward cuprates (see section 6.2.2).

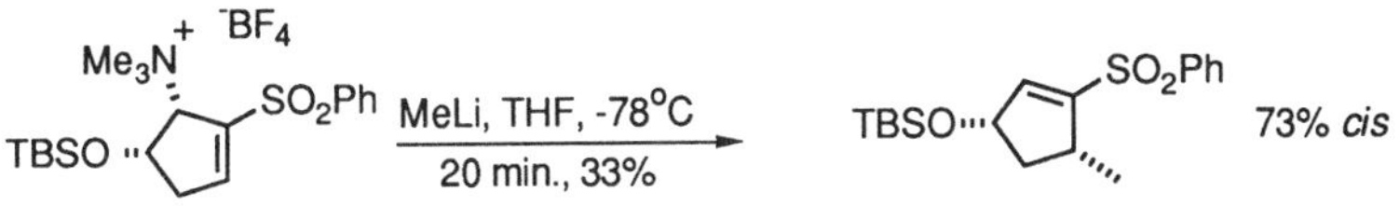

Pietrusiewicz's group has applied Luche's ultrasonication method, adding alkyl zinc reagents to conjugate acceptors in aqueous media, to alkenylphosphine oxides.[78]

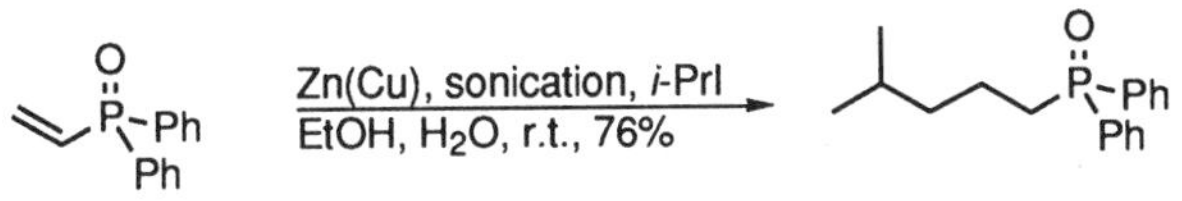

6.2.4   Heteronucleophiles

6.2.4.1   *Nitrogen*

Benzylamine adds with good stereoselectivity to isoborneol-derived chiral alkenylsulfoxides.[79,80] Only the Z-isomer gave good stereoselectivity.

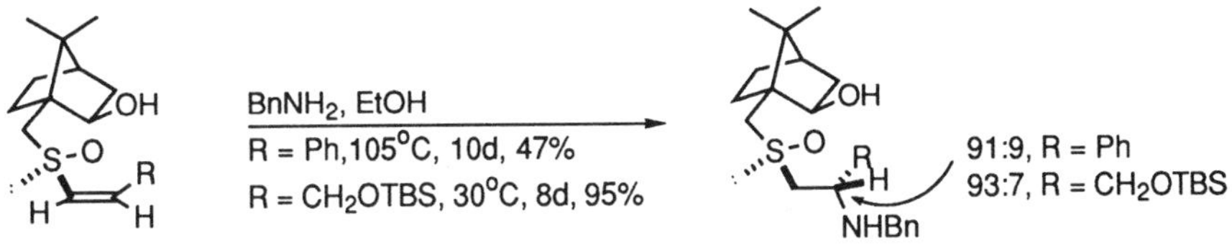

The authors proposed that addition occurred, preferentially, to one of two hydrogen-bonded conformations, as shown below.

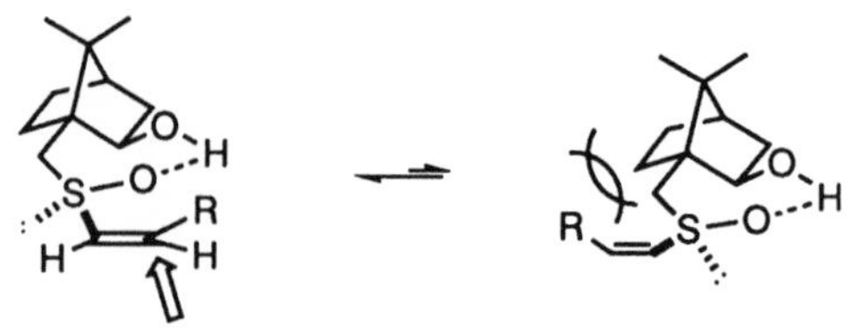

The same group has also shown that addition of amines to either the E- or Z-alkenylsulfoxides is also stereoselective.[81]

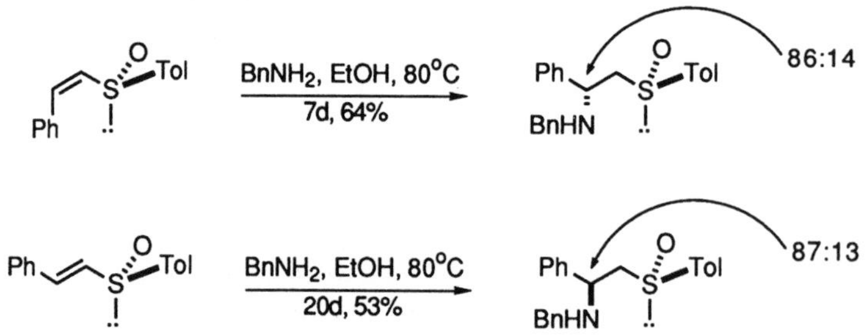

Several studies on intramolecular additions of amides to alkenylsulfoxides and sulfones have been reported, with the following transformation in mind:

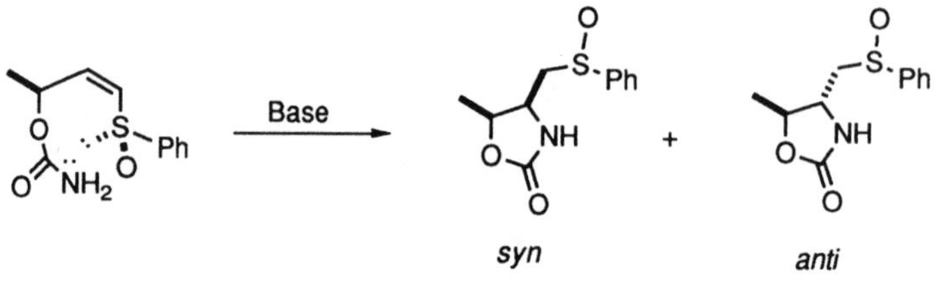

Excellent *anti*-selectivity was obtained and the authors accounted for this by modelling the transition state, with the assumption that all low energy conformations would have the sulfur lone pair in plane with the alkene.[82] Their calculations showed that the difference in energies between the two low energy conformations for each of the matched and mis-matched cases was significantly greater in the matched case.

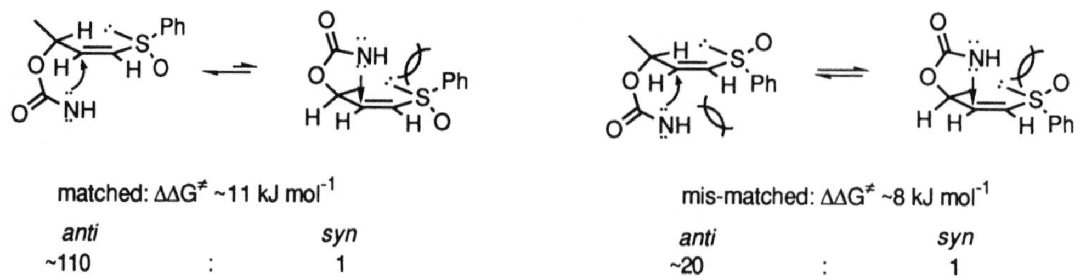

matched: $\Delta\Delta G^{\neq}$ ~11 kJ mol^{-1}

anti		syn
~110	:	1

mis-matched: $\Delta\Delta G^{\neq}$ ~8 kJ mol^{-1}

anti		syn
~20	:	1

They also found that sulfones give excellent selectivity, thus obviating the need for separating diastereomeric sulfoxides.[83] This led to a new synthesis of homochiral serine derivatives.

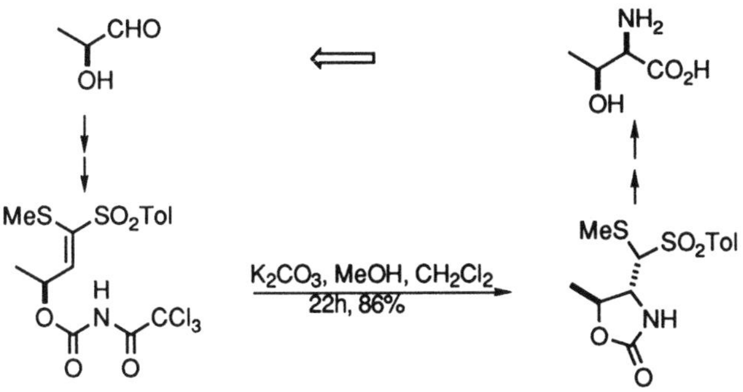

6.2.4.2   *Oxygen and sulfur*

A highly stereoselective synthesis of spiroketals has been developed based on an intramolecular conjugate addition of alkoxide to diastereomeric alkenylsulfoxides.[84] Oxidation of a 3-phenylthio-4,5-dihydrofuran (**17**) yields an easily separable diastereomeric mixture of sulfoxides. Each of these was then subjected to base-catalyzed ring closure followed by desulfurization. The isomeric spiroketals were produced in high yield and again with excellent stereoselectivity.

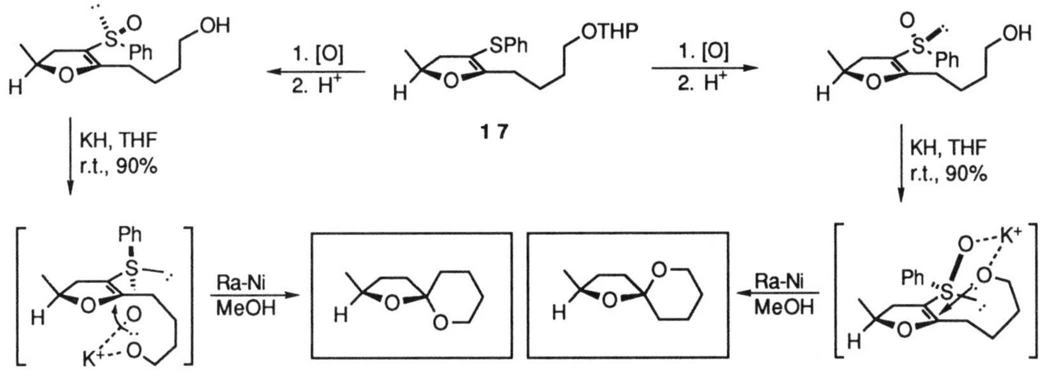

Thiocresol itself does not add to alkenylsulfones, however its sodium salt does.[50]

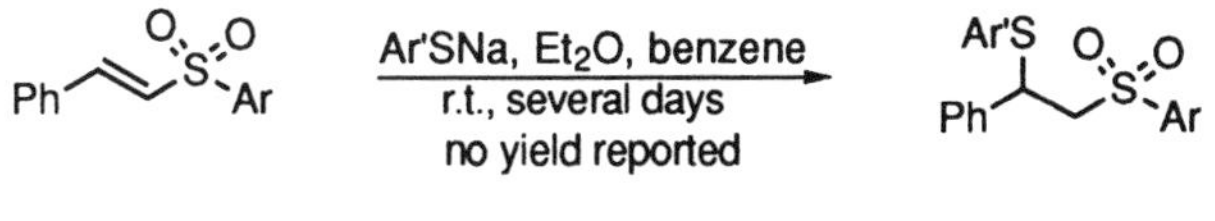

6.2.5.3   Silicon

The addition of LO silylcuprates to alkenylsulfoxides can be highly stereoselective, particularly in additions to the Z-isomer.[58] The use of HO silylcuprates gave better yields. However, the selectivities were much reduced. Quenching with iodomethane failed and quenching with deuterium oxide gave a mixture of all four diastereomers. Other electrophiles may have successfully reacted with the intermediate, but they were not mentioned in this study.

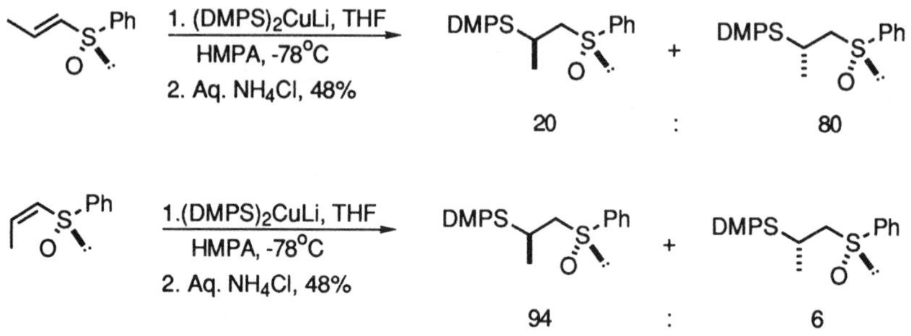

The authors used Hehre's conformational analysis (see section 1.3.7.1) to account for the observed stereoselectivity.

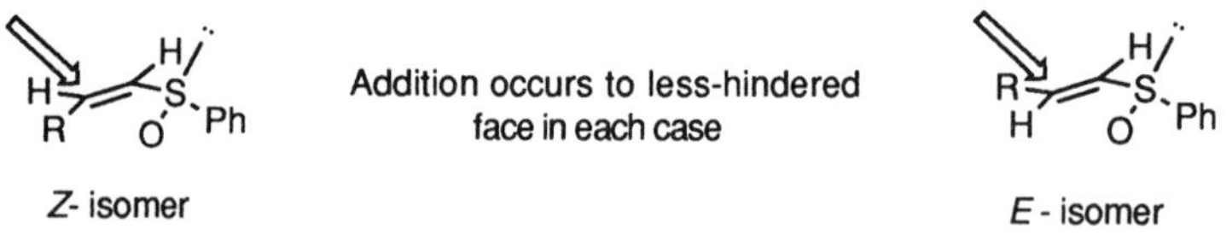

Addition occurs to less-hindered face in each case

Z- isomer   *E* - isomer

Alkenylsulfones may be desulfonated via conjugate addition of tributylstannyllithium followed by treatment with silica gel.[85]

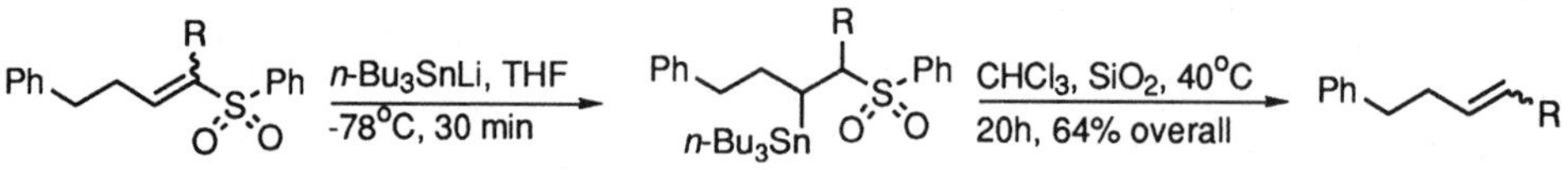

6.3   Quinones

Additions to quinones form one of the oldest areas of conjugate addition chemistry.[86] This is in large part due to the importance of naphthoquinones and anthraquinones to the natural and synthetic dyestuff industries. However, more recently this chemistry has found application in the synthesis of anthracyclines and related drugs.

   Conjugate additions to quinones are unique in that, overall, they often replace a hydride, i.e. they are oxidative additions. However, the mechanism is not one of direct displacement of hydride, but, rather, involves an aromatic intermediate (a quinol) which is usually reoxidized, either during the reaction or in a separate step. The oxidant may be either air (i.e. oxygen), a chemical oxidant or a second mole of the starting quinone. The overall process is summarized on the next page.

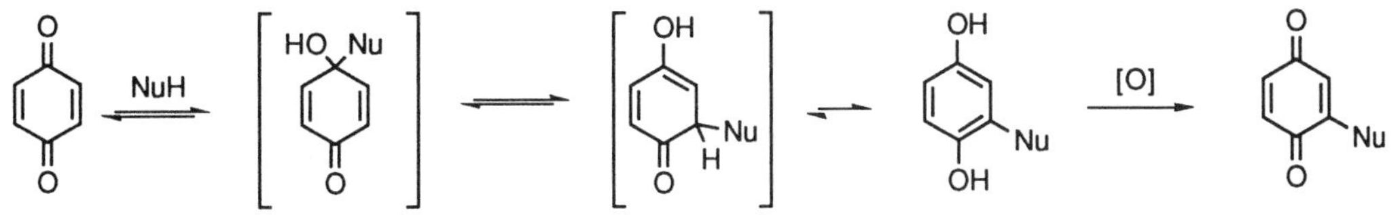

   Recently, clear evidence for initial addition to the carbonyl carbon, rather than to the double bond, has been presented for additions of ammonia to 1,4-benzoquinones.[87]

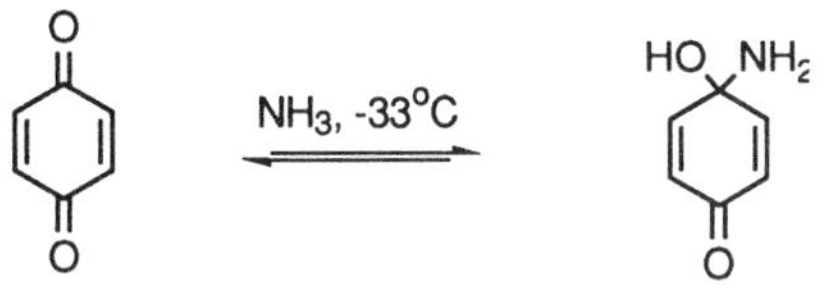

18

   These 1,2 adducts, e.g. (**18**), are stable for several hours to several days, at the low temperatures used in this study, ~-33°C. The quinone may be regenerated by simply pumping off the ammonia under vacuum. It was also demonstrated that the additions are quite selective for unsymmetrically-substituted 1,4-benzoquinones. Addition occurs to the carbonyl flanked by the greater number of *ortho*-substituents. This is probably due to selective solvation of the less-hindered carbonyl, leaving the other carbonyl relatively exposed to nucleophilic attack.

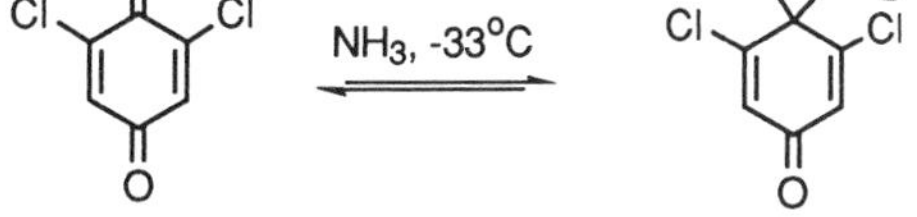

Regiochemistry of additions

Hard nucleophiles, i.e. non-stabilized carbanions and organometallic reagents, generally add to the carbonyl group of quinones. Advantage was taken of this in an interesting method for additions to quinones which involves 1,2 addition followed by a chelation-controlled 1,4 addition.[88]

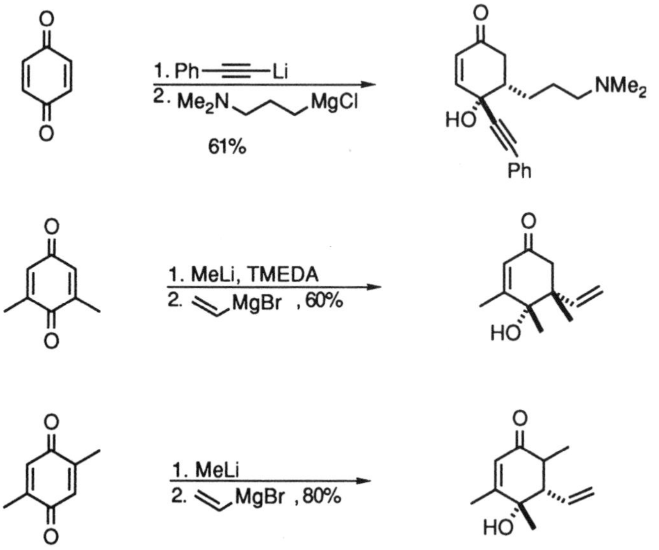

Chelation is proposed to occur to the lithium alkoxide (formed by the first addition) thus directing the incoming nucleophile to the same face as that bearing the alkoxide. (For a related process see pp 347 to 348).

As hard nucleophiles generally give carbonyl addition, the remainder of this section will focus on the addition of softer nucleophiles to 1,4-quinones. Several important guidelines now exist for predicting the site of addition to a variety of substituted 1,4-quinones. It should be stressed that these generalizations apply to ionic and not free radical reactions. Where free radicals have been added, mention is also made of the regiochemical outcome (usually the opposite to that for ionic reactions).

The Nenitzescu reaction is an important method for the synthesis of 5-hydroxyindoles.[89,90] It relies upon the addition of an enamine (usually an enaminoester) to a benzoquinone. The original procedure consisted of heating an acetone solution of benzoquinone with an enaminoester.[91] Since then it has been applied to many substituted quinones.

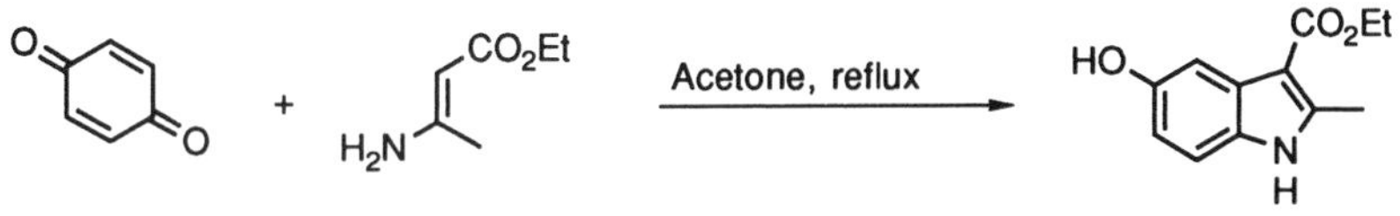

Another important reaction, which has spawned numerous approaches to the total synthesis of naturally-occurring quinones, is the addition of ketene acetals to quinones. This reaction was originally studied by McElvain and Engelhardt.[92] Some thirty years later it was demonstrated that two equivalents of a ketene acetal could add to the quinone in what constituted a new annulation procedure.[93]

Conditions were sought which could control the stoichiometry of the reaction and it was soon established that the reaction was highly solvent-dependent, as the following examples demonstrate.[94] Although this reaction has proved extremely useful in the synthesis of many naturally-occurring quinones, its importance lies more in the extension of this approach to the cycloadditions of nucleophilic, oxygenated butadienes.[95,96]

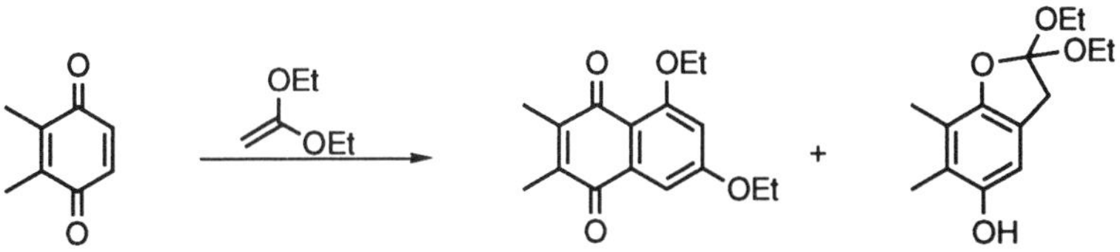

neat, heat	9%	63%
DMSO, r.t.	56%	0%

2-Acyl-1,4-quinones

Additions to 2-acyl-1,4-quinones occur at C3 when that position is unsubstituted. In fact this type of quinone is one of the most reactive known, often reacting exothermically with nucleophiles. After initial addition to the quinone nucleus, condensation between the acyl carbonyl and a functional group in the nucleophile can sometimes occur.[97] In the case shown below, a Nenitzescu-type reaction is diverted to produce an isoquinoline instead of an indole.

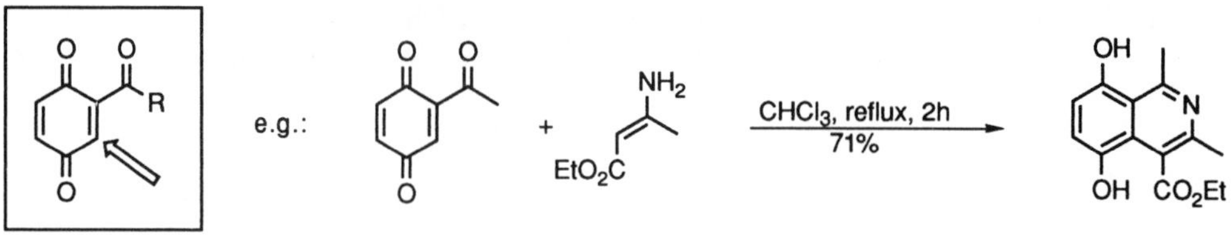

The tetracyclic skeleton of 11-deoxydaunomycinone has been synthesized from an acylnaphthoquinone by the following sequential intermolecular conjugate addition/intramolecular cycloaddition.[98] Note that the acyl group overrides any directing effect of the *peri*-methoxy substituent (see below).

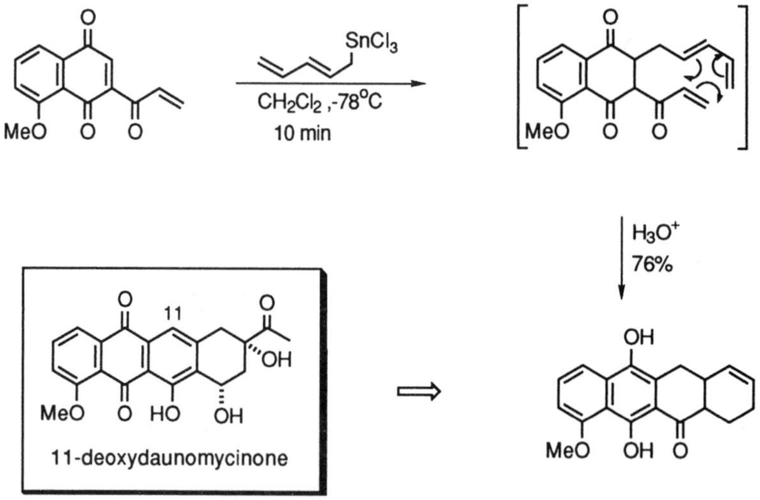

11-deoxydaunomycinone

Alkoxy, alkylthio or amino-1,4-quinones

Additions to 2-alkoxybenzo-1,4-quinones occur at C5. This is due to the deactivation of the carbonyl at C4 by the interaction of an oxygen lone pair of the alkoxy substituent.

When an alkoxy group is "in opposition" to an acyl group, as in 2-acetyl-5-methoxybenzo-1,4-quinone (19), then the influence of the acyl group dominates and addition occurs at C3.

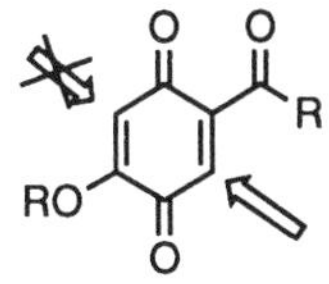

19

Another example of the regio-directing power of these substituents is in the Nenitzescu reaction. In the following reactions only one isomer is formed in each case.

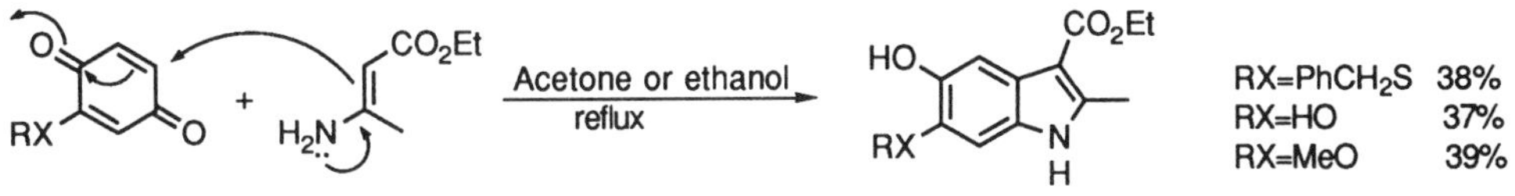

Acetone or ethanol
reflux

RX=PhCH$_2$S 38%
RX=HO   37%
RX=MeO   39%

A key step in Kishi's total synthesis of mitomycins was an intramolecular conjugate addition of a primary amine to a trisubstituted benzo-1,4-quinone (**20**).[99] The amine could not be isolated but added as soon as it was generated *in situ*. Only conjugate addition to the more activated system occurred, generating a seven-membered ring.

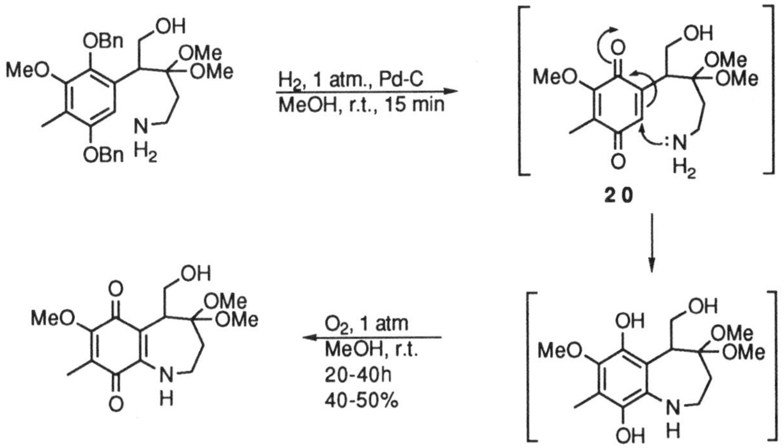

H$_2$, 1 atm., Pd-C
MeOH, r.t., 15 min

20

O$_2$, 1 atm
MeOH, r.t.
20-40h
40-50%

Substituents attached to the aromatic ring of naphtho- and anthraquinones can also have a profound influence on regiochemistry. As shown below, hydrogen bonding between a "peri" hydroxyl group[100] and its adjacent carbonyl leads to exclusive addition at the more activated carbon of the quinonoid double bond. An alkoxy group at C5 or C6 deactivates the C4 and C1 carbonyls, respectively.

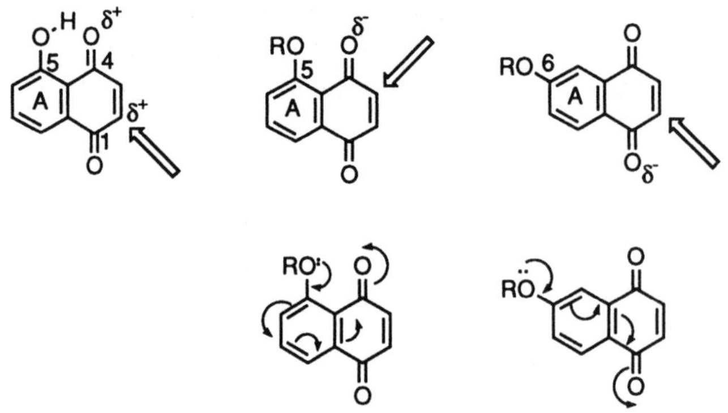

For example, addition of two equivalents of 1,1-dimethoxyethene to juglone (5-hydroxy-1,4-naphthoquinone) yields only one regioisomer.[101]

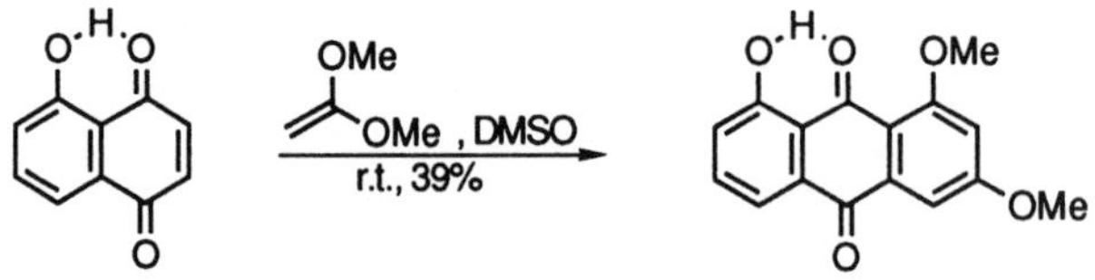

This is in sharp contrast to the addition to 5-methoxy-1,4-naphthoquinone, which yields, exclusively, the alternative regioisomer.

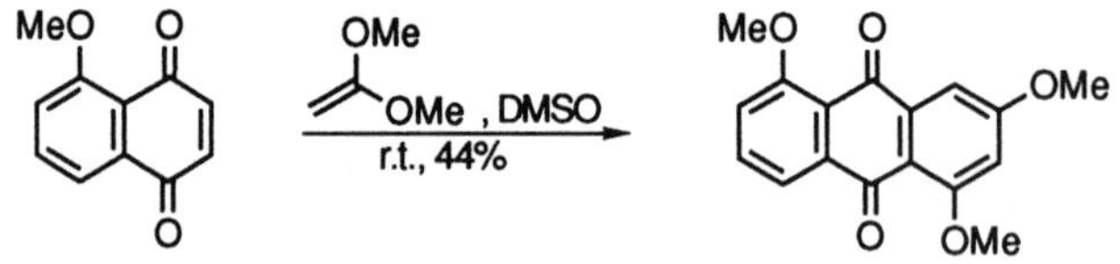

These are clear examples of the use of remote substituents in naphthoquinones to control the regioselectivity in these conjugate additions.

Alkyl-1,4-quinones

An alkyl substituent at C2 of 1,4-benzoquinones also deactivates the carbonyl at C4. However, this is not as effective as an alkoxy or amino substituent and, consequently, regioisomeric mixtures of addition products are often obtained. Remarkably, the alkyl group is occasionally *displaced* by a nucleophile.[102] The mechanism is thought to involve a reverse Mannich reaction.

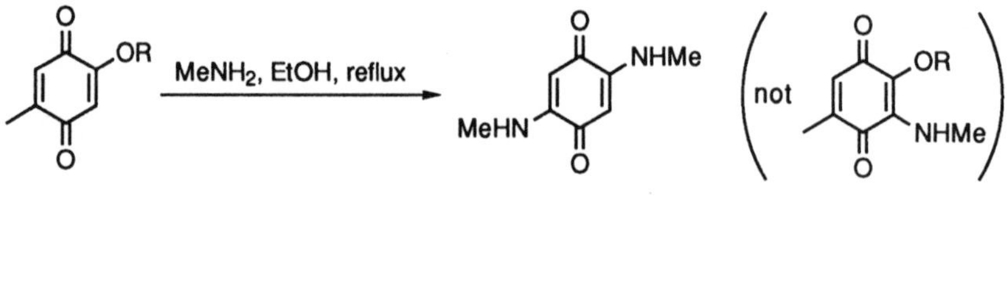

An attempt to use 2-methylbenzo-1,4-quinone in the Nenitzescu reaction led to isomeric mixture of products, although the 2,6-dimethyl derivative separated cleanly from the reaction mixture.[103] By contrast, the 2-ethyl derivative provides the 2-methyl-6-ethylindole isomer exclusively.[104]

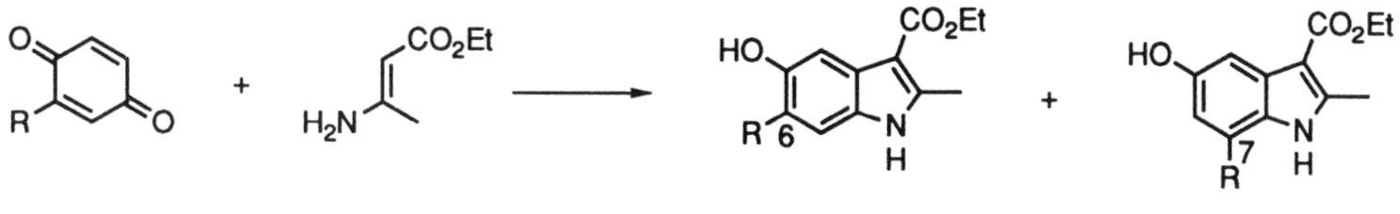

R = Me	1	:	1	
R = Et	1	:	0	

The addition of *cyclo*enamino esters to 2-alkylbenzo-1,4-quinones, another variant of the Nenitzescu reaction, provides a very direct method for preparing the pyrroloindole nucleus of the mitomycin antibiotics.[105]

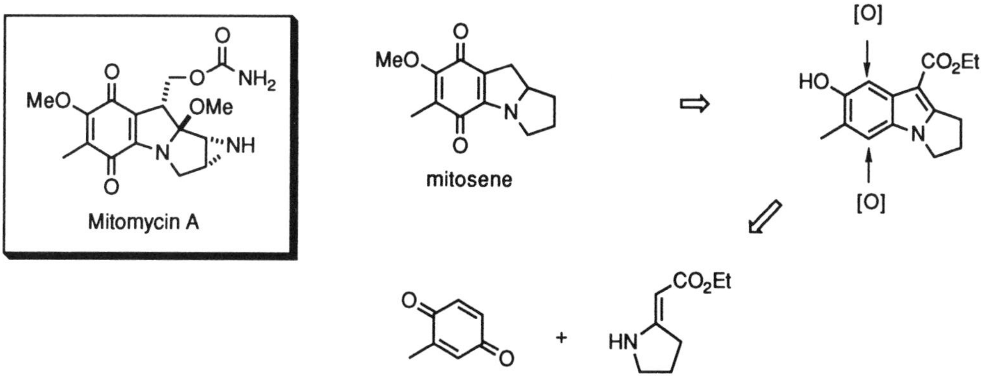

Mitomycin A

mitosene

This idea has been extended to reactions with benzo-1,4-quinone monoacetals.[106]

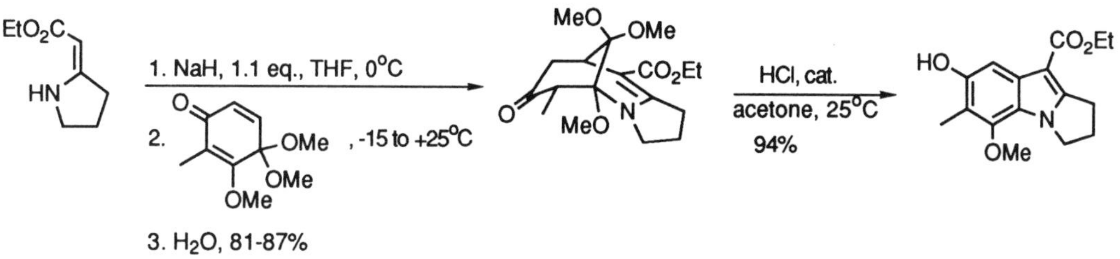

1. NaH, 1.1 eq., THF, 0°C
2. -15 to +25°C
3. H₂O, 81-87%

HCl, cat.
acetone, 25°C
94%

2-Halobenzo-1,4-quinones

The chemistry of halobenzo-1,4-quinones is largely that of addition/elimination reactions.[107] However, occasionally addition occurs at the adjacent carbon, as in the following reaction.[108]

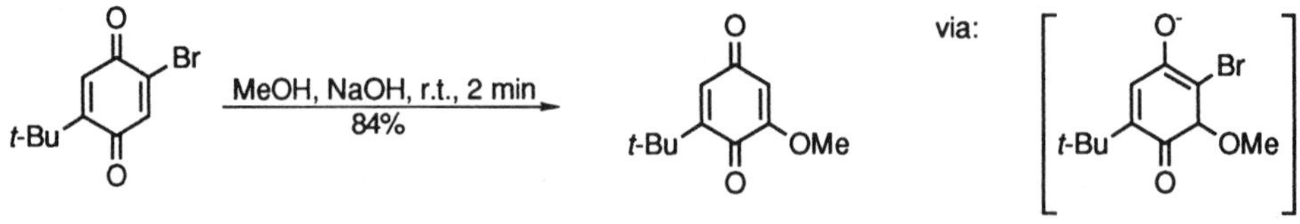

6.4   Alkenylsilanes, sulfides and selenides

Additions to these conjugate acceptors have been dominated by reactions with organometallic reagents. For example, organolithium reagents add smoothly to alkenylsilanes. In the original work on this reaction, Cason and Brooks found that, in addition to the expected triphenylsilylethene, a second compound, identified as 2-phenyl-1-(triphenylsilyl)ethane, was produced when phenyllithium was added to trichlorosilylethene in diethyl ether.[109] They postulated that this unexpected product came from the addition of phenyllithium to triphenylsilylethene which had been generated *in situ*. They then showed that this was the case by isolating the product alkene and showing that phenyllithium adds quite efficiently to it. This was quite an early example of the addition of an organometallic to an alkene activated by something other than the more conventional activating groups, such as a carbonyl or nitrile. Their results are summarized in the following scheme.

Seebach's group has developed a ketone synthesis based upon organolithium addition to 1,1-bis-(trimethylsilyl)alkenes, e.g. (21), quenching with an aldehyde and epoxidation/rearrangement.[110]

SiMe₃ / SiMe₃ (structure **21**)

1. RLi, 1.1 eq., THF, hexane,
-78°C, then r.t., 6h

→

SiMe₃
R⌣SiMe₃
Li

2 1

R = n-Bu, s-Bu, t-Bu not Me
R' = H, Ph or PhCHCH

1. R'CHO, THF, 12h
2. H₂O, 61-81%

O
R⌣R'

1. MCPBA, CH₂Cl₂, r.t., 2.5h
2. Aq. NaHCO₃, 76%

←

SiMe₃
R⌣R'

Alkenylsulfides and selenides also react well with organometallic reagents. The intermediate carbanion may be quenched with a variety of electrophiles. In the case of alkenylselenides, α-deprotonation or C-Se bond cleavage are also possible, hence the choice of reaction conditions is critical (Table 6.6, entry 3). The use of diethyl ether at 0°C works quite well, however lower temperatures lead to poorer yields and the use of tetrahydrofuran, especially at -78°C, yielded mainly α-deprotonation.

Table 6.6 Conjugate additions of organolithiums to alkenylsilanes, sulfides and selenides[111]

Entry	Substrate	Conditions	Product
1[112]	SiPh₃	1. n-BuLi, 1 eq., Et₂O, r.t. 2. PhCHO 3. Dil. HCl, 0°C, "good" yield	(chain) Ph
2[112]	SiMe₃	1. PhLi, Et₂O, TMEDA, 1 eq.,0°C, 1h 2. PhSSPh, 55%	SiMe₃ Ph⌣SPh
3[112]	SPh	1. MeLi, Et₂O, TMEDA, 1 eq.,0°C, 1h 2. TMSCl, 81%	SPh ⌣SiMe₃
4[110]	SePh	1. n-BuLi, hexane, DMM, 0°C, 10 min 2. Acetone, 1.5 eq., -78°C, 1h 3. Aq. NH₄Cl, 60%	SePh (chain) OH
5[70]	SPh SiMe₃	1. n-BuLi, Et₂O, TMEDA, 1 eq.,0°C, 1h 2. H₂O, 85%	SPh (chain) SiMe₃
6[70]	SPh SiMe₃	1. t-BuLi, 1.05 eq., THF, TMEDA, 2 eq., -78°C, then -40°C, 15 min 2. SO₂, -78°C 3. Aq. NH₄Cl, 73%	SPh ⌣S=O

Yoshida's group has applied its electro-initiated oxygenation of alkenes to the transformation of alkenylsilanes into α-thiophenyl carbonyl compounds.[113] For example, treatment of an acetic acid solution of a γ-oxoalkenylsilane (see section 3.1.2 for their preparation) with oxygen and thiophenol, whilst intermittently passing a current through the solution, leads to the α-thiophenyl carbonyl derivative in good yield. The process is believed to involve a radical chain mechanism as shown below.

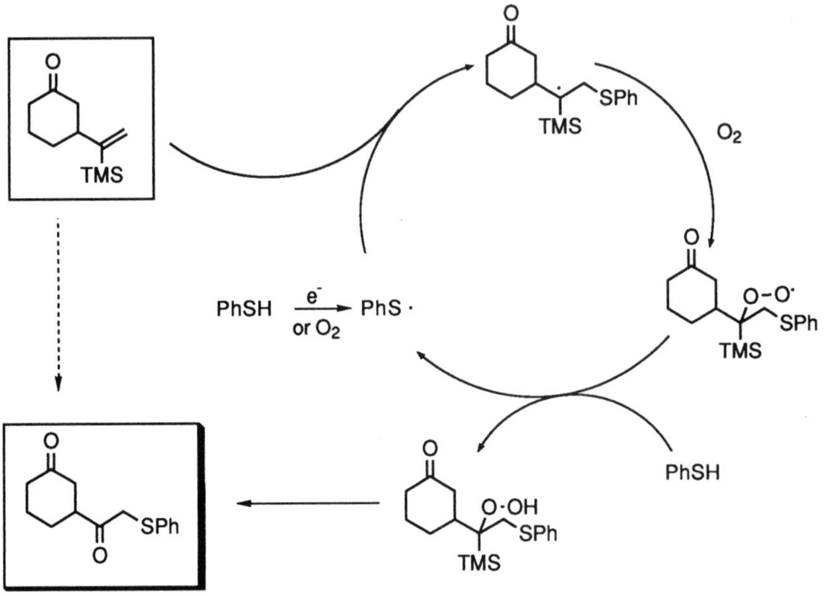

6.5   Ketene acetals

Quite a large variety of ketene dithioacetals undergo conjugate additions with organometallic reagents. These additions provide an alternative method for preparing 2-substituted 1,3-dithianes and dithiolanes. Some of these are shown in Table 6.7.

Table 6.7 Conjugate additions of organolithiums to dithioketene acetals

1[114]	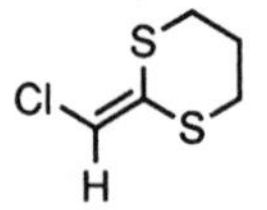	MeMgBr, 3 eq., THF, 55°C, 65h, 85% or MeLi, THF,-78°C, 3 min, 20%	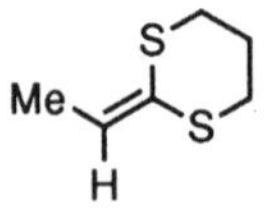
2[114]	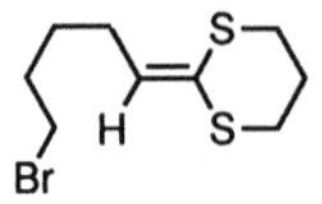	Mg,THF, 50°C,5h, 74%	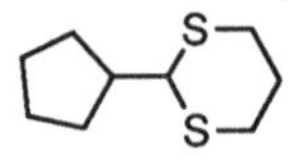
3[115,116]	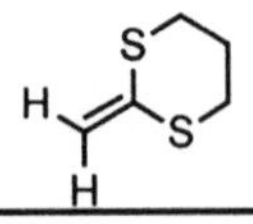	1. n-BuLi, THF,-30°C, 2.5h 2. H₂O, 89%	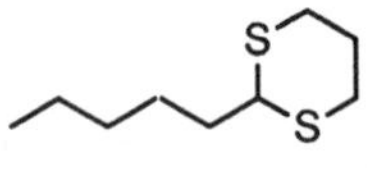

In general, organomagnesium reagents add well, although only with heating. Organolithiums react extremely rapidly (entry 1) but sometimes participate in side reactions when deprotonation or rearrangement are possible.[114,115]

6.6   Alkenylphosphonium salts

Cory's method for preparing tricyclic ketones (see section 6.1) culminated in a total synthesis of the sunflower constituent, trachylobanic-19-oic acid.[117]

trachylobanic-19-oic acid

pyridine, r.t., 3h
Aq. NaHCO₃, 23%

LDA, THF, hexane
0°C, 45 min

References

1.   For a review of the addition of stabilized carbanions to nitroalkenes, see A. Yoshikoshi and M. Miyashita, *Acc. Chem. Res.*, 1985, **18**, 284

2.   M. Miyashita, R. Yamaguchi and A. Yoshikoshi, *J. Org. Chem.*, 1984, **49**, 2857

3.   M. Miyashita, R. Yamaguchi and A. Yoshikoshi, *Chem. Lett.*, 1982, 1505

4.   E. J. Corey and H. Estreicher, *J. Amer. Chem. Soc.*, 1978, **100**, 6294

5.   I. V. Troitskaya, M. D. Boldyrev and B. V. Gidaspov, *Russ. J. Org. Chem.*, 1974, 1641

6.   S. Hoz, M. Albeck and Z. Rappoport, *Synthesis*, 1978, 162

7.   D. Seebach and P. Knochel, *Helv. Chem. Acta*, 1984, **67**, 261

8.   A. G. M. Barrett, G. G. Graboski and M. A. Russell, *J. Org. Chem.*, 1986, **51**, 1012

9.   R. M. Cory, P. C. Anderson, M. D. Bailey, F. R. McLaren, R. M. Renneboog and B. R. Yamamoto, *Can. J. Chem.*, 1985, **63**, 2618

10.   R. M. Cory and R. M. Renneboog, *J. Org. Chem.*, 1984, **49**, 3898

11.   R. M. Cory and R. M. Renneboog, *J. Chem. Soc., Chem. Commun.*, 1980, 1081

12.   F. Boberg, K-H. Garburg, K-J. Gorlich, E. Pipereit and M. Ruhr, *Liebig's Ann. Chem.*, 1985, 239

13.   See also, A. G. Gomez-Sanchez, B. M. Stictel, R. Férnandez-Férnandez, C. Pascval and J. Ballanoto, *J. Chem. Soc., Perkin Trans. I*, 1982, 441

14.	D. H. R. Barton and S. Zard, *J. Chem. Soc., Chem. Commun.*, 1985, 1098
15.	M. Miyashita, B. Z. E. Awen and A. Yoshikoshi, *J. Chem. Soc., Chem. Commun.*, 1989, 841
16.	F. C. Escribano, M. P. Alcantara and A. Gomez-Sanchez, *Tetrahedron Lett.*, 1988, **29**, 6001
17.	D. Seebach and P. Knochel, *Helv. Chem. Acta*, 1984, **67**, 261
18.	S. B. Bowlus, *Tetrahedron Lett.*, 1975, 3591
19.	A-T. Hansson and M. Nilsson, *Tetrahedron*, 1982, **38**, 389
20.	See reference 8
21.	D. Seebach and P. Knochel, *Helv. Chem. Acta*, 1984, **67**, 261
22.	D. H. R. Barton, H. Togo and S. Z. Zard, *Tetrahedron*, 1985, **41**, 5507
23.	M. E. Kuehne and L. Foley, *J. Amer Chem. Soc.*, 1965, **30**, 4280
24.	D. Seebach and J. Golinski, *Helv. Chim. Acta*, 1981, **64**, 1413
25.	(a) A. Risalti, M. Forchiassin, and E. Valentin, *Tetrahedron*, 1968, **24**, 1889; (b) F. P. Colonna, E. Valentin, G. Pitacco and A. Risalti, ibid., 1973, **29**, 3011; (c) E. Valentin, G. Pitacco, F. P. Colonna, and A. Risalti, ibid., 1974, **30**, 2741; (d) M. Calligaris, G. Manzini, G. Pitacco and E. Valentin, ibid., 1975, **31**, 1501; (e) S. Fabrissin, S. Fatutta, N. Malusa and A. Risalti, *J. Chem. Soc. Perkin I*, 1980, 686; (f) S. Fabrissin, S. Fatutta, and A. Risalti, ibid., 1981, 109
26.	J. W. Patterson and J. E. McMurry, *J. Chem. Soc., Chem. Commun.*, 1971, 488
27.	P. Bradamante, G. Pitacco, A. Risalti and E. Valentin, *Tetrahedron Lett.*, 1982, **23**, 2683
28.	A. Mezzetti, P. Nitti, G. Pitacco and E. Valentin, *Tetrahedron*, 1985, **41**, 1415
29.	M. Miyashita, T. Yanami, T. Kumazawa and A. Yoshikoshi, *J. Amer. Chem. Soc.*, 1984, **10**, 2149; see also M. Miyashita, T. Yanami and A. Yoshikoshi, *Org. Synth.*, 1981, **60**, 117
30.	I. Fleming and T. W. Newton, *J. Chem. Soc., Perkin Trans. I*, 1984, 119
31.	E. J. Corey and H/ Estreicher, *J. Amer. Chem. Soc.*, 1978, **100**, 6294
32.	A. G. M. Barrett, G. G. Graboski and M. A. Russell, *J. Org. Chem.*, 1985, **50**, 2603
33.	See also, M. Shibaya, M. Kuretani and J. Kubota, *Tetrahedron Lett.*, 1981, **22**, 4453
34.	D. Seebach and P. Knochel, *Helv. Chem. Acta*, 1984, **67**, 261
35.	M. Yamada, M. Yamashita and S. Inokawa, *Synthesis*, 1982, 1026
36.	See reference 8
37.	N. Ono, A. Kamimura and A. Kaji, *J. Org. Chem.*, 1986, **51**, 2139
38.	A. Kamimura and N. Ono, *J. Chem. Soc., Chem. Commun.*, 1988, 1278
39.	A. Kamimura and N. Ono, *Tetrahedron Lett.*, 1989, **30**, 731
40.	D. Seebach and P. Knochel, *Helv. Chem. Acta*, 1984, **67**, 261
41.	See reference 8
42.	N. Kobayashi and K. Iwai, *J. Org. Chem.*, 1981, **46**, 1823
43.	J. R. Hwu and N. Wang, *J. Chem. Soc., Chem. Commun.*, 1987, 427
44.	A. Padwa, U. Chiacchio, D. C. Dean, A. M. Schoffstall, A. Hassner and K. S. K. Murthy, *Tetrahedron Lett.*, 1988, **29**, 4169
45.	P. P. Wadgaonkar and G. W. Kabalka, *Syn. Comm.*, 1989, **19**, 805
46.	See also R. S. Varma and G. W. Kabalka, *Syn. Comm.*, 1985, **15**, 151
47.	H. Ohta, N. Kobayashi and K. Ozaki, *J. Org. Chem.*, 1989, **54**, 1802
48.	P. J. Brown,, D. N. Jones, M. A. Khan and N. A. Meanwell, *Tetrahedron Lett.*, 1983, **24**, 405
49.	I. Hori and T. Oishi, *Tetrahedron Lett.*, 1979, **20**, 4087

50.   E. P. Kohler and H. Potter, *J. Amer. Chem. Soc.*, 1935, **57**, 1316
51.   N. Ono, H. Miyake, A. Kamimura, N. Tsukui and A. Kaji, *Tetrahedron Lett.*, 1982, **23**, 2957
52.   N. Ono, A. Kamimura, H. Miyake, I. Hamamoto and A. Kaji, *J. Org. Chem.*, 1985, **50**, 3692
53.   N. Ono, H. Miyake, R. Tanikaga and A. Kaji, *J. Org. Chem.*, 1982, **47**, 5017
54.   T. Ohnuma, T. Oishi and Y. Ban, *J. Chem. Soc., Chem. Commun.*, 1973, 301
55.   I. M. Dawson, J. A. Gregory, R. B. Herbert and P. G. Sammes, *J. Chem. Soc., Chem. Commun.*, 1986, 620
56.   T. B. R. A. Chen, J. J. Burger and E. R. de Waard, *Tetrahedron Lett.*, 1977, **18**, 4527
57.   H. Sugihara, R. Tanikaga, K. Tanaka and A. Kaji, *Bull. Soc. Chem. Jap.*, 1978, **51**, 655
58.   K. Takaki, T. Maeda and M. Ishikawa, *J. Org. Chem.*, 1989, **54**, 58
59.   S. G. Pyne, *J. Org. Chem.*, 1986, **51**, 81
60.   F. Nicotra, L. Ponza and G. Russo, *J. Chem. Soc., Chem. Commun.*, 1984, 5
61.   K. M. Pietrusiewicz, M. Zablocka and J. Monkiewicz, *J. Org. Chem.*, 1984, **49**, 1522
62.   R. Bodalski, T. Michalski and K. M. Pietrusiewicz, *ACS Symp. Ser.*, 1981, **171**, 243
63.   P. C. Conrad and P. L. Fuchs, *J. Amer. Chem. Soc.*, 1978, **100**, 346
64.   J. C. Saddler, P. C. Conrad and P. L. Fuchs, *Tetrahedron Lett.*, 1978, **19**, 5079
65.   M. Isobe. M. Kitamura and T. Goto, *Tetrahedron Lett.*, 1980, **21**, 4727
66.   M. Isobe, Y. Ichikawa, Y. Funabashi, S. Mio and T. Goto, *Tetrahedron*, 1986, **42**, 2863
67.   M. Isobe, Y. Funabashi, Y. Ichikawa, S. Mio and T. Goto, *Tetrahedron Lett.*, 1984, **25**, 2021
68.   See also, M. Isobe, J. Obeyama, Y. Funabashi and T. Goto, *Tetrahedron Lett.*, 1988, **29**, 4773
69.   See also, R. Annunziata, M. Cinquini and S. Colonna, *J. Chem. Soc., Perkin Trans. I*, 1980, 2422
70.   M. van der Leij and B. Zwanenburg, *Tetrahedron Lett.*, 1978, 3383
71.   D. L. Barton, P. C. Conrad and P. L. Fuchs, *Tetrahedron Lett.*, 1980, **21**, 1811
72.   For a review of these methods, see P. L. Fuchs and T. F. Braish, *Chem. Rev.*, 1986, **86**, 903
73.   J. C. Saddler and P. L. Fuchs, *J. Amer. Chem. Soc.*, 1981, **103**, 2112
74.   R. E. Donaldson and P. L. Fuchs, *J. Amer. Chem. Soc.*, 1981, **103**, 2108
75.   J. C. Saddler, R. E. Donaldson and P. L. Fuchs, *J. Amer. Chem. Soc.*, 1981, **103**, 2110
76.   D. K. Hutchinson and P. L. Fuchs, *J. Amer. Chem. Soc.*, 1985, **107**, 6137
77.   R. E. Donaldson, A. McKenzie, S. Byrn and P. L. Fuchs, *J. Org. Chem.*, 1983, **48**, 2167
78.   K. M. Pietrusiewicz and M. Zablocka, *Tetrahedron Lett.*, 1988, **29**, 937
79.   S. G. Pyne, P. Bloem and R. Griffith, *Tetrahedron*, 1989, **45**, 7013
80.   For early work on amine additions to alkenylsulfoxides, see D. J. Abbott, S. Colonna and C. J. M. Stirling, *J. Chem. Soc., Chem. Commun.*, 1971, 471 and *J. Chem. Soc., Perkin Trans. I*, 1976, 49
81.   S. G. Pyne, R. Griffith and M. Edwards, *Tetrahedron Lett.*, 1988, **29**, 2089
82.   M. Hirama, H. Hioki, S. Ito and C. Kabuto, *Tetrahedron Lett.*, 1988, **29**, 3121
83.   M. Hirama, H. Hioki and S. Ito, *Tetrahedron Lett.*, 1988, **29**, 3125

84.   C. Iwaka, K. Hattori, S. Uchida and T. Imanishi, *Tetrahedron Lett.*, 1984, **25**, 2995

85.   M. Ochiai, T. Ukita and E. Fujita, *J. Chem. Soc., Chem. Commun.*, 1983, 619

86.   For reviews, see "Methoden der Organischen Chemie", E. Muller (ed.), Georg Thieme Verlag, Stuttgart, 1977, Vol. 7, Part 3a and K. T. Finley in "The Chemistry of the Quinonoid Compounds", S. Patai (ed.), Wiley, London, 1974, 877

87.   J. A. Chudek, R. Foster and F. J. Reid, *J. Chem. Soc., Perkin Trans. II*, 1984, 287

88.   M. Solomon, W. C. L. Jamison, M. McCormick, D. Liotta, D. A. Cherry, J. E. Mills, R. D. Shah, J. D. Rodgers and C. A. Maryanoff, *J. Amer. Chem. Soc.*, 1988, **110**, 3702

89.   G. R. Allen, Jr., *Org. React.*, 1973, **20**, 337

90.   For related studies using quinone diimines, see D. L. Boger and H. Zarrinmayeh, *J. Org. Chem.*, 1990, **55**, 1379 and references cited therein

91.   C. D. Nenitzescu, *Bull. Soc. Chim. Romania.*, 1929, **11**, 37 (*Chem. Abstr.*, 1930, **24**, 110)

92.   S. M. McElvain and E. L. Engelhardt, *J. Amer. Chem. Soc.*, 1944, **66**, 1077

93.   J. Banville, J. Grandmaison, G. Lang and P. Brassard, *Can. J. Chem.*, 1974, **52**, 80

94.   D. W. Cameron, M. J. Crossley and G. I. Feutrill, *J. Chem. Soc., Chem. Commun.*, 1976, 275

95.   This work lies beyond the scope of this book. However, see for example, D. W. Cameron, G. I. Feutrill, P. G. Griffiths and B. K. Merrett, *Tetrahedron Lett.*, 1986, **27**, 2421 and references cited therein

96.   See also W. Carruthers, "Cycloaddition Reactions in Organic Synthesis", Pergamon Press, Oxford, 1990

97.   G. R. Allen, Jr. and M. J. Weiss, *J. Org. Chem.*, 1968, **33**, 198

98.   Y. Naruta, Y. Nishigaichi and K. Maruyama, *Chem. Lett.*, 1986, 1703

99.   Y. Kishi, *J. Nat. Prod.*, 179, **42**, 549. See also, F. Nakatsubo, A. J. Cocuzza, D. E. Keely and Y. Kishi, *J. Amer. Chem. Soc.*, 1977, **99**, 4836 and F. Nakatsubo, T. Fukuyama, A. J. Cocuzza and Y. Kishi, *J. Amer. Chem. Soc.*, 1977, **99**, 8115

100.   R. G. Cooke and W. Seigel, *Aust. J. Sci. Res.*, 1950, **A3**, 628

101.   D. W. Cameron, M. J. Crossley, G. I. Feutrill and P. G. Griffiths, *Aust. J. Chem.*, 1978, **31**, 1363

102.   W. K. Anslow and H. Raistrick, *J. Chem. Soc.*, 1939, 1446

103.   R. J. S. Beer, K. Clarke, H. F. Davenport and A. Robertson, *J. Chem. Soc.*, 1951, 2029

104.   G. R. Allen, Jr., *Org. React.*, 1973, **20**, 344

105.   Y. Yamada and M. Matsui, *Agr. Biol. Chem.*, 1971, **35**, 282

106.   R. M. Coates and P. A. MacManus, *J. Org. Chem.*, 1982, **47**, 4822

107.   For a review, see "Methoden der Organischen Chemie", E. Muller (ed.), Georg Thieme Verlag, Stuttgart, 1977, Vol. 7, Part 3a

108.   F. R. Hewgill and L. R. Mullings, *J. Chem. Soc. B*, 1969, 1155

109.   L. F. Cason and H. G. Brooks, *J. Amer. Chem. Soc.*, 1952, **74**, 4582

110.   D. Seebach, R. Bürstinghaus, B-T. Gröbel and M.Kölb, *Liebig's Ann. Chem.*, 1977, 830

111.   Entries 2 to 5: S. Raucher and G. A. Koolpe, *J. Org. Chem.*, 1978, **43**, 4252

112.   T. H. Chan, E. Chang and E. Vinokur, *Tetrahedron Lett.*, 1970, 1137

113.   J. Yoshida, S. Nakatani and S. Isoe, *J. Org. Chem.*, 1989, **54**, 5655

114.   N. H. Andersen, P. F. Duffy, A. D. Denniston and D. B. Grotjahn,
        Tetrahedron Lett., 1978, **19**, 4315
115.   D. Seebach, *Synthesis*, 1969, 17
116.   R. M. Carlson and P. M. Helquist, *Tetrahedron Lett.*, 1969, 173
117.   R. M. Cory, D. M. T. Chan, Y. M. A. Naguib, M. H. Rastall and R. M. Renneboog,
        J. Org. Chem., 1980, **45**, 1852

Chapter Seven Activated alkynes

7.1 Alkynones

7.1.1 Stabilized carbanions

7.1.1.1 *Carbanions stabilized by π-conjugation with one heteroatom*

Attempts to extend the Robinson annulation procedure to include additions to alkynones have been reported. The earliest report came from Woodward's group in their work on the naturally-occurring cyclohexadienone, santonin.[1] However, the yield was poor. Several other attempts to improve this reaction met with little success.[2]

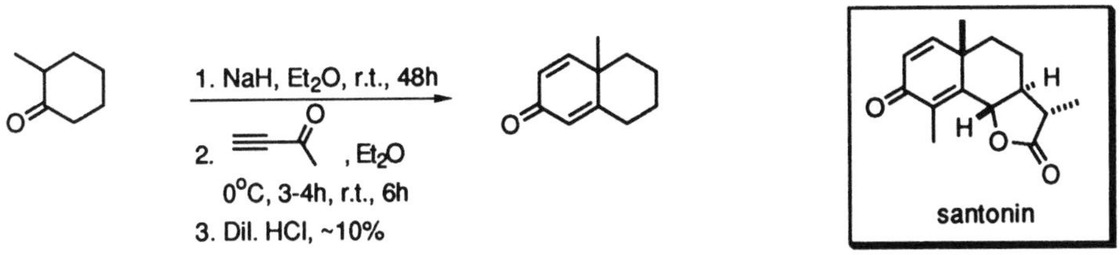

A chemoselective, as well as remarkably stereoselective, double addition was devised by Stork's group in its total synthesis of the orally active antifungal antibiotic, (±)-griseofulvin.[3] The reaction must be under kinetic control, as griseofulvin is quite readily equilibrated to a 60:40 mixture of griseofulvin and *epi*-griseofulvin.[4]

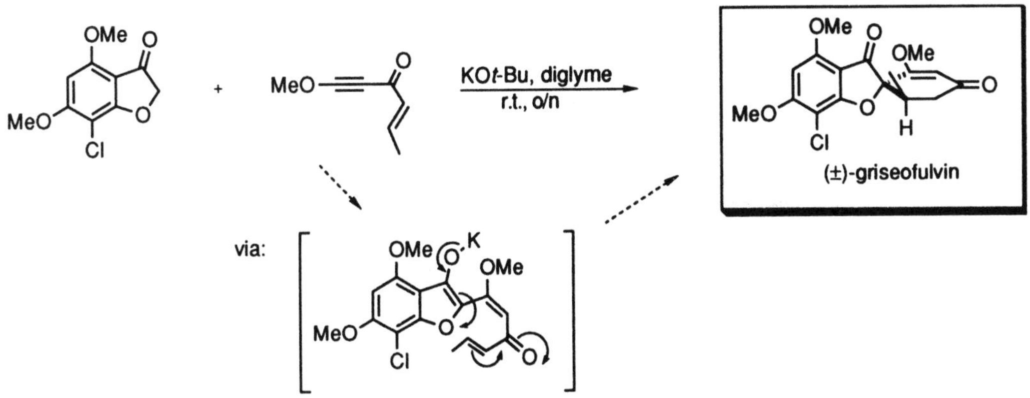

Intramolecular ring closures onto alkynones have only recently begun to be investigated. Corey's group has developed the following method in their total synthesis of the C15 gingkolide, bilobalide in both racemic[5] and enantiomerically-pure[6] forms. There are several points worth noting in this synthesis. Firstly, all the fifteen carbons of bilobalide are assembled in this coupling procedure. Second, the authors demonstrated that the reaction proceeds via a Claisen condensation. Treatment of the diester with one equivalent of base, followed by addition of the alkynone, gave the Claisen product in high yield (see below). They also offer the suggestion that the ring closure proceeds via electron transfer rather than a direct 5-*endo*-dig process.

The fact that the Claisen condensation occurs first is critical to the success of their enantioselective synthesis. Starting with the readily available bis-menthyl ester, Claisen condensation produces one enantiomerically-pure diastereomer. (Presumably, treatment with two equivalents of base would have led to the dianion and complete loss of stereoselection in the desired sense). The ring closure is then constrained to give the *cis*-fused product.

Nitroalkanes add efficiently to alkynones using PTC as shown in the following examples.[7] The stereoselectivity of the reaction varied with reaction time. For example, after one hour the addition of 2-nitropropane to 3-butyn-2-one gave a 62:38 ratio of E/Z isomers. Increasing the reaction time to 72 hours gave the E isomer exclusively.

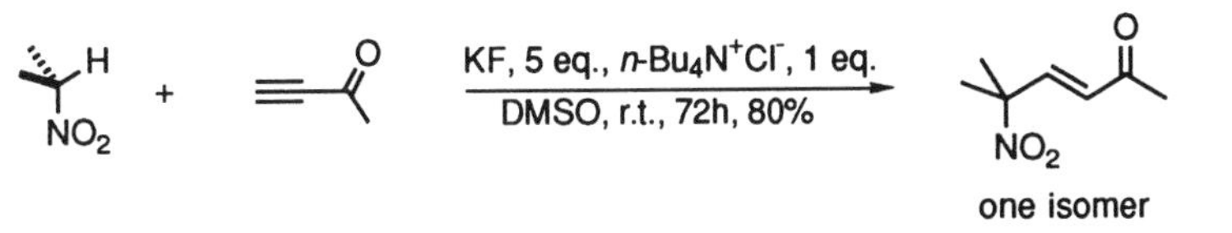

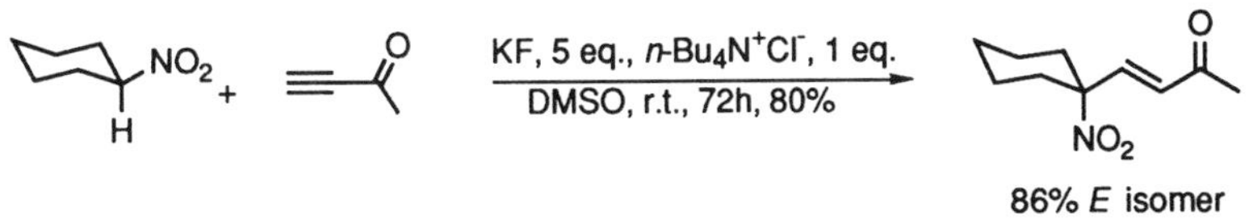

one isomer

86% *E* isomer

Tandem additions were also examined and the resulting method was applied to a simple synthesis of norsolanadione, a terpene isolated from tobacco leaves. Norsolanadione is an attractant for *tabakoshibanmushi*, a pest insect which infests food and tobacco.[8] It is also a flavourant and flavour enhancer for tobacco.[9]

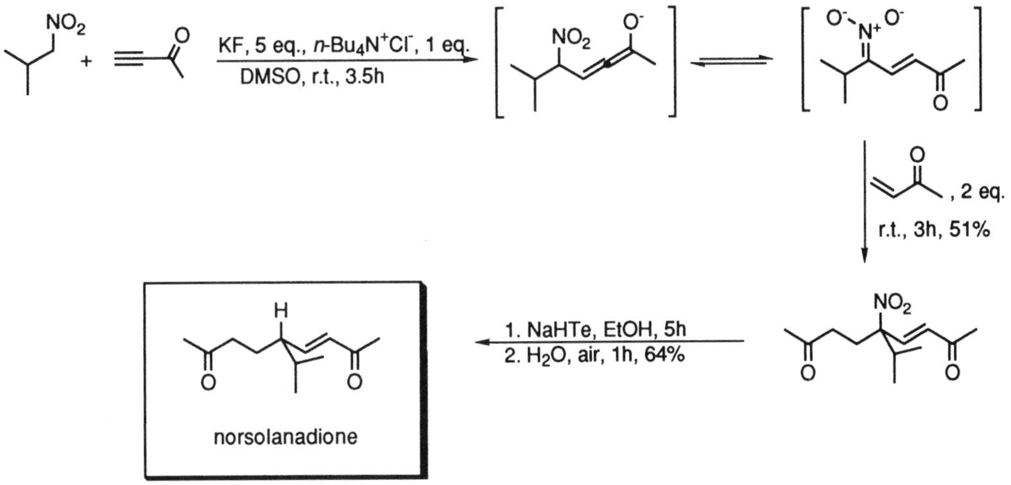

norsolanadione

7.1.1.2   *Carbanions stabilized by π-conjugation with more than one heteroatom*

The product from the addition of malonates to alkynones is often an α-pyrone.[10]

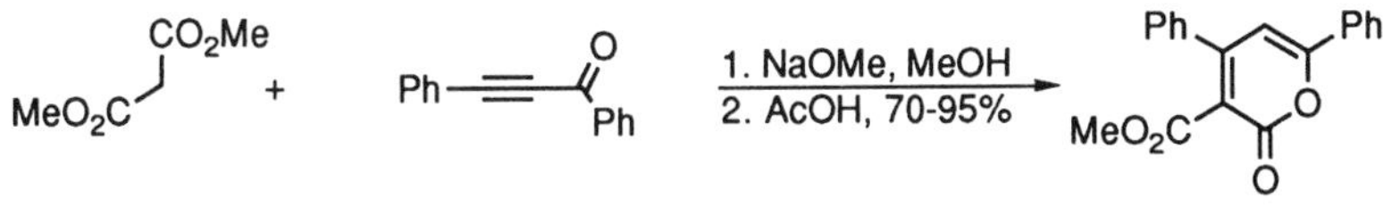

As part of their studies on the total synthesis of (±)-griseofulvin (see section 7.1.1.1), Stork's group added diethyl malonate to the enynone shown below.[3]

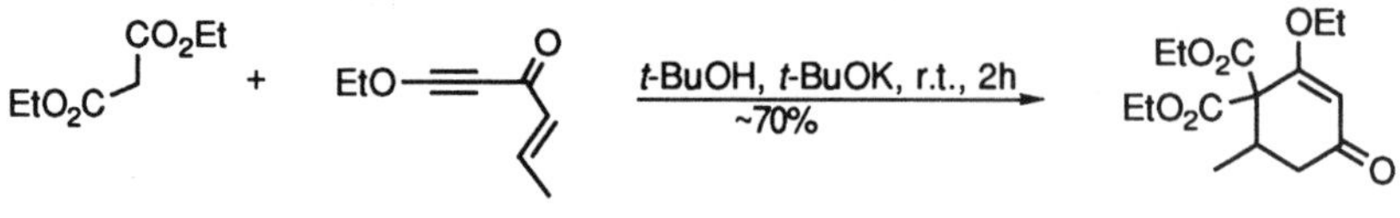

Small rings may be formed by closure of highly-stabilized carbanions onto alkynones.[11]

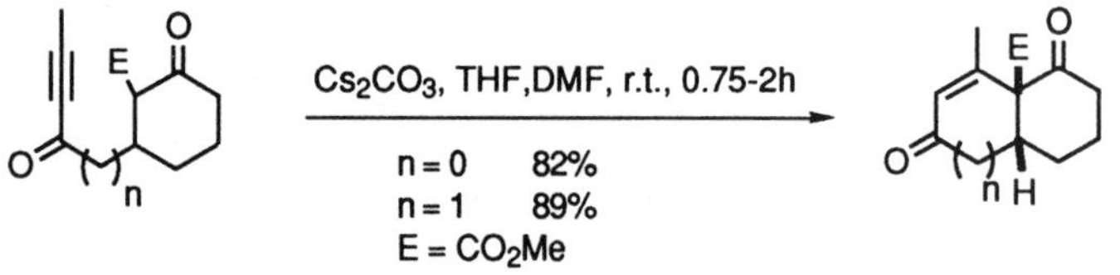

Cs₂CO₃, THF,DMF, r.t., 0.75-2h

$$Cs_2CO_3, \text{ THF,DMF, r.t., 0.75-2h}$$

n = 0   82%
n = 1   89%
E = CO₂Me

In this case, addition to the π bond which is conjugated to the carbonyl would lead to an intermediate allenolate. However, it seems unlikely that an allenolate could form in a five-membered ring. Alternatively, addition could occur to the non-conjugated π-bond leading to a higher energy (~71 kJ mol⁻¹) intermediate alkenyl anion. The authors proposed that this must be the situation where the alternative (allenolate) is too strained.

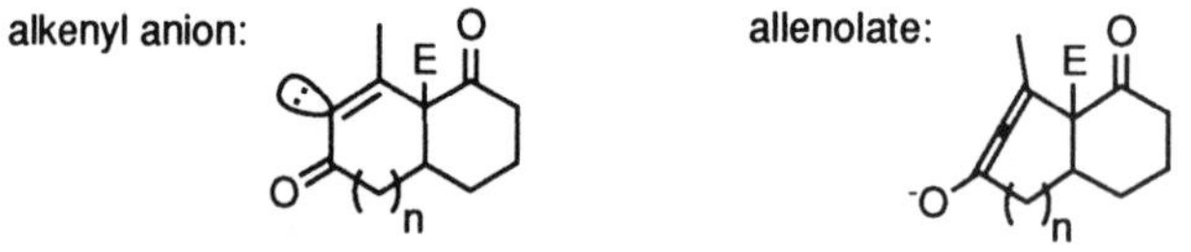

alkenyl anion: allenolate:

7.1.2   Organocopper reagents

Compared with other conjugate acceptors, there have been relatively few studies published on the addition of organocopper reagents to alkynones. Those that have been reported mostly deal with the problem of controlling the alkene stereochemistry. For example, it was found that the product E-Z isomer ratio from additions to trifluoromethylalkynones was quite sensitive to the type of organocopper reagent used.[12] Thus HO cuprates give considerably improved yields compared to LO cuprates.

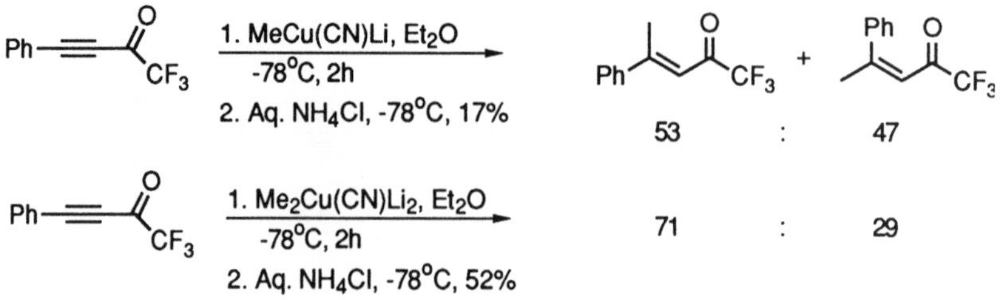

1. MeCu(CN)Li, Et₂O
   -78°C, 2h
2. Aq. NH₄Cl, -78°C, 17%

53   :   47

1. Me₂Cu(CN)Li₂, Et₂O
   -78°C, 2h
2. Aq. NH₄Cl, -78°C, 52%

71   :   29

Other reagents were less successful. Organomagnesium-derived copper reagents gave mainly reduction and/or 1,2-addition products. An attempt to use a residual ligand (1-hexynyl) led to transfer of this ligand to the carbonyl. Not unexpectedly, trifluoromethyl ketones are so electrophilic that they can provide these unusual results.

Inclusion of several equivalents of chlorotrimethylsilane increased the amount of the Z-isomer without changing the chemical yield.

LO cuprates add to 3-silylalkynones with excellent stereoselectivity and, depending on the type of quench used, either the alkenone or the allenol ether may be obtained in excellent yields.[13]

An addition of dimethylcopperlithium to a cycloalkynone, followed by equilibration, was used in a total synthesis of a cembranolide.[14,15]

7.1.3   Heteronucleophiles

A new method for preparing tetrahydroazocines involves conjugate addition of a 2-vinylazetidine to an alkynone. The adducts are then converted to the tetrahydroazocines by a thermal aza-Cope rearrangement.[16,17]

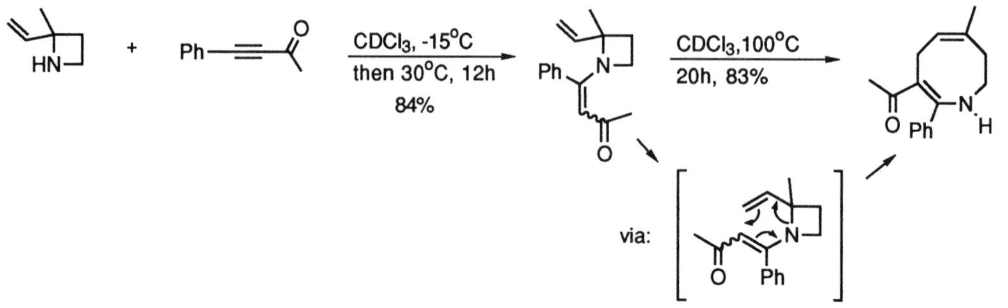

N-Trialkylstannyl tetrazoles, e.g. (**1**), react with alkynones, however a mixture of products is usually formed.[18]

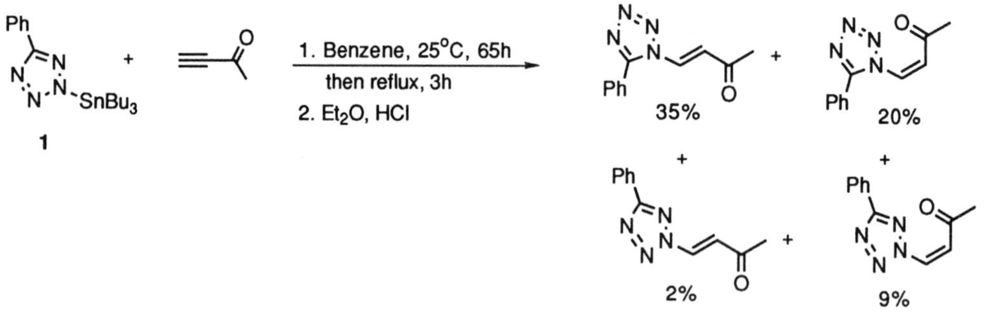

Tris(trimethylsilyl)alane delivers a trimethylsilyl group to alkynones efficiently.[19]

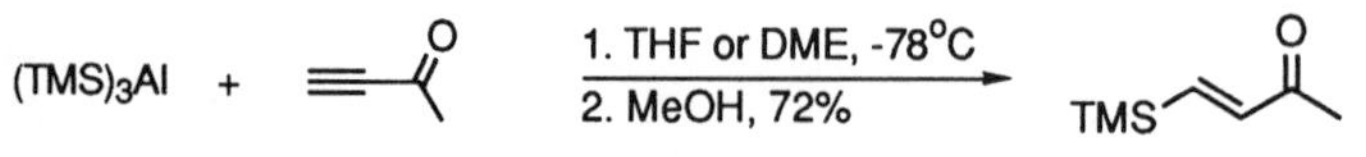

Iodotrimethylsilane adds to alkynones with excellent *syn* selectivity. The reaction gives a mixture of mono and bis addition.[20] Treatment of the reaction mixture with diisopropylethylamine eliminates HI from any di-iodo adduct.

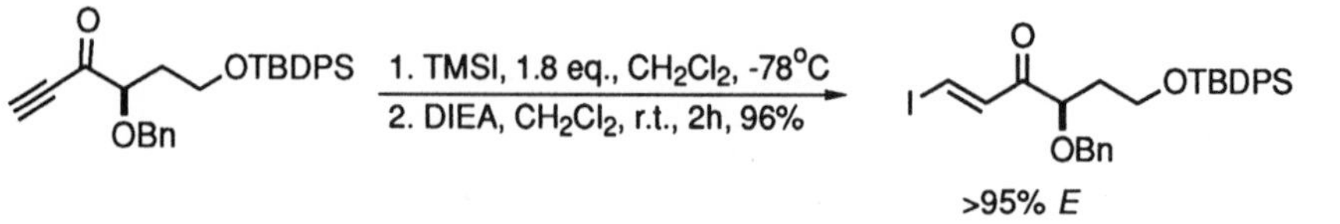

If a Lewis acid is included in the above reaction, then an intermediate iodoallenolate is formed which can be quenched with aldehydes.[21]

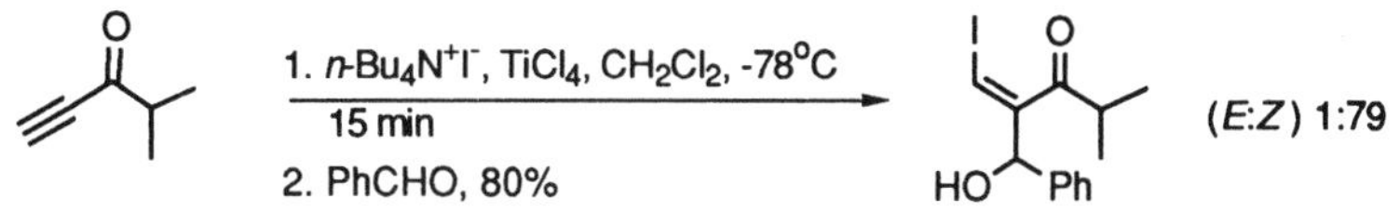
1. n-Bu$_4$N$^+$I$^-$, TiCl$_4$, CH$_2$Cl$_2$, -78°C
15 min
2. PhCHO, 80%
HO
Ph
(E:Z) 1:79

7.2 Alkynoic acids and their derivatives

7.2.1 Stabilized carbanions

7.2.1.1 *Carbanions stabilized by π-conjugation with one heteroatom*

Only a relatively few examples of conjugate additions of ketone enolates have been reported. In each of the cases shown below only the *E*-isomer was isolated.[22,23]

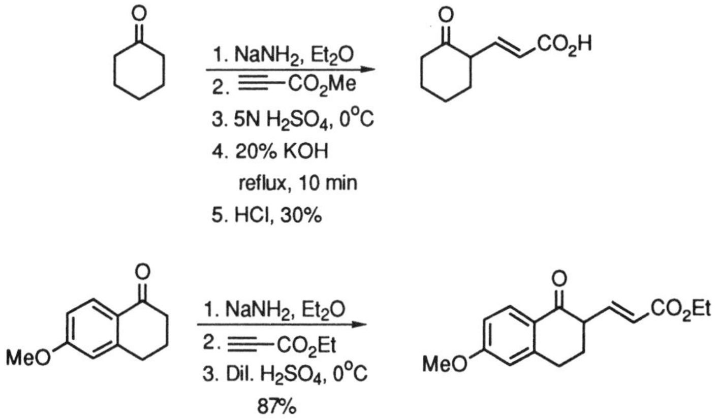
1. NaNH$_2$, Et$_2$O
2. ≡—CO$_2$Me
3. 5N H$_2$SO$_4$, 0°C
4. 20% KOH
reflux, 10 min
5. HCl, 30%
CO$_2$H
1. NaNH$_2$, Et$_2$O
2. ≡—CO$_2$Et
3. Dil. H$_2$SO$_4$, 0°C
87%
MeO
CO$_2$Et

Trost's group required the bicyclic cyanoketone (**2**) shown below in their total synthesis of (±)-hirsutic acid.[24] This proved to be especially useful as attempts to close the ring onto an alk*en*oate were unsuccessful.

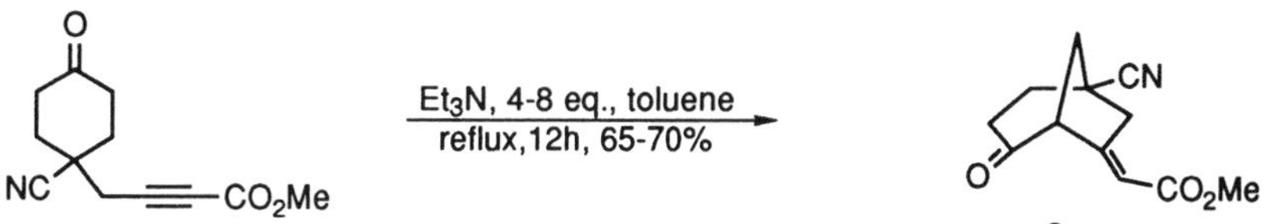
NC
CO$_2$Me
Et$_3$N, 4-8 eq., toluene
reflux, 12h, 65-70%
CN
CO$_2$Me

2

As with alkynones, PTC conjugate addition of nitroalkanes to alkynoates is an efficient reaction.[7] However, unlike additions to alkynones, the stereoselectivity does not change with reaction time, with the product E/Z ratios remaining constant at ~4:1. Some examples are given on the next page.

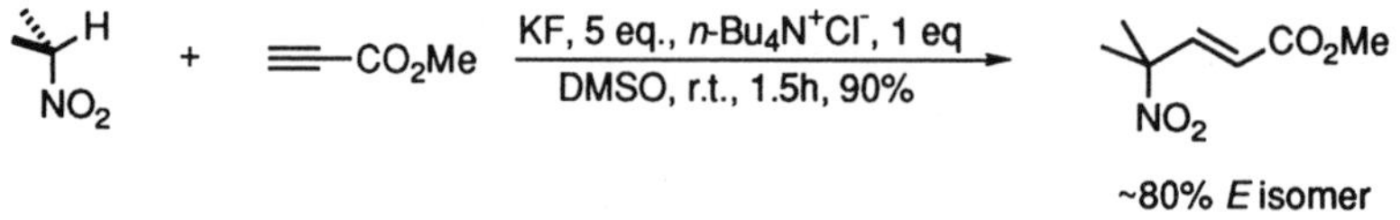

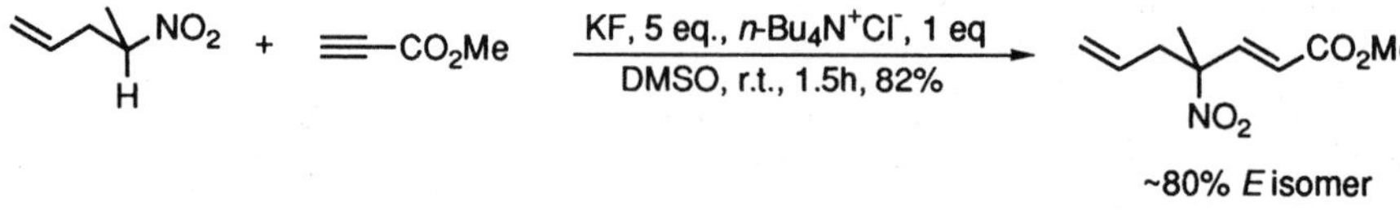

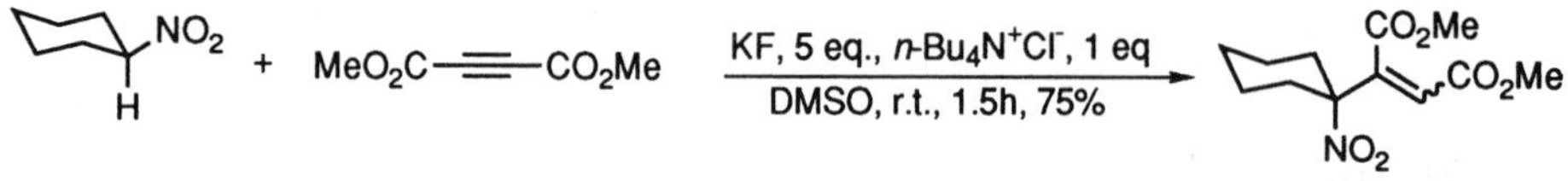

~80% E isomer

~80% E isomer

Simply using an excess of the propynoate gave products arising from the addition of the initial adduct to a second molecule of propynoate.

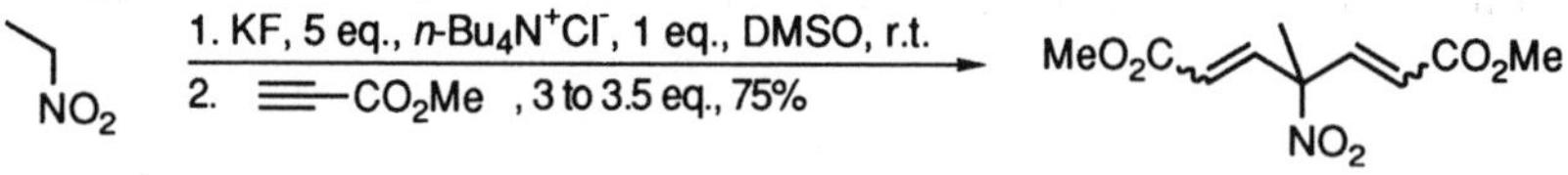

7.2.1.2   Carbanions stabilized by π-conjugation to more than one heteroatom

Doubly-stabilized carbanions also add to alkynoates. Some of the pioneering work on these additions was carried out by Kon in the 1930's.[25]

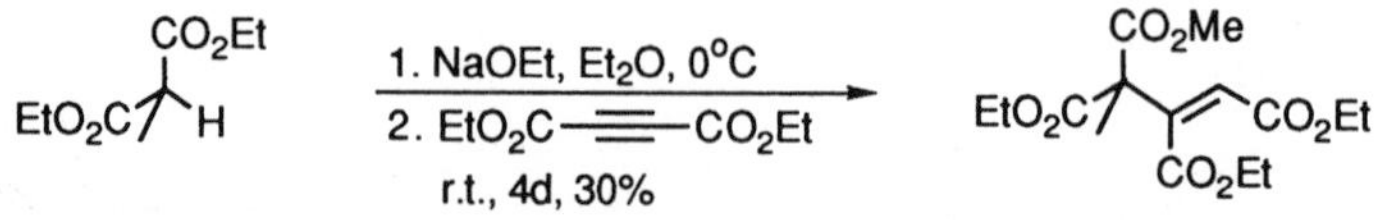

Apart from some early attempts by Kon's group to alkylate the intermediate allenolates formed initially on addition, successful trapping experiments have only recently been reported (see also the next section). Stirling's group has shown that the initial allenolate can be trapped by aldehydes.[26]

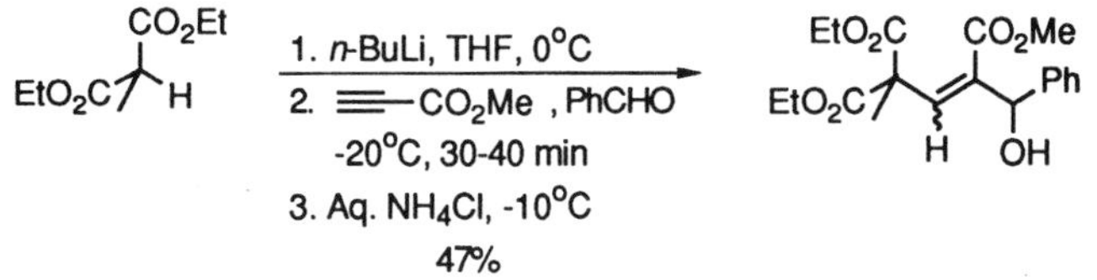

A ruthenium hydride-triphenylphosphine complex readily catalyzes additions of stabilized carbanions to alkynoates.[27]

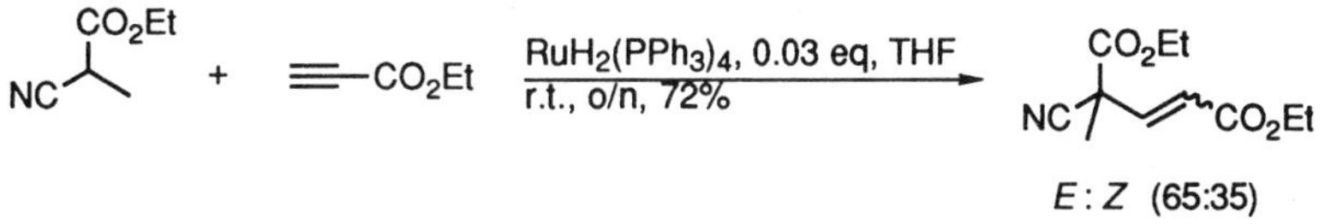

$E:Z$ (65:35)

7.2.2   Organocopper reagents

LO cuprates add efficiently to alkynoates and their derivatives. Regioselectivity is not a problem. However, stereoselectivity can be difficult to control. Corey's group demonstrated that very high overall *syn* selectivity was possible when the intermediate was trapped with either a proton or iodine.[28] It is important that these reactions are carried out in tetrahydrofuran. The use of diethyl ether leads to loss of stereocontrol, even at -70°C.

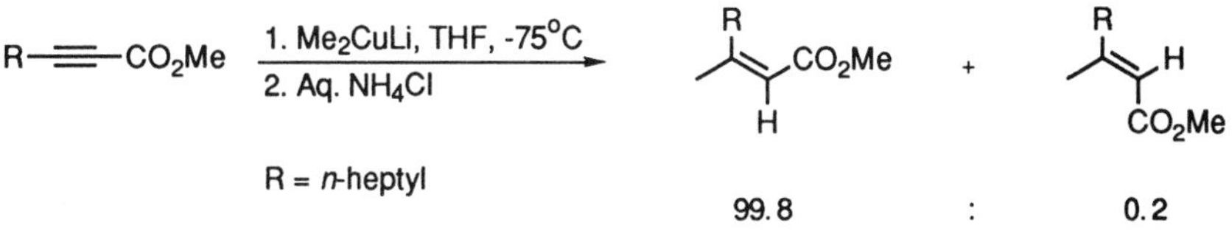

A nice example of these additions can be found in Corey's total synthesis of bongkrekic acid.[29] In this case addition is to an alkynylnitrile.

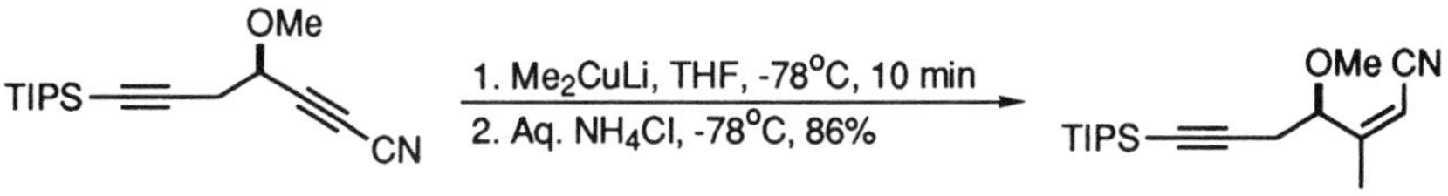

Good chemo- as well as stereoselectivity has been demonstrated for such additions in a total synthesis of some sex pheromones of the green stink bug.[30] In the following example no side reactions with the epoxide were observed.

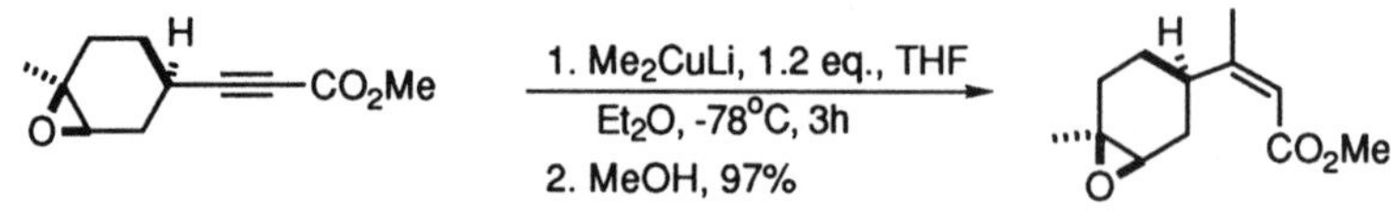

In their studies on the synthesis of juvenile hormones, Henrick's group discovered that *both* ligands on LO homocuprates could be transferred.[31] This, of course, means that only 0.5 molar equivalents of cuprate are required. They also found that mono-organocoppers (the authors called these "polymeric complexes") gave the highest stereoselectivity, as well as being relatively insensitive to the solvent used.

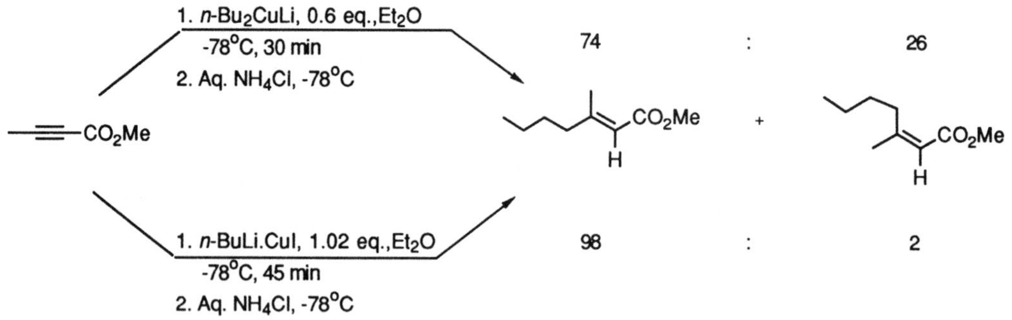

The same group showed that, in a mixed homocuprate, a tertiary ligand transfers to an alkynoate much more readily than a primary ligand.

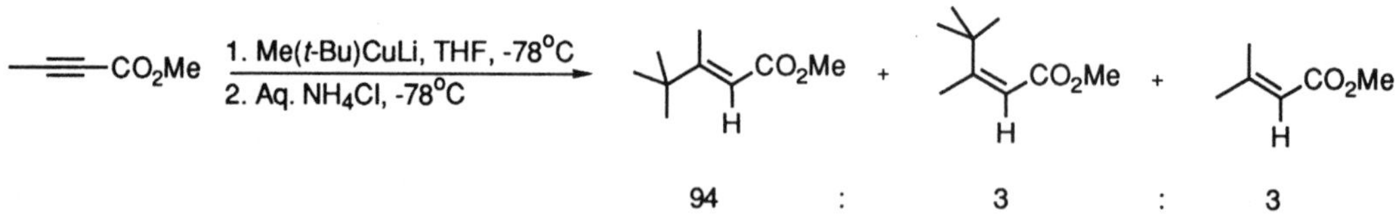

Attempts to trap the intermediates with alkyl halides have not been very successful, producing isomeric mixtures in poor yields. Other traps have been much more successful. For example, Marino and Linderman have used the addition of an LO cuprate, followed by trapping with an acid chloride as the first of a two step annulation procedure.[32]

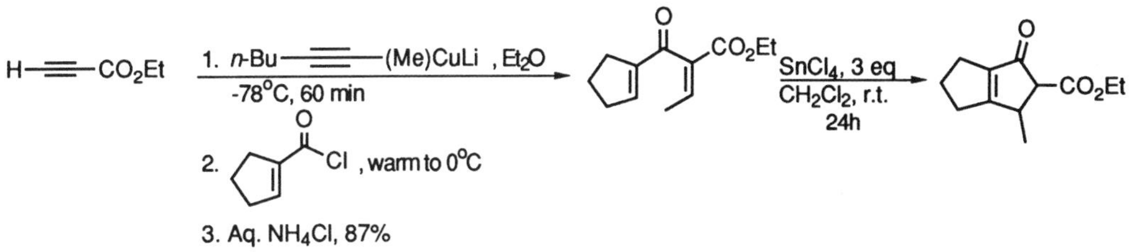

$H\!\!-\!\!\equiv\!\!-CO_2Et$ 1. n-Bu$-\!\!\equiv\!\!-$(Me)CuLi , Et$_2$O
                              $\overline{-78°C, 60\ min}$

2. [cyclopentene acid chloride structure] Cl , warm to 0°C

3. Aq. NH$_4$Cl, 87%

$\xrightarrow{\text{SnCl}_4,\ 3\ eq}$ [bicyclic product structure] $-CO_2Et$
$\quad$ CH$_2$Cl$_2$, r.t.
$\quad\quad$ 24h

Many electrophiles can be used to quench the intermediate, which they postulate is an allenolate.[33]

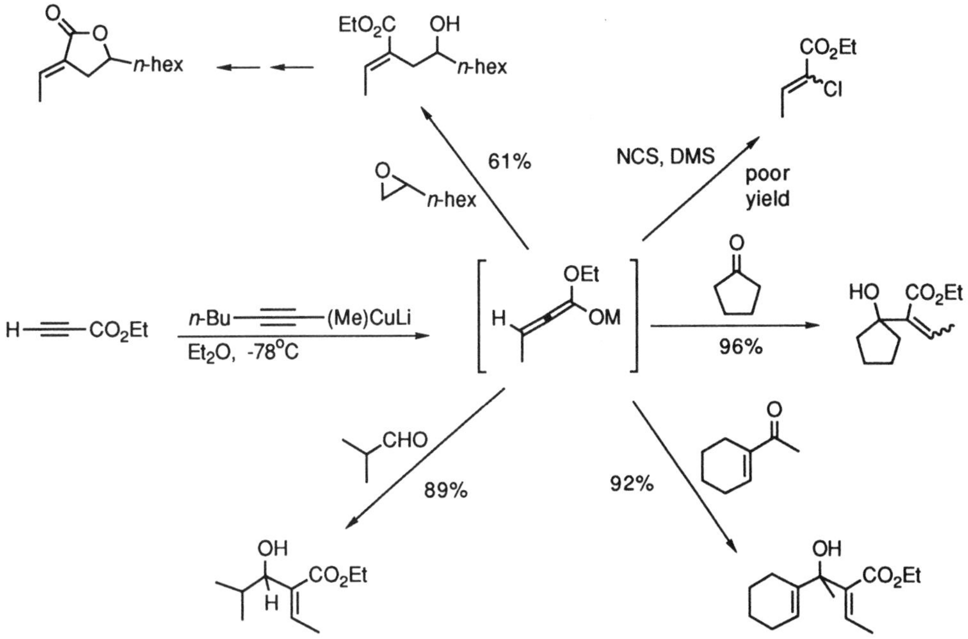

$H\!\!-\!\!\equiv\!\!-CO_2Et$

Intramolecular traps can sometimes succeed. Interestingly, in the following example, the reactivity of the intermediate from addition of an HO cuprate was sufficient to attack an epoxide in the same molecule.[34] The corresponding intermediate from an LO cuprate addition did not close onto the epoxide. The authors attributed this to the allenolate structure of the intermediate from HO cuprate additions.

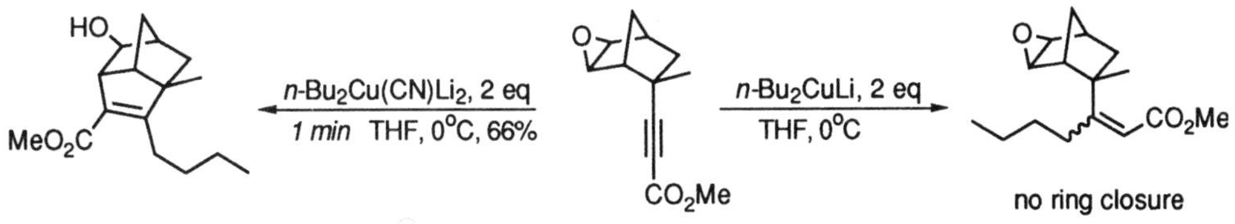

n-Bu$_2$Cu(CN)Li$_2$, 2 eq
$\overline{1\ min\ \ THF, 0°C, 66\%}$

n-Bu$_2$CuLi, 2 eq
$\overline{THF, 0°C}$

no ring closure

It is also possible to add LO cuprates to an alkynoate-iron cationic complex.[35]

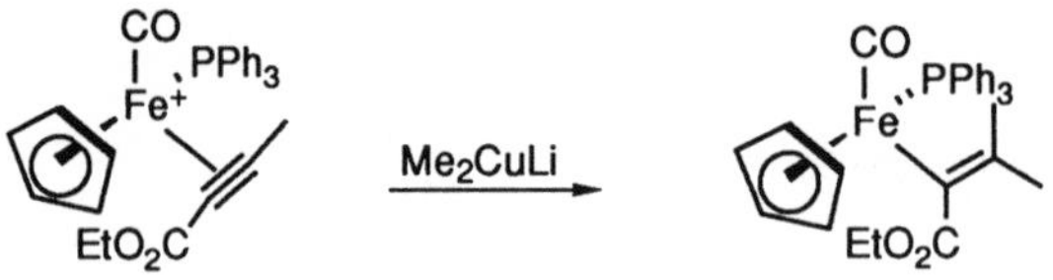

The addition of LO *alkenyl*cuprates gives a mixture of stereoisomers.[36]

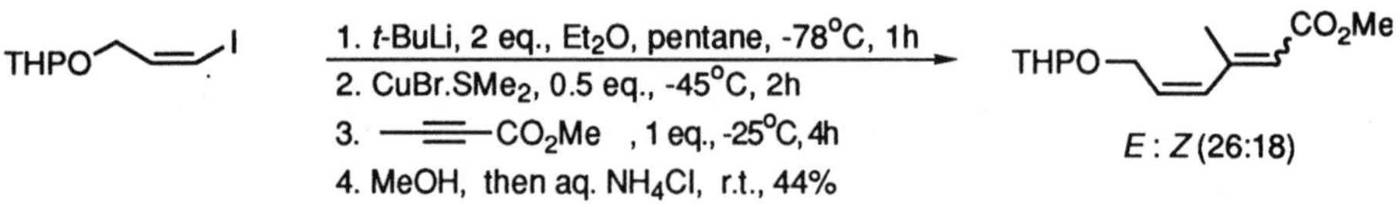

7.2.3   Enamines and enol ethers

7.2.3.1   *Enamines*

Excellent reactivity and regioselectivity is obtained simply by heating an enamine with an alkynoate. The additions are assumed to proceed via a cyclobutene intermediate.[37]

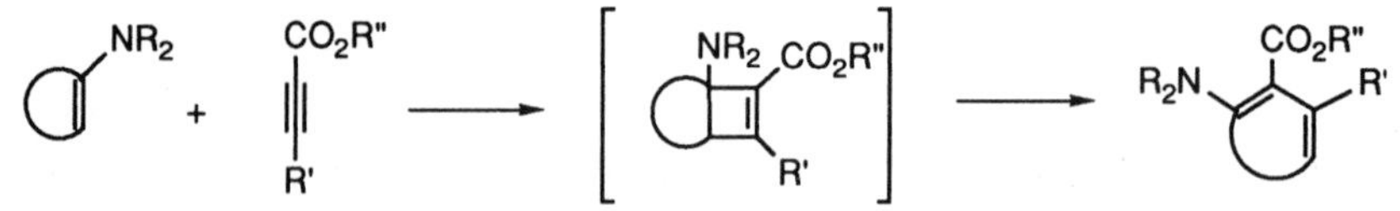

If the enamine is part of a ring, the overall result is a two-atom ring expansion. This methodology was applied to a synthesis of, for example, some sesquiterpene fungal metabolites such as velleral, from basidiomycetes of the genus *Lactarius*.[38]

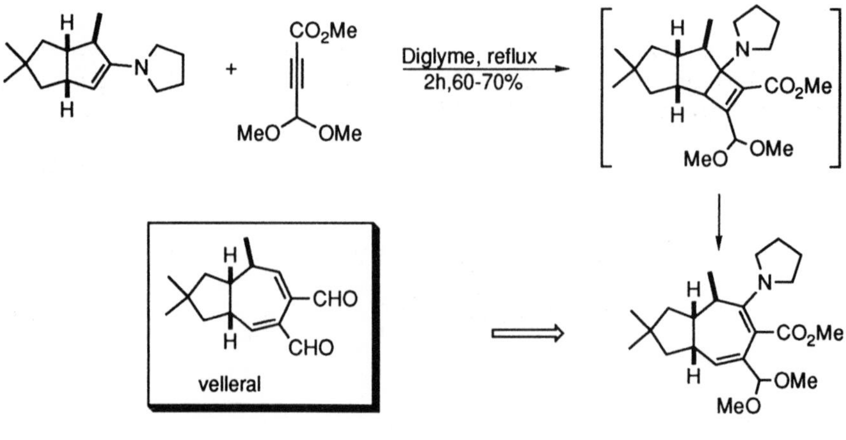

The nature of the product, i.e. cyclobutene or conjugate adduct, is highly solvent dependent.[39] In the following example only conjugate adducts were obtained.

Me₂NHN   O   +   MeO₂C—≡—CO₂Me   EtOH, r.t.   →   Me₂NHN   O
                                       12h, 44%   MeO₂C   CO₂Me

A rather interesting example of an ene-diaminoester addition to an alkynoate is shown below.[40] The E-isomer is produced exclusively.[41]

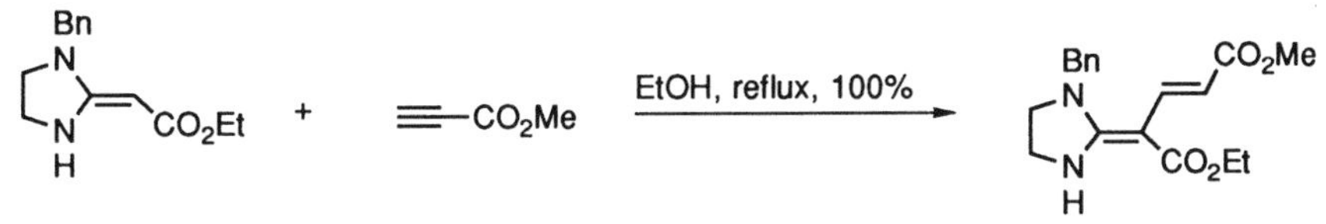

Bn
N
  ⟩=⟨CO₂Et   +   ≡—CO₂Me   EtOH, reflux, 100%   →   Bn   CO₂Me
N   N
H   ⟩   CO₂Et
                                                     N
                                                     H

Many studies have been carried out on the addition of indole derivatives to alkynoates.[42] Where a 2-methyl substituent is present cyclization to a cyclohexadienone system is possible.

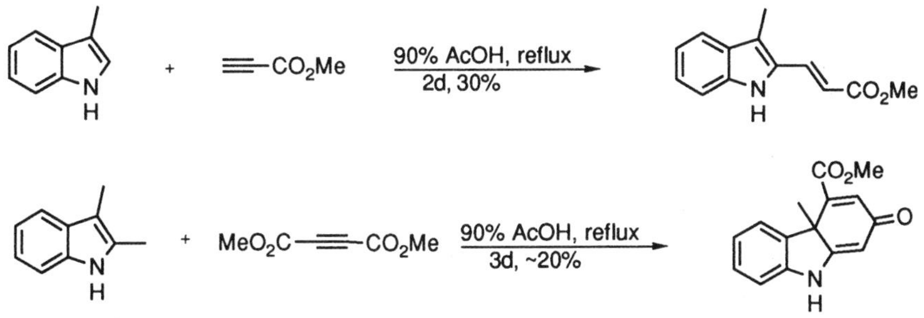

(indole with 3-methyl)   +   ≡—CO₂Me   90% AcOH, reflux   →   (indole)—CO₂Me
N   2d, 30%   N
H   H

(indole with 3-methyl)   +   MeO₂C—≡—CO₂Me   90% AcOH, reflux   →   CO₂Me
N   3d, ~20%   ⟩=O
H   N
                                                                     H

7.2.3.2   Enol ethers and ketene acetals

Lewis acid-catalyzed reactions of enol ethers to alkynoates give only [2+2] cycloadditions.[43] Results from the addition of ketene acetals to alkynoates have been reported. Rousseau's group found that titanium tetrachloride-catalyzed additions were quite efficient.[44]

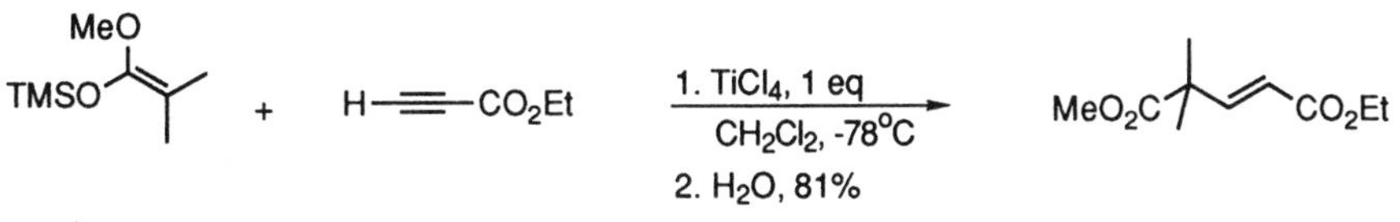

MeO
TMSO   ⟩   +   H—≡—CO₂Et   1. TiCl₄, 1 eq   MeO₂C   CO₂Et
                           CH₂Cl₂, -78°C   →
                           2. H₂O, 81%

Changing the Lewis acid to zinc iodide led to excellent yields of stereochemically pure E-isomers of adducts bearing a C2-silyl substituent. This product is probably formed by silylation of an intermediate allenolate.

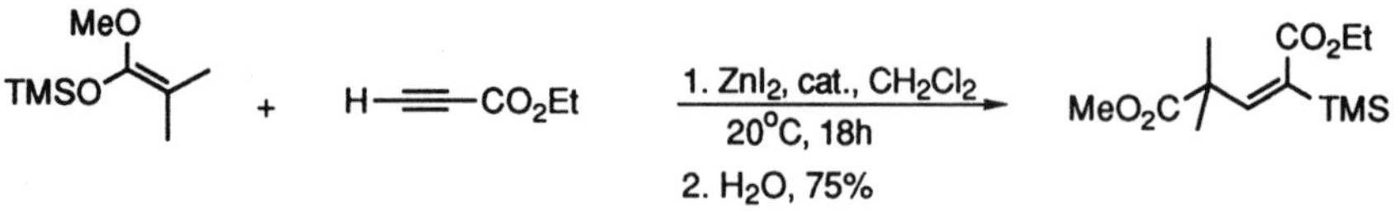

Perhaps even more remarkable is their finding that changing to a third type of Lewic acid, zirconium tetrachloride, leads to isolation of the (formally) [2+2] cycloaddition product (3).

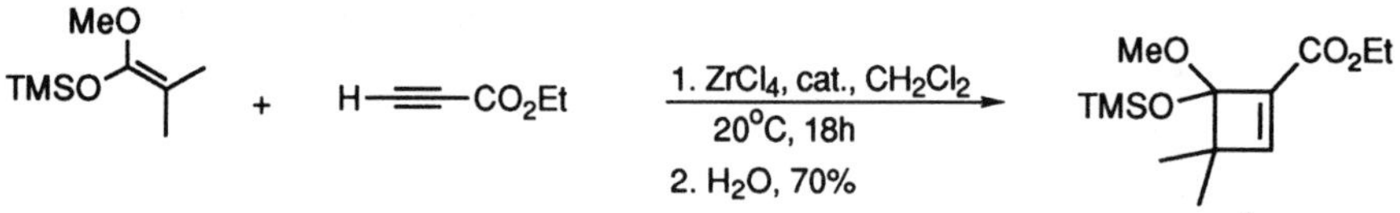

7.2.4   Heteronucleophiles

7.2.4.1   *Nitrogen*

Primary and secondary alkylamines and aromatic amines add readily to alkynoates in diethyl ether or benzene.[45] The product stereochemistry is often dependent upon the type of amine added. For example, primary alkylamines tend to give mainly Z-isomers, whereas cyclic secondary amines (and aromatic amines) give the E-isomers. Additions to acetylene dicarboxylate are similarly variable.

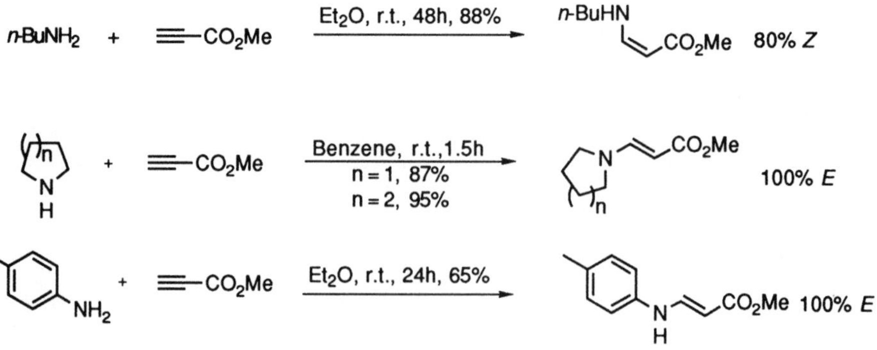

Heating acylimidazoles with acetylenedicarboxylate in acetonitrile provides direct access to the imidazo[1,2-a]pyridine ring system.[46]

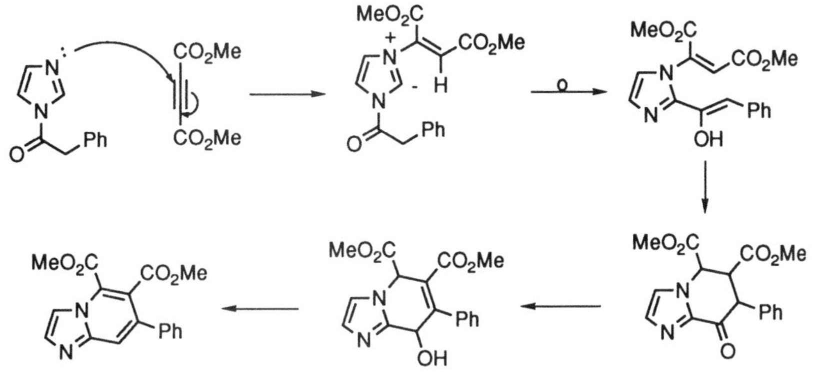

Additions of pyrazolidinones bearing a functionalized side chain lead to annulated products.[47]

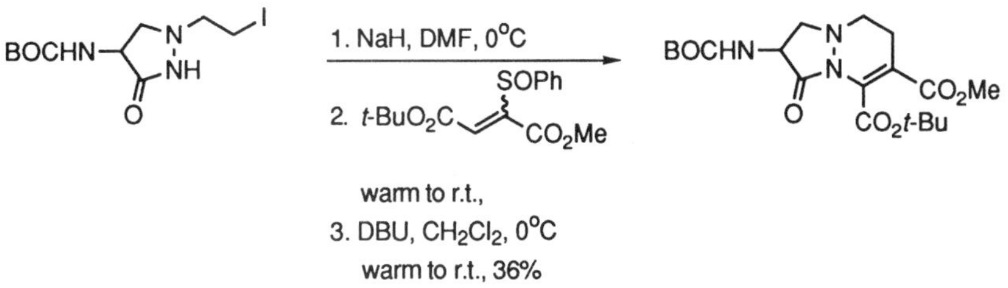

7.2.4.2   Sulfur and selenium

Quite a large number of methods exists for the addition of sulfur and selenium nucleophiles. Excellent yields are usually obtained. The stereoselectivity is often high and varies with the nature of the thiol and the alkynoate. For example, addition of methanethiol to methyl propynoate (Table 7.1 entry 1) gives 4:1 selectivity in favour of the Z-isomer. However, base-catalyzed addition of thiophenol to propynoic *acid* (presumably to the carboxylate anion) is completely Z-selective, although in half the chemical yield.

Mukaiyama has developed a useful method for synthesizing 3,3-disubstituted alkenoates by (i) conjugate addition of thiophenoxide to a 2-butynoate, followed by (ii) addition/elimination of an organomagnesium reagent, with overall retention.[48] This method is one of the few that exists for adding (overall) an R group and a proton in *anti* fashion across an alkynoate triple bond. For example, Cha's group has used this in their studies on the potent neurotoxin, verrucosin.[49]

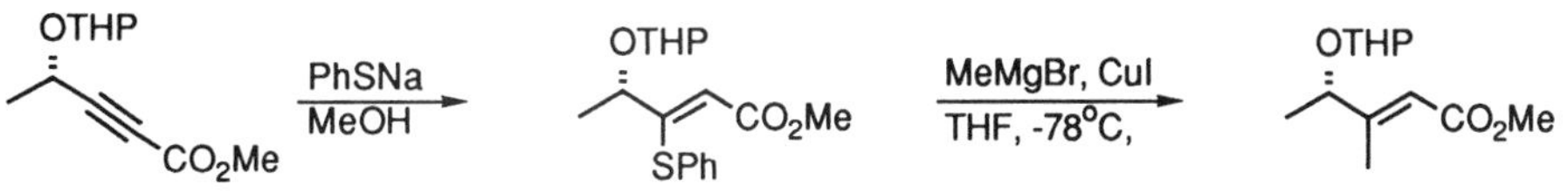

Table 7.1   Addition of sulfur, selenium and tellurium nucleophiles to alkynoates

1[50]	MeSH + ≡–CO$_2$Me	$\dfrac{\text{CDCl}_3, 21°C, 1h}{100\%}$	MeS⌒CO$_2$Me 80% Z
2[51]	R–≡–CO$_2$Me	$\dfrac{\text{Bu}_4\text{NSCN, H}_2\text{SO}_4}{\text{CH}_2\text{Cl}_2, 40°C, 8h}$	NCS—C(R)=CH–CO$_2$Me R = H, 85%% E R = CO$_2$Me, 80% E
3[52]	o-(NH$_2$)(SH)C$_6$H$_4$ + Ph–≡–CN	$\dfrac{\text{neat, 32°C}}{95\%}$	o-(NH$_2$)C$_6$H$_4$–S–C(Ph)=CH–CN
4[53]	PhSH	1. NaOEt, EtOH, 0°C 2. ≡–CO$_2$H , 100°C, 3h 3.10% H$_2$SO$_4$, 50%	PhS⌒CO$_2$H
5[54,55]	PhXXPh	1. NaBH$_4$, Et$_2$O, THF (1:1), r.t. 2. R–≡–CO$_2$Et , r.t., 0.5h 3. AcOH	PhX–C=C(R)–CO$_2$Et X = Se, R = Me 83% X = Te, R = Ph 85%

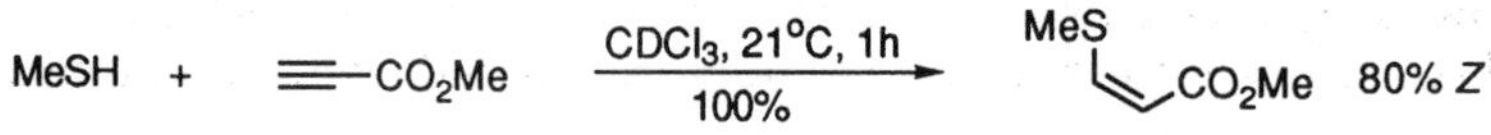

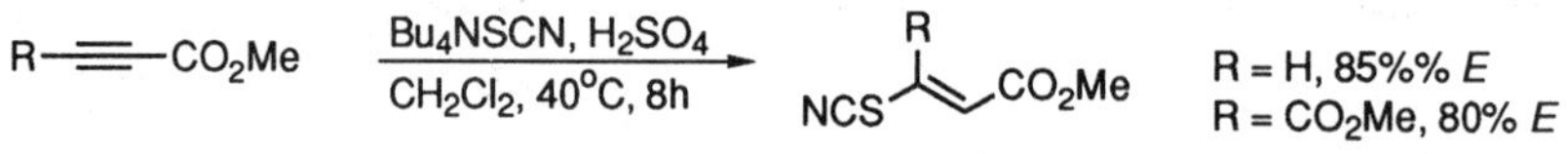

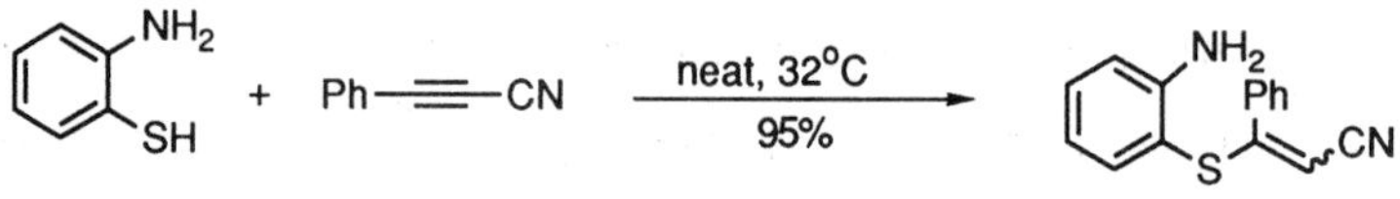

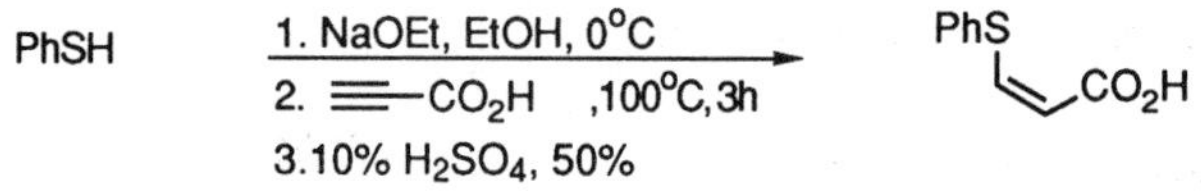

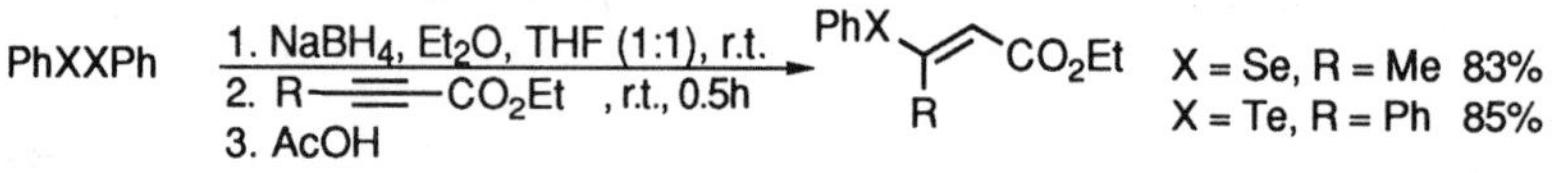

When an allyl sulfide is added to an alkynoate in the presence of a Lewis acid, the product is effectively that which would be obtained from the conjugate addition of a sulfide anion, followed by trapping with, say, allyl bromide.[56] As well as this, changing the Lewis acid has a dramatic influence on the stereochemical outcome of the reaction, as shown below.

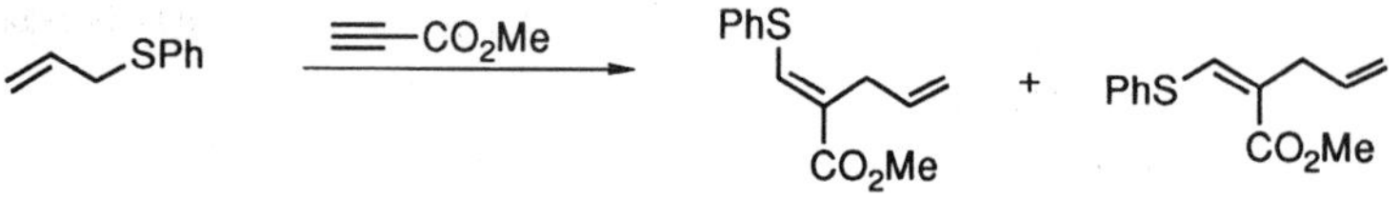

⌒SPh   ≡–CO$_2$Me → PhS–C(CH$_2$CH=CH$_2$)=CH–CO$_2$Me + PhS–C(=CH$_2$)(CH$_2$CH=CH$_2$)–CO$_2$Me

CH$_2$Cl$_2$, AlCl$_3$, 1.1 eq., 25°C, 71%	72	:	28
Neat, ZnCl$_2$,	92% 3	:	97

7.2.4.3   Tin

Piers' group has designed reagents which add stereoselectively to alkynoates to give either the E- or Z- isomers.[57] Precise reaction conditions were also defined for these reactions. As pointed out above there are no direct methods currently available for the selective *anti* addition of an R group and a proton across the triple bonds of alkynoates. The following reagents were studied:

1. Me$_3$SnCu.SMe$_2$ 2. [Me$_3$SnCuY]Li   Y = SPh, SnMe$_3$ or   ≡–C(Me)=CH–OMe

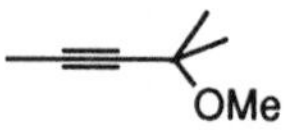

It was found that if the additions were run in the presence of 1.7 equivalents of methanol good yields of essentially pure *E*-isomers could be obtained. On the other hand, addition followed by quenching with methanol in a separate step led to the *Z*-isomers selectively. These results were general for additions to a variety of 3-substituted alkynoates.

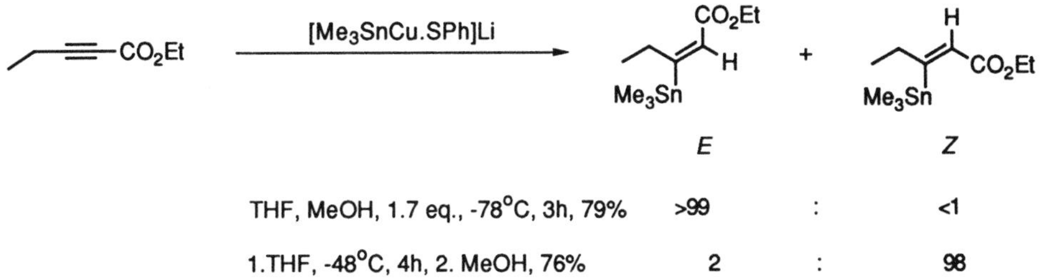

		E		*Z*
THF, MeOH, 1.7 eq., -78°C, 3h, 79%		>99	:	<1
1.THF, -48°C, 4h, 2. MeOH, 76%		2	:	98

In only one case, that shown below, was the yield poor and this was found to be due to competitive addition of the phenylthio ligand.

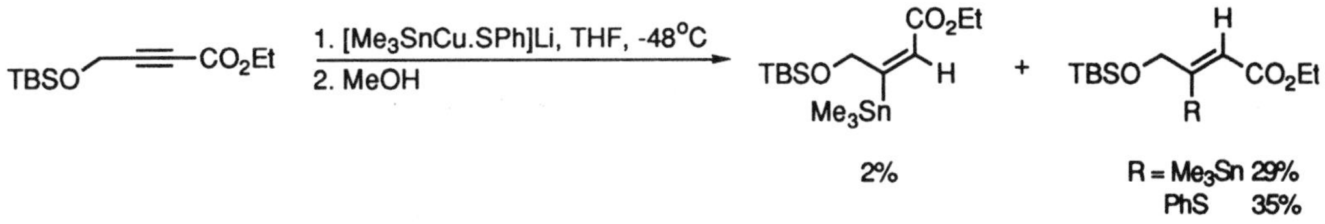

7.2.4.4   *Conjugate reduction*

Conjugate reduction of alkynoates is possible using Semmelhack's copper based reagents. However stereochemical mixtures, as well as some saturated ester, are produced.[58]

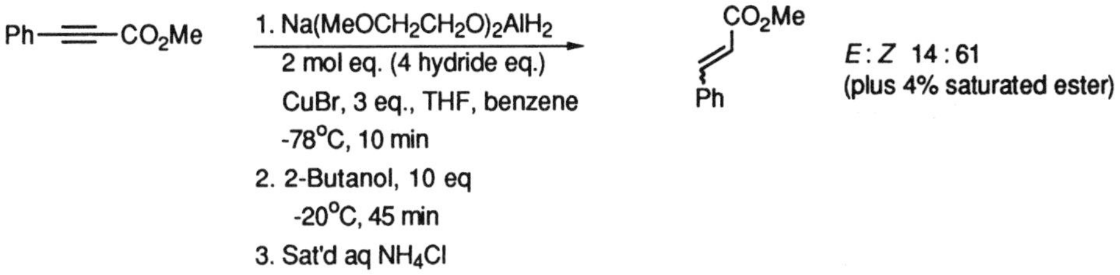

7.3   **Alkynyl sulfoxides and related systems**

As with alkynoates (mentioned in section 7.2.1), addition/trapping reactions are possible with alkynylsulfones.[26]

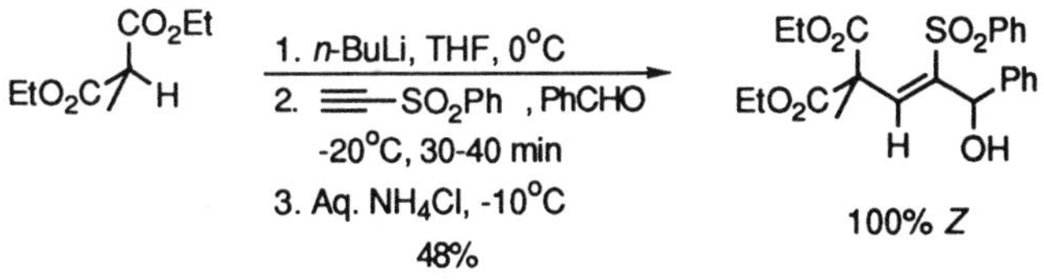

Mono-organocopper reagents add easily, and with excellent *syn* selectivity, to alkynyl sulfoxides.[59]

They also add well to alkynyl(thio)phosphine oxides.[60]

References

1. R. B. Woodward and T. Singh, *J. Amer. Chem. Soc.*, 1950, **72**, 494
2. For a discussion of these, see E. D. Bergman, D. Ginsburg and R. Pappo, *Org. React.*, 1959, **10**, 213-216
3. G. Stork and M. Tomasz, *J. Amer. Chem. Soc.*, 1962, **84**, 310
4. K. Nakanishi, T. Goto, S. Ito, S. Natori and S. Nozoe (eds), "Natural Products Chemistry, Vol. 2, Kodanshi Ltd. and Academic Press, New York, 1975, 192
5. E. J. Corey and W-g. Su, *J. Amer. Chem. Soc.*, 1987, **109**, 7534
6. E. J. Corey and W-g. Su, *Tetrahedron Lett.*, 1988, **29**, 3423
7. D. A. Anderson and J. R. Hwu, *J. Org. Chem.*, 1990, **55**, 511
8. Japan Tobacco and Salt Public Corp. Jpn. Patent 82 72 901, 1982; *Chem. Abstr.*, 1982, 97:105613
9. International Flavors and Fragrances Inc. U.S. Patent 4,210,158, 1980; *Chem. Abstr.*, 1980, 93:183048
10. E. P. Kohler, *J. Amer. Chem. Soc.*, 1922, **44**, 379
11. J-F. Lavallee, G. Berthiaume and P. Deslongchamps, *Tetrahedron Lett.*, 1986, **27**, 5455
12. R. J. Linderman and M. S. Lonikar, *J. Org. Chem.*, 1988, **53**, 6013
13. I. Fleming and D. A. Perry, *Tetrahedron*, 1981, **37**, 4027
14. J. A. Marshall, S. L. Croos and B. S. DeHoff, *J. Org. Chem.*, 1988, **53**, 1616 and J.A. Marshall and W. Y Gung, *Tetrahedron Lett.*, 1988, **29**, 3899
15. See also, J. A. Marshall, *Tetrahedron Lett.*, 1986, **27**, 4845
16. A. Hassner and N. Wiegard, *J. Org. Chem.*, 1986, **51**, 3652
17. For a related preparation of dihydroazepines from vinylazirines, see A. Hassner and W. Chan, *Tetrahedron Lett.*, 1982, **23**, 1989
18. M. Casey, C. J. Moody, C. W. Rees and R. G. Young, *J. Chem. Soc., Perkin Trans. I,* 1985, 741
19. G. Altnau and L. Rosch, *Tetrahedron Lett.*, 1983, **24**, 45
20. S. H. Cheon, W. J. Christ, L. D. Hawkins, H. Jin, Y. Kishi and M. Taniguchi, *Tetrahedron Lett.*, 1986, **27**, 4759
21. M. Taniguchi, T. Hino and Y. Kishi, *Tetrahedron Lett.*, 1986, **27**, 4767
22. W. E. Bachmann, and E. K. Raunio, *J. Amer. Chem. Soc.*, 1950, **72**, 2530

23. W. E. Bachmann, G. I. Fujimoto and E. K. Raunio, *J. Amer. Chem. Soc.*, 1950, **72**, 2533

24. B. M. Trost, C. D. Shuey and F. DiNinno, Jr., *J. Amer. Chem. Soc.*, 1979, **101**, 1284

25. E. H. Farmer, S. C. Ghosal and G. A. R. Kon, *J. Chem. Soc.*, 1936, 1804

26. A. Bury, S. D. Joag and C. J. M. Stirling, *J. Chem. Soc., Chem. Commun.*, 1986, 124

27. T. Naota, H. Taki, M. Mizuno and S-I. Murahashi, *J. Amer. Chem. Soc.*, 1989, **111**, 5954

28. E. J. Corey and J. A. Katzenellenbogen, *J. Amer. Chem. Soc.*, 1969, **91**, 1851

29. E. J. Corey and A. Tramontano, *J. Amer. Chem. Soc.*, 1984, **106**, 462

30. B. E. Marron and K. C. Nicolau, *Synthesis*, 1989, 537

31. R. J. Anderson, V. L. Corbin, G. Cotterrell, G. R. Cox, C. A. Henrick, F. Schaub and J. B. Siddall, *J. Amer. Chem. Soc.*, 1975, **97**, 1197

32. J. P. Marino and R. J. Linderman, *J. Org. Chem.*, 1981, **46**, 3696

33. J. P. Marino and R. J. Linderman, *J. Org. Chem.*, 1983, **48**, 4621

34. D. E. Lewis and H. L. Rigby, *Tetrahedron Lett.*, 1985, **26**, 347

35. D. L. Reger, P. J. Mc Elligot, N. G. Charles, E. A. Griffith and E. L. Amma, *Organometallics*, 1982, **1**, 443

36. R. J. Sundberg and B. C. Pearce, *J. Org. Chem.*, 1982, **47**, 725

37. G. H. Alt and A. G. Cook in A. G. Cook, (ed.), "Enamines", Marcel Dekker, New York, 1988, Chapter 4

38. J. Froborg and G. Magnusson, *J. Amer. Chem. Soc.*, 1978, **100**, 6728

39. W. Verboom, G. W. Visser, W. P. Trompenaars, D. N. Réinhoudt, S. Harkema and G. J. van Hummel, *Tetrahedron*, 1981, **20**, 3525

40. R. C. F. Jones and M. J. Smallridge, *Tetrahedron Lett.*, 1988, **29**, 5005

41. For similar additions, see Z-t. Huang and L-h. Tzai, *Chem. Ber.*, 1986, **119**, 2208

42. R. M. Letcher, M. C. K. Choi, R. M. Acheson and R. J. Pronce, *J. Chem. Soc., Perkin Trans. I*, 1983, 501

43. R. D. Clark and K. G. Untch, *J. Org. Chem.*, 1979, **44**, 248

44. A. Quendo and G. Rousseau, *Tetrahedron Lett.*, 1988, **29**, 6443

45. R. Huisgen, K. Herbig, A. Siegl and H. Huber, *Chem. Ber.*, 1966, **99**, 2526

46. H-J. Knölker and R. Boese, *J. Chem. Soc., Chem. Commun.*, 1988, 1151

47. R. J. Ternansky and S. E. Draheim, *Tetrahedron Lett.*, 1988, **29**, 6569

48. S. Kobayashi and T. Mukaiyama, *Chem. Lett.*, 1974, 705

49. J. K. Cha and R. J. Cooke, *Tetrahedron Lett.*, 1987, **28**, 5473

50. M. Renard and L Hevesi, *Tetrahedron*, 1985, **41**, 5939

51. M. Giffard, J. Cousseau, L. Gouin and M-R. Crahe, *Tetrahedron*, 1985, **41**, 801

52. S. R. Landor, P. D. Landor, Z. T. Fomum, J. T. Mbafor and G. W. B. Mpango, *Tetrahedron*, 1984, **40**, 2141

53. S. M. Proust and D. D. Ridley, *Aust. J. Chem.*, 1984, **37**, 1677

54. X=Te, M. R. Detty, B. J. Murray, D. L. Smith and N. Zumbulyadis, *J. Amer. Chem. Soc.*, 1983, **105**, 875

55. X=Se, M. R. Detty and B. J. Murray, *J. Amer. Chem. Soc.*, 1983, **105**, 883

56. K. Hayakawa, Y. Kamakawaji, A. Wakita and K. Kanematsu, *J. Org. Chem.*, 1984, **49**, 1985

57. E. Piers, J. M. Chong and H. E. Morton, *Tetrahedron*, 1989, **45**, 363

58. M. F. Semmelhack, R. D. Stauffer and A. Yamashita, *J. Org. Chem.*, 1977, **42**, 3180

59. W. E. Truce and M. J. Lusch, *J. Org. Chem.*, 1978, **43**, 2252

60.   A. M. Aguiar and J. R. S. Irelan, *J. Org. Chem.*, 1969, **34**, 4030

Index

* Entries in the "conjugate addition" section of this index which are preceded by "/" indicate a quenching reagent, other than a proton source, for that particular conjugate addition